Modern Technologies and Tools Supporting the Development of Industry 5.0

In an era where technological advancements are not just tools but partners in our workspaces, *Modern Technologies and Tools Supporting the Development of Industry 5.0* emerges as a seminal guide to understanding and navigating the complexities of the Fifth Industrial Revolution. This book, a collective work of expert authors, delves into the heart of Industry 5.0, exploring how it synergizes human creativity with robotic precision to redefine industrial landscapes. From collaborative robotics to sustainable development, each chapter unfolds layers of knowledge essential for professionals, academics, and students alike.

Features

- Covers modern technologies including artificial intelligence, robotics, and the Internet of Everything for modernizing Industry 5.0 and the transformative role of collaborative robots in the workplace and how they are changing the dynamics of human labour.
- Focuses on technologies mimicking human behaviour and reasoning to solve complex problems and explores the evolving role of human expertise in an increasingly automated world and the competencies needed to thrive in this new era.
- Showcases the impact of Industry 5.0 on the environment, and industry commitment to sustainable development by laying a map to understand how Industry 5.0 is steering industries towards sustainable practices, focusing on green supply chains, reverse logistics, and the critical role of internal audits.
- Highlights future perspectives such as smart manufacturing and the Industrial Internet of Things (IIoT) for Industry 5.0 manufacturing processes and provides insights into the challenges and security concerns as industries prepare to adopt Industry 5.0, offering foresight into its long-term impacts on global markets and societies.
- Presents real-time case studies on tools, technologies, architecture, and product outcomes for Industry 5.0.

Modern Technologies and Tools Supporting the Development of Industry 5.0 is more than a book; it's a roadmap for the future, guiding readers through the intricacies of industrial evolution. It is primarily written for senior undergraduate and graduate students and academic researchers in the fields of industrial engineering, production engineering, mechanical engineering, and aerospace engineering.

Modern Technologies and Tools Supporting the Development of Industry 5.0

Edited by
Justyna Żywiołek
Joanna Rosak-Szyrocka
Anand Nayyar
Mohd Naved

CRC Press
Taylor & Francis Group
Boca Raton London New York

CRC Press is an imprint of the
Taylor & Francis Group, an **informa** business

First edition published 2024
by CRC Press
2385 NW Executive Center Drive, Suite 320, Boca Raton FL 33431

and by CRC Press
4 Park Square, Milton Park, Abingdon, Oxon, OX14 4RN

CRC Press is an imprint of Taylor & Francis Group, LLC

© 2024 selection and editorial matter, Justyna Żywiołek, Joanna Rosak-Szyrocka, Anand Nayyar and Mohd Naved; individual chapters, the contributors

ISBN: 978-1-032-53104-5 (hbk)
ISBN: 978-1-032-78757-2 (pbk)
ISBN: 978-1-003-48926-9 (ebk)

DOI: 10.1201/9781003489269

Typeset in Sabon
by SPi Technologies India Pvt Ltd (Straive)

Contents

Preface

The factory floors of our ancestors were fortresses of sweat and steel, where human toil met the unyielding rhythm of machines. This was Industry 1.0, the era of mechanized muscle, laying the groundwork for the mass production that defined the 20th century. But progress, like a restless river, never stays still. Industry 2.0 brought electricity, powering the assembly line and ushering in an age of standardized efficiency. Then came the digital revolution of Industry 3.0, where computers took the reins, weaving automation into the fabric of production. And now, on the horizon, a new dawn breaks: Industry 5.0.

This book, *Modern Technologies and Tools Supporting the Development of Industry 5.0*, stands as a map for navigating this new landscape. It is more than just a collection of technical treatises; it is a compass guiding us through the intricate network of emerging technologies and transformative tools reshaping the very fabric of industry.

Industry 5.0 is not merely an incremental upgrade. It is a paradigm shift, a fundamental reimagining of the relationship between humans, machines, and the industrial ecosystem. It is about collaboration, not competition, where humans and AI work in tandem, leveraging their unique strengths to push the boundaries of innovation. It is about sustainability, where factories become closed-loop systems, minimizing environmental impact and maximizing resource efficiency. It is about personalization, where products and services adapt to individual needs, reflecting the diverse tapestry of our global society.

This journey towards Industry 5.0 is paved with a myriad of modern technologies and tools. Artificial intelligence, the once-mythical whisper, now roars with the potential to optimize every facet of production, from predictive maintenance to self-learning robots. Big Data, the silent observer, now holds the key to unlocking hidden patterns and insights, enabling data-driven decision-making. The Internet of Things (IoT), an invisible web of connected devices, promises to orchestrate a symphony of real-time data, ensuring seamless communication and intelligent adaptation. These are just a glimpse of the tools at our disposal, each a brushstroke in the grand canvas of Industry 5.0.

But technology, like any powerful tool, demands thoughtful stewardship. We must ensure that the benefits of Industry 5.0 are shared equitably, bridging the digital divide and empowering individuals with the skills and knowledge to thrive in this evolving landscape. We must address ethical concerns, ensuring that AI serves humanity, not the other way around. And we must remain vigilant, protecting our data and infrastructure from the ever-evolving threats of the digital age. The road ahead is not without its challenges. But within these challenges lie opportunities, waiting to be seized by those bold enough to embrace the future. This book is not just a chronicle of technology; it is a call to action. It is an invitation to join the vanguard of those who are shaping the future of industry, who are building a world where humans and machines dance in harmony, where innovation fuels sustainability, and where progress serves the needs of all.

The book comprises 12 chapters enlisting the development of Industry 5.0 in a strong methodological approach. Chapter 1 titled "Industry 5.0: Employee Readiness for Co-Bot Workspace" enlightens the current level of employee readiness for co-bot workspace in Industry 5.0, stresses on HR requirements and adaptation-related challenges faced by industries to build a sustainable co-bot workspace, highlights key skills of employees to work in a co-bot workspace, as well as HR preparedness and adaptation-related challenges faced by industries for building a co-bot workspace. Chapter 2 titled "Automation and Robotization: An Aerial View of Industry 5.0 on Employment and Employees" illustrates various technology enablers for Industry 5.0 like the IoT, AI, ML, and Blockchain and also determines the effects of technology on employee's psychology at the workplace. Chapter 3 titled "Structural Dimensions and Measurement of Readiness for Industry 5.0 Implementation: A Fresher Insight from SMEs in Developing Country" explores the readiness for Industry 5.0 implementation (RII) which is defined from the perspectives of SMEs and formulates a measurement scale utilizing mixed methods in two phases. In the first phase, following Churchill's scale development process, three dimensions are determined—the Technology aspect, the Organizational aspect, and the Environmental aspect—with eight factors that have appropriate convergent validity and discriminant validity. In the second phase, the nomological validity of the RII scale is evaluated by investigating its interconnection with other theoretical constructs to offer managerial and operational pointers and suggests a roadmap, at the policy degree, regarding Industry 5.0 implementation to policymakers, at the industrial and national levels. Chapter 4 titled "Thriving in Industry 5.0: The Evolving Role of Employees in the Age of Automation and Innovation" provides a comprehensive analysis of the evolving role of employees in Industry 5.0, including its impact on employment, the key competencies required for career development, the challenges faced by employees, strategies to overcome these challenges, and the management of human–robot collaboration. The chapter sheds light on how employees can not only acclimatise but also thrive in Industry 5.0, and contribute to the ongoing dialogue about the changing role of employees in the age of automation and innovation. Chapter 5 titled "Exploring the Impact of Collaborative Robots on Human–Machine Cooperation in the Era of Industry 5.0" explores the impact of cobots on human–machine cooperation and the workforce in the era of Industry 5.0. This chapter presents a comprehensive review of the existing literature on cobots, including their development, implementation, and applications in various industries and investigates the differences between human and machine work, including the defects of each, and how cobots can be used to compensate for these defects. Additionally, the chapter examines the social perception of Industry 5.0 and the impact of cobots on workforce readiness and job quality. The chapter also presents a case study of a manufacturing company that has recently integrated cobots into its production process. The case study investigates the impact of cobots on the company's workforce, including changes in job responsibilities and perceptions of job quality. Chapter 6 titled "Harmonising Human and Robotic Workforces in Industry 5.0: Creating a Resilient, Human-Centric, and Sustainable Organizational Ecosystem" evaluates each manager's competence in an automotive organization embracing Industry 5.0, enlightens human-centric values managers perceive as crucial to involvement in Industry 5.0 and also illustrates the managers' role in encouraging team members' resilience termed as crucial to involvement in Industry 5.0. Chapter 7 titled "Impact of Industry 5.0 on Sustainable Development with Machine Learning and the Role of Internal Audit: Literature Review" analyses the impact of Industry 5.0 on sustainable development and examines the role of ML and IA function with the application of Stakeholder Theory, Resource-Based View (RBV) of the industry, and Ethical Decision-Making theory. Chapter 8 titled "Analysis of Industry 5.0 Contributions to Sustainable Development: From Welfare to Wellbeing" illustrates the relationship between Industry 4.0 and Industry 5.0, demonstrates the Industry 4.0 to Industry 5.0 transition, and stresses the significance of sustainability in Industry 5.0. Chapter 9 titled "Reverse Logistics and Green Supply Chain on Terms of Sustainability: An Application in an Enterprise" identifies the product-recovering options and factors of RLs and GCSC towards attaining sustainability and sustainable development of companies, finds the inter-relationship among the identified potential drivers, to evaluate the success of GSCM in any industry or marketplace and chapter analyses the obstacles to RLs and GSCM for a mathematical model approach to decision-making for the sustainability of a business. Chapter 10 titled "Influence of Internal Audit on Implementation of Environment, Social and Governance (ESG) Factor: Case of Public-Listed Companies in Oman" examines the influence of internal audit (IA) on Environmental, Social, and Governance (ESG) practices and emphasizes the importance of integrating ESG practices into IA for effective corporate governance and also explores the relationship between IA, ESG, and Industry 5.0. Chapter 11 titled "Foresight the Future Consequences of Industry 5.0 in Sustainable

Services Production and Consumption in Developing Countries" illustrates the future consequences of Industry 5.0 in the context of sustainable services, analyses the role of Industry 5.0 in sustainable service production and consumption in developing countries, and also highlights the significance of Industry 5.0 in the attainment of sustainable development goals. Chapter 12 titled "Readiness of Enterprises to Implement Industry 5.0: Challenges and Security Concerns" enlightens a meta-analysis of the literature review to describe the benefits of the 5.0 Industry along with several components and related technologies of Industry 5.0. In addition, diverse challenges of this technology to enterprises, such as human–robot interaction, human factors, and security concerns, are discussed to raise awareness and minimize the gaps between Industry 4.0 and 5.0. The chapter aims to provide helpful information about Industry 5.0 via several aspects: main components of Industry 5.0, challenges of Industry 5.0 application for enterprise, outstanding opportunities of Industry 5.0, human and Industry 5.0, and potential contribution of Industry 5.0 to future production and business.

"As we stand at the threshold of Industry 5.0," writes Alan Kay, "the best way to predict the future is to invent it." Let this book be your compass, your blueprint, and your spark of inspiration. Let us invent a future where technology amplifies human potential, where progress serves all, and where Industry 5.0 becomes not just a revolution, but a renaissance.

Justyna Żywiołek
Joanna Rosak-Szyrocka
Anand Nayyar
Mohd Naved

Editors

Justyna Żywiołek completed her master's studies in the field of Metallurgy in 2010. In 2014, she defended her doctoral dissertation with honors in the field of management science, specializing in security science, entitled: *The impact of information process management on the security of knowledge resources in the enterprise*, at the Faculty of Organization and Management of the Lodz University of Technology. Since January 2015, she has been an assistant professor in the Faculty of Management at the Technical University of Częstochowa. She is a respected researcher in Poland and abroad for her experience in management issues: knowledge and information management and their security. She has published over 160 articles and is the author/co-author/editor of seven books. She presented research results at various international conferences. She deals with data protection, is an inspector of personal data protection and a leading ISO 27001:2018 auditor. Phd. Eng. Justyna Żywiołek is a member of ICAA (Intelectual Capital Association) and EU-OSHA, as well as organizations operating in Poland promoting information and knowledge security. She is a guest editor of Energies MdPI, International Journal of Environmental Research and Public Health MdPI Journal, Computer Modeling in Engineering & Sciences. During her early career, she served on the editorial board of The Journal of Strategic Information Systems IF 14.682 and the Editorial Advisory Board of the International Journal of Management and Applied Research IF 3.508. She was a participant in multiple Erasmus+ teacher mobility: Slovenia, Hungary, Czech Republic, Slovakia and France.

Joanna Rosak-Szyrocka is an Assistant Professor, Erasmus+ coordinator at the Faculty of Management, Czestochowa University of Technology, Poland. She specialized in the fields of digitalization, industry 5.0, quality 4.0, education, IoT, AI, and quality management. She was a participant in multiple Erasmus+ teacher mobility programs: Italy, UK, Slovenia, Hungary, Czech Republic, Slovakia, and France. She cooperates with many universities both in her country and abroad. She is on the editorial boards of Plos One Journal, PeerJ Journal and on the Advisory Board of Heliyon Journal. She is the Associate Editor for Cogent Business and Management, Taylor & Francis, Guest Editor of Entertainment Computing Journal, Elsevier Journal, Resources MdPI, IJERPH MdPI, Energies MdPI, Sustainability MdPI, Springer Discover Sustainability Journal, Frontiers and Elsevier Measurement Journal. She is reviewer for a number of prestigious journals such as IEEE, Elsevier, MdPI, Frontiers, Sage, Springer and Emerald, and a member of the research team at the Faculty of Management of the Częstochowa University of Technology 2022 and the interdisciplinary team 2022. Additionally, she is a member of the team for surveying deans' offices and the course of studies in the field of occupational health and safety and a member of the technical team for evaluation at the Faculty of Management of the Częstochowa University of Technology. She was awarded a diploma of recognition from the Dean of the Faculty of Management of the Częstochowa University of Technology for her publishing activities, as well as a medal from the Dean of the Faculty of Management of the Częstochowa University of Technology in gratitude for long-term cooperation with the Faculty of Management (2017). In 2023, she was awarded the medal of the President of the Republic of Poland for long-term service.

Dr. Anand Nayyar earned a Ph.D (Computer Science) from Desh Bhagat University in 2017 in the area of Wireless Sensor Networks, Swarm Intelligence and Network Simulation. He is currently working in the School of Computer Science-Duy Tan University, Da Nang, Vietnam as Professor, Scientist, Vice-Chairman of Research, and Director- IoT and Intelligent Systems Lab. He is a Certified Professional with 125+ Professional certifications from CISCO, Microsoft, Amazon, EC-Council, Oracle, Google, Beingcert, EXIN, GAQM, Cyberoam among others. He has published more than 200+ Research Papers in various High-Quality ISI-SCI/SCIE/SSCI Impact

Factor- Q1, Q2, Q3, Q4 Journals cum Scopus/ESCI indexed Journals, 80+ Papers in International Conferences indexed with Springer, IEEE and ACM Digital Library, 50+ Book Chapters in various SCOPUS/WEB OF SCIENCE Indexed Books with Springer, CRC Press, Wiley, IET, Elsevier with Citations: (Google Scholar): 14000+, H-Index: 62 and I-Index: 225; (Scopus): 7000+; H-Index: 45. Member of more than 60+ Associations as Senior and Life Member such as, IEEE (Senior Member) and ACM (Senior Member). He has authored/co-authored cum Edited 65+ Books of Computer Science and is associated with more than 600+ International Conferences as Programme Committee/Chair/Advisory Board/Review Board member.

He is Associate Editor for Wireless Networks (Springer), Computer Communications (Elsevier), International Journal of Sensor Networks (IJSNET) (Inderscience), Computers Materials and Continua (CMC), Tech Science Press- IASC, Cogent Engineering, Human Centric Computing and Information Sciences (HCIS), PeerJ Computer Science, IET-Quantum Communications, IET Networks, IEEE Transactions on Artificial Intelligence (IEEE TAI), Indonesian Journal of Electrical Engineering and Computer Science, IJFC, IJISP, IJDST, IJCINI, IJGC, IJSIR, IJBDCN, IJNR, IJSI. He is Managing Editor of IGI-Global Journal, USA titled "*International Journal of Knowledge and Systems Science (IJKSS)*" and Editor-in-Chief of IGI-Global, USA Journal titled "*International Journal of Smart Vehicles and Smart Transportation (IJSVST)*". He has reviewed more than 2600+ Articles for diverse Web of Science and Scopus Indexed Journals. He is currently researching in the areas of Wireless Sensor Networks, Internet of Things, Swarm Intelligence, Cloud Computing, Artificial Intelligence, Drones, Blockchain, Cyber Security, Healthcare Informatics, Big Data and Wireless Communications.

Dr. Mohd Naved is an Associate Professor in Jaipuria Institute of Management, Noida, India. He has an impressive career spanning over a decade in the fields of Business Analytics, Data Science, and Artificial Intelligence. As an educator, Dr. Naved has consistently demonstrated a commitment to the highest standards of teaching and mentoring, ensuring that his students receive an education that is both cutting-edge and grounded in real-world experience. His dedication to helping students achieve their full potential extends beyond the classroom, as he has been an active participant in the university's Mentor-Mentee Program, providing guidance and support to over 150 undergraduate and postgraduate students. In addition to his teaching prowess, Dr. Naved has excelled in the areas of education management, research, and curriculum development. He has served on various committees and led initiatives related to curriculum development, faculty recruitment and retention, and accreditation, contributing to the institutions he has worked with becoming centers of academic excellence in their respective fields. He has also successfully led the launch of several BBA/MBA programs, resulting in increased admissions and student satisfaction.

As a researcher, Dr. Naved has made significant contributions to the fields of Business Analytics, Data Science, and Artificial Intelligence, with over 80+ publications in reputed scholarly journals and books. His research focuses on the applications of these disciplines in various industries, and he has supervised numerous research projects and dissertations, guiding students to successful outcomes.

Contributors

Nahad Al-Maskari
Finance Department
A'Sharqiyah University
Ibra, Oman

Dusmanta Behera
Department of Commerce
Vyasanagar Autonomous College
Jajpur, India

T. Blažauskas
Department of Software
 Engineering
Kaunas University of Technology
Kaunas, Lithuania

R. Damaševičius
Department of Applied Informatics
Vytautas Magnus University
Kaunas, Lithuania

Nguyen Huu Phuoc Dai
Bánki Donát Faculty
Óbuda University, Hungary

Girish Prasad Das
Department of Humanities and
 Social Sciences
Parala Maharaja Engineering College
Berhampur, India

P. Deepika
Department of Management Studies
Gayatri Vidya Parishad College of
 Engineering (A)
Visakhapatnam, India

Ayşenur Erdil
Department of Business
 Administration
Istanbul Medeniyet University
Turkey

B. Udaya Bhaskar Ganesh
Mittal School of Management Studies
Lovely Professional University
Phagwara, India

Yuvaraj Ganesan
Graduate School of Business
Universiti Sains Malaysia
Penang, Malaysia

Pham Phi Giang
Can Tho Technical Economics
 College
Vietnam

V. Harish
PSG Institute of Management
PSG College of Technology
Coimbatore, India

Fathyah Hashim
Graduate School of Business
Universiti Sains Malaysia
Penang, Malaysia

Pham Quang Huy
School of Accounting
University of Economics Ho Chi
 Minh City (UEH)
Vietnam

Mugdha Shailendra Kulkarni
Symbiosis Centre for Information
  Technology (SCIT)
Symbiosis International (Deemed
  University)
Pune, India

Jyoti Kushwaha
Department of Home Management
KRG Post Graduate College
Jiwaji University
Gwalior, India

Ruchi Kushwaha
School of Studies in Management
Jiwaji University
Gwalior, India

Narayana Maharana
Department of Management Studies
Gayatri Vidya Parishad College of
  Engineering (A)
Visakhapatnam, India

Jyotirmayee Mohanty
Department of Economics
Kendrapada Autonomous College
Kendrapada, India

Vrinda V. Nair
RV University
Bengaluru, India

L. Narbutaitė
Department of Software
  Engineering
Kaunas University of Technology
Kaunas, Lithuania

Anand Nayyar
Graduate School
Faculty of Information Technology
Duy Tan University
Da Nang, Vietnam

Kanchan Pranay Patil
Symbiosis Centre for Information
  Technology (SCIT)
Symbiosis International (Deemed
  University)
Pune, India

MA. Vu Kien Phuc
School of Accounting
University of Economics Ho Chi
  Minh City (UEH)
Vietnam

Ali Rehman
Internal Audit Department
A'Sharqiyah University
Ibra, Oman

Renuka Sharma
Amity College of Commerce
Amity University
Gurugram, India

Pankaj Singh
Institute of Business Management
GLA University
Mathura, India

Tariq Umar
School of Architecture and
  Environment
University of the West England
Bristol, United Kingdom

M. Vasiljevas
Department of Software
  Engineering
Kaunas University of Technology
Kaunas, Lithuania

Truong Hong Vo Tuan Kiet
FPT University
Vietnam

Chapter 1

Industry 5.0

Employee readiness for co-bot workspace

Narayana Maharana
Gayatri Vidya Parishad College of Engineering (A), Visakhapatnam, India

Girish Prasad Das
Parala Maharaja Engineering College, Berhampur, India

B. Udaya Bhaskar Ganesh
Lovely Professional University, Phagwara, India

Jyotirmayee Mohanty
Kendrapada Autonomous College, Kendrapada, India

P. Deepika
Gayatri Vidya Parishad College of Engineering (A), Visakhapatnam, India

Dusmanta Behera
Vyasanagar Autonomous College, Jajpur, India

1.1 INTRODUCTION

Industry 5.0, also known as the 'human-centred industry', is the next stage of the industrial revolution that emphasizes the integration of human capabilities with advanced technologies such as co-bots, augmented reality, and artificial intelligence. Technology is certainly the backbone of the futuristic knowledge economy of Industry 5.0. In the coming days, robots are going to play an important role in almost every sector of the economy by integrating the physical world with the digital world (Culot et al., 2020; Ghobakhloo, 2020). Many academicians and industrial experts are anticipating and discussing how robotic technology is going to change the work environment, productivity, and competition (Breque et al., 2021; Cassioli et al., 2021; Demir et al., 2019; Jafari et al., 2022). In the past couple of years, researchers have been trying to figure out the pros and cons of the industrial transformation in Industry 5.0, including its impact on employees, employment, and the work environment (Briggs & Scheutz, 2014; Kwanya, 2023;

Lu et al., 2022; Min et al., 2019; Østergaard, 2018; Schrum et al., 2021; Tamers et al., 2020; Vermeulen et al., 2018). Besides, some scholars also highlighted the risks of unemployment in areas where the work is more routine and repetitive since these works can be easily replaced by robots (Dalenogare et al., 2018; Frey & Osborne, 2013) and most of the existing jobs are going to be automated in the coming few years (Hammershøj, 2019). This in turn would create a new generation of employees with an entirely novel job profile, skill set, and abilities (Demir et al., 2019). Some have pointed out that the advent of technology has made salary disparity worse by excessively favouring jobs that are inclined towards information flows. Essentially, the digitalization of society has exacerbated the information and resource gap in support of inherently strong professions that are strategically positioned and technologically capable of reorganizing, aggregating, and transferring data as well as translating, interpreting, and manipulating the data (Kristal, 2020). Some researchers also countered by arguing that machines cannot outsmart intricate knowledge systems, and, on the other hand, the adverse ripples of the digital revolution on employment have been overstated (Pettersen, 2019).

A study by Shirokii et al. (2018) during the era of Industry 4.0 claimed that automation of tasks replaced around 47% of the employees in Germany and over 60% of the jobs are at risk (Min et al., 2019; Vermeulen et al., 2018). Contrarily, it is also argued by many that we should embrace new technology since it has the potential to drive the economy and society towards a better future rather than to fear it just because it can abolish some routine and manual jobs (Kristal, 2020; Spencer, 2018; Vermeulen et al., 2018). Very recently, many scholars and policymakers from Europe and Japan also claimed that the emergence of Industry 5.0 would bring a drastic change in three areas: human-centricity, sustainability, and resilience (Breque et al., 2021; Choi et al., 2022). Hence, it is high time to contemplate the essential attributes that employees must possess to thrive in a technology-driven work environment, where the learning capabilities of robots surpass those of humans. Simultaneously, it is imperative for industries to proactively equip themselves to integrate such technologies seamlessly, fostering a harmonious relationship between human workers and robots within the workplace. The objective of this chapter is to analyse different theoretical perspectives and frameworks concerning key findings, with the aim of providing insight for future academic inquiry and policy interventions.

To achieve this goal, two primary research questions are posed as follows:

a. In what ways advanced AI technologies transformed industries around the world and what kind of skill and knowledge an employee needs to harvest to fit into such an economy? This research question focuses on the preparation of the labour market for the digital innovations that are going to change the job profile of next-generation employees.

b. How do the industries need to prepare themselves while adopting and calibrating new-age technologies and employees in Industry 5.0? This research question addresses the requirements that industries need to cultivate to adopt the drastic change brought by the use of AI and robotic technology at every level of the workplace.

1.1.1 Objectives of the chapter

The following are the objectives of the chapter:

- To identify the current level of employee readiness for co-bot workspace in Industry 5.0;
- To identify the HR requirements and adaptation-related challenges faced by industries to build a sustainable co-bot workspace;
- To highlight and prioritize the key skills of employees to work in a co-bot workspace;
- To highlight and prioritize the HR preparedness and adaptation-related challenges faced by industries for building a co-bot workspace;
- And, to suggest future research areas and initiatives to improve employee readiness for co-bot workspaces.

1.1.2 Organization of the chapter

The rest of the chapter is organized as follows: Section 1.2 presents a detailed review and analysis of the existing literature to explore how the future of Industry 5.0 would be and what kind of changes it brings to the labour market and industry. This section also explores the requirements that an employee needs to have to perform in a co-bot workspace and highlights what changes and capabilities an industry needs to have to sustain such an AI-driven economy. Section 1.3 deals with the methodology adopted for the study by discussing survey design, data collection, and estimation in AHP. Section 1.4 enlists the analysis, followed by observations and discussions in Section 1.5. Section 1.6 highlights the implications and limitations of the study conducted. Finally, Section 1.6 concludes the chapter with the future scope.

1.2 LITERATURE REVIEW

Industry 5.0, also referred to as the 'human-centred industry', is the beginning of a new era in manufacturing. It is built upon the advancements made in Industry 4.0, where automation and advanced technologies were integrated into the manufacturing processes. However, Industry 5.0 takes it a step further by placing humans at the forefront. It envisions a future where advanced technologies and human capabilities work hand-in-hand, leveraging the strengths of both to create more efficient, sustainable, and personalized manufacturing systems (Kumar & Nayyar, 2020; Nayyar et al., 2020). Industry 5.0 represents a paradigm shift that places a strong emphasis on human–machine collaboration, customization, sustainability, and digitalization (Jafari et al., 2022). It heralds a new era of manufacturing that embraces the potential of technology while prioritizing the well-being and empowerment of the workforce. In this section, we will delve into the key concepts, potential benefits, and challenges of Industry 5.0, exploring how it is shaping the future of manufacturing and redefining the way we approach different aspects of a co-bot workspace. With advances in robotics and automation, co-bots are becoming increasingly prevalent in various industries, revolutionizing the way we work. Unlike traditional robots that operate in isolation, co-bots are designed to work alongside humans, enhancing their capabilities and productivity (Sherwani et al., 2020). Co-bots are transforming work environments by creating a synergy between humans and machines, where they collaborate and coexist seamlessly and safely.

There are a lot of uncertainties in the future where robots will be playing significant roles in industries. As we navigate the early days of the Industrial Revolution, what direction it will take remains uncertain, except for some hypotheses and studies providing very limited insights. Østergaard (2018) highlighted the need for the future industrial revolution is certainly due to increasing consumer demand for customization. This idea was further supported by a study published in Bloomberg, which noted that a German car manufacturer is prioritizing people in production plants, recognizing that adaptability is a crucial factor for modern consumers (Mattioli et al., 2020). A few such studies suggested that Industry 5.0 can be characterized by 'intensified collaboration' between intelligent systems and people, in contrast to Industry 4.0 where the focus was primarily on 'automation' rather than collaboration. The shift towards Industry 5.0 reflects the growing recognition of the importance of human–machine collaboration in shaping the future of manufacturing and meeting evolving consumer demands. In 2021, the European Commission released a report that introduced the concept of Industry 5.0 as a means of addressing the economic and environmental challenges facing the industry. The goal of Industry 5.0 is to achieve green production, which refers to the creation of a carbon-neutral and energy-efficient industry (Solanki & Nayyar, 2019). This approach is aimed at reducing the environmental impact of industrial processes while also promoting sustainable economic growth. By adopting Industry 5.0 principles, companies can create innovative and sustainable solutions that benefit both the economy and the environment (Breque et al., 2021). The report highlights that Industry 4.0, over the past decade, has focused primarily on digitization and AI-driven technologies to enhance production efficiency and flexibility, while deviating from the novel objectives of social justice and sustainability (Nayyar & Kumar, 2020; Sahoo et al., 2021). In contrast, Industry 5.0

places a renewed emphasis on research and innovation while supporting the long-term objectives of service to humanity and sustainability (Breque et al., 2021). Another study by Maddikunta et al. (2022) characterized Industry 5.0 as more intriguing, agile, and scalable than its predecessors, owing to the advanced technologies at its disposal. The emergence of Industry 5.0 reflects a shifting paradigm in manufacturing, aiming to address pressing environmental concerns while driving innovation and sustainable practices in the industry (Kumar & Nayyar, 2020; Paschek et al., 2022).

At present, there are two distinct visions that are being proposed for Industry 5.0. The first vision centred around 'human–robot co-working', where humans and robots collaborate closely, leveraging their respective strengths. Humans would focus on tasks that require creativity and innovation, while robots would handle other routine or repetitive tasks. The second vision for Industry 5.0 is centred around bio-economy, which involves the intelligent utilization of biological resources for industrial purposes. This approach aims to strike a balance between ecology, industry, and economy, by harnessing the potential of bio-based materials and processes in manufacturing and production (Aaltonen & Salmi, 2019). These two visions reflect different perspectives on how Industry 5.0 may unfold, with one emphasizing human–robot collaboration and the other focusing on sustainable bio-economy principles. According to the European Commission, bio-economy is

> The production of renewable biological resources and the conversion of these resources and waste streams into value-added products, such as food, feed, bio-based products, and bioenergy. It includes agriculture, forestry, fisheries, food, pulp and paper production, as well as parts of the chemical, biotechnological and energy industries. Its sectors have a strong innovation potential due to their use of a wide range of sciences (life sciences, agronomy, ecology, food science, and social sciences), enabling and industrial technologies (biotechnology, nanotechnology, information and communication technologies (ICT), and engineering, and local and tacit knowledge.
>
> Demir et al. (2019)

The principle of biologization,[1] which entails the intelligent use of biological resources, has the potential to bring about a profound transformation in various industries (Schütte, 2018). As such, bio-economy could potentially serve as the overarching theme or a significant component of the next industrial revolution. It's important to note that Industry 5.0 may encompass both human–robot co-working and bio-economy, while also potentially incorporating other themes, such as space life, space industries, and space mining, as part of the next revolution. Scientists are already urging caution in the utilization of space resources (Elvis & Milligan, 2019), as space mining could potentially become the next 'gold rush'.

1.2.1 Co-bots in the manufacturing industries

The use of co-bots has seen a significant increase in recent years, with the market size estimated to be $649 million in 2019 and projected to grow at a rate of 45% between 2019 and 2025 (Bi et al., 2021). Manufacturing companies have introduced co-bots into their production environments to enhance productivity, usability, and flexibility (Calitz et al., 2017). Compared to manual workers, co-bots can minimize the risk of errors in operations, and the production process requires special personnel to supervise each step of the process (Aaltonen & Salmi, 2019; Calitz et al., 2017). Co-bots are robots designed to work alongside human workers in the same workspace, performing different tasks (Simões et al., 2022). This allows human workers to focus on developing innovative solutions and taking on decision-making roles, while the co-bots assist with various tasks (Calitz et al., 2017; Longo et al., 2020). In terms of technical aspects, co-bots offer several advantages over traditional industrial robots, especially in terms of user-friendliness and flexibility. Co-bots are capable of handling repetitive, tedious, and physically demanding tasks, and are equipped with sensors that can detect contact with their surroundings, whether it is intentional or accidental. These factors are crucial for ensuring safety and ease of use, and for promoting natural interactions between co-bots and human workers. Several studies have highlighted the significant role of co-bots in the manufacturing industry. For example, Calitz et al. (2017) conducted a study on African workers to examine the current and future implementation of co-bots and their impact to suggest that co-bot solutions can provide mobility and flexibility, thereby creating a sustainable work environment for the future (Batth et al., 2018).

Few other studies have also examined expectations and success factors in adopting co-bots from both industry and academic perspectives (Aaltonen & Salmi, 2019; Kildal et al., 2022). Like, for instance, Simões et al. (2022) had identified 39 factors that influence the intention to adopt co-bots, based on a conceptual framework integrating the 'Diffusion of Innovation Theory', 'Technology Organization-Environment Theory', and the 'Institutional Theory'. According to Aaltonen and Salmi (2019), the sustainable development of co-bots lies in the successful allocation of work among human workers and co-bots. Additionally, Bi et al. (2021) identified some notable co-bot developments in various applications. Most previous studies focused on the technical challenges of implementing co-bots in manufacturing processes. However, the actual use of co-bots in collaborative work environments can pose different challenges that are crucial for determining the success of human-co-bot collaboration in a smart working environment (Calitz et al., 2017). For instance, the acceptance of co-bots by human workers during the initial stages of implementation can only be achieved if they are involved in defining the work conditions. The use of co-bots can potentially improve the design of work systems and how work is organized (Simões et al., 2022). To achieve this, manufacturing companies must consider task requirements, safety and trust assurance, ergonomics, and the flexibility to adapt to changes (Bi et al., 2021; Maharana, 2020).

1.2.2 Issues in human–robot collaboration

Human–robot co-working is a concept that is becoming increasingly relevant with the rise of Industry 5.0. However, several issues need to be addressed for the successful implementation of this co-working model. One major issue is the need for training and upskilling of the human workforce to adapt to the new technologies and work processes (Agarwal & Chauhan, 2022). There is also a need for effective communication and collaboration between humans and robots to ensure safe and efficient work environments. Another issue is the potential loss of jobs and the need for re-skilling and re-employment programs for workers who may be displaced by automation. In addition, there is a need for standardization and regulation of the co-working model to ensure the ethical and fair treatment of both humans and robots. Overall, while human–robot co-working offers several benefits such as increased productivity and efficiency, it also presents various challenges that need to be carefully addressed for successful implementation. On the other hand, there are a lot of changes the organization need to undergo including evolution in organizational behaviour, ethics, work culture, organizational structure, workflow, and work environment. Additionally, some important factors to consider in a human–robot co-working environment include the acceptance of robots in the workplace, potential discrimination towards either robots or human workers, privacy concerns, building trust between human and robot workers, the redesign of workspaces to accommodate robots, and the need for education and training (Huang et al., 2022; Pizoń & Gola, 2023). The study of Buchner et al. (2013) discussed three major factors that undergo a drastic change in Industry 5.0, including individuals, organizations, and the robotic agent, where, individual factors include the employee's age, gender, educational level, technical experience, expectations, and social skills. Organizational factors include workflow, physical environment, social and emotional context, training, and alignment of employee–robot goals. Robotic agent factors encompass appearance, behaviour, interaction capabilities, and safety. The study of Demir et al. (2019) was based on the legal and regulatory issues due to human–robot coworking that any organization can face. Authors suggested that a clear and legal definition of a robot is essential to address the issues arising from human–robot co-working. Without a legally binding definition of a robot, human–robot co-working will create numerous issues. Therefore, there is a need to consider various factors such as individual, organizational, and robotic agents while designing human–robot co-working environments. The necessary regulations for human–robot co-working should include a clear definition of what a robot is, as well as distinctions between robots, drones, and cyborgs. The law and related regulations need to define the categories of robots that are appropriate for use in work environments, their designated functions and obligations, the scope of decisions that robots are authorized to make with regard to humans, the allocation of liability in cases of robot malfunctions, and whether the robot software operates on pre-programmed rules or has the ability to learn and adapt. Regulations should also cover robot development, manufacturing, and certifications to ensure safety and responsibility (Senna, 2022).

On the other hand, employee preference for working in a technology-driven environment varies from person to person. Employees who are incompatible with the co-bot environment will not be able to perform better. Similarly, psychological issues resulting from working with robots would impart certain disorders and may increase the stress level. Sometimes, it may result in some sort of addiction or phobia like 'robophobia' among the employees with increased use of robots. It is quite possible that advanced technology would bring some unprecedented social and psychological changes that we are not ready to face. The increase in the use of robots in workplaces may lead to a decrease in social interactions between humans (EunJeong et al., 2022; Smids et al., 2020). Employees may have different opinions on how to interact with robots socially, especially if the robots are in a management position. The issue is complicated by the fact that humans tend to show respect to their superiors or managers, but respect means nothing to a robot. Furthermore, some humans may prefer robots that exhibit social behaviour, while others may think it is unreal and fake. Studies have shown that children treat robots like living pets, indicating that social studies will be an essential part of researching human–robot co-working (Piçarra & Giger, 2018).

The human resource department has to incorporate a lot of changes in its functioning from work analysis to the job description, recruitment, training, and development. They have to fix new challenges arising out of human–robot co-working in addition to the current issues. Eventually, in a highly robot-dominated workplace, the importance given to humans would be minimal; therefore, human resource management may also take a different form to cater to such human-related issues. On the other hand, the introduction of robots into the workplace raises questions about their ethical status and how they may affect current work ethics. Robots that are selfless, do not know laziness, and cannot lie will challenge human employees expected ethical behaviour. It is difficult to predict the development of work ethics in an environment where humans and robots collaborate. A set of ethical standards will probably be established to govern the interactions between humans and robots and to define the ethical standing of robots in the workplace (Briggs & Scheutz, 2014). Moreover, in human–robot co-working environments, employees may need to learn how to behave towards robots and understand their different capabilities. Nonverbal communication may be important, but robots may or may not understand it. Employees may need time to get used to communicating properly with robots, and robot manufacturers will need to develop robots that can inform about their capabilities without causing confusion.

Whether or not Industry 5.0 will focus on human–robot co-working, it will still represent a significant shift for organizations. In reality, having machines in our lives will probably cause a big shift for humanity. This will mimic, excite, and inspire some people. Some people will think it's outlandish, annoying, or even dangerous for humanity. The media actively promotes this anti-robot sentiment. A few surveys on how people feel about machines in society will be skewed by this disapproval (Huang et al., 2022; Martynov et al., 2019; Paschek et al., 2022). One of the first measures designed to gauge attitudes towards machines takes a pessimistic stance. We are not sure how people will respond to robots until they truly live and interact with robots. As people interact with robots, their outlook towards them is likely to change. Children of today might respond differently from how members of our age did because they will develop alongside this or comparable technologies. Robotic civilization will exist for their offspring. In order to create a robotic society where humans can fully profit from technology while attempting to limit negative effects, it is important to be conscious of generational differences. A study by Matheson et al. (2019) on human–robot collaboration in the context of manufacturing applications analysed more than 40 papers discussing the control mechanism, and collaboration strategies for improving production, performance, and flexibility. Authors observed that most of the studies were carried out in the automotive and electronics industries where the co-bots are easily integrated into the workplace and also anticipated that the Small- and Medium-Scale Enterprises (SMEs) in future are the early adopters of such technological changes whereas the upcoming challenges are not quite clear at this point and need more in-depth study.

1.2.3 Employee and organizational readiness for Industry 5.0

The readiness of employees to work alongside co-bots is influenced by various factors, including their skills, knowledge, attitudes, and perceptions. One critical factor is the level of technology literacy among employees (Cheon et al., 2022; Kwanya, 2023; Schrum et al., 2021). Studies have shown that employees with higher levels of technology literacy are more comfortable working with co-bots than those with lower levels of technology literacy. Technology literacy includes the ability to use digital devices, software, and other technologies effectively (Davies, 2011). Employees who lack technology literacy may struggle to understand how to operate and interact with co-bots, leading to reduced efficiency and productivity (Tamers et al., 2020).

Another critical factor is the perception of co-bots among employees. Employees' perceptions of co-bots can be influenced by various factors, including their previous experiences with technology, their knowledge of co-bots, and their attitudes toward automation (Kwanya, 2023). Studies have shown that employees who have positive perceptions of co-bots are more likely to be comfortable working with them (Kwanya, 2023; Lambrechts et al., 2021). On the other hand, employees who view co-bots negatively may be resistant to working with them, leading to reduced productivity and efficiency (Lambrechts et al., 2021).

Training is critical to ensure that employees are ready to work with co-bots effectively. The training should focus on providing employees with the knowledge and skills necessary to operate and interact with co-bots safely and efficiently (Calitz et al., 2017). The training should also address any concerns or misconceptions that employees may have about co-bots. Studies have shown that employees who receive comprehensive training on co-bots are more comfortable and confident working with them. Training should also be ongoing to ensure that employees remain up-to-date with the latest co-bot technologies and best practices (Cassioli et al., 2021).

Industry 5.0 requires a significant shift in organizational mindset and the adoption of new technologies, business models, and work processes. This literature review examines the preparedness of industries to adopt the changes associated with Industry 5.0. The readiness of industries to adopt Industry 5.0 depends largely on their technological maturity. Industry 5.0 requires advanced automation, machine learning, and artificial intelligence technologies that are not yet widely available. Therefore, industries that have already adopted Industry 4.0 technologies and have a culture of innovation are more likely to be prepared for the changes of Industry 5.0. Similarly, the organizational culture of industries can significantly affect their preparedness for Industry 5.0. Industries that have a culture of innovation, adaptability, and flexibility are more likely to be prepared for the changes of Industry 5.0 (Lachvajderová & Kádárová, 2022; Romero & Stahre, 2021).

On the other hand, industries with a rigid and hierarchical organizational culture may struggle to adapt to the changes brought about by Industry 5.0. Along with this, the preparedness of the existing workforce to adopt the changes of Industry 5.0 is another essential factor (Dwivedi et al., 2023). Industries need to ensure that their workforce has the necessary skills and knowledge to operate and maintain Industry 5.0 technologies. They also need to invest in upskilling and reskilling programs to ensure that their employees are ready for the new work processes and business models associated with Industry 5.0 (Agarwal & Chauhan, 2022; Doyle-Kent & Shanahan, 2022; Leopold et al., 2017). As such, the regulatory framework can significantly affect the adoption of Industry 5.0. Governments need to create an enabling environment for the development and adoption of Industry 5.0 technologies (Işcan, 2021). This includes creating supportive policies, funding research and development, and providing incentives for industries to invest in Industry 5.0. Moreover, there is a sheer necessity for collaboration between industries, which plays a crucial role in the adoption of Industry 5.0. Collaboration can help to create economies of scale, promote knowledge sharing, and accelerate the development and adoption of Industry 5.0 technologies (Senna, 2022).

The adoption of Industry 5.0 poses several challenges for human resource (HR) professionals. This literature review also examines the challenges faced by HR while adapting to Industry 5.0. The adoption of advanced technologies requires a new set of skills and competencies, which are different from those required in traditional industries (Agarwal & Chauhan, 2022; Doyle-Kent & Shanahan, 2022). HR professionals face the challenge of identifying and recruiting individuals with the necessary skills and knowledge to work in the Industry 5.0 environment (Maddikunta et al., 2022; Paschek et al., 2022). This can be a daunting task as the demand for skilled workers is high, and the competition for talent is intense. In Industry 5.0, employees need to be equipped with new skills and knowledge to operate and manage advanced technologies such as AI, IoT, and robotics (Spencer, 2018). HR professionals are responsible for providing training and development opportunities to help employees acquire new competencies (Orlova, 2021). However, this can be a challenge as the training programs need to be designed to meet the specific needs of the Industry 5.0 workforce, which requires a significant investment of time and resources (Doyle-Kent & Shanahan, 2022). Moreover, there is a need for significant changes in the work environment, which can be unsettling for some employees. HR professionals need to develop retention strategies to ensure that employees feel valued and engaged (Cillo et al., 2022; Orlova, 2021). This can involve creating a supportive work culture, providing opportunities for career development, and offering competitive compensation and benefits packages. At the same time, another critical challenge faced by HRM in Industry 5.0 is the shortage of skilled workers who can work effectively with the latest technologies. With Industry 5.0, organizations require employees who possess a unique set of skills that combines technical expertise with social and emotional intelligence (Cillo et al., 2022; Martynov et al., 2019). However, many organizations struggle to find employees with these skills, and the shortage of skilled workers can impede the adoption of Industry 5.0. On the other hand, resistance to change also brings a lot of complications for HR (Romero & Stahre, 2021). The implementation of Industry 5.0 requires a significant shift in the way organizations operate, and many employees may be resistant to this change. Some employees may fear that their jobs will be replaced by machines, while others may feel overwhelmed by the new technologies and processes. HRM needs to address these concerns and provide employees with the necessary training and support to adapt to the changes brought by Industry 5.0.

Industry 5.0 necessitates a diverse and inclusive workforce that can collaborate effectively. HR professionals need to develop strategies to recruit and retain individuals from diverse backgrounds and create an inclusive work environment that values and respects individual differences. In addition to

this, it also brings a lot of compliance with new legal and regulatory requirements such as data privacy and security laws (Leng et al., 2022). HR professionals need to be aware of these requirements and ensure that the organization
is compliant to avoid legal and financial risks. Where organizations are
increasingly relying on interconnected systems and devices to automate processes and gather data, the importance of security cannot be ignored. While
robots can enhance productivity and efficiency, they also create new kinds of
cybersecurity risks about which we are quite clueless (Özsungur, 2021; Venaik
et al., 2023). HRM needs to ensure that employees are trained to identify and
respond to cybersecurity threats to prevent data breaches and protect sensitive information. More importantly, it also creates several ethical concerns
that HRM needs to address. For example, the use of AI and automation may
lead to job displacement and inequality, and organizations need to ensure
that their technology and practices are ethical and inclusive. HRM needs to
establish clear ethical guidelines and ensure that employees are trained to
uphold these standards.

Working in a collaborative robot or 'co-bot' workspace requires certain
skills, knowledge, and qualities from employees. Based on the previous
studies and discussions with industrial experts, the following points highlight the things that an employee may need to work in a co-bot workspace
(Figure 1.1).

a) **Adaptability:** Working in a co-bot workspace may require employees to adapt to new and changing work environments. Co-bots may
be reprogrammed or moved to different locations in the workspace,
which may require employees to adjust their workflows and work
methods accordingly.

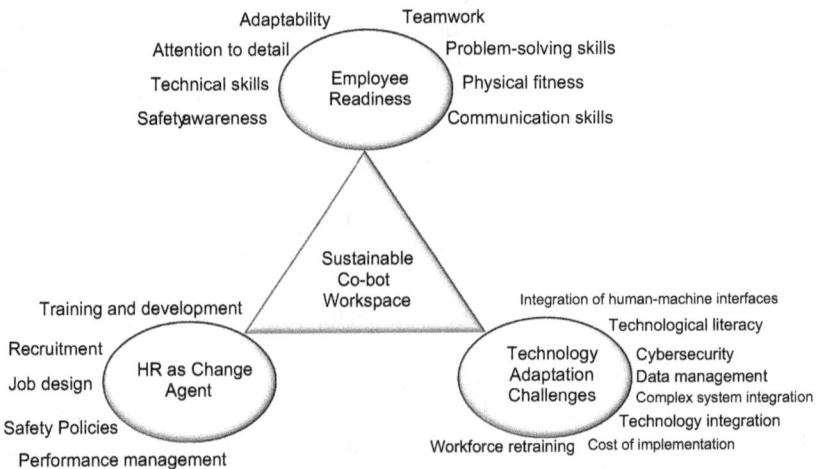

Figure 1.1 Employee and industry requirements for sustainable co-bot workspace.

b) **Attention to detail:** The precision and accuracy required in a co-bot workspace mean that employees need to be detail-oriented. They should be able to follow instructions carefully and ensure that their work is accurate and error-free.

c) **Communication skills:** Communication is key in a co-bot workspace, as employees must work closely with the co-bots and with each other. They should be able to communicate effectively and clearly, both verbally and in writing.

d) **Physical fitness:** Depending on the tasks being performed in the co-bot workspace, employees may need to be physically fit and able to perform tasks such as lifting, standing for extended periods of time, or working in awkward positions.

e) **Problem-solving skills:** Co-bots are designed to work alongside humans and assist them with tasks, but they may encounter unexpected situations or errors. Employees should be able to think critically and creatively to solve problems as they arise.

f) **Safety awareness:** Safety is of utmost importance in a co-bot workspace. Employees should be aware of the safety protocols and procedures in place, and understand the risks associated with working with co-bots. They should also be able to identify and respond to potential safety hazards.

g) **Teamwork:** Collaboration is a key aspect of working in a co-bot workspace. Employees should be able to work well with others and be willing to help and support their colleagues as needed.

h) **Technical skills:** Employees working in a co-bot workspace should have a good understanding of the technology behind the co-bots and how they operate. This may include programming skills, knowledge of sensors and automation systems, and the ability to troubleshoot technical issues.

Overall, employees working in a co-bot workspace should have a combination of technical skills, safety awareness, communication skills, adaptability, attention to detail, teamwork, and problem-solving skills.

To make the co-bot environment effective, the HR department in an organization may need to make the following changes.

a) **Job design:** The HR department may need to redesign job roles to incorporate co-bots into work processes. This could include identifying tasks that co-bots can perform and redefining the roles of employees to focus on more complex or strategic tasks that require human expertise.

b) **Performance management:** The HR department may need to review and update performance management systems to account for the use of co-bots. This could include setting performance targets for both employees and co-bots, evaluating the efficiency of the co-bot environment, and assessing the performance of employees working with co-bots.

c) **Recruitment:** The HR department may need to recruit employees with the necessary skills and knowledge to work in a co-bot environment. This could include recruiting individuals with engineering, robotics, or computer science backgrounds to operate and program the co-bots.

d) **Safety policies:** The HR department may need to review and update safety policies to ensure that employees can work safely with co-bots. This could include implementing safety protocols to prevent accidents and ensuring that employees are trained to operate co-bots safely.

e) **Training and development:** The HR department may need to provide employees with training and development programs to enhance their skills and knowledge of co-bots. This could include technical training, safety training, and collaboration skills training to ensure that employees can work efficiently with co-bots.

Overall, the HR department may need to provide training and development programs, recruit the right employees, redesign job roles, review and update safety policies, and adapt performance management systems to make the co-bot environment effective.

The technological adaptation challenges that organizations may face while implementing Industry 5.0 are enlisted as follows.

a) **Complex system integration:** Integrating various technologies such as co-bots, sensors, and artificial intelligence to work together seamlessly can be a complex process. It requires a high level of technical expertise to integrate these systems effectively.

b) **Technology integration:** Integrating advanced technologies like robotics, artificial intelligence, and IoT into existing manufacturing systems can be a significant challenge. Organizations need to ensure that these technologies are compatible with existing systems and processes, and they may need to invest in additional infrastructure and tools to facilitate this integration.

c) **Cost of implementation:** Implementing advanced technologies required for Industry 5.0 can be expensive, and it may be a challenge for some companies to justify the cost of implementation. They need to ensure that the benefits, such as increased productivity and efficiency, outweigh the costs of implementation.

d) **Cybersecurity:** As Industry 5.0 relies heavily on connected devices and technologies, cybersecurity is a significant challenge. There is a need to ensure that the systems and technologies used are secure from cyber-attacks and data breaches. The integration of advanced technologies into manufacturing processes creates new cybersecurity challenges. Organizations need to ensure that their systems are secure and protected from cyber threats, and they need to invest in cybersecurity tools and practices to mitigate these risks (Venaik et al., 2023).

e) **Data management:** With the proliferation of sensors and IoT devices, the amount of data generated by Industry 5.0 systems can be overwhelming. Organizations need to develop efficient data management strategies to collect, store, process, and analyse this data to derive insights and make informed decisions.

f) **Integration of human–machine interfaces:** In Industry 5.0, humans and machines will work together seamlessly. However, integrating human–machine interfaces will require a significant technological adaptation as machines and humans have different ways of interacting and communicating.

g) **Technological literacy:** Industry 5.0 requires a highly skilled workforce that is comfortable with advanced technologies such as artificial intelligence, augmented reality, and co-bots. There is a need for technological literacy among the workforce to ensure the effective integration and use of these technologies.

h) **Workforce retraining:** The shift to Industry 5.0 requires organizations to retrain their workforce to work collaboratively with machines. Employees need to be equipped with new skills such as data analytics, machine learning, and robotics to effectively collaborate with the machines.

Overall, organizations may face several technological adaptation challenges while implementing Industry 5.0. They need to ensure that they have the right infrastructure, tools, and skills to successfully integrate advanced technologies into their manufacturing processes.

The emergence of Industry 5.0, which involves human collaboration with specialized robots, has prompted inquiries into the necessary skills for successful co-working with co-bots in manufacturing settings. This research chapter aims to investigate the essential employability skills required for such collaboration, as well as the strategies organizations can use to foster a collaborative environment in the era of Industry 5.0. To identify these skills, it is first necessary to examine the key factors influencing the co-working relationship between humans and robots (refer to Figure 1.2). Based on a review of relevant literature and expert interviews, this study identifies three major factors: employee readiness, HR as a change agent, and challenges in adapting to new technologies.

1.3 METHODOLOGY

Saaty (1980) introduced the Analytic Hierarchy Process (AHP), which is a structured approach to organizing and analysing the elements of a problem in a hierarchical manner. AHP is a user-friendly method for making and

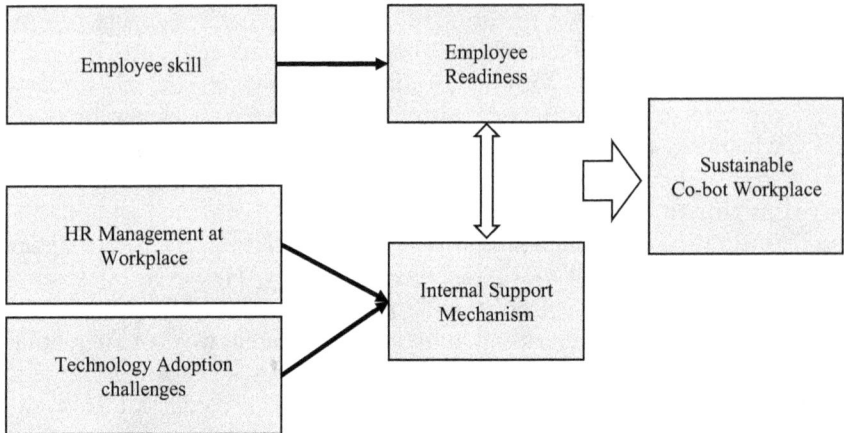

Figure 1.2 Conceptual model for building a sustainable co-bot workplace.

evaluating decisions (Millet & Saaty, 2000; Saaty, 1980), particularly when there are a limited number of choices with multiple attributes. AHP involves comparing objects concerning a common goal or criterion, with weight scores (preference scores) assigned to the factors being compared to help inform the decision. The primary criteria are compared in paired groups, while the sub-criteria are compared within those groups. The preference response scale (Hafeez et al., 2002; Saaty, 1980) used during the comparison of criteria is presented in Table 1.1.

1.3.1 Survey design

For this study, we designed a survey tool that contains three independent variables that lead to the goal of attaining a sustainable co-bot workplace,

Table 1.1 Rating of preferences in pairwise comparison of sub-criteria in AHP

Preference weight	Meaning	Rating
Equally preferred	Both elements contribute equally to the objective	1
Moderately preferred	Weak importance of one over the other	3
Strongly preferred	One element is strongly favoured over the other	5
Very strongly preferred	One element is very strongly preferred over the other	7
Extremely preferred	The evidence favouring one element is the highest possible order of affirmation	9
Intermediate values	When compromise is needed between two adjacent judgments	2, 4, 6, 8

Source: Bhutta and Huq (2002); Saighal et al. (2015)

which is our dependent variable. We identified a total of 28 items for the survey through a literature review and consultation with key employees from selected organizations. The survey instrument was developed in consultation with experts in the field and academic professionals. The primary goal of the survey tool was to assess the priority of employee readiness and industry preparedness for creating an effective co-bot workplace. The readiness of HR personnel and the challenges associated with adopting new technologies served as key indicators of industrial preparedness.

1.3.2 Data collection

Responses were collected from the 14 industrial experts working in the seven selected industries online by asking them to respond to the AHP pairwise comparison using the website https://bpmsg.com/ahp/ahp-calc.php. We contacted each selected employee to respond to the AHP comparison matrix for each criterion online and also ensured that the consistency ratio (CR) value was within the acceptable limit. Further, the validity of the instrument was also ensured using both expert opinion and statistical tools.

1.3.3 Estimations in AHP

The process of AHP starts with the collection of pairwise comparison data in a matrix form of size $n \times n$ (n = number of sub-criteria). The upper trigonal comparison matrix values are represented as 'a_{ij}' and the lower trigonal comparison matrix value can be determined by getting the reciprocal of 'a_{ij}':

$$a_{ij} = \frac{1}{a_{ij}}$$

where i and $j = 1, 2, ..., 9$.

In the second step, we calculate the average of all response matrices and then normalized using the following formula

$$a_{ik} = \frac{a_{ij}}{\sum\limits_{i=1}^{n} a_{ij}}$$

Calculating the priority vector for a criterion W_{ik} we can use the formula given below:

$$W_{ik} = \frac{\sum\limits_{i=1}^{n} a_{ij}}{n}$$

To weight the Eigenvectors based on the criteria weights, hierarchical synthesis is employed, where the sum of all weighted Eigenvector entries corresponding to those in the next lower level of the hierarchy is calculated.

Now the Weighted Vector (W_v) can be estimated as

$$W_V = \sum_{i=1}^{n} a_{ij}W_{ik}$$

where 'i' is the row number and 'k' is the column number.

Now let

$$AW = \begin{bmatrix} X_1 \\ X_2 \\ X_3 \\ . \\ . \\ . \\ X_n \end{bmatrix}$$

Then, we can calculate λ_{max}

$$\lambda_{max} = \frac{1}{n}\left[\frac{X_1}{W_1} + \frac{X_2}{W_2} + \frac{X_3}{W_3} + \ldots + \frac{X_n}{W_n}\right]$$

Eigenvalues can be calculated using the formula

$$E_v = \frac{W_{ik}}{RI}$$

Having made all the pairwise comparisons, the consistency is determined by using the Eigenvalue, λ_{max}, to calculate the consistency index (CI):

$$CI = \frac{\lambda_{max} - n}{n - 1}$$

where 'n' is the matrix size or the number of criteria.

> *Consistency Ratio*: The consistency ratio (CR) was calculated using Saaty's random index table (ref. Table 1.2). An inconsistency of 10% or less implies that the adjustment is small as compared to the actual values of the Eigenvector entries. Further, a higher value of CR is

Table 1.2 Saaty's random index for different numbers of elements

N	1	2	3	4	5	6	7	8	9	10	11	12
RI	0	0	0.52	0.89	1.11	1.25	1.35	1.40	1.45	1.49	1.51	1.54

Source: Saaty (1980)

considered unworthy since those responses would be random and comparison is not reliable.

$$CR = \frac{CI}{RI}$$

1.4 ANALYSIS

1.4.1 Demographic profile

Table 1.3 provides the demographic profile of the participants of the survey. We selected 34 employees from seven manufacturing industries where they are using artificial intelligence, robotics, and other advanced techniques for their manufacturing and other operational activities. Since the introduction of artificial intelligence in manufacturing is in the early stage in India; therefore, the availability of samples is very limited. In this particular study, we collected data from 34 employees out of which about 56% are male and

Table 1.3 Demographic profile of the respondents (N = 34)

Demographics	Number of respondents	Percentage (%)
Gender		
Male	19	55.88
Female	15	44.12
Age		
Less than 35 years	16	47.06
36–45 years	9	26.47
Above 45 years	9	26.47
Department		
Production	12	35.29
Quality control and R&D	8	23.53
Human resource	13	38.24
Position		
Managerial/supervisor	18	52.94
Executive	16	47.06

Source: Interpretation of primary data

44% are females. The majority (47%) of the respondents who participated are at a young age (less than 35 years) whereas 26% of the respondents are in the middle age group (35–45 years) and another 26% of respondents are aged above 45 years. Therefore, we maintained uniformity among customers of different age groups to get an unbiased result that may arise due to the respondents of similar age groups. We also selected respondents from three prominent departments in a manufacturing industry where AI is already in use; 35% of respondents are from the production department, 24% are from the Quality Control and R&D (Research and Development) department, and 38% are from the HR department. Out of the total respondents, 53% are holding managerial positions in the selected organizations whereas 47% are in executive positions.

1.4.2 Development of a hierarchical model

To build the hierarchical structure, first, a graphical model needs to be planned. The graphical presentation acts as a guide map for achieving the overall objective of the AHP process. The factors that are identified by reviewing the prior studies and from expert opinion are now considered as the sub-criteria for the AHP model. Figure 1.3 provides the criteria and sub-criteria of the AHP model. Usually, all the identified elements or criteria under a factor can be considered in the AHP model since AHP allows pairwise comparisons of all the available options. However, with the increase in the number of criteria, the level of consistency in response would be critical

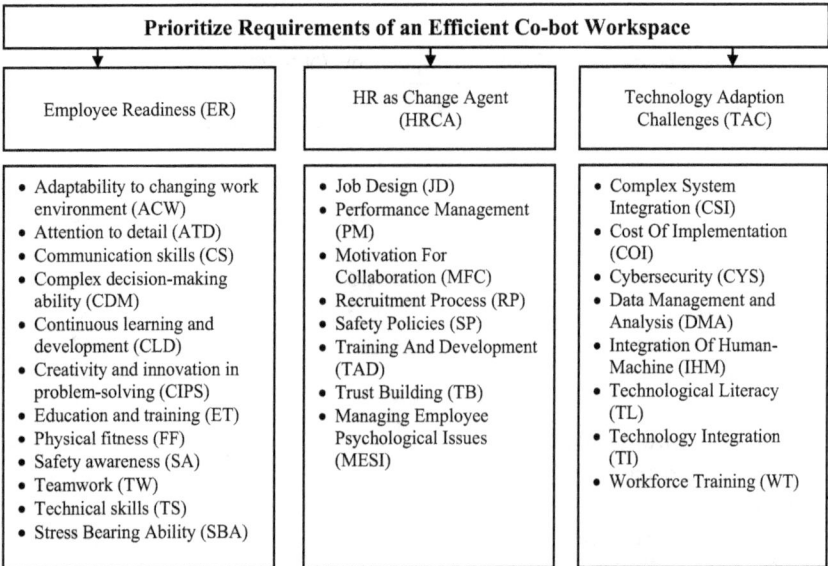

Prioritize Requirements of an Efficient Co-bot Workspace		
Employee Readiness (ER)	HR as Change Agent (HRCA)	Technology Adaption Challenges (TAC)
• Adaptability to changing work environment (ACW) • Attention to detail (ATD) • Communication skills (CS) • Complex decision-making ability (CDM) • Continuous learning and development (CLD) • Creativity and innovation in problem-solving (CIPS) • Education and training (ET) • Physical fitness (FF) • Safety awareness (SA) • Teamwork (TW) • Technical skills (TS) • Stress Bearing Ability (SBA)	• Job Design (JD) • Performance Management (PM) • Motivation For Collaboration (MFC) • Recruitment Process (RP) • Safety Policies (SP) • Training And Development (TAD) • Trust Building (TB) • Managing Employee Psychological Issues (MESI)	• Complex System Integration (CSI) • Cost Of Implementation (COI) • Cybersecurity (CYS) • Data Management and Analysis (DMA) • Integration Of Human-Machine (IHM) • Technological Literacy (TL) • Technology Integration (TI) • Workforce Training (WT)

Figure 1.3 Factors and parameters for AHP.

to achieve. We have 12 sub-criteria under the factor of 'employee readiness', and eight sub-criteria under the factor of 'HR as a Change Agent' and 'technology adoption challenges' each. Further, the assessment bias of the respondents needs to be looked into while collecting pairwise comparison data from the respondents. By using the online AHP pairwise comparisons tool, we collected data from the respective respondents personally after explaining how to respond to the AHP pairwise comparisons data sheet.

After the collection of pairwise comparison data in a matrix form for each factor, we used all 34 responses to find out the average response pairwise comparisons matrix. Then, we obtained the column total in the average response pairwise comparisons matrix, which is represented in the first table of the AHP outcome of each factor (ref. Tables 1.4, 1.7, and 1.10). In the next tables (ref. Tables 1.5, 1.8, and 1.11), we normalized the average responses by dividing each value in the column by the respective column total. After getting the normalized values, we estimated the average of each row called the prioritized vector (PV) and ranked them to get erstwhile information about the priorities of the sub-criteria. In the next step, we estimated the weighted Vectors and Eigenvalues by using formulas stated in the methodology. We then estimated the Average Lambda (λ_{max}), Consistency Index (CI), and Consistency Ratio (CR) for each factor to ensure the reliability of the model outcome, the results of which are given in Tables 1.6, 1.9, and 1.12.

1.5 OBSERVATION AND DISCUSSION

The final AHP ranking outcome for the factor Employee Readiness given in Table 1.5 indicates that the sub-criteria 'Creativity and innovation to problem solving' is the most important element for the employees to make them suitable for working in a co-bot workplace. Since the working of robots along with humans will be guided by artificial intelligence but needs to be supported by the innovativeness of the worker which would be missing in the case of robots. This quality in employees should be improved followed by 'Technical Skill' and 'Complex Problem-solving ability' which were ranked two and three, respectively. The least preferred skill that an employee may require for working in the co-bot workspace is 'Physical fitness'. 'Attention to detail' was ranked 11th and one of the least preferred characteristics of an employee may be due to the fact that robots can do the job of attention to detail more efficiently than humans. Further, Table 1.6 provides the reliability measures for the hierarchical model for the factor of 'employee readiness'. The reliability of the responses and the AHP ranking of the sub-criteria can be ensured by looking at the consistency index and consistency ratios estimated thereof. We can see that the CI of the AHP model is estimated to be 0.0882 (8.82%), which is below 0.10, indicating the consistency of the responses and reliability of the model.

Table 1.4 Average responses of pairwise comparison matrix for employee readiness

	ACW	ATD	CS	CDM	CLD	CIPS	ET	FF	SA	TW	TS	SBA
ACW	1.0000	6.6000	1.0000	0.1580	0.3680	0.1520	0.3980	5.4000	1.0000	2.2000	0.1640	1.0000
ATD	0.1520	1.0000	0.2820	0.1820	0.1580	0.1580	0.1820	1.6000	0.2200	1.6000	0.1520	0.2200
CS	1.0000	3.6000	1.0000	0.2200	0.2820	0.1520	0.7000	6.6000	1.0000	3.8000	0.1880	0.5000
CDM	6.4000	5.6000	4.6000	1.0000	1.0000	1.0000	4.0000	7.6000	6.6000	6.6000	0.7320	5.6000
CLD	3.6000	6.4000	3.6000	1.0000	1.0000	1.0000	1.0000	7.0000	3.6000	4.0000	0.4400	3.4000
CIPS	6.6000	6.4000	6.6000	1.0000	1.0000	1.0000	5.0000	6.6000	8.6000	7.6000	2.6000	5.6000
ET	2.6000	5.6000	1.6000	0.2500	1.0000	0.2000	1.0000	5.6000	4.6000	4.0000	1.0000	1.8000
FF	0.1880	0.7000	0.1520	0.1280	0.1400	0.1580	0.1820	1.0000	0.3500	0.3460	0.1520	0.1880
SA	1.0000	4.6000	1.0000	0.1520	0.2820	0.1140	0.2200	3.2000	1.0000	2.2000	0.1400	0.8500
TW	0.5980	0.7000	0.2780	0.1520	0.2620	0.1280	0.2500	3.4000	0.6400	1.0000	0.1860	0.3720
TS	6.2000	6.6000	5.4000	1.8000	3.2000	0.6800	1.0000	6.8000	7.0000	5.6000	1.0000	5.0000
SBA	1.0000	4.6000	2.0000	0.1820	0.4000	0.2240	0.6660	5.4000	1.6000	3.0000	0.2000	1.0000
Sum	30.3380	52.4000	27.5120	6.2240	9.0920	4.9660	14.5980	60.2000	36.2100	41.9460	6.9540	25.5300

Table 1.5 Normalized pairwise comparison matrix for employee readiness

	ACW	ATD	CS	CDM	CLD	CIPS	ET	FF	SA	TW	TS	SBA	Prioritized vector (PV)	Rank
ACW	0.03296	0.12595	0.03635	0.02539	0.04048	0.03061	0.02726	0.08970	0.02762	0.05245	0.02358	0.03917	0.04596	8
ATD	0.00501	0.01908	0.01025	0.02924	0.01738	0.03182	0.01247	0.02658	0.00608	0.03814	0.02186	0.00862	0.01888	11
CS	0.03296	0.06870	0.03635	0.03535	0.03102	0.03061	0.04795	0.10963	0.02762	0.09059	0.02703	0.01958	0.04645	7
CDM	0.21096	0.10687	0.16720	0.16067	0.10999	0.20137	0.27401	0.12625	0.18227	0.15735	0.10526	0.21935	0.16846	3
CLD	0.11866	0.12214	0.13085	0.16067	0.10999	0.20137	0.06850	0.11628	0.09942	0.09536	0.06327	0.13318	0.11831	4
CIPS	0.21755	0.12214	0.23990	0.16067	0.10999	0.20137	0.34251	0.10963	0.23750	0.18119	0.37389	0.21935	0.20964	1
ET	0.08570	0.10687	0.05816	0.04017	0.10999	0.04027	0.06850	0.09302	0.12704	0.09536	0.14380	0.07051	0.08662	5
FF	0.00620	0.01336	0.00552	0.02057	0.01540	0.03182	0.01247	0.01661	0.00967	0.00825	0.02186	0.00736	0.01409	12
SA	0.03296	0.08779	0.03635	0.02442	0.03102	0.02296	0.01507	0.05316	0.02762	0.05245	0.02013	0.03329	0.03643	9
TW	0.01971	0.01336	0.01010	0.02442	0.02882	0.02578	0.01713	0.05648	0.01767	0.02384	0.02675	0.01457	0.02322	10
TS	0.20436	0.12595	0.19628	0.28920	0.35196	0.13693	0.06850	0.11296	0.19332	0.13350	0.14380	0.19585	0.17938	2
SBA	0.03296	0.08779	0.07270	0.02924	0.04399	0.04511	0.04562	0.08970	0.04419	0.07152	0.02876	0.03917	0.05256	6
	1.00000	1.00000	1.00000	1.00000	1.00000	1.00000	1.00000	1.00000	1.00000	1.00000	1.00000	1.00000		

Table 1.6 Reliability test matrix for employee readiness

	ACW	ATD	CS	CDM	CLD	CIPS	ET	FF	SA	TW	TS	SBA	PV	Weighted vector	Eigenvalue
ACW	1.0000	6.6000	1.0000	0.1580	0.3680	0.1520	0.3980	5.4000	1.0000	2.2000	0.1640	1.0000	0.04596	0.599	13.035
ATD	0.1520	1.0000	0.2820	0.1820	0.1580	0.1580	0.1820	1.6000	0.2200	1.6000	0.1520	0.2200	0.01888	0.244	12.912
CS	1.0000	3.6000	1.0000	0.2200	0.2820	0.1520	0.7000	6.6000	1.0000	3.8000	0.1880	0.5000	0.04645	0.601	12.938
CDM	6.4000	5.6000	4.6000	1.0000	1.0000	1.0000	4.0000	7.6000	6.6000	6.6000	0.7320	5.6000	0.16846	2.383	14.145
CLD	3.6000	6.4000	3.6000	1.0000	1.0000	1.0000	1.0000	7.0000	3.6000	4.0000	0.4400	3.4000	0.11831	1.617	13.666
CIPS	6.6000	6.4000	6.6000	1.0000	1.0000	1.0000	5.0000	6.6000	8.6000	7.6000	2.6000	5.6000	0.20964	3.004	14.328
ET	2.6000	5.6000	1.6000	0.2500	1.0000	0.2000	1.0000	5.6000	4.6000	4.0000	1.0000	1.8000	0.08662	1.202	13.876
FF	0.1880	0.7000	0.1520	0.1280	0.1400	0.1580	0.1820	1.0000	0.3500	0.3460	0.1520	0.1880	0.01409	0.188	13.340
SA	1.0000	4.6000	1.0000	0.1520	0.2820	0.1140	0.2200	3.2000	1.0000	2.2000	0.1400	0.8500	0.03643	0.484	13.272
TW	0.5980	0.7000	0.2780	0.1520	0.2620	0.1280	0.2500	3.4000	0.6400	1.0000	0.1860	0.3720	0.02322	0.306	13.182
TS	6.2000	6.6000	5.4000	1.8000	3.2000	0.6800	1.0000	6.8000	7.0000	5.6000	1.0000	5.0000	0.17938	2.494	13.905
SBA	1.0000	4.6000	2.0000	0.1820	0.4000	0.2240	0.6660	5.4000	1.6000	3.0000	0.2000	1.0000	0.05256	0.701	13.333

Average Lambda = **13.494**

N = **12**

Consistency Index = **0.136**

Random Index = **1.54**

Consistency Ratio = **0.088**

Table 1.7 Average responses of pairwise comparison matrix for HR as a change agent

	JD	PM	MFC	RP	SP	TAD	TB	MESI
JD	1.0000	0.8400	0.2240	0.8660	0.1760	0.7660	1.4400	0.1760
PM	1.0000	1.0000	0.1800	0.7075	0.1400	2.5000	1.2275	0.2050
MFC	5.2500	5.7500	1.0000	4.0000	2.0000	5.2500	6.2500	1.0000
RP	1.0000	1.7500	0.2575	1.0000	0.2025	1.7500	1.0000	0.1425
SP	5.7500	7.0000	0.8000	5.2500	1.0000	5.7500	6.2500	1.2575
TAD	1.2500	0.6250	0.1975	0.7075	0.1775	1.0000	0.3725	0.1775
TB	1.1650	2.5625	0.1650	1.0000	0.1625	2.7500	1.0000	0.1775
MESI	5.5000	5.0000	1.0000	7.0000	2.5625	5.7500	5.7500	1.0000
Sum	**21.9150**	**24.5275**	**3.8240**	**20.5310**	**6.4210**	**25.5160**	**23.2900**	**4.1360**

Again, the second AHP model for the factor 'HR as a Change Agent' is given in Table 1.8. It can be observed that 'Managing Employee Psychological Issues' and 'Safety Policies' are the two most important aspects that human resource management needs to focus on a priority basis. Out of eight different elements recognized in the literature review, employee psychology is going to be affected very severely due to a fully automated robotic and AI-driven workplace (Ljungholm, 2022). The psychological aspects may create unnecessary havoc and stress among the employees due to work-related uncertainties. The lack of clarity about future developments, particularly in relation to the evolving role of automation and robotics, leaves many employees feeling uncertain and insecure. It is imperative to promptly address these concerns to foster a conducive work environment that promotes productivity and well-being among employees. The second most important aspect of the 'safety policy' is a matter of concern when more machines are working and communicating with each other in a manufacturing facility. The more the number of machines, the more stringent the safety measures should be. Therefore, Industry 5.0 should also focus on safety measures to ensure harmony in the co-bot workplace. On the other hand, the least preferred elements that the HR needs to focus on are 'Job Design' and 'Training and Development' which were ranked seven and eight, respectively. A few studies on Industry 5.0 stressed the importance of job design and training and development as they are important to select the right candidate and induce the required skill and expertise to perform well (Cillo et al., 2022; Longo et al., 2020; Lu et al., 2022). However, in Industry 5.0, the situation is completely different where job design and training and development are given the least priority because robots and AI are going to be doing more than 50% of the routine work of any organization. At the same time, these HR activities in an organization can also be managed well by robots with AI technology (Arslan et al., 2022; Vrontis et al., 2022). For this AHP result, we can see the reliability measures given in Table 1.9, where

Table 1.8 Normalized pairwise comparison matrix for HR as a change agent

	JD	PM	MFC	RP	SP	TAD	TB	MESI	Prioritized vector (PV)	Rank
JD	0.04563	0.03425	0.05858	0.04218	0.02741	0.03002	0.06183	0.04255	0.04281	7
PM	0.04563	0.04077	0.04707	0.03446	0.02180	0.09798	0.05271	0.04956	0.04875	6
MFC	0.23956	0.23443	0.26151	0.19483	0.31148	0.20575	0.26836	0.24178	0.24471	3
RP	0.04563	0.07135	0.06734	0.04871	0.03154	0.06858	0.04294	0.03445	0.05132	5
SP	0.26238	0.28539	0.20921	0.25571	0.15574	0.22535	0.26836	0.30404	0.24577	2
TAD	0.05704	0.02548	0.05165	0.03446	0.02764	0.03919	0.01599	0.04292	0.03680	8
TB	0.05316	0.10447	0.04315	0.04871	0.02531	0.10778	0.04294	0.04292	0.05855	4
MESI	0.25097	0.20385	0.26151	0.34095	0.39908	0.22535	0.24689	0.24178	0.27130	1
	1.00000	1.00000	1.00000	1.00000	1.00000	1.00000	1.00000	1.00000		

Table 1.9 Reliability test matrix for HR as a change agent

	JD	PM	MFC	RP	SP	TAD	TB	MESI	PV	Weighted vector	Eigenvalue
JD	1.0000	0.8400	0.2240	0.8660	0.1760	0.7660	1.4400	0.1760	0.04281	0.387	9.029
PM	1.0000	1.0000	0.1800	0.7075	0.1400	2.5000	1.2275	0.2050	0.04875	0.426	8.735
MFC	5.2500	5.7500	1.0000	4.0000	2.0000	5.2500	6.2500	1.0000	0.24471	2.277	9.305
RP	1.0000	1.7500	0.2575	1.0000	0.2025	1.7500	1.0000	0.1425	0.05132	0.454	8.843
SP	5.7500	7.0000	0.8000	5.2500	1.0000	5.7500	6.2500	1.2575	0.24577	2.217	9.021
TAD	1.2500	0.6250	0.1975	0.7075	0.1775	1.0000	0.3725	0.1775	0.03680	0.319	8.669
TB	1.1650	2.5625	0.1650	1.0000	0.1625	2.7500	1.0000	0.1775	0.05855	0.514	8.784
MESI	5.5000	5.0000	1.0000	7.0000	2.5625	5.7500	5.7500	1.0000	0.27130	2.532	9.335

Average Lambda = **8.965**

N = **8**

Consistency Index = **0.138**

Random Index = **1.400**

Consistency Ratio = **0.098**

Table 1.10 Average responses of pairwise comparison matrix for technology adaption challenges

	CSI	COI	CYS	DMA	IHM	TL	TI	WT
CSI	1.0000	6.2000	1.8000	7.0000	0.3500	4.4000	2.0000	5.4000
COI	0.1640	1.0000	0.2620	4.2000	0.1360	1.0000	0.6660	2.8000
CYS	0.7320	4.0000	1.0000	5.0000	0.4320	4.8000	3.4000	6.6000
DMA	0.1400	0.2700	0.2080	1.0000	0.1520	0.2860	0.2020	2.0000
IHM	3.2000	7.2000	2.4000	6.6000	1.0000	4.6000	5.6000	9.0000
TL	0.2300	1.0000	0.2140	4.0000	0.2200	1.0000	0.4000	4.8000
TI	0.7400	1.8000	0.3500	5.2000	0.1820	2.8000	1.0000	5.8000
WT	0.1880	0.5560	0.1520	0.6320	0.1100	0.2140	0.1760	1.0000
Sum	**6.3940**	**22.0260**	**6.3860**	**33.6320**	**2.5820**	**19.1000**	**13.4440**	**37.4000**

the consistency ratio is 0.0984 (or 9.84%), which is less than 0.10 (10%), indicating good consistency in the responses and results.

In Table 1.11, we can observe the ranking of the sub-criteria for the factor 'Technology Adaption Challenges'. 'Integration of Human–Machine' and 'Complex System Integration' are the two major challenges that need to be addressed on a priority basis. In a co-bot workplace, integration of advanced technology with the existing operations that require complex integration of various systems, processes, and departments is a very critical task. Further, 'cybersecurity' is the third important aspect and the proper adoption of which is critical for the sustainability of a co-bot workplace. With the increased use of technology, the risk of cyber-attacks, phishing, and many other unknown threats would increase. Therefore, integrating, updating, and protecting the technology can only ensure sustainability. Besides, the 'integration of humans and machines' is in fact a more important task and organizations need to make this possible on a priority basis. Better coexistence, communication, and collaboration are very important to building a sustainable co-bot work environment (Hanna et al., 2020; Jost et al., 2020). In the adaptability-related issues, the least preferred criteria are 'Workforce Training' and 'Data Management and Analysis' which are ranked seven and eight, respectively. These two elements are not very imperative to worry about since these jobs can be done easily with technology. Moreover, many companies are managing these two aspects very efficiently with the use of various advanced techniques. Some authors also highlighted that employee training and management of big data can easily be done with the use of cutting-edge database management techniques where they can use three-dimensional dummy models, virtual reality, and advanced database management software (Chiang et al., 2022; Ljungholm, 2022; Wellsandt et al., 2022). The reliability of AHP results for the factor 'Technology Adaption Challenges' is given in Table 1.12, where we can see the value of the consistency ratio is 0.0949 (or 9.49%) which is within the acceptable limit of 0.10 (10%).

Table 1.11 Normalized pairwise comparison matrix of technology adaption challenges

	CSI	COI	CYS	DMA	IHM	TL	TI	WT	Prioritized vector (PV)	Rank
CSI	0.15640	0.28149	0.28187	0.20814	0.13555	0.23037	0.14877	0.14439	0.19837	2
COI	0.02565	0.04540	0.04103	0.12488	0.05267	0.05236	0.04954	0.07487	0.05830	6
CYS	0.11448	0.18160	0.15659	0.14867	0.16731	0.25131	0.25290	0.17647	0.18117	3
DMA	0.02190	0.01226	0.03257	0.02973	0.05887	0.01497	0.01503	0.05348	0.02985	7
IHM	0.50047	0.32689	0.37582	0.19624	0.38730	0.24084	0.41654	0.24064	0.33559	1
TL	0.03597	0.04540	0.03351	0.11893	0.08521	0.05236	0.02975	0.12834	0.06618	5
TI	0.11573	0.08172	0.05481	0.15461	0.07049	0.14660	0.07438	0.15508	0.10668	4
WT	0.02940	0.02524	0.02380	0.01879	0.04260	0.01120	0.01309	0.02674	0.02386	8
	1.00000	1.00000	1.00000	1.00000	1.00000	1.00000	1.00000	1.00000		

Table 1.12 Reliability test matrix for technology adaption challenges

	CSI	COI	CYS	DMA	IHM	TL	TI	WT	PV	Weighted vector	Eigenvalue
CSI	1.0000	6.2000	1.8000	7.0000	0.3500	4.4000	2.0000	5.4000	0.19837	1.846	9.305
COI	0.1640	1.0000	0.2620	4.2000	0.1360	1.0000	0.6660	2.8000	0.05830	0.513	8.805
CYS	0.7320	4.0000	1.0000	5.0000	0.4320	4.8000	3.4000	6.6000	0.18117	1.692	9.338
DMA	0.1400	0.2700	0.2080	1.0000	0.1520	0.2860	0.2020	2.0000	0.02985	0.250	8.384
IHM	3.2000	7.2000	2.4000	6.6000	1.0000	4.6000	5.6000	9.0000	0.33559	3.139	9.352
TL	0.2300	1.0000	0.2140	4.0000	0.2200	1.0000	0.4000	4.8000	0.06618	0.559	8.451
TI	0.7400	1.8000	0.3500	5.2000	0.1820	2.8000	1.0000	5.8000	0.10668	0.962	9.016
WT	0.1880	0.5560	0.1520	0.6320	0.1100	0.2140	0.1760	1.0000	0.02386	0.210	8.794

Average Lambda = 8.931

N = 8

Consistency Index = 0.133

Random Index = 1.4

Consistency Ratio = 0.095

Table 1.13 The result of AHP priority ranking of individual elements under a factor and combined ranking

Factor	Subfactor	Local weight	Local rank	Consistency		Combined weight	Ranking
Employee Readiness	Adaptability to changing work environment (ACW)	0.04596	8	CI	0.1358	0.0153	9
	Attention to detail (ATD)	0.01888	11	RI	1.54	0.0063	27
	Communication skills (CS)	0.04645	7	CR	0.0882	0.0155	19
	Complex decision-making ability (CDM)	0.16846	3			0.0562	9
	Continuous learning and development (CLD)	0.11831	4			0.0394	10
	Creativity and innovation in problem-solving (CIPS)	**0.20964**	**1**			**0.0699**	**5**
	Education and training (ET)	0.08662	5			0.0289	12
	Physical fitness (FF)	0.01409	12			0.0047	28
	Safety awareness (SA)	0.03643	9			0.0121	23
	Teamwork (TW)	0.02322	10			0.0077	26
	Technical skills (TS)	0.17938	2			0.0598	8
	Stress bearing ability (SBA)	0.05256	6			0.0175	16
HR as a Change Agent	Job design (JD)	0.04281	7	CI	0.1378	0.0143	21
	Performance management (PM)	0.04875	6	RI	1.40	0.0162	18
	Motivation for collaboration (MFC)	0.24471	3	CR	0.0984	**0.0816**	**4**
	Recruitment process (RP)	0.05132	5			0.0171	17
	Safety policies (SP)	0.24577	2			**0.0819**	**3**
	Training and development (TAD)	0.03680	8			0.0123	22
	Trust building (TB)	0.05855	4			0.0195	14
	Managing Employee Psychological Issues (MESI)	**0.27130**	**1**			**0.0904**	**2**

(Continued)

Table 1.13 (Continued) The result of AHP priority ranking of individual elements under a factor and combined ranking

Factor	Subfactor	Local weight	Local rank	Consistency		Combined weight	Ranking
Technology Adaption Challenges	Complex system integration (CSI)	0.19837	2	CI	0.1329	0.0661	6
	Cost of implementation (COI)	0.05830	6	RI	1.40	0.0194	15
	Cybersecurity (CYS)	0.18117	3	CR	0.0949	0.0604	7
	Data management and analysis (DMA)	0.02985	7			0.0100	24
	Integration of Human–Machine (IHM)	**0.33559**	1			**0.1119**	1
	Technological literacy (TL)	0.06618	5			0.0221	13
	Technology integration (TI)	0.10668	4			0.0356	11
	Workforce training (WT)	0.02386	8			0.0080	25
	Total	3					

Table 1.13 presents a combined view of the individual ranking called the local ranking of the sub-criteria under each of the three variables and a universal ranking or 'combined weight and ranking' of all the criteria to find out which criteria are more important. It was observed that 'Integration of Human–Machine' under the 'Technology Adaption Challenges' factor is ranked one and supposed to be the most important element that needs primary focus followed by 'Managing Employee Psychological Issues' which is ranked second under the 'HR as a Change Agent' factor. The third and fourth important criteria 'Safety Policies' and 'Motivation for Collaboration', respectively, are again under the 'HR as a Change Agent' factor. The fifth-ranked criteria 'Creativity and Innovation to Problem Solving' comes much after the adaptation-related challenges and HR preparedness. On the contrary, the least preferred criteria are 'attention to detail' and 'physical fitness' which are also the lowest locally ranked elements under the employee readiness factor.

1.6 IMPLICATIONS AND LIMITATIONS

The study on employee readiness for co-bot workplaces has several implications for employees, organizations and HRM. By understanding the factors that influence employee readiness, organizations can take steps to ensure the successful integration of co-bots into the workplace. First of all, the study highlighted the importance of providing employees with what skills they need to develop in order to get employment and survive in a co-bot technology-driven economy. The study provides the employees with a list of qualities that need to be developed on a priority basis so that it will help save time and make them more apt for future-ready organizations in Industry 5.0. This will enable employees to gain confidence in working with co-bots and enhance their readiness.

Second, organizations need to communicate the potential advantages of co-bots, such as increased productivity, efficiency, and safety, to their employees. By doing so, employees are more likely to be receptive to the introduction of co-bots into their work environment. For HR managers, it is more important to understand the psychology of the employees so that their productivity can be ensured. Further, organizations need to address any concerns relating to both psychological issues and security-related aspects that may bother an employee while working with co-bots. Employees may be worried about job displacement or the loss of control over their work tasks. Organizations need to address these concerns and provide reassurance that the introduction of co-bots is not intended to replace human labour, but rather to augment it.

Third, the organizations which are entering into Industry 5.0 are going to face tremendous challenges while adopting new technology and coping with the changing work environment. The study also highlighted what the adaptation-related challenges that an organization may face and be supposed to handle on a priority basis would help devise better strategies and save time in the adaptation process. Overall, the study highlights the importance of creating a work environment that supports employee readiness for co-bot workplaces. Organizations that take the necessary steps to ensure employee readiness will be better equipped to reap the benefits of co-bot technology and achieve greater efficiency and productivity. Moreover, by addressing these factors, organizations can maximize the benefits of co-bots while minimizing potential risks and challenges associated with their introduction.

The study has also a few limitations, first, the sample size of the study is small, which could limit the generalizability of the observation. Second, the study only focussed on the manufacturing sector which limits the applicability of the findings to the manufacturing sector only. Therefore, the selection of such sectors in further study may bring some novel outcomes. Third, the study did not take into account the contextual factors that could influence employee readiness for co-bot workplaces, such as the type of industry or the organizational culture which may impact results. Finally, Industry 5.0 is in the budding stage so, it is very difficult to list the factors that affect an employee's preparedness and the organization's preparedness for such a radical change, thus, the results may not have long-term applicability.

1.7 CONCLUSION AND FUTURE SCOPE

In conclusion, employee readiness along with the role of human resource management and the adaptation-related challenges are very crucial for the successful integration of co-bots into the workplace. As per the findings of the study, in order to ensure employee collaboration with robots, organizations should invest more time and effort in understanding employee psychology, motivating them while ensuring safety, by promoting a positive organizational culture that values innovation and collaboration. On the other hand, the survival of an employee necessitates the acquisition of certain qualities on a priority basis including innovation and creativity in problem-solving, technical skill, complex decision-making ability along with an attitude of continuous learning and development. Further, it was observed that organizations may also face adaptation-related challenges that need to be addressed first, including proper integration of humans and machines, complex system integration, and good cybersecurity. By doing so, organizations can maximize the benefits of co-bots, including increased productivity, efficiency, and safety, while creating a positive and inclusive work environment that supports employee well-being and job satisfaction.

Future research could involve longitudinal studies that track changes in employee attitudes and behaviours over time, providing insights into the factors that influence employee readiness in the long term. Second, the study can be extended to different other sectors, for instance, the healthcare sector, research and development organizations, and services sector where many jobs presently performed by humans may be taken over by robots with Industry 5.0. Third, to gain a better understanding of the factors that influence employee readiness for co-bot workplaces, comparative studies could be conducted across different industries, organizations, and countries. This would allow researchers to identify commonalities and differences in employee attitudes and behaviours and to develop strategies that are tailored to specific contexts. Fourth, employee readiness for co-bot workplaces is a complex phenomenon that involves factors related to technology, psychology, and organizational behaviour. Future research could involve multidisciplinary studies that bring together experts from different fields to develop a more comprehensive understanding of the phenomenon. Finally, employee readiness for co-bot workplaces is a complex phenomenon that involves factors related to technology, psychology, and organizational behaviour. Future research could involve multidisciplinary studies that bring together experts from different fields to develop a more comprehensive understanding of the phenomenon.

NOTE

1 To engage in biological investigations, especially superficially or amateurishly.

REFERENCES

Aaltonen, I., & Salmi, T. (2019). Experiences and expectations of collaborative robots in industry and academia: Barriers and development needs. *Procedia Manufacturing, 38*, 1151–1158.

Agarwal, N., & Chauhan, S. (2022). Amplifying employability skills to create co-working space for human and cobots in the E-Commerce industry. *Procedia Computer Science, 214*, 1040–1048.

Arslan, A., Cooper, C., Khan, Z., Golgeci, I., & Ali, I. (2022). Artificial intelligence and human workers interaction at team level: A conceptual assessment of the challenges and potential HRM strategies. *International Journal of Manpower, 43*(1), 75–88.

Batth, R. S., Nayyar, A., & Nagpal, A. (2018). Internet of robotic things: Driving intelligent robotics of future-concept, architecture, applications and technologies. *2018 4th International Conference on Computing Sciences (ICCS),*

Bhutta, K. S., & Huq, F. (2002). Supplier selection problem: A comparison of the total cost of ownership and analytic hierarchy process approaches. *Supply Chain Management: An International Journal, 7*(3), 126–135. https://doi.org/10.1108/13598540210436586

Bi, Z. M., Luo, C., Miao, Z., Zhang, B., Zhang, W., & Wang, L. (2021). Safety assurance mechanisms of collaborative robotic systems in manufacturing. *Robotics and Computer-Integrated Manufacturing, 67*, 102022. https://doi.org/10.1016/j.rcim.2020.102022

Breque, M., De Nul, L., & Petridis, A. (2021). *Industry 5.0: Towards a sustainable, human-centric and resilient European industry.* Publications Office of the European Union. https://doi.org/doi/10.2777/308407

Briggs, G., & Scheutz, M. (2014). How robots can affect human behavior: Investigating the effects of robotic displays of protest and distress. *International Journal of Social Robotics, 6,* 343–355.

Buchner, R., Wurhofer, D., Weiss, A., & Tscheligi, M. (2013). Robots in time: How user experience in human-robot interaction changes over time. *Social Robotics: 5th International Conference,* ICSR 2013, Bristol, UK, October 27–29, 2013, Proceedings 5,

Calitz, A. P., Poisat, P., & Cullen, M. (2017). The future African workplace: The use of collaborative robots in manufacturing. *SA Journal of Human Resource Management, 15*(1), 1–11.

Cassioli, F., Fronda, G., & Balconi, M. (2021). Human–co-bot interaction and neuroergonomics: Co-botic vs. robotic systems. *Frontiers in Robotics and AI, 8,* 659319. https://doi.org/10.3389/frobt.2021.659319

Cheon, E., Schneiders, E., & Skov, M. B. (2022). Working with bounded collaboration: A qualitative study on how collaboration is co-constructed around collaborative robots in industry. *Proceedings of the ACM on Human-Computer Interaction, 6*(CSCW2), 1–34.

Chiang, F.-K., Shang, X., & Qiao, L. (2022). Augmented reality in vocational training: A systematic review of research and applications. *Computers in Human Behavior, 129,* 107125.

Choi, T. M., Kumar, S., Yue, X., & Chan, H. L. (2022). Disruptive technologies and operations management in the Industry 4.0 era and beyond. *Production and Operations Management, 31*(1), 9–31.

Cillo, V., Gregori, G. L., Daniele, L. M., Caputo, F., & Bitbol-Saba, N. (2022). Rethinking companies' culture through knowledge management lens during Industry 5.0 transition. *Journal of Knowledge Management, 26*(10), 2485–2498.

Culot, G., Orzes, G., & Sartor, M. (2020). The 4.0 industrial revolution. *International Journal of Production Economics, 226,* 107617.

Dalenogare, L. S., Benitez, G. B., Ayala, N. F., & Frank, A. G. (2018). The expected contribution of Industry 4.0 technologies for industrial performance. *International Journal of Production Economics, 204,* 383–394.

Davies, R. S. (2011). Understanding technology literacy: A framework for evaluating educational technology integration. *TechTrends, 55,* 45–52.

Demir, K. A., Döven, G., & Sezen, B. (2019). Industry 5.0 and human-robot co-working. *Procedia Computer Science, 158,* 688–695.

Doyle-Kent, M., & Shanahan, B. W. (2022). The development of a novel educational model to successfully upskill technical workers for Industry 5.0: Ireland a case study. *IFAC-PapersOnLine, 55*(39), 425–430.

Dwivedi, A., Agrawal, D., Jha, A., & Mathiyazhagan, K. (2023). Studying the interactions among Industry 5.0 and circular supply chain: Towards attaining sustainable development. *Computers & Industrial Engineering, 176,* 108927.

Elvis, M., & Milligan, T. (2019). How much of the solar system should we leave as wilderness? *Acta Astronautica, 162,* 574–580.

EunJeong, C., Eike, S., Kristina, D., & Mikael, B. S. (2022). Robots as a Place for Socializing: Influences of Collaborative Robots on Social Dynamics In- and

Outside the Production Cells. *Proceedings of the ACM on Human-Computer Interaction, 6*(CSCW2), Article 457. https://doi.org/10.1145/3555558

Frey, C. B., & Osborne, M. A. (2013). The future of employment: How susceptible are jobs to computerisation. *Technological Forecasting and Social Change, 114,* 254–280.

Ghobakhloo, M. (2020). Industry 4.0, digitization, and opportunities for sustainability. *Journal of Cleaner Production, 252,* 119869.

Hafeez, K., Zhang, Y., & Malak, N. (2002). Determining key capabilities of a firm using analytic hierarchy process. *International Journal of Production Economics, 76*(1), 39–51.

Hammershøj, L. G. (2019). The new division of labor between human and machine and its educational implications. *Technology in Society, 59,* 101142.

Hanna, A., Bengtsson, K., Götvall, P.-L., & Ekström, M. (2020). Towards safe human robot collaboration-Risk assessment of intelligent automation. *2020 25th IEEE International Conference on Emerging Technologies and Factory Automation (ETFA),* Vienna, Austria, 2020, pp. 424–431. https://doi.org/10.1109/ETFA46521.2020.9212127

Huang, S., Wang, B., Li, X., Zheng, P., Mourtzis, D., & Wang, L. (2022). Industry 5.0 and Society 5.0—Comparison, complementation and co-evolution. *Journal of Manufacturing Systems, 64,* 424–428.

Işcan, E. (2021). An old problem in the new era: Effects of artificial intelligence to unemployment on the way to Industry 5.0. *Yaşar Üniversitesi E-Dergisi, 16*(61), 77–94.

Jafari, N., Azarian, M., & Yu, H. (2022). Moving from Industry 4.0 to Industry 5.0: What are the implications for smart logistics? *Logistics, 6*(2), 26.

Jost, C., Le Pévédic, B., Belpaeme, T., Bethel, C., Chrysostomou, D., Crook, N., Grandgeorge, M., & Mirnig, N. (2020). *Human-robot interaction.* Springer.

Kildal, J., Molina, J., & Andrés, U. (2022). User-centred human-robot collaborative handling of small parts in a MIM process. In M. Shafik and K. Case (Eds.), *Advances in manufacturing technology XXXV* (pp. 113–118). IOS Press. https://doi.org/10.3233/ATDE220576

Kristal, T. (2020). Why has computerization increased wage inequality? Information, occupational structural power, and wage inequality. *Work and Occupations, 47*(4), 466–503.

Kumar, A., & Nayyar, A. (2020). si 3-Industry: A sustainable, intelligent, innovative, internet-of-things industry. In Nayyar, A., Kumar, A. (eds) *A roadmap to Industry 4.0: Smart production, sharp business and sustainable development* (pp. 1–21). Springer, Cham. https://doi.org/10.1007/978-3-030-14544-6_1

Kwanya, T. (2023). Working with robots as colleagues: Kenyan perspectives of ethical concerns on possible integration of co-bots in workplaces. In *Responsible AI in Africa: Challenges and opportunities* (pp. 65–99). Springer International Publishing Cham. https://doi.org/10.1007/978-3-031-08215-3_4

Lachvajderová, L., & Kádárová, J. (2022). Industry 4.0 Implementation and Industry 5.0 Readiness in Industrial Enterprises. *Management and Production Engineering Review, 13*(3), 102–109. https://doi.org/10.24425/mper.2022.142387

Lambrechts, W., Klaver, J. S., Koudijzer, L., & Semeijn, J. (2021). Human factors influencing the implementation of Cobots in high volume distribution centres. *Logistics, 5*(2), 32.

Leng, J., Chen, Z., Huang, Z., Zhu, X., Su, H., Lin, Z., & Zhang, D. (2022). Secure blockchain middleware for decentralized IIoT towards Industry 5.0: A review of architecture, enablers, challenges, and directions. *Machines*, *10*(10), 858.

Leopold, T. A., Ratcheva, V., Zahidi, S., & Samans, R. (2017). *The future of jobs and skills in Africa: Preparing the region for the fourth industrial revolution*. World Economic Forum.

Ljungholm, D. P. (2022). Metaverse-based 3D visual modeling, virtual reality training experiences, and wearable biological measuring devices in immersive workplaces. *Psychosociological Issues in Human Resource Management*, *10*(1), 64–77.

Longo, F., Padovano, A., & Umbrello, S. (2020). Value-oriented and ethical technology engineering in industry 5.0: A human-centric perspective for the design of the factory of the future. *Applied Sciences*, *10*(12), 4182.

Lu, Y., Zheng, H., Chand, S., Xia, W., Liu, Z., Xu, X., Wang, L., Qin, Z., & Bao, J. (2022). Outlook on human-centric manufacturing towards Industry 5.0. *Journal of Manufacturing Systems*, *62*, 612–627.

Maddikunta, P. K. R., Pham, Q.-V., Prabadevi, B., Deepa, N., Dev, K., Gadekallu, T. R., Ruby, R., & Liyanage, M. (2022). Industry 5.0: A survey on enabling technologies and potential applications. *Journal of Industrial Information Integration*, *26*, 100257.

Maharana, N. (2020). Entrepreneurial Lessons from the Bhagavad Gita. *PURUSHARTHA-A journal of ....* http://journals.smsvaranasi.com/index.php/purushartha/article/view/822

Martynov, V. V., Shavaleeva, D. N., & Zaytseva, A. A. (2019). Information technology as the basis for transformation into a digital society and industry 5.0. *2019 International Conference"Quality Management, Transport and Information Security, Information Technologies"(IT&QM&IS)*, Sochi, Russia, 2019, pp. 539–543. https://doi.org/10.1109/ITQMIS.2019.8928305

Matheson, E., Minto, R., Zampieri, E. G., Faccio, M., & Rosati, G. (2019). Human–robot collaboration in manufacturing applications: A review. *Robotics*, *8*(4), 100.

Mattioli, G., Roberts, C., Steinberger, J. K., & Brown, A. (2020). The political economy of car dependence: A systems of provision approach. *Energy Research & Social Science*, *66*, 101486.

Millet, I., & Saaty, T. L. (2000). On the relativity of relative measures–accommodating both rank preservation and rank reversals in the AHP. *European Journal of Operational Research*, *121*(1), 205–212.

Min, J., Kim, Y., Lee, S., Jang, T.-W., Kim, I., & Song, J. (2019). The fourth industrial revolution and its impact on occupational health and safety, worker's compensation and labor conditions. *Safety and Health at Work*, *10*(4), 400–408.

Nayyar, A., & Kumar, A. (2020). *A roadmap to industry 4.0: Smart production, sharp business and sustainable development*. Springer.

Nayyar, A., Rameshwar, R., & Solanki, A. (2020). Internet of Things (IoT) and the digital business environment: A standpoint inclusive cyber space, cyber crimes, and cybersecurity. In *The evolution of business in the cyber age* (pp. 111–152). Apple Academic Press.

Orlova, E. V. (2021). Design of personal trajectories for employees' professional development in the knowledge society under Industry 5.0. *Social Sciences*, *10*(11), 427.

Østergaard, E. H. (2018). Welcome to industry 5.0. *Retrieved Febr, 5,* 2020.

Özsungur, F. (2021). Business management and strategy in cybersecurity for digital transformation. In K. Sandhu (Ed.), *Handbook of research on advancing cybersecurity for digital transformation* (pp. 144–162). IGI Global. https://doi.org/10.4018/978-1-7998-6975-7.ch008

Paschek, D., Luminosu, C.-T., & Ocakci, E. (2022). Industry 5.0 challenges and perspectives for manufacturing systems in the Society 5.0. *Sustainability and Innovation in Manufacturing Enterprises: Indicators, Models and Assessment for Industry 5.0*, 17–63.

Pettersen, L. (2019). Why artificial intelligence will not outsmart complex knowledge work. *Work, Employment and Society*, 33(6), 1058–1067.

Piçarra, N., & Giger, J.-C. (2018). Predicting intention to work with social robots at anticipation stage: Assessing the role of behavioral desire and anticipated emotions. *Computers in Human Behavior*, 86, 129–146.

Pizoń, J., & Gola, A. (2023). Human–machine relationship—Perspective and future roadmap for Industry 5.0 solutions. *Machines*, 11(2), 203.

Romero, D., & Stahre, J. (2021). Towards the resilient operator 5.0: The future of work in smart resilient manufacturing systems. *Procedia CIRP*, 104, 1089–1094.

Saaty, T. (1980). *The analytic hierarchy process (AHP) for decision making*. Japan: Kobe.

Sahoo, K. S., Tiwary, M., Luhach, A. K., Nayyar, A., Choo, K.-K. R., & Bilal, M. (2021). Demand–Supply-based economic model for resource provisioning in industrial IoT traffic. *IEEE Internet of Things Journal*, 9(13), 10529–10538.

Saighal, B., Bhat, M., & Gupta, A. (2015). Analytical study: Workplace environment in IT companies and ranking of companies using AHP–topsis method. *IOSR Journal of Business and Management (IOSR-JBM)*, 17(5), 11–22. https://doi.org/10.9790/487X-17511122

Schrum, M. L., Neville, G., Johnson, M., Moorman, N., Paleja, R., Feigh, K. M., & Gombolay, M. C. (2021). Effects of social factors and team dynamics on adoption of collaborative robot autonomy. *Proceedings of the 2021 ACM/IEEE International Conference on Human-Robot Interaction (HRI '21). Association for Computing Machinery*, New York, NY, pp. 149–157. https://doi.org/10.1145/3434073.3444649

Schütte, G. (2018). What kind of innovation policy does the bioeconomy need? *New Biotechnology*, 40, 82–86.

Senna, P. P. (2022). *Policies for the digital era: Digital technology adoption*, Implementation and the Role of Governments.

Sherwani, F., Asad, M. M., & Ibrahim, B. S. K. K. (2020). Collaborative robots and industrial revolution 4.0 (ir 4.0). *2020 International Conference on Emerging Trends in Smart Technologies (ICETST)*.

Shirokii, V., Batusov, R., Chubarov, A., Dolenko, S., & Samsonovich, A. (2018). Patterns of cognitive activity in a human vs collaborative robot interactive game. *Procedia Computer Science*, 145, 495–499.

Simões, A. C., Pinto, A., Santos, J., Pinheiro, S., & Romero, D. (2022). Designing human-robot collaboration (HRC) workspaces in industrial settings: A systematic literature review. *Journal of Manufacturing Systems*, 62, 28–43. https://doi.org/10.1016/j.jmsy.2021.11.007

Smids, J., Nyholm, S., & Berkers, H. (2020). Robots in the workplace: A threat to—or opportunity for—meaningful work? *Philosophy & Technology*, 33(3), 503–522.

Solanki, A., & Nayyar, A. (2019). Green internet of things (G-IoT): ICT technologies, principles, applications, projects, and challenges. In G. Kaur & P. Tomar (Eds.), *Handbook of research on big data and the IoT* (pp. 379–405). IGI Global. https://doi.org/10.4018/978-1-5225-7432-3.ch021

Spencer, D. A. (2018). Fear and hope in an age of mass automation: Debating the future of work. *New Technology, Work and Employment, 33*(1), 1–12. https://doi.org/10.1111/ntwe.12105

Tamers, S. L., Streit, J., Pana-Cryan, R., Ray, T., Syron, L., Flynn, M. A., Castillo, D., Roth, G., Geraci, C., & Guerin, R. (2020). Envisioning the future of work to safeguard the safety, health, and well-being of the workforce: A perspective from the CDC's National Institute for Occupational Safety and Health. *American Journal of Industrial Medicine, 63*(12), 1065–1084. https://doi.org/10.1002/ajim.23183

Venaik, A., Jain, S., & Nayyar, A. (2023). Industry 4.0—Its Advancement and Effects on Security of Whistle-Blowers on Dark Web. In Singh, G., Goel, R., & Garg, V. (eds) *Industry 4.0 and the Digital Transformation of International Business* (pp. 103–121). Springer. https://doi.org/10.1007/978-981-19-7880-7_6

Vermeulen, B., Kesselhut, J., Pyka, A., & Saviotti, P. P. (2018). The impact of automation on employment: Just the usual structural change? *Sustainability, 10*(5), 1661. https://doi.org/10.3390/su10051661

Vrontis, D., Christofi, M., Pereira, V., Tarba, S., Makrides, A., & Trichina, E. (2022). Artificial intelligence, robotics, advanced technologies and human resource management: A systematic review. *The International Journal of Human Resource Management, 33*(6), 1237–1266. https://doi.org/10.1080/09585192.2020.1871398

Wellsandt, S., Klein, K., Hribernik, K., Lewandowski, M., Bousdekis, A., Mentzas, G., & Thoben, K.-D. (2022). Hybrid-augmented intelligence in predictive maintenance with digital intelligent assistants. *Annual Reviews in Control, 53*(2022), 382–390. https://doi.org/10.1016/j.arcontrol.2022.04.001

Chapter 2

Automation and robotization

An aerial view of Industry 5.0 on employment and employees

Vrinda V. Nair

RV University, Bengaluru, India

2.1 INTRODUCTION

Industrial Revolution refers to the historical period in which there is a significant shift in the way goods are produced and services are delivered. There have been five distinct phases of the Industrial Revolution, each with its own unique characteristics and technological advancements. Industry 5.0 sets the stage when Industry 4.0 is still at large, establishing its popularity. Industry 5.0 is known more for its revolution in which consumers could satisfy their requirements as per their tastes and expectations. In Industry 4.0, mass customization is achieved with the help of robots, while Industry 5.0 aims for personalization with the help of artificial intelligence or AI (Demir and Cicibas, 2017). Industry 5.0 is expected to bring in a sea change in the production process. It provides higher autonomy to collaborative robots. It brings in more creativity and innovation in the production process by allowing robots to perform repetitive tasks (Demir and Cicibaş, 2018). Industry 5.0 is futuristic in nature, bringing creativity and innovation to the production process through the creative intellectual ability of humans in the most optimized manner (Demir et al., 2019b). Industry 5.0 is the trend of moving from mass production to custom manufacturing and techniques, where the production system is moving from digitalization to intelligentization (Vaidya et al., 2018).

Since the First Industrial Revolution, successive developments have led in manufacturing, from water and steam-driven devices to electrical and digital electronic production, making manufacturing processes more complex, automatic, and sustainable so that machines can be operated with simplicity, efficiency, and persistence (Qin et al., 2016). Technological advances in manufacturing processes fuel all Industrial revolutions. One of the critical things that can be noticed is the reduced timeline between the Second Industrial Revolution and the Third Industrial Revolution. Industry 4.0, is characterized by the integration of automation, data exchange, and other digital technologies into manufacturing processes (Dalenogare et al., 2018; Culot et al., 2020). Industry 5.0, on the other hand, is a concept that describes the next phase of industrial development. Industry 5.0 builds on

DOI: 10.1201/9781003489269-2

this foundation and emphasizes the importance of human collaboration and creativity in manufacturing. People nowadays need the human touch of mass personalization, so Industry 5.0 helps them change from mass manufacturing to mass personalization (Oks et al., 2022). Industry 5.0 is making mass customization a reality and, the fast advances in manufacturing techniques, production system digitization, and intelligence are the needs of today's sector (Vaidya et al., 2018). Industry 4.0 previously activated mass customization, which was not sufficient. For instance, type-1 diabetes is difficult to manage because people have distinct levels of metabolism and distinct dimensions with distinct skin thicknesses, behaviors, lifestyles, etc. Shift to industry 5.0 allows providing people with an application that follows their habits and routine, producing a diabetes control production method and eventually a lower, more discreet, and reliable device tailored to the person. The capacity to produce an Industry 5.0 method would therefore be totally life-changing in every aspect.

In Industry 4.0, CPPS (Cyber-Physical Production System) helps make intelligent decisions in "manufacturing things" through real-time communication and cooperation, enabling flexible production of personalized products at mass efficiency and augmenting high quality (Kachiche et al., 2023). AI uses various technological blends enabling software and machines to comprehend, sense, and act and also to learn operations done by human operations. This process enables the industrial production system to be more efficient and ensure more precision. "Artificial intelligence is not a substitute for human intelligence; it is a tool to amplify human creativity and ingenuity", states Fei-Fei Li, Co-Director of the Stanford Institute for Human-Centered Artificial Intelligence.

AI can be applied in various industries, from healthcare to finance to manufacturing. Along with machine learning (ML), generative and discriminative AI can bring massive value to technological advancement in both Industry 4.0 and Industry 5.0. AI at this stage brings in a human-like level in software which raises the bar in the area of automation. This leads to a reduction in the need for manpower to evaluate information and helps in better decision-making in manufacturing and other sectors, reducing costs and improving productivity. Machine learning can help in guiding robotic operation which helps in better precision and better quality.

The Fifth Industrial Revolution (Industry 5.0) goes beyond jobs and growth. It aims to focus on achieving societal goals and expects to become a buoyant provider of prosperity. One of the agenda is to make the production process respect the boundaries of the planet and place the well-being of the employees working in the organization as an essential criterion in the production process. Industry 4.0 focuses more on digitalization and AI-driven technologies for increasing efficiency and allowing flexibility in the organization while Industry 5.0 focuses on the importance of research and innovation which helps and supports the industry in the long term and helps serve humanity and sustain planet boundaries (Dengler and Matthes, 2018).

It is an "Age of Augmentation" where humans and machines reconcile and work in a symbiosis. AI can play a significant role in Industry 5.0 by helping to improve efficiency and productivity, but it is just one part of a larger shift toward more human-centered manufacturing processes. So, while they are related, AI and Industry 5.0 are not the same.

2.1.1 Evolution of Industry 5.0

The horse-drawn steam engine and horse wagon clamored out, moving into the reign of a motorized coach. To understand on which era of the industry we are currently in, let us understand the evolution of Industry 5.0.

Figure 2.1 highlights the evolution of Industry 5.0.

> **The First Industrial Revolution**, also known as Industry 1.0, began in the late 18th century in Britain and lasted until the mid-19th century. The period was marked by the introduction of steam power, the development of new textile machines, and the use of iron and coal as key raw materials. Factories were built and mass production methods were introduced, leading to significant increases in productivity and economic growth.
>
> **The Second Industrial Revolution**, or Industry 2.0, occurred during the late 19th and early 20th centuries. This period was characterized by the widespread use of electricity and the development of new technologies such as the telephone, radio, and the internal combustion engine. Assembly lines and other mass production methods were further refined, leading to even greater increases in productivity.
>
> **The Third Industrial Revolution**, or Industry 3.0, took place during the late 20th century and was marked by the introduction of computer technology and automation computer/information revolution. This

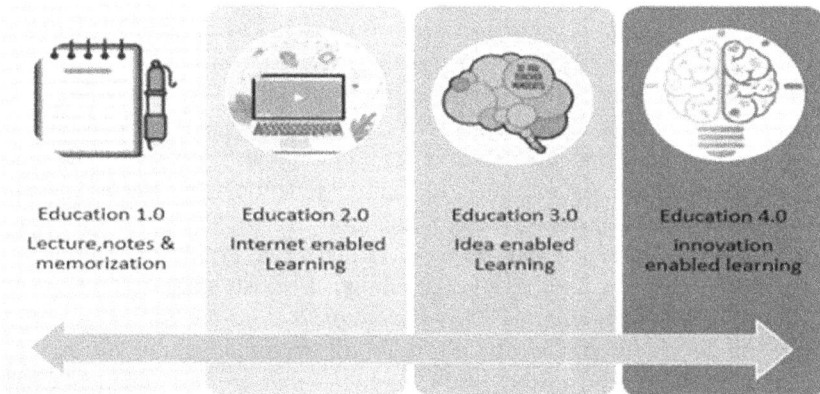

Figure 2.1 Evolution of Industry 5.0.

led to the development of new manufacturing processes and the widespread adoption of information technology, including the use of personal computers and the internet.

Industry 4.0 is the current phase of the industrial revolution and is characterized by the integration of digital technologies such as artificial intelligence, the Internet of Things (IoT), and robotics. These technologies are changing how we work and live, increasing efficiency, productivity, and innovation. Today, we use handheld devices that use large amounts of cloud computing, machine learning devices, and Big Data through 4G/5G to cloud infrastructure. These data derive analytics that the average human cannot produce. The machine learning models are presented to the user with magical accuracy and a shorter time. For example, the application or ability maps will help you to tell how long it will take to commute home if you leave 30–60 minutes in the future or suggest alternative routes when there is a traffic jam. These are specific examples of the convergence of technology and application of technology in Industry 5.0.

Industry 4.0 relies on various technologies like robotics/cobotics, the IoT, additive manufacturing, digital twinning, 3D printing, and analytics (Haleem and Javaid, 2019). A digital factory is a combination of intelligent instrumentation that watches and controls every aspect of production on one end and has minute data on material quality and millisecond updates on the machinery (Dalenogare et al., 2018). Beyond automation in manufacturing sectors, AI can be used for various purposes. It can aid in constructing digital twins, a mechanism for speeding up the product development cycle. It is widely used in Industrial IoT or IIoT infrastructure. A typical example of this is the filtering of event information both at the spot and predicting potential production problems based on sensor data. It can help identify unseen patterns in production and usage data which helps suggest design or process change.

Industry 5.0 is the next stage of the industrial revolution, and its main focus is on integrating human-centric approaches with advanced digital technologies. This includes using collaborative robots, augmented reality, and virtual reality to create a more seamless and interactive working environment. The goal of Industry 5.0 is to create a more sustainable and inclusive society by placing greater emphasis on human creativity and social responsibility Beier et al., 2020). The exponential advancement from Industry 4.0 to 5.0 has occurred quickly in one to two years. The differentiating factor between 4.0 and 5.0 requires deliberate evaluation and is confined to the given technology and capability area. An analysis done in the future may help in evaluating if we have moved from one industrial era to another especially from Industry 4.0 to 5.0. To make it concise, Industry 5.0 is a combination of advanced 4.0 technologies that are assembled to create a digital assistant to support humans in various activities at the workplace.

Unlike Industry 4.0, Industry 5.0 creates more skilled jobs as it requires intellectual professionals working with machines. In Industry 4.0, robots are actively engaged in large-scale production but Industry 5.0 focuses on mass customization and is designed for enhancing customer satisfaction. Humans will be guiding robots. Cyber-Physical System (CPS) connectivity is the main focus of Industry 4.0, whereas Industry 5.0 links the application of Industry 4.0 to establish a relationship with collaborative robots (cobots). Industry 5.0 focuses on greener and more sustainable solutions compared to existing industrial transformations which lack focus on protecting the natural environment. The major aims of Industry 5.0 are to model and use predictive analytics and operating intelligence to create more accurate and unstable decisions for organizations. Industry 5.0 helps in obtaining real-time data through the process of production automation and collaborating with specialized and equipped human resources.

2.1.2 Objectives of the chapter

The objectives of the chapter are:

(1) To identify the various technology enablers for Industry 5.0;
(2) To explain the application of technologies (IoT, artificial intelligence, ML, and blockchain) at the workplace;
(3) To determine the effects of technology on the employee's psychology at the workplace;
(4) And, to analyze the impact of technology on future employment and employees.

2.1.3 Organization of the chapter

The rest of the chapter is organized as follows: Section 2.2 elaborates related terminologies. Section 2.3 stresses deep insights into Industry 5.0 specifying how technology enables Industry 4.0 and 4.0 applications and applications of technology in Industry 5.0. Section 2.4 signifies opportunities of Industry 5.0. Section 2.5 highlights remonstrance under Industry 5.0. And, finally, Section 2.6 concludes the chapter.

2.2 RELATED TERMINOLOGIES

A theoretical framework enables to shape the contribution of the scholars. It provides a window to understanding relatively new issues highlighted by some scholars. There are many perspectives related to the future of work and a few of them are mentioned below (Kolade and Owoseni, 2022).

2.2.1 Cyber-physical system (CPS)

CPS integrates the physical system with computer-based control and communication systems. CPS adopts a close integration of advanced technologies like IoT, AI, and ML with traditional manufacturing processes to create a more efficient and flexible system of production. It is an engineered system that arrays sensing, computation, control, networking, and analysis to interact with the physical world which includes humans. They enable a safe, real-time, secure, resilient, adaptable, and reliable performance (Cropley, 2020). Figure 2.2 highlights CPS.

CPS processes manage and optimize physical outcomes from individual processes to entire eco systems. Examples include robotic arms, gas and oil processing technology, water purification systems, meat cutting, gain processing, baggage handling, and building management systems.

2.2.2 Human-centric system (HCS)

The HCS model adopts a methodology of designing technologies and systems that are intuitive and user-friendly as mentioned in Figure 2.3. In the context of Industry 5.0, this model integrates advanced technologies that support and enhance human functions at the workplace rather than displacing them. This model uses designing technologies that are human-friendly, provide timely feedback, and support decision-making in integrating a sustainable model.

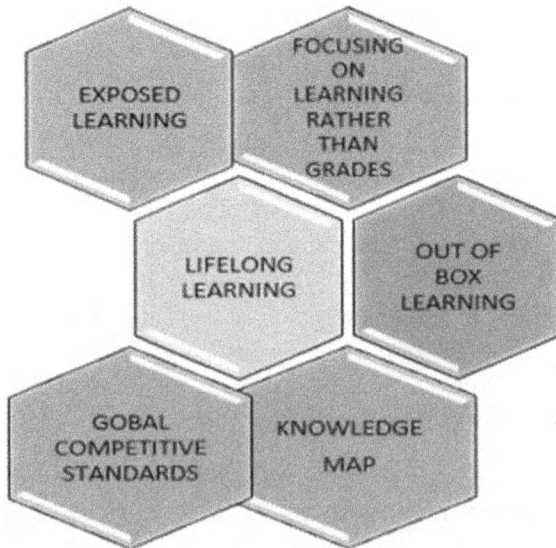

Figure 2.2 Cyber-physical system.

Traditional Learning to Lifelong Learning

| Mentor's or Teacher's Specific | Campus based -out off packet expense | Learner have less exposed to real-time problem | Performance based Learning |

Lifelong Learning

| Globally competitive standards | High cognitive Thinking Skills And patterns | Flexible in Time and cost effective | Exposed Learning with wide curriculum | Learner learn by self / often in groups |

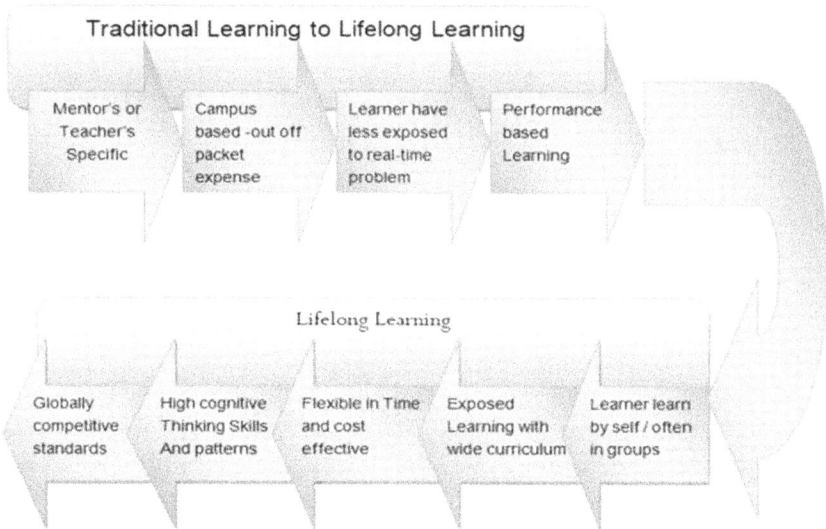

Figure 2.3 Human-centric system (HCS).

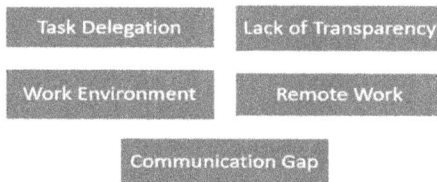

Task Delegation	Lack of Transparency
Work Environment	Remote Work
Communication Gap	

Figure 2.4 Lean manufacturing system.

2.2.3 Lean manufacturing perspective

Lean manufacturing is a methodology that focuses on eliminating waste in the production process. This perspective emphasizes the importance of integrating advanced technologies to streamline the manufacturing process and reduce waste. This includes designing flexible and agile systems, allowing for quick adaptation to changing market demands. Figure 2.4 shows the lean manufacturing perspective.

2.2.4 Socio-technical systems

Kolade and Owoseni (2022) highlighted the gripping change at the workplace and how technology is changing the needs of human factors (Kadir et al., 2019). It is defined as "the participative, multidisciplinary study and improvement of how jobs, single organisations, networks and ecosystem

Social Media Application	Face Book	Communication Information or data
	Twitter	
	Blog	
	You Tube	Sharing Content Analysis
	Forums	
		Interpretation Result

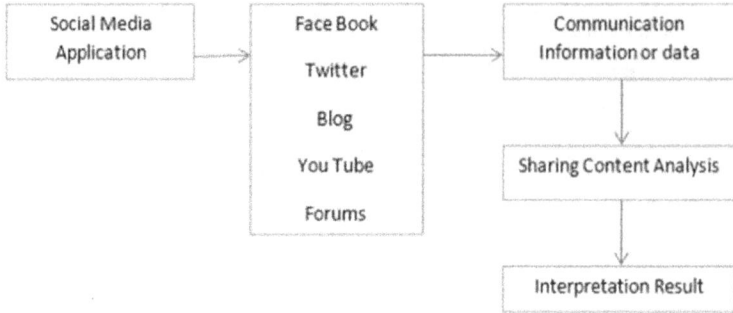

Figure 2.5 Socio-technical system.

functions internally and externally in relation to their environment context, with special focus on the mutual interactions of entity's value creation process" (Davis et al., 2014). Figure 2.5 highlights the socio-technical system.

The assumption underlying the emergence of Industry 4.0 and Industry 5.0 is that it will precipitate an increasing degree of automation that in turn would lead to less and less need for human interaction in the future of worker-less production (Mohr, 2016). However, the emergence of a new system will not reduce the need for human interaction but change how humans interact with computers – both at accomplishing the task or while using products and services in the future, as certain other scholars claim. Thus, socio-technical theory was proposed as a framework to analyze the complexity of the human and technical association and create performance (Sony and Naik, 2020, Beier et al., 2020).

This system is a six-dimensional structure comprising (1) People, (2) Infrastructure, (3) Technology, (4) Culture, (5) Process and Procedure, and (6) Goals. This offers an organizational-centric view of how these components interact in a complex system. This approach is participative and multi-disciplinary and improves how jobs function, organizations network and create an ecosystem that functions internally to meet the external environment with a focus on value creation. This perspective helps in analyzing technological innovation for organizations and also social innovation to optimize the benefits and to compact the possible side effects arising out of regulating growth. For example, the technological development around IoT has been significant in the last decade, equal importance should be given in terms of human components with reference to security, safety, and privacy. The dynamics and human–machine interaction in Industry 5.0 has led to changes in tasks and new profiles in skills leading to skill upgradation along with skills to use technology to communicate between humans, machines, and products. These include new communication roles to support the integration of customers with digital products, and more mentally resilient people who are doing less physical work (Fareri et al., 2020). In short, this approach provides information on employment challenges and opportunities and the guidelines to be made in an era of digital transformation (Cirillo et al., 2020).

2.2.5 Skill-based technology

Frey and Osborne (2017) explained how the swift adoption of computing technology changes the worker's task and the demand for human skills. It maps the changes required in the skill profile in the current and future occupation due to digital transformation. This theory is based on two related actualities: Computers are a substitution for workers following routine tasks and following certain explicit rules at work. Computers can complement workers involved in tasks with no explicit rules but require more flexibility and creativity. The framework classifies tasks into routine and non-routine time tasks and further classifies them into manual and non-manual, and cognitive and interactive tasks. The major challenge emerging in this context along with the productivity-enhancing mechanisms is the polarization and wage inequality across occupations and tasks (Dengler and Matthes (2018).

The skill-based model classifies the task into three categories: (1) perception and manipulation task, (2) creative intelligence, and (3) social intelligence. While the perception and manipulation task requires manual and physical dexterity of fingers, in awkward positions, and in cramped work spaces, creative intelligence requires originality and the ability to come up with new or unusual ideas to solve problems at work. The examples range from fine arts to music, mainly composing, performing, visual arts, etc. Social intelligence comprises the art of negotiation, persuasion, and perceptiveness, among many others. In the era of automation, the perception and manipulation task fall under the high-risk category which can be easily substituted (Frey and Osborne, 2017). Tasks related to creative and social intelligence are less likely to be substituted and replaced as they tend to be non-routine and less amenable to coding on computers. However, the pace of technological progress does indicate the risk factor and is not completely immune to replacement (Novakova, 2020). Sophisticated algorithms have emerged that use pattern recognition to do non-routine tasks and robots with enabled senses and adroitness. There is a proposal for routine-based technological change that distinguishes routines from skills which enable locating specific points and tasks even for low-skilled jobs where automation replacements are high (Fernández-Macías and Hurley, 2017).

Another modification of skill-based technological change (SBTC) is class-based technological change (CBTC). In this model, computer-based technology is used as a key driver that transforms work into an activity that is more knowledge-intensive (Caines et al., 2017). While SBTC focuses on productivity enhancement leading to job polarization and wage inequality across tasks and occupations, the CBTC model is focused on a power-enhancing mechanism explaining the occupations with respect to access, control, and information on the production process (Kristal, 2020).

The SBTC framework states that occupational earning results from the social relationship among the occupations. It has posited a power struggle between occupation and education where education has failed to keep pace with the technology-driven skill requirement and this has resulted in higher wages only for selected more educated groups. According to this model, there are two powerful occupations that emerge (Abduljabbar et al., 2019). Computer programmers or information specialists who have expertise in technology reorganize, aggregate, and transfer knowledge to managers and engineers who are experts in translating, managing, and interpreting the data (Adel, 2022; Kristal, 2020; Tomaskovic-Devey and Avent-Holt, 2019).

2.2.6 The political economy of automation and transformation

Spencer (2017) enlightened the political and policy factors related to technology, ownership, the gig economy, and work-less future issues. Digital technology leads to unequal power between the capitalist and labor. The main objective of technological advancement is to create surplus value. In the process, human workers are either reduced, replaced, or pushed down in the ranking order in terms of wages and remuneration. On one side, there is a huge expansion in production and consumption, but on the other side, it is alienation and exploitation of the workforce. This perspective advocates collective ownership and looks at technology as a "Positive" force that helps liberate rather than enslave the workforce. Automation should be embraced willingly as a step toward better work and more leisure. There is a need for fundamental reforms, as the need for technological changes is empirical (Caines et al., 2017). It should move into a collectivist ownership rather than capital ownership. The future of work should be based on production, forms, and relationships based on a social context.

Hence, the possible Industrial 5.0 can be depicted as shown in Figure 2.6.

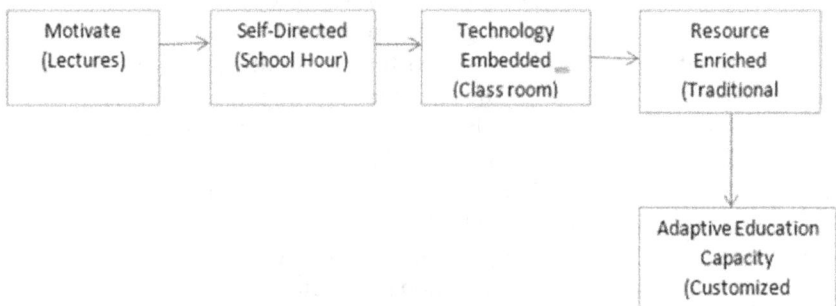

Figure 2.6 Critical factors for Industry 5.0.

2.3 DEEP INSIGHT INTO INDUSTRY 5.0

Industry 5.0 is connected with three core values – human centricity, sustainability, and resilience – as shown in Figure 2.7. The human-centric approach shows interest in putting humans at the center of the production process. The approach is to shift progress using technology for a human-centric and societal benefit. The new approach shifts the perception of industry workers from "cost" to "investment" through new roles in the organization. Technology is seen as an enabler to serve people wherein there is a shift to a sustainable production process, which includes the diversity of the industry workers. The intention is to create a safe work environment with priority given to the physical and mental health of the employees, focusing on the employees' holistic well-being. It is also essential to guard their fundamental rights, i.e., autonomy, human dignity, and privacy. Employees will have to keep reskilling and upskilling themselves to have better career opportunities and work–life balance (Garcia-Murillo et al., 2018).

2.3.1 Technology enablers of Industry 5.0

Various technologies that enable the implementation of Industry 5.0 are highlighted as follows:

1. **Cloud Computing**
   This technology helps in the delivery of computing services that include databases, software, intelligence analytics, networks, and others. Cloud computing offers economies of scale and innovation efficiently.

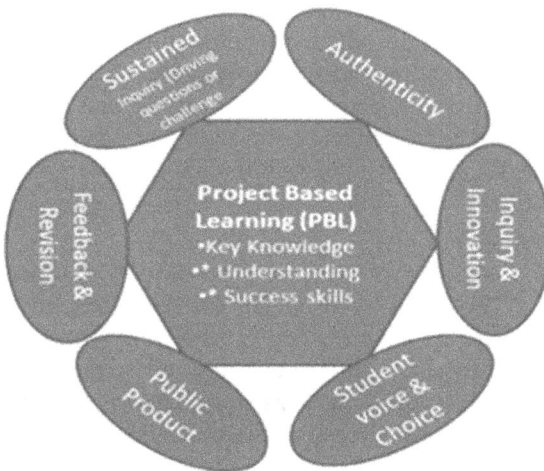

Figure 2.7 Industry 5.0 model.

Data is stored in remote servers. The internet is used to store, manage, and access data.

The industrial cloud is a virtual environment that provides a supportive environment for industry applications, for example, IoT monitoring tools for mobile and web usage, and automating data normalization from diverse production data sources. Autonomous robots and diverse robots are managed through these platforms. This helps determine the machine defects or failures and mitigate them using more workforce (Bekey, 2005).

Scalable infrastructure to support data edge devices is provided by cloud computing. The IoT platforms are primarily supported by cloud computing. It is used in managing autonomous and diverse robots mainly deployed in shop floor plants of manufacturing companies.

2. **Collaborative Robots**

Industry 5.0 aims to provide a human touch in the production units for development. Collaborative robots represent a new age in robotics-aided production. Cobots are the technology that combines the innate ability of fine craftsmanship and creativity of people with high efficiency and consistency. Industry 5.0 focuses on highly skilled people along with robots to create individualized products ranging from smart devices to cars for consumers (Cormier et al., 2013). Robots accomplish heavy tasks while skilled humans provide the cognitive skills of the craftsman (Adel, 2022). Robots help humans work better via IoT and thus increase automation and business efficiency. For example, it provides human operators the benefit of robots with more technical precision and heavy lifting abilities. Robots can be leveraged using advanced technologies using the IoT. It uses a network of physical objects and things embedded with sensors, software, and other technologies (Bagdasarov et al., 2020). It helps in connecting and exchanging ideas using devices plus internet systems. IoT is an important component in a smart city. It enables IoT-incorporated home appliances, heating systems, and security systems for better comfort and reduced energy. Authentication and access control are issues in the IoT. It is a dynamic approach to data-centric applications using a cloud platform.

3. **Big Data Analytics**

Big Data analysis is an intricate procedure to examine Big Data to uncover certain trends, and hidden patterns using advanced analytical methods for decision-making processes. In Industry 5.0, Big Data plays a vital role. It helps in utilizing 3D symmetry in innovation ecosystem design. Big Data helps store massive amounts of data and processes through traditional methods. Technology helps in analyzing these data, which could be diverse, semi-structured unstructured data using innovative analytical methods and disseminating information that leverages a list of root causes and helps better decision-making. This lays the foundation for better customer experience and building strong customer relationships.

Huge data that is stored requires time and storage space. Big Data analytics bring value by helping disseminate the data for industries. This is mainly achieved through connecting with physical devices to the internet using novel technologies. This analyzed data helps in providing systematic guidance in manufacturing units, related to production activities within the entire product life cycle, a process of cost-efficient fault-free running of units, and helping in decision-making and solving problems related to the operation.

Big Data is running on information. For a better characterization, the dimension can be classified into Veracity, Vision, Volatility, Verification, Validation, Variability, and Value. To explore data, advanced analytics is used, for example, machine learning and forecasting models. Knowledge is extracted, enabling manufacturers to understand the product life cycle allowing them to make rational, informed, and responsive decision-making. IoT converges with Big Data being an enabler for further growth.

4. **Blockchain Technology**
   Complete information and data are stored in an immutable ledger which is decentralized and distributed containing a digital records ledger. Through blockchain, the members can access the data. This information includes orders, payments, and production information wherein the customers can track the information.

   The blockchain is a digital ledger of any transaction or contract that must be recorded independently. This is encoded and dispersed across several hundred and thousands of computers which is distrusted across various locations (Lage et al., 2022). The digital ledger is accessible across various locations and remains a critical feature of blockchain. This technology has disrupted the financial sector and underpins the digital currency. Bitcoin is the finest example of the same.

   This has become a boon to the financial sector. Using blockchain technology, the financial sector customers make transactions across the internet without any third-party interference. The transaction records are created and encrypted for security (Bhushan et al., 2021). This also reduces the possibility of data breaches. This advanced technology can potentially revolutionize various business sectors in the future.

2.3.2 Industry 4.0 and 5.0 applications

The usage of advanced technologies like AI, machine learning, and robotics is clearly visible in various retail, healthcare, finance, manufacturing, and other industries (Adel, 2022).

2.3.2.1 AI and its applications in Industry 4.0

AI can be widely used in manufacturing in both Industry 4.0 and Industry 5.0 environments. In the first product life phase of designing and prototyping, 3D printing, computer-controlled machining, or additive manufacturing is done using AI. Generative AI helps in optimizing the design and making it more efficient in terms of materials. For example, the cutting layout of clothing panels on the cloth's bolts can be controlled and fabric waste minimized using AI (Zawacki-Richter et al., 2019).

In other manufacturing processes, the AI system helps incorporate simplicity and optimize the number of parts required for a design (Muro et al., 2019). For example, AI helps in concocting a design and speeds up production by reducing the number of separate cuts in the making of a finished chair leg in an automatic lathe machine.

A few other uses of AI in factories are as follows:

- Use of generative AI by Nutella in designing packages for its products which run to millions
- 3D printing: In ADDMAN, hybrid modeling tools along with AI are used to design and prototype machine parts.
- Use of computerized numeric control machine tools by FANUC which learns from errors and improvises the control as they operate.
- In an oxygen-free environment or extreme temperature, where it is difficult for humans to be physically present, cobots are used for plant control in a production line.
- In 3B Fibreglass, AI monitors the performance of machinery by speeding and slowing lines or operations based on real-time events.
- Finally, in BMW, to save time and energy, AI uses cameras and other sensors to review the product quality and material.

AI algorithms and techniques are dynamic and keep growing. Computing is available both on-site and in the cloud. The manufacturers are looking for efficiency and responsiveness of these latest tools to compete in the business world. As new manufacturers emerge, old factories are retooling to cope with the advancement and hence AI will continue to expand from Industry 4.0 to 5.0.

2.3.3 Applications of technology in Industry 5.0

2.3.3.1 Healthcare

By 2023, AI had made considerable advances in the healthcare sectors, especially in the areas of diagnostics and personal medicine. It has helped accelerate the discovery of new drugs and has emerged as a new paradigm in telemedicine. Machine learning algorithms are used for early disease detection and accurate diagnoses. It has helped in personalized healthcare and treatment plans based on the unique genetic makeup of the patient.

AI has also made a substantial impact through wearable devices and IoT-enabled health monitoring systems. These technologies have enabled the collection of patient data like the heartbeat, blood pressure, and sugar levels to monitor and manage chronic medical conditions. Integration of AI into the healthcare sector is going to make a dynamic change in diagnosis and health management.

These advanced technologies using AI have created a significant impact in the area of mental improvement by creating accessible personalized support systems. Powered by natural language processing and machine learning, chatbots and virtual therapists can engage users in therapeutic conversations. This has become highly effective in alleviating symptoms of anxiety, depression, and other mental health issues especially during and post-COVID-19.

Big Data analytics helps healthcare providers with information related to patient care, treatment effectiveness, prediction of diseases, and effective resource allocation. Big Data is widely used in clinical trial optimizations, improving operational efficiencies in the hospital, drug discovery, and genomics research.

Blockchain technology helps to enhance the security, privacy, and interoperability of data related to healthcare. It enables the sharing of medical records, streamlining the claiming process, and managing the pharmaceutical supply chain effectively.

2.3.3.2 Smart hospital

Industry 5.0 has enabled the creation of real-time capability for smart hospitals by providing remote monitoring systems using technology. Smart healthcare technologies enable doctors to monitor and focus on patients using efficient data and provide better treatment. It helps in providing customized implants and performing surgery with more precision.

1. **Early Detection of Sepsis**: At Humber River Hospital in Toronto, Canada, the usage of AI-powered algorithms helps to identify early signs of sepsis, a potentially life-threatening infection. This system has led to a reduction in sepsis-related deaths by 25% and helped cut the average length of stay in the hospital by one day.
2. **Image Analysis in Radiology**: Mount Sinai Hospital in New York City uses AI systems to analyze breast cancer scans and detect malignancies wherein the accuracy is 94.5%. These processes are done by using AI to scan medical images such as X-rays, CT scans, and MRIs.
3. **Predictive Analytics**: By analyzing a large amount of patient data, patterns and predictions are made to identify patients with diabetes and heart disease. This also helps in predicting patient outcomes and readmission.
4. **Personalized Treatment Plan**: At Mayo Clinic, the researcher used an AI system to analyze the genetic data of patients, their medical history, and current symptoms to recommend targeted therapies for their cancer.

5. **Virtual Assistants for Patients:** AI-powered virtual assistants such as chatbots help patients navigate the healthcare system and quick answers to their questions. The patients' scheduled visits and reminders for taking their prescriptions and keeping an eye on health-related activities are done through this technology. AI is specifically useful for old care. Virtual assistants help seniors in their daily routines. For instance, the virtual assistant greets an elderly person who says "Good Morning", prompts them to take medicine and have breakfast, and reminds them of their health visits. This enables elderly users to take care of their activities and themselves independently. This virtual assistant by having a conversation with these older users helps in alleviating loneliness to a large extent.

2.3.3.3 Manufacturing industry

Applications of AI in manufacturing range from quality control, predictive maintenance, optimization, and supply chain to robotics. The usage of advanced algorithms helps in detecting quality issues in products while predictive maintenance ensures minimizing equipment downtime. The supply chain can be optimized, allowing resource allocation more efficiently. Robotics helps increase productivity and precision in manufacturing processes.

Companies use digital twins by creating replicas of physical items, processes, or systems virtually and these digital representations enable them to manage real-time through simulation, monitoring, and optimizing the performance. These are specifically applicable to manufacturing industries.

Big Data helps optimize the production process, reduce downtime, minimize maintenance costs, and improve quality. It enables predictive maintenance, supply chain optimization, inventory management, and real-time monitoring.

Blockchain can enhance the security and interoperability of IoT devices. Provides a secure and authenticated decentralized communication and data exchange between devices in a network which is an essential requirement in an automated environment.

The manufacturing system is shifting in the era of Industry 5.0. The mundane, repetitive tasks done by humans are getting automized (Autor et al., 2003). It is a collaboration combining the accuracy of machines and human beings' creative and innovative potential. Cloud-based platforms help in controlling manufacturing services and provide cost optimization.

A few of the industry-based applications are as follows:

1. **General Electric:** GE has implemented AI in its manufacturing plants to improve productivity and reduce costs and developed Brilliant Manufacturing which uses AI and machine learning to optimize production processes, reducing waste and increasing efficiency. The system collects data from machines, sensors, and other resources to identify patterns and anomalies that could indicate problems or opportunities for improvement.

2. **BMW:** BMW uses AI to improve its production line and streamline the manufacturing process and developed an AI system called BMW AI assistant which uses natural language processing to understand and respond to commands from workers. This system helps in optimizing production schedules and identifying potential bottlenecks in the manufacturing process.

3. **Foxconn:** A Taiwanese electronics manufacturer, has implemented AI and automation in its factories called Foxconn Intelligent Robotics Initiative which is mainly used to do repetitive tasks and improve productivity. They also use machine learning to optimize production schedules and to identify potential issues before they occur.

4. **Intel:** AI is implemented in the manufacturing process to improve product quality and reduce defects. The data generated from sensors and other sources are analyzed to identify patterns and anomalies that could indicate problems with a particular component or process. The significant benefit is to identify issues before they become substantial.

5. **Siemens:** AI and machine learning are used for optimizing manufacturing processes and improving product quality. Siemens developed a system "Siemens Digital Enterprise" which enables production schedules, reduces waste and increases efficiency along with analyzing data patterns to identify potential issues in the manufacturing process.

These are a few examples of how AI and machine learning are used to transform the way products are designed and delivered along with enhanced efficiency.

2.3.3.4 Futuristic human-centric assembly

The approach in component assembly is changing rapidly. A few of the repeated tasks have been automated. The last decade saw the collaboration of humans and robots (HRC). This has helped in retaining quality in mundane tasks and improving work conditions. This model of HR combines qualities like agility, accuracy, and strength along with cognition, flexibility, and adaptability in robots to maintain the quality of the products. Industry 5.0 aims for sustainability and resilience. The futuristic robots are facilitated by four qualities of humans such as augmented thinking, cognition, mixed reality, and co-intelligence. This quality enables the robot to co-work with humans by energizing, advising, supporting, and empowering in their work. They are modeled as super operators that help in applying knowledge to perform various tasks assigned (Figure 2.8).

1. **Augmented Robot:** Robots are made with augmented strength which has more sustainable muscle power than human beings. So this factor eliminates fatigue at the workplace. Hence, with the exoskeleton, augmented robots can be used to energize with active or passive actuation. Human operators and the exoskeleton can perform tasks beyond

Figure 2.8 Human-centric assembly in future factories (Wang, 2022).

Source: A futuristic perspective on human-centric assembly. *Journal of Manufacturing Systems.*

normal physical limits along with consistency, accuracy, and quality (Wang et al., 2019).

2. **Cognitive System:** A high level of cognitive ability is possessed by human beings within an allotted time at times with a restricted holistic view. To enhance this ability, an intelligent multi-cognitive system is added that helps advise operators in making optimized decision-making. This also leads to a collaboration of humans and robots while together (Monostori et al., 2006; Kemény et al., 2021).

3. **Mixed Reality:** Humans rely on senses to make decisions. At times with perception and fading memories, they cannot foresee the future. Mixed reality is a combination of Augmented Reality (AR) and digital twin, which helps humans to see what is invisible now and presumed to happen in the future, for making informed decisions for organizational effectiveness.

4. **Co-Intelligence:** Collaborative intelligence is based on assistance provided to robots, i.e., to train, explain, and sustain by amplifying, interacting, and embodying the primitives. Humans need to apply competence to robots, train on what to do, and use sustainable decision-making. Robots must amplify humans' self-adaptably so that humans are empowered to interact with work using the full potential of (artificial and anthropological) co-intelligence.

5. **Brain Robotic:** Brainwaves are used to control robots, another modality of HRC in production assembly. Here, human thought control commands are sent to the robot for the desired action. This process is done by controlling commands through deep learning. This process aligns well with human centricity and complements modalities like voice, and gesture for effective communication between humans and robots.

6. **Human-Centric Assembly:** This system enables the human operator to do this task along with the function of instructing the cobots to co-work with him. This is done through contact-awareness-enabled real-time sensing systems (Wang et al., 2019).

This is definitely the reality for tomorrow, wherein humans and robots work together to meet the common objective of the organization effectively maximizing the growth and development of the organization (Wang, 2022).

2.3.3.5 Customer service

In the realm of customer service, AI has taken over in a big way. AI-powered virtual assistants and chatbots provide service 24/7 to cater to various customer queries. Automation in call centers has been able to capture customer emotions through sentiment analysis. This has enabled increased productivity by tailoring the responses of the customers.

AI is used to analyze consumer data, including buyer behavior patterns, preferences, and purchase history. This data is used to provide hyper-personalized customer experiences by businesses (Morgan-Thomas et al., 2020). Algorithms help in creating content, promotions, and personalized recommendations for consumers.

2.3.3.6 Finance

The usage of AI in the finance domain is becoming prominent. Professionals in finance are employing AI to detect fraud, use algorithms in trading and credit scoring, and analyze customers' risk assessments. AI tools like machine learning are used to identify suspicious transactions and execute faster trading with accuracy. With AI, financial institutions can more accurately assess risk so they can improve loan decisions and investment strategies.

Finance is a highly regulated industry and needs to maintain compliance and regulations. AI helps financial institutions to simplify regulatory compliance. The documents and records are automated and AI helps in monitoring transactions for violations and deviations in compliance.

AI is also revolutionizing the area of financial and wealth management and planning by creating robo-advisors which is intelligent enough to cater to client requirements who range from seasoned to novice professionals. These platforms use advanced algorithms to assess the trends in the market and risk tolerance and provide recommendations at personalized levels.

Big Data analytics enables real-time monitoring, anomaly detection, and fraud prevention. It helps financial institutions analyze vast amounts of data to detect fraudulent activities, assess credit risk, improve customer experience, optimize trading strategies, and comply with regulatory requirements.

With blockchain technology in the financial sector, this facility has enabled cross-border payment and remittance with ease. The receiver has direct access to the payment with no delays, unnecessary fees, or remittances involved from any part of the globe and can trade finance at their fingertips. Participants can interact directly and make transactions across the internet without the interference of a third party. Helps in cross-border payments, remittances, and trade finance. Through blockchain technology, personal information and transaction records are encrypted to maintain safety and security. Blockchain technology has helped in reducing data breaches to a large extent.

2.3.3.7 Transportation

The automobile industry has introduced self-driving cars and trucks using advanced technology like AI. This helps in reducing human errors and enhances safety. Also, using technology, intelligent traffic management helps in reducing congestion. It also helps in route optimization and saving time and fuel which is in alignment with sustainable goals (Bednar and Welch, 2020). Also, drone delivery is picking up as an alternative, which is more eco-friendly than the traditional drop-out method.

Predicting passenger demand and optimizing schedules is one of the key components for which AI is used – to enhance and improve public transportation systems (Abduljabbar et al., 2019).

Logistics companies leverage Big Data analytics to optimize routes, reduce delivery times, manage fleet operations, and improve supply chain visibility. It enables real-time tracking, demand forecasting, and efficient inventory management.

2.3.3.8 Agriculture

The innovation in technology is helping the agriculture world too. Farmers and scientists are using AI technology to monitor crops, keep pests at bay, and predict farm yield. The technological precision aided by AI is helping to improve fertilization, optimize irrigation, and aid farmers reduce waste through data-driven informed decision-making.

The agriculture sector has seen farmers using tractors and machinery and have revolutionized the traditional agriculture practices followed. Equipped with sensors, GPS, and control systems, self-driven tractors have started performing tasks like plowing, seeding, and spraying, leading to increased output due to precision and efficiency.

Equipped with AI, ML, and remote sensing features, drone technology is growing at a rapid pace. There is a rising demand for the usage of drones in the future in this field due to their immense advantage. It is increasing performance and solving assorted barriers through precision farming. The market for agriculture drones has increased to a whopping $1.3 billion. The Unmanned Ariel Vehicles (UAVs) are filling the gap and inefficiencies of traditional farming. The primary aim of adopting drone technology is to avoid guesswork or ambiguity and focus on precise accurate information before making decisions.

Machine learning and remote sensing help provide information related to weather, soil conditions, and temperature, which plays a critical role in farming. Agriculture drone provides input to farmers helping to decide on the optimum choice of farming based on the environmental conditions. The data obtained through drones are used for regulating crop health, treatment of crops, scouting, irrigation, analyzing the soil, and assessment of crop damage. The surveys conducted through drones provide information on crop yield and help in minimizing time and expenses.

2.3.3.9 Retail and supply chain management

AI application is widely used in retail for inventory management and targeted marketing. They leverage artificial intelligence through chatbots and create personalized recommendations based on customer preferences and buying history.

Retailers are also integrating visual search technologies into their online stores, so customers can find products by uploading images instead of relying on text-based queries. These types of AI-powered visual search engines can analyze the features of the uploaded image and provide a list of similar products available for purchase.

Visual search technologies are adopted in online stores by retailers to help customers find products by uploading images. The earlier method mainly was text-based queries. AI-powered visual engines analyze the features of the product and provide similar products available in the store.

Big Data analytics is used to understand customer behavior, preferences, and buying patterns. It helps in personalized marketing, targeted advertising, inventory optimization, demand forecasting, and recommendation systems.

Blockchain technology also provides end-to-end visibility and traceability of products. It enables tracking and verification, ensuring authenticity and improving efficiency. It also helps reduce counterfeiting.

The digital supply chain is the next-generation supply chain solution. This technology integrates human intelligence with cognitive computing and intelligent automation provides the opportunity for hyper-personalization (Krzywdzinski, 2017). Machine learning, robotics, and other automated technologies help organizations and employees increase business proficiency and provide immense value to customers.

AI is widely used in supply chain management to optimize processes, reduce costs, and increase customer satisfaction. A few of the areas where it is widely used are as follows.

1. **Demand Forecasting:** AI algorithms analyze extensive data of customer orders, social media trends, and weather patterns to predict future demand. This helps in optimizing inventory levels, production planning, and logistics.

2. **Inventory Management:** Helps in analyzing the inventory levels, lead times, and demand to optimize inventory levels and reduce the risk of stockouts or overstocking.

3. **Logistics Optimization:** Optimization of transportation routes (Abduljabbar et al., 2019) and schedules based on real-time data such as traffic conditions and weather forecasts, reducing delivery times and costs.

4. **Supplier Selection:** Identifying potential suppliers based on criteria like pricing, quality, and delivery times helps companies make informed supplier selection decisions.

5. **Quality Control:** Sensors' data from production lines are used to identify potential quality issues and alert operations before they become more significant problems.

6. **Risk Management:** Systems that help in identifying potential risks in supply chain management, disruption in transportation, and bankruptcies of suppliers allowing companies to develop contingency plans and mitigate risks.

A few of the industry examples are as follows:

1. **Walmart:** Walmart has been experimenting with AI for long. They use machine learning algorithms to optimize the supply chain including forecasting demand, tracking inventory levels, and predicting the product demands. They use AI-powered robots to scan store shelves and track inventory levels in real time.

2. **Amazon:** Amazon is well known for its usage of AI and machine learning in recommending products based on the browsing and purchase history of the customers. Alexa Voice Assistant is also AI-enabled to help customers order through voice commands.

3. **Sephora:** AI-powered chatbot provides personalized beauty product recommendations to customers. The chatbot uses natural language to understand customers' questions and preferences to recommend beauty products. Sephora has developed a color-matching AI tool which provides inputs for selecting the correct type of foundation based on skin tone.

4. **H&M:** AI is used to optimize the supply chain, predict demand, and optimize inventory levels. They also help to analyze customer reviews to identify trends and improve product design.

5. **Starbucks:** Starbucks uses AI to optimize its mobile ordering system. The system uses a machine learning algorithm to predict when customers will arrive and prepare orders accordingly, reducing waiting time and improving customer satisfaction.

2.3.3.10 Education

AI has made an impact in the education field as well. AI-powered learning techniques are adaptive and can be tailored to meet customized student needs. The use of AI in academics helps in predicting the performance of the students and helps in early intervention to address this issue. This is specifically useful in cases of students dealing with learning challenges (Zawacki-Richter et al., 2019).

These advanced technologies are playing a significant role as enablers in democratizing access to education for lesser privileged. AI tools help in language translation and real-time transcription services. This helps in breaking language barriers enabling wide access to students on educational content sitting in any part of the globe enabling students worldwide to access educational content from anywhere in the world. One-to-one support and guidance are provided by supplementing regular classroom instructions making quality education accessible to a broader audience. This technology helps in the evaluation of the performance of teachers and professional development (Akgunduz and Mesutoglu, 2021).

2.3.3.11 Energy

Energy management is also significantly impacted in a positive way through the usage of AI in various applications like smart grid management, forecasting demand, optimization of renewable energy, and conservation of energy. The usage of AI-driven systems helps in predicting customer usage patterns, helps in balancing supply and demand, recommends energy-saving measures, and optimizes renewable energy resources, which is the main aim of Industry 5.0, sustainability and resilience (Ghobakhloo, 2020).

Apart from energy management, it is used in improving the performance of energy storage systems like thermal and battery storage. By analyzing historical performance metrics and data from sensors, AI-powered algorithms can optimize charging and discharging cycles and extend the life of energy storage systems.

Big Data analytics help provide inputs to companies the consumption patterns, predict demand, optimize energy distribution, and predict anomalies in the power grid. This data helps in energy integration, equipment maintenance, and initiatives for efficient energy utilization.

Blockchain technology helps in energy trading and grid management, especially peer-to-peer energy trading, where organizations and individuals can trade electricity. It helps in energy distribution and encourages renewable energy via transparent tracking of production of energy production and consumption.

2.3.3.12 Human resources

Organizations are using AI tools in Human Resource Management as well. The time-consuming tasks like resume screening and workforce planning are being done using automation services. AI is helping HR professionals in assessing employee training needs and monitoring employee performance.

Many diversity and inclusion initiatives are supported through artificial intelligence tools. Algorithms are used to assess job postings and for hiring to attract a diverse pool of candidates. These technologies have helped ensure fair and smooth HR processes for candidates.

Artificial intelligence tools are increasingly acting as gatekeepers between the job seeker and job provider. More companies are using chatbots to screen candidates and video interviewing that evaluates body language to judge personality, skills, and aptitude.

2.3.3.13 Environment

Big Data and AI are helping address environmental challenges by providing inputs related to climate modeling, wildlife conservation, prediction of natural disasters, protecting endangered species, trailing pollution norms, and addressing various environmental challenges. Advanced technologies adopted help give insights into predicting climate patterns, and pollution norms and help in taking measures to avoid mass calamities to a large extent.

2.3.3.14 Security

AI is a game-changer in the security world. They are widely used in facial recognition, surveillance, and threat detection by law enforcement agencies. Cyber security firms are using these advanced technologies to combat cybercrime. AI helps in identifying and neutralizing potential threats in virtual and real-time situations.

Advanced biometric authentications like fingerprint, iris, and voice recognition are developed using AI. Technological support helps a high level of security unlike traditional methods using passwords and access cards, and also relying on biological features that are difficult to forge or replicate.

Blockchain-based voting systems can provide transparent and tamper-resistant voting processes. It ensures the integrity of the voting process, enhances trust, and eliminates the possibility of fraud or manipulation.

2.3.3.15 Entertainment

AI is the real "name of the game" in the entertainment industry. Game designers are creating an immersive experience for the players, which includes virtual and augmented reality elements. Players also get AI-powered recommendations that help personalize and provide customized, engaging entertainment. With AI-powered recommendation systems, companies can curate content for users so they have more personalized and engaging entertainment experiences.

AI has also entered into the space of generative art, virtual concerts, and interactive installations. This has created an enhanced experience for the customers in art and music and changed the way we interact with the art forms.

2.3.3.16 Law and legal services

Big Data is emerging as a boon to law and legal services. This technology helps in storing and collecting data related to clients, contracts, and legal history and provides access anytime and anywhere, including courtrooms. AI-driven tools help in processing large volumes of legal documents, expedite legal research from extensive data, and identify issues in legal contracts. AI is also used in dispute resolution and to streamline mediation and arbitration. This also helps in cost reduction.

Blockchain can establish proof of existence and ownership for intellectual property rights. It provides a secure and timestamped record of creations, patents, and copyrights.

2.3.3.17 Space exploration

The advent of the application of AI is visible in space exploration as well. Scientists are using this technology for spacecraft navigation, imaging satellites, and mission planning. This technology is immensely helpful in astronomy, for example, in identifying a new astronomical phenomenon.

AI plays a critical role in the detection and tracking of asteroids and comets (near-Earth objects). Big Data helps in analyzing vast amounts of astronomical data; machine learning is of immense use to identify and predict the trajectory of these objects to make decisions and provide alerts to avoid any disaster.

2.4 OPPORTUNITIES OF INDUSTRY 5.0

The following points enlighten diverse opportunities of Industry 5.0:

- Automation will positively impact many sectors through the deployment of next-generation technology.
- It creates opportunities for higher customization of products to the customers in a highly automated production system.
- Industry 5.0 provides greater opportunities for creative people enabling optimization of human efficiency.
- The versatility of the machines provides employees with the choice of creating a high level of customization for the customers. Also, it helps in managing repeated or follow-up assignments through digital mode, making work much easier (Haleem and Javaid, 2019).

- Industry 5.0 demands highly competent skills leading to higher-value employment. This also provides people the opportunity to be creative and innovative (Rossi et al., 2018).
- In Industry 5.0, the operation in the production unit can engage themselves in the planning and execution than in a less automated production unit. This also has a positive impact on the well-being of the employees as work that leads to fatigue is mostly done by robots (Spencer, 2017).
- In Industry 5.0, Automation provides opportunities for designing tailor-made and personalized products especially related to medicine and prosthetics for physically challenged people.
- Automation helps in providing real-time information from various sectors helping for informed decision-making. This is specifically applicable to the manufacturing and service sectors.
- The introduction of cobots on the floor helps in taking up jobs or work that are more hazardous and dangerous in nature. This leads to better safety and quality at the workplace.
- More personalized products and services increase customer satisfaction, and loyalty and attract new customers which results in increased profit and market share for the companies.
- AI and Big Data provide great opportunities for start-ups and entrepreneurs to be in creative and innovative spheres and come up with unique products and services allied to the concept of sustainability. It also helps look at opportunities in a holistic manner.
- Increasing human–machine interaction is also providing opportunities in Industry 5.0 for future research and development which adds greater value for society and organizations (Mohr, 2016).
- Quality services can be provided at remote locations with the help of Industry 5.0, especially in the healthcare industry such as robots' medical surgeries in rural areas.
- Technology is appealing as it is cost-effective, especially in the arena of HR where thousands of resumes can be screened and filtered to get the right candidates.
- The cobots in HR can help in emotional analysis and performance rating. It helps in analyzing human behavior related to burnout stress syndrome and lack of commitment.
- Robots can be used for working in high-risk environments which may be dangerous or difficult for humans to work. They can be used in situations involving hazardous material, nuclear facilities, deep sea, or space exploration. Industrial robots possess significant physical strength and endurance allowing them to perform heavy lifting and repetitive physical tasks for extended periods. This is highly beneficial in industries like construction, logistics, and manufacturing.
- Since communication and collaboration are key factors in Industry 5.0, engineering education should focus on innovations in human–technology interaction in a futuristic manner.

2.5 REMONSTRANCE UNDER INDUSTRY 5.0

- Industry 5.0 comes up with enormous challenges. High automation leads to an increase in work polarization. On one end, there are highly trained and qualified professionals and, on the other end, there are low-paid, unqualified employees. This may augment the skilled and unskilled divide in society (Boyd and Huettinger, 2019; Kolade and Owoseni, 2022).
- Due to highly automated manufacturing systems, skill development is a humongous task such as training the workforce in advanced and cutting-edge technologies. Along with this, it is essential to provide behavioral training to interact with robots.
- Collaborative robotics has its own safety issues, as there are instances where robots have emerged as dangerous for human beings co-working with them (Rossi et al., 2018).
- Smart manufacturing systems demand higher autonomy as they have enabled robots' cognitive capabilities. The shift from the present context to Industry 5.0 is difficult due to the lack of autonomy in the present systems, such as integrated decision-making (Thoben et al., 2017).
- Even through automation, accommodating diverse data repositories and acquiring high quality and integrity of the data is a concern, especially for manufacturing sectors.
- Increased connectivity and standard communication protocols have led to issues about cyber security threats in industrial and manufacturing lines.
- Industry 5.0 demands a huge amount of investment to fully implement all its pillars which is difficult for industry, especially for the SMEs to adopt.
- For instance, Industry 5.0 offers great potential in the healthcare industry but a high degree of precision and accuracy is needed. The research on this front is still in a nascent stage and demands a high number of investments and infrastructure.
- It is challenging for startups and entrepreneurs since Industry 5.0 demands high investments and infrastructure with cutting-edge technology requirements.
- Challenging to draw regulatory mechanisms in Industry 5.0 due to the high amount of automation presence. For example, whom to be held accountable in case of failures and to what extent.
- Due to higher levels of automation in the industries, the existing business strategy and business models have to be modified and customized to meet the requirements of Industry 5.0. Due to mass personalization, business strategy will be focusing more on customer-centric operations. Customer subjectivity changes over time and it is difficult to change business strategies and business models frequently.
- Business strategies in Industry 5.0 demand a higher level of dynamism to sustain competition due to differential customer preferences.

2.5.1 Challenges emerging from the co-work of human–robots

It is not easy to overlook the challenges with respect to Industry 5.0 in the organizational behavior, structure, workflow, ethics, and environment.

A few of the major challenges are for Industry 5.0 developments to succeed in the business (Tomaskovic-Devey and Avent-Holt, 2019).

1. Human–robot co-working is expected to create many challenges. It is essential to legally define what a robot is (Demir et al., 2019a). This should include what are the distinctions of robots that can be used in the workplace, the role and responsibilities of robots, and the types of decisions that can be made by the robot regarding the work.
2. People need to develop competency skills; employees are required to upskill and reskill to collaborate and work with advanced robots and smart machines. Along with the soft skills required, gaining technical skills is a key component to remain relevant for the employees (Goos et al., 2010). Programming the industrial robot and managing translation in the new jobs leads to an upgrade to a high level of technical skills (Gorman and Sandefur, 2011).
3. Customized software-connected factories, artificial intelligence, real-time information, and the IoT, collaborative robotics, are widely used for Industry 5.0. Employees need to continuously adopt advanced technology which requires more time and effort from the side of the human workers. Personal preference for not using a particular technology varies from human to human. Some people would readily accept or strongly oppose it. Hence, change management becomes a crucial factor that needs to be addressed.
4. Advanced technologies require heavy investments. Implementing Industry 5.0 is expensive as it requires smart machines and highly skilled employees to increase productivity and efficiency. Cobots are expensive too. Training the employees for new jobs brings extra costs. The companies need to upgrade the production lines for Industry 5.0.
5. Advanced technology issues may have certain psychological effects on employees like "robophobia" or "nonrobophobia" (Briggs and Scheutz, 2014). Humans are social creatures. Social interactions are essential for effective work performance. When the number of robots increases, it could lead to unprecedented effects in the workplace (Bartlett et al., 2004).
6. The Human Resources department may face new challenges. Nowadays, most organizations are advertising themselves as "Green". They claim to be sensitive to the environment. Hence, an automized environment may contradict the green philosophy (Kovacs, 2018). Some organizations may go for a "human workforce" to prove they are socially responsible.

7. To establish trust in the ecosystem, it is necessary to have a fully proofed security. The authentication used in the industry is the scale to interact with various devices, to stand against the future quantum computing applications to deploy nodes of IoT. Artificial intelligence and automation in Industry 5.0 come with their own challenges due to security. Various strategies need to be adopted to maintain cyber security as it may prove dangerous to organizations, including nations, as it contains data that could be misused.

8. Robots are selfless, have no ambition, and do not have laziness. All these factors have an impact on how the work is perceived. Robots do not require breaks, sleep, or rest, enabling them to operate continuously. In industries where constant operation is critical, such as customer support, surveillance, or autonomous vehicles, robots can provide non-stop service, ensuring uninterrupted operations (Bekey, 2005). Humans may not be able to compete with robots. It is hard to predict how work ethics evolve in a human–robot co-working environment (Demir, 2017).

9. Robots if not properly designed, can cause physical harm to the human. A malfunctioning industrial robot may inadvertently injure a worker by striking or pinning them against the surface. Accidents can occur when humans make errors in programming or designing. Human operators also make mistakes while interacting with robots leading to accidents. There are possible accidents from collaborative robots working alongside humans due to miscommunication or failure in sensors. Software bugs and glitches can lead to unexpected behavior causing unsafe situations or accidents (Roboticsbiz, 2022; Winfield et al., 2021).

A few examples are as follows:

- The robot stopped because of some technical issue with the locating switch. The operator entered the cell and attempted to locate the part. The robot which got activated in the process crushed the employee.
- An electrician who entered a robot cell to reposition a part was leaving back passed through an envelope of an adjacent robot which had completed a task and moving back to its home position to pick another part. This robot pinned the electrician with its end effector against a parts feed conveyor.
- Four workers entered a restricted space to troubleshoot a robot with power on. One worker puts the part to the fixture back on the live robot. The command the robot took was "Go". The worker was pinned to the fixture.
- An employee who was cleaning up was crushed between the "safety post" and the back end.

- When attempting to load parts into empty parts on conveyors, the robot pinned an employee against one of the conveyors causing severe injuries to the employee.
- It was necessary to wipe the dust off two photoelectric eyes of the robot of a cell design to perform the next task. The robot had stopped and the employee entered through the opening to fix the issue. While dusting, it signalled the robot to move crushing the employee.
- A technician was working on a robot-conveyor arm which dips part into solution every 15 minutes. The operator forgot about the 15-minute cycle and was caught between the conveyor and the robot arm.

There are multiple reports of workers sustaining injuries while working alongside robotic systems in the automotive industry. These incidents often involve collisions between workers and robots or the unintended activation of robotic equipment.

To mitigate the risks associated with robot usage, safety standards and regulations should be in place in industries. These standards aim to ensure the proper design, implementation, and operation of robots, as well as the training of personnel involved in their use. Continued research and development in the field of robotics are also focused on improving safety features and creating a more robust and reliable system.

It's important to note that while robots may outperform humans in certain tasks, they lack the adaptability, creativity, emotional intelligence, and complex decision-making abilities that humans possess. The ideal scenario often involves a collaboration between humans and robots, where each can leverage their strengths to achieve optimal results.

2.6 CONCLUSION

Industry 5.0 is mainly an HCS which helps human–machine interaction making jobs easier and quicker. Personalization and customization top the agenda by providing greater effectiveness along with building virtual environments, advanced computers, and information technologies. They integrate it through technologies such as Big Data, artificial intelligence, IoT, cloud computing, and cobots. This is an era of high innovation and creativity. Industry 5.0 is expected to create a huge transformation in the work systems for industries. It is anticipated to create higher-value employment with a huge focus on design thinking and creativity. It is bound to increase the productivity of labor and greater opportunities for customization in various spheres. The technologies are not just going to revolutionize the production system, it is going to collaborate between humans and robots to provide tailor-made products to customers.

On the flip side, high automation in manufacturing and other domains may create the need for skill development for the workforce at a constant pace which can be a gargantuan task. The increased connectivity and standard communication protocols lead to increased security and cyber threats for industrial and manufacturing lines (Qin et al., 2016).

Even though Industry 5.0 provides greater autonomy to robots, important and moral-based decision-making is vested with humans. Human centricity is a way of looking at the lasting existence and well-being of employees but challenges loom large. On one end, there exist humans with limited physical strength and restricted intellectual capacity in prediction, and reasoning for optimal decision-making in a short time, automation should be used as an adaptive support for enhancing and empowering their work. The futuristic perspective should be based on four EHAs (enhanced human abilities) to EASE (energize, advise, support, empower) human TASK (human thoughts, robot assistance, real-time sensing, and human knowledge) in human-centric assembly.

Einstein had said after the Second World War "I know not with what weapons World War III will be fought, but World War IV will be fought with sticks and Stones" (Smith and Browne, 2021). The aspect that needs to be looked into is "Is artificial intelligence making people uneasy"? Computers are endowed with the ability to learn and make decisions increasingly free from human interventions. How would they make these decisions? Would it reflect best on humanity? AI technologies need to be guided by solid ethical principles if they are to serve society well.

REFERENCES

Abduljabbar, R., Dia, H., Liyanage, S., & Bagloee, S. A. (2019). Applications of artificial intelligence in transport: An overview. *Sustainability*, 11(1), 189.

Adel, A. (2022). Future of industry 5.0 in society: Human-centric solutions, challenges and prospective research areas. *Journal of Cloud Computing*, 11(1), 1–15.

Akgunduz, D., & Mesutoglu, C. (2021). STEM education for Industry 4.0 in technical and vocational high schools: Investigation of teacher professional development. *Science Education International*, 32(2), pp. 172–181. doi:10.33828/sei.v32.i2.11

Autor, D., Levy, F., & Murnane, RJ (2003). The skill content of recent technological change: An empirical exploration. *The Quarterly Journal of Economics*, 118(4), 1279–1333.

Bagdasarov, Z., Martin, A. A., & Buckley, M. R. (2020). Working with robots: Organizational considerations. *Organizational Dynamics*, 49(2), 100679.

Bartlett, B., Estivill-Castro, V., & Seymon, S. (2004, January). Dogs or robots: Why do children see them as robotic pets rather than canine machines? In *Proceedings of the fifth conference on Australasian user interface-Volume 28* (pp. 7–14).

Bednar, P. M., & Welch, C. (2020). Socio-technical perspectives on smart working: Creating meaningful and sustainable systems. *Information Systems Frontiers*, 22(2), 281–298.

Beier, G., Ullrich, A., Niehoff, S., Reißig, M., & Habich, M. (2020). Industry 4.0: How it is defined from a sociotechnical perspective and how much sustainability it includes–A literature review. *Journal of Cleaner Production*, 259, 120856.

Bekey, G. A. (2005). *Autonomous robots: From biological inspiration to implementation and control.* MIT Press.

Bhushan, B., Sinha, P., Sagayam, K. M., & Andrew, J. (2021). Untangling blockchain technology: A survey on state of the art, security threats, privacy services, applications and future research directions. *Computers & Electrical Engineering, 90,* 106897.

Boyd, J. A., & Huettinger, M. (2019). Smithian insights on automation and the future of work. *Futures, 111,* 104–115.

Briggs, G., & Scheutz, M. (2014). How robots can affect human behavior: Investigating the effects of robotic displays of protest and distress. *International Journal of Social Robotics, 6,* 343–355.

Caines, C., Hoffmann, F., & Kambourov, G. (2017). Complex-task biased technological change and the labor market. *Review of Economic Dynamics, 25,* 298–319.

Cirillo, V., Evangelista, R., Guarascio, D., & Sostero, M. (2020). Digitalization, routineness and employment: An exploration on Italian task-based data. *Research Policy,* July, 104079. https://doi.org/10.1016/j.respol.2020.104079

Cormier, D., Newman, G., Nakane, M., Young, J. E., & Durocher, S. (2013, August). Would you do as a robot commands? An obedience study for human-robot interaction. In *International Conference on Human-Agent Interaction* (pp. 1–3).

Cropley, A. (2020). Creativity-focused technology education in the age of industry 4.0. *Creativity Research Journal, 32*(2), 184–191.

Culot, G., Nassimbeni, G., Orzes, G., & Sartor, M. (2020). Behind the definition of Industry 4.0: Analysis and open questions *International Journal of Production Economics.* press, corrected proof, Available online, 10.

Dalenogare, L. S., Benitez, G. B., Ayala, N. F., & Frank, A. G. (2018). The expected contribution of Industry 4.0 technologies for industrial performance. *International Journal of Production Economics, 204,* 383–394.

Davis, M. C., Rose Challenger, D. N. Jayewardene, & Chris W. Clegg. 2014. Advancing socio-technical systems thinking: A call for bravery. *Applied Ergonomics, 45*(2), 171–180.

Demir, K. A. (2017). Research questions in roboethics. *Mugla Journal of Science and Technology, 3*(2), 160–165.

Demir, K. A., & Cicibas, H. (2017, October). Industry 5.0 and a critique of Industry 4.0. In *Proceedings of the 4th international management information systems conference,* Istanbul, Turkey (pp. 17–20).

Demir, K. A., & Cicibaş, H. (2018). The next industrial revolution: Industry 5.0 and discussions on industry 4.0, industry 4.0 from the management information systems perspectives.

Demir, K. A., Döven, G., & Sezen, B. (2019a). Industry 5.0 and human-robot co-working. *Procedia Computer Science, 158,* 688–695.

Demir, K. A., Turan, B., Onel, T., Ekin, T., & Demir, S. (2019b). Ambient intelligence in business environments and internet of things transformation guidelines. *Guide to Ambient Intelligence in the IoT Environment: Principles, Technologies and Applications,* 39–67.

Dengler, K., & Matthes, B. (2018). The impacts of digital transformation on the labour market: Substitution potentials of occupations in Germany. *Technological Forecasting and Social Change, 137,* 304–316.

Fareri, S., Fantoni, G., Chiarello, F., Coli, E., & Binda, A. (2020). Estimating Industry 4.0 impact on job profiles and skills using text mining. *Computers in Industry, 118,* 103222.

Fernández-Macías, E., & Hurley, J. (2017). Routine-biased technical change and job polarization in Europe. *Socio-Economic Review*, 15(3), 563–585.

Frey, C. B., & Osborne, M. A. (2017). The future of employment: How susceptible are jobs to computerisation *Technological Forecasting & Social Change*, 114, 254–280.

Garcia-Murillo, M., MacInnes, I., & Bauer, J. M. (2018). Techno-unemployment: A framework for assessing the effects of information and communication technologies on work. *Telematics and Informatics*, 35(7), 1863–1876. https://doi.org/10.1016/j.tele.2018.05.013

Ghobakhloo, M. (2020). Industry 4.0, digitization, and opportunities for sustainability. *Journal of Cleaner Production*, 252, 119869.

Goos, M., Manning, A., & Salomons, A. (2010). *Explaining job polarization in Europe: The roles of technology and globalization.* University of Maastricht, mimeo.

Gorman, E. H., & Sandefur, R. L. (2011). "Golden age," quiescence, and revival: How the sociology of professions became the study of knowledge-based work. *Work and Occupations*, 38(3), 275–302.

Haleem, A., & Javaid, M. (2019). Additive manufacturing applications in industry 4.0: A review. *Journal of Industrial Integration and Management*, 4(04), 1930001. https://roboticsbiz.com/robot-safety-types-and-sources-of-accidents/

Kachiche, S., Gahi, Y., Gharib, J. (2023). An efficient framework for the implementation of sustainable Industry 4.0. In: Ben Ahmed, M., Boudhir, A.A., Santos, D., Dionisio, R., Benaya, N. (eds) *Innovations in smart cities applications*, Vol. 6. SCA 2022. Lecture Notes in Networks and Systems, vol. 629. Cham: Springer. https://doi.org/10.1007/978-3-031-26852-6_74

Kadir, B. A., Broberg, O., & da Conceicao, C. S. (2019). Current research and future perspectives on human factors and ergonomics in Industry 4.0. *Computers & Industrial Engineering*, 137, 106004.

Kemény, Z., Váncza, J., Wang, L., & Wang, X. V. (2021). Human–robot collaboration in manufacturing: A multi-agent view. In Wang, L., Wang, X. V., Váncza, J., & Kemény, Z. (eds.) *Human–robot collaboration in manufacturing: A multi-agent view* (pp. 3–41). Cham: Springer International Publishing. https://doi.org/10.1007/978-3-030-69178-3_1

Kolade, O., & Owoseni, A. (2022). Employment 5.0: The work of the future and the future of work. *Technology in Society*, 71, 102086. https://doi.org/10.1016/j.techsoc.2022.102086

Kovacs, O. (2018). The dark corners of industry 4.0–Grounding economic governance 2.0. *Technology in society*, 55, 140–145.

Kristal, T. (2020). Why has computerization increased wage inequality? Information, occupational structural power, and wage inequality. *Work and Occupations*, 47(4), 466–503. https://doi.org/10.1177/0730888420941031

Krzywdzinski, M. (2017). Automation, skill requirements and labour-use strategies: High-wage and low-wage approaches to high-tech manufacturing in the automotive industry. *New Technology, Work and Employment*, 32(3), 247–267.

Lage, O., Saiz-Santos, M., & Zarzuelo, J. M. (2022). Decentralized platform economy: Emerging blockchain-based decentralized platform business models. *Electronic Markets*, 32(3), 1707–1723.

Mohr, B. J. (2016). Co-Creating Humane and Innovative Organizations. Creating high-performing organizations: The North American open socio-technical systems design approach. Global STS-D Network Press, pp. 16–33.

Monostori, L., Váncza, J., & Kumara, S. R. (2006). Agent-based systems for manufacturing. *CIRP Annals*, 55(2), 697–720.

Morgan-Thomas, A., Dessart, L., & Veloutsou, C. (2020). Digital ecosystem and consumer engagement: A socio-technical perspective. *Journal of Business Research*, 121, 713–723.

Muro, M., Maxim, R., & Whiton, J. (2019). Automation and artificial intelligence: How machines are affecting people and places. Brookings Institute, https://www.brookings.edu/research/automation-and-artificial-intelligence-howmachines-affect-people-and-places/

Novakova, L. (2020). The impact of technology development on the future of the labour market in the Slovak Republic. *Technology in Society*, 62, 101256.

Oks, S. J., Jalowski, M., Lechner, M. et al. (2022). Cyber-physical systems in the context of Industry 4.0: A review, categorization and outlook. *Information Systems Frontiers*.

Qin, J., Liu, Y., & Grosvenor, R. (2016). A categorical framework of manufacturing for industry 4.0 and beyond. *Procedia CIRP*, 52, 173–178.

Roboticsbiz, 2022 https://roboticsbiz.com/robot-safety-types-and-sources-of-accidents/#:~:text=Types%20of%20accidents,four%20categories%20of%20robotic%20incidents

Rossi, S., Staffa, M. & Tamburro, A. (2018). Socially assistive robot for providing recommendations: Comparing a humanoid robot with a mobile application. *International Journal of Social Robotics*, 10, 265–278. https://doi.org/10.1007/s12369-0Fift

Smith, B., & Browne, C. A. (2021). *Tools and weapons: The promise and the peril of the digital age*. Penguin.

Sony, M., & Naik, S. (2020). Industry 4.0 integration with socio-technical systems theory: A systematic review and proposed theoretical model. *Technology in Society*, 61, 101248.

Spencer, D. (2017). Work in and beyond the Second Machine Age: The politics of production and digital technologies. *Work, Employment and Society*, 31(1), 142–152. https://doi.org/10.1177/0950017016645571

Thoben, K. D., Wiesner, S., & Wuest, T. (2017). "Industrie 4.0" and smart manufacturing-a review of research issues and application examples. *International Journal of Automation Technology*, 11(1), 4–16.

Tomaskovic-Devey, D., & Avent-Holt, D. (2019). *Relational inequalities: An organizational approach*. USA: Oxford University Press.

Vaidya, S., Ambad, P., & Bhosle, S. (2018). Industry 4.0–a glimpse. *Procedia Manufacturing*, 20, 233–238.

Wang, L. (2022). A futuristic perspective on human-centric assembly. *Journal of Manufacturing Systems*, 62, 199–201.

Wang, L., Gao, R., Váncza, J., Krüger, J., Wang, X. V., Makris, S., & Chryssolouris, G. (2019). Symbiotic human-robot collaborative assembly. *CIRP Annals*, 68(2), 701–726.

Winfield, A. F., Winkle, K., Webb, H., Lyngs, U., Jirotka, M., & Macrae, C. (2021). Robot accident investigation: A case study in responsible robotics. *Software Engineering for Robotics*, 165–187.

Zawacki-Richter, O., Marín, V.I., Bond, M. (2019). Systematic review of research on artificial intelligence applications in higher education – where are the educators? *International Journal of Higher Education* 16(1), 1–27.

Chapter 3

Structural dimensions and measurement of readiness for Industry 5.0 implementation
A fresher insight from SMEs in developing country

Pham Quang Huy and MA. Vu Kien Phuc
University of Economics Ho Chi Minh City (UEH), Vietnam

3.1 INTRODUCTION

3.1.1 Background and research motivation

While the world is still attempting to adjust to Industry 4.0 and realize its promise, numerous industrialists and academics have begun to imagine and explore the next Industrial Revolution, known as Industry 5.0. While Industry 4.0 is about digitally connecting machines to enable a seamless flow of data and the highest possible optimization, Industry 5.0 is thought to focus on sustainable manufacturing while also bringing humans back into the game for collaboration and adding the human touch to manufactured products (Nahavandi, 2019; Demir et al., 2019). Nevertheless, because a considerable portion of businesses are micro-, small-, and medium-sized firms, such transformations in Asian countries face a significant problem of industry structure (Madhavan et al., 2022). The majority of countries' gross domestic products are contributed by small and medium enterprises (SMEs), which are regarded as the foundation of an economy (Robu, 2013). They are one of the fundamental forces behind the development of competence because they make up the bulk of all apprenticeships.

Therefore, radical changes to Industry 5.0 will not be feasible in terms of fundamental values without taking the SMEs into account (Madhavan et al., 2022). In fact, manufacturing companies in developing nations are far from implementing Industry 5.0 (Xu et al., 2021), and many sectors of emerging economies have yet to embrace new technologies like them (Sharma et al., 2022; Raj et al., 2020). The ability of industries to undergo this digital transition is also not fully understood; it is not simple to convert all processes to digital because it requires a lot of time, money, and labor. Indeed, SMEs' transition to Industry 5.0 from Industry 4.0 faces a number of difficulties, including a lack of tools, motivation, and expertise; a lack of open innovation and strategies; and a concern for sustainability. Even though the readiness of SMEs for digital transformation is critical, firms have a limited understanding of the required future skills and experience and their readiness to embark on the Industry 5.0 transformation. Thus, it is crucial to investigate SMEs' readiness for Industry 5.0 implementation (RII) because implementation efforts regarding new technology processes or organizational change frequently fail when leaders do not establish adequate organizational readiness for change. Despite the importance of SMEs' readiness for the digital transformation, firms' knowledge of the necessary future knowledge and experience and their readiness for the Industry 5.0 transition is low. Investigating SMEs' RII is thus essential since leaders frequently fail to build enough organizational readiness for change when implementing new technological processes or organizational transformation.

With this agenda in consideration, this study seeks to develop a thorough and useful model that can identify the RII level, allowing businesses to deploy and utilize the technology components of this. To that end, this work bridges this knowledge gap by tackling the research question as follows.

> **RQ1.** *How can the measurement model for RII be created?*
> RQ2. *How can the measurement model for RII be standardized?*

3.1.2 Research contribution

The current study provides contributions to the academic and practitioner communities by filling several gaps based on the examination of the overall findings and key insights. The majority of the current research on the adoption of Industry 5.0 focuses on the many technologies needed for its integration (Özdemir& Hekim, 2018) and its prospective applications (Sachsenmeier, 2016; Ciasullo et al., 2022). The bulk of previous research has only conducted descriptive analyses to examine specific hurdles to the adoption of Industry 5.0 (Maddikunta et al., 2022; Huang et al., 2022). Given the significance and wide-ranging applications of Industry 5.0 in numerous organizations and industries, as well as the numerous advantages of this digital transformation framework, organizations need to assess their RII as a starting point and take action to fulfill the strategic objectives of Industry 5.0. From a theoretical perspective, this chapter represents likely the first attempt to pinpoint the RII that SMEs in emerging nations should achieve. In particular, this research is one of the first empirical ones to investigate structural dimensions and measures of RII. Accordingly, the obtained findings in this chapter will enable academicians to gain a good grasp of how Industry 5.0 is being adopted in emerging markets. From the standpoint of practitioners, the in-depth analysis of the interconnection between RII and expected benefits from the Industry 5.0 implementation will allow the SMEs to sense and seize which issue should be addressed first. Besides, this research will assist policymakers, governmental influencers, and regulatory authorities in using the findings and insights to guide their decisions and action plans about the adoption of Industry 5.0.

3.1.2.1 Organization of chapter

The rest of the chapter is organized as: Section 3.2 tackles the existing literature on the theoretical underpinning and Industry 5.0. Section 3.3 delineates
the data collection process and analyzed as well as outlines the main results.
And, finally, Section 3.4 concludes the chapter with implications and future
scope.

3.2 LITERATURE REVIEW

3.2.1 The technology–organization–environment (TOE) framework

The TOE which Tornatzky and Fleischer first introduced in 1990, has
become a well-known theoretical perspective that integrates both human
and non-human actors into the network. This approach helps manage the
illusion of accumulated traditions and techno-centric predictions that are
found to be the weakness of other technology adoption frameworks like
TAM, TRA, UTAUT, and TPB (Awa et al., 2016). By describing how a firm's
technological, organizational, and environmental contexts determine the
adoption, implementation, and probability of technological innovations
(Masood & Egger, 2019; Abed, 2020), the TOE framework more specifically states the affecting factor for the acceptance of new technologies at
the organizational level (Rajbhandari et al., 2022). Since its inception to
the present, TOE has been one of the theories of organizational technology
adoption that scholars and practitioners have used the most (Muhamad et
al., 2021). Since innovation adoption decisions must be researched within
appropriate contexts and with variables adapted to the peculiarity of the
invention, it is appropriate to adapt the TOE framework to fit into a different study context (Chau & Tam, 1997). The study as a whole uses the TOE
framework as a theoretical foundation to explain the institutionalization
of Industry 5.0 technologies among SMEs and to further build a structured
classification of adoption drivers.

3.2.2 Industry 5.0

Although numerous academic notes have deepened analyses on Industry
4.0 and its advancement and effect on smart production, sharp business,
and sustainable development (i.e., Kumar and Nayyar (2020); Nayyar
and Kumar (2020); Singh et al. (2021)), security of whistle-blowers on
Dark Web (Venaik et al., 2023), Industry 4.0 is primarily focused on a
shift in the industrial paradigm that is driven by technology, but society
and human factors have received less attention (Jafari et al., 2022). The
potential for job loss and job security with the rising use of autonomous
systems is one worry associated with this industrial revolution (Doyle-Kent

& Kopacek, 2019). Several scholars like Nahavandi (2019) and Leng et al. (2022) criticized Industry 4.0's focus on productivity. Therefore, it is crucial that the technological transition be made in a sustainable manner and in line with the objectives of socioeconomic growth (Nahavandi, 2019). The concerns of people and society throughout the industrial transition gave rise to Industry 5.0, which was established in 2015 to advance the idea of "Industrial Upcycling" (Jafari et al., 2022). In the interim, Özdemir and Hekim (2018) identified Industry 5.0 as the incremental progression of Industry 4.0 that can support symmetric innovation to overcome the flaws of the Industry 4.0 innovation ecosystem. In order words, Industry 5.0 expands on the principles and ideas of Industry 4.0 in a more comprehensive and futuristic way. Industry 5.0 has been the subject of substantial research since 2017 (Nahavandi, 2019; Paschek et al., 2019; Demir et al., 2019). Due to the fact that Industry 5.0 has not yet fully evolved, a number of industry practitioners and scholars have offered a variety of definitions (Maddikunta et al., 2022; Mehdiabadi et al., 2022). Based on the perspectives of Nahavandi (2019) and Leng et al. (2022), Industry 5.0 represents the start of human-focused industrial sector activities brought on by recent technological advancements that encourage joint participation between humans and machines.

According to Mehdiabadi et al. (2022), everything in Industry 5.0 revolves around the deployment of digitalization and intelligent technologies to boost human productivity. Industry 5.0 has been divided into distinct segments by experts, including edge computing, big data, the Internet of Things, autonomous robots, artificial intelligence, augmented reality, and innovation system integration (Maddikunta et al., 2022). By integrating technology and people, it seeks to serve humanity by putting artificial intelligence to use (Xu et al., 2021). Industry 5.0, as defined by Espina-Romero et al. (2023), is the most recent technological revolution that aims to transform industries into intelligent structures supported by cognitive computing and the Internet of Things.

As humans and machines work together to improve process efficiency by harnessing human creativity and brainpower through the integration of workflows with intelligent systems, Industry 5.0 brings back the human workforce to the factory (Nahavandi, 2019). According to the European Economic and Social Committee, Industry 5.0 is a new revolutionary wave that combines the varying powers of human intelligence and cyber-physical production systems to build synergistic factories (Longo et al., 2020). Furthermore, policymakers are searching for creative, moral, and human-centered designs to overcome the workforce weakness brought on by Industry 4.0. Industry 5.0 is the era of the socially intelligent industry, in which Cobots converse with people (Koch et al., 2017). Given that this industrial model encourages the economic productivity of the commercial sector while ensuring environmental sustainability, the high energy consumption of this sector as a result of the use of novel technologies necessitates a progressive display of rational and advantageous parameters within this medium (Ghobakhloo et al., 2022). Therefore, Industry 5.0 aims to create a comprehensive framework by implementing disruptive technologies and creative solutions to address the societal and human challenges that are currently emerging (Jafari et al., 2022).

3.3 METHODOLOGY

3.3.1 Phase 1: qualitative scale development

3.3.1.1 Domain of construct specification and content analysis

This study gathered the scale items through an analysis of Industry 5.0 and RII by examining the pertinent literature in order to acquire the initial item pool. Alternatively, Churchill (1979) also advocated doing qualitative interviews to gather knowledge from experts and create better survey items for the target topic. In the current study, semi-structured interviews were used since they allowed for the collection of open-ended data, respondent thought exploration, topic-specific feelings, and beliefs, as well as in-depth personal and sensitive issue examination (DeJonckheere & Vaughn, 2019). In order to find the right set of field participants, purposive sampling was chosen since it allowed for the collection of a variety of opinions and the acquisition of fresh ideas for discussion (Kummer et al., 2021). A professional team of business elites who are knowledgeable about the issue, have the necessary tools, appreciate the value of working, and have the ability to set aside enough time has been chosen. Table 3.1 lists the traits of this group of experts.

Table 3.1 Profile of panel of expert

Item	Frequency	Valid (%)	Item	Frequency	Valid (%)
Gender			Career		
Female	18	50	Manager	8	22.22
Male	18	50	Employee	8	22.22
			University Lecturer	20	55.56
Experience (Years)			Education		
5–10	15	41.67	Associate Professor	2	5.56
11–16	8	22.22	PhD degree	11	30.56
17–21	8	22.22	Master degree	23	63.88
22–27	1	2.78			
28–33	3	8.33			
Over 33	1	2.78			

A series of questions were left open-ended, leading to a number of follow-up questions based on the respondents' responses because open-ended responses were a useful tool for examining and comprehending participants' experiences and perspectives (Feng & Behar-Horenstein, 2019). This allowed the respondents to simply and fully express their various perceptions of RII. In order for participants to fully understand what we discussed because these terms are theoretical, the researchers clarified the definitions of Industry 5.0 and RII before the interview began. Every interview lasted 45–60 minutes.

3.3.1.2 Analysis strategy

The qualitative data was then analyzed using NVivo version 12.2 to extract all of the responses and record the most common keywords. The content was analyzed from the feedback of 36 respondents,1,341 words in total, and only words that appeared four times or more were retained (Fetscherin & Stephano, 2016). This standard assisted in removing unimportant data. For further assessment and selection, the remaining terms were ordered in descending order by frequency of occurrence.

3.3.1.3 Peer review

The assertions in the experimental form, which had 83 statements, were assessed in the following phase. These questions were given to subject matter experts for feedback on content validity, clarity, and developmental relevance in accordance with DeVellis's (2017) recommendations. The expert panel, made up of ten respondents including five lecturers from the information technology field and five practitioners from SMEs, was requested to rate each item as "prerequisite," "helpful and indispensable," or "not imperative" based on how well it adhered to the definition. Items that received more than 90% of positive votes (either "prerequisite" or "helpful and not indispensable") would be retained.

The items were improved, developed, and combined based on the evaluation process to create a preliminary scale with 35 candidate items for each sub-scale. This led to the creation of the first pilot questionnaire, which contained 35 items, based on the statements discovered through the conceptual analysis evaluations.

3.3.1.4 Pilot survey study

A five-point Likert-type scale, ranging from 1 (extensively disagree) to 5 (extensively agree) was employed for all the measurements. In order to test for internal consistency and to whittle down the number of items to a tolerable quantity, the first questionnaire was created. The small-scale pilot test required participation from 300 respondents in order to maintain the input corroboration. Information technology middle and senior managers make up the bulk of this stage's audience. The questionnaires were sent using the convenience sampling method to those who were reachable. The survey's implementation period was from April 2022 to June 2022. The Cronbach's α value was employed to examine the degree of internal consistency of each construct of the pilot questionnaire (Dunn et al., 2013) and expected to be at 0.7 or more to generate an acceptable reliability coefficient (Hair et al., 2011). In this regard, the measurement scales with the Cronbach's α value reported to be greater than 0.7 would be retained. In contrast, the measurement scales with the Cronbach's α value reported to be below 0.7 would be eliminated from the dataset. In addition, the questionnaire and several items were later updated and rewritten where necessary based on their feedback and observations. In doing so, a total of 32 items were used in the questionnaire.

3.3.1.5 Discussion

Hinged on the scale construction phase recommendations from DeVellis and Thorpe (2021) and Churchill (1979), Phase 1 in this research sought to produce a suitable item pool that accurately matched the construct of RII. The methods used in this study created a clear, logical, and systematic process that guaranteed the quality of the first item pool. In a nutshell, the initial item pool was improved in Phase 1 using several techniques, yielding 32 items for additional quantitative scale validation in Study 2.

3.3.2 Phase 2: quantitative scale validation

3.3.2.1 Scale purification and refinement

This study gathered quantitative survey data from accountants in SMEs to further analyze the scale's validity and reliability. The SMEs in the Southern regions in Vietnam made up the sample unit of analysis for this study, and the accountants in SMEs were selected as the study's target participants. These businesses were frequently chosen in a number of earlier studies on the application of innovation (e.g., Nguyen & Wongsurawat, 2012; Hoang & Otake, 2014), as the southern regions of Vietnam have been regarded as the most dynamic regions (Pham & Vu, 2022). In addition, SMEs in Southern Vietnam tended to implement innovations more quickly and successfully than SMEs in other parts of Vietnam, according to Ha et al. (2022). Incredibly, in view of the rapidly spreading use of digital technologies, accountants would become essential to a successful outcome (Zybery & Rova, 2014). In this study, convenience sampling and snowball sampling were used in combination. The paper-and-pencil questionnaire survey was employed for data collection. The primary data collection lasted from mid-May to mid-September 2022. After screening and examination of the questionnaires, a final sample size of 723 cases with a data loss rate of 12.89 percent was left for analysis. A detailed description of demographic data collected from the survey is demonstrated in Table 3.2.

Table 3.2 Distribution of participants based on demographic characteristics

Item	Contents	Frequency	Valid (%)
Gender			
	Male	452	62.52
	Female	271	37.48
Age (years)			
	20 to Under 30	39	5.39
	30 to Under 40	364	50.35
	40 to Under 50	256	35.41
	50 to Under 60	64	8.85
Education			
	Undergraduate	712	98.48
	Postgraduate	11	1.52
Experience (years)			
	Under 10	39	5.39
	10 to Under 20	358	49.52
	20 to Under 30	262	36.24
	30 to Under 40	64	8.85
Business type			
	Production of food items	284	39.28
	Production of other machinery and equipment	66	9.13
	Production of fabricated metal products	42	5.81
	Production of electrical equipment	39	5.39
	Manufacture of electronic components, computers, visual, sounding, and communication equipment	25	3.46
	Production of goods made of rubber and plastic	71	9.82
	Production of medical, precision, and optical instruments, watches, and clocks	12	1.66
	Production of pharmaceuticals, medicinal chemicals	29	4.01
	Production of coke, briquettes, and refined petroleum products	14	1.94
	Production of goods made of fur, clothes, and accessories	23	3.18
	Production of fundamental metals	66	9.13
	Production of leather, luggage, and footwear	52	7.19

As the goal of exploratory factor analysis (EFA) is to identify the quantity and type of common factors in a group of variables in order to decrease their factor structure, EFA is appropriate for use in an exploratory investigation, especially when there is no prior theory regarding the interactions between variables (Stevens, 2002). By using the Kaiser–Meyer–Olkin (KMO) test and Bartlett's test of sphericity, it is determined whether the factor analysis of the variables under study can be used and whether the model's joint significance is valid. Since the KMO value was 0.812, it was clear that the data set was appropriate for EFA. Additionally, Bartlett's test of sphericity was performed to examine the possibility of multiple co-relationships among variables. The result was highly significant [$\chi^2(666)$ = 18.020, p = 0.000], suggesting that the correlation matrix demonstrated appropriate linearity between the variables. A structure of eight components that accounted for about 58.60% of the overall variation was revealed by the initial study. The outcomes of factor loads are displayed in Table 3.3.

3.3.2.2 Collecting information and reexamining scale characteristics

Confirmatory factor analysis (CFA) was performed following the EFA to determine how measured RII factors contribute to the conceptual framework (Harrington, 2009). CFA was applied using structural equation modeling (SEM) using the maximum likelihood method.

Using Kaynak's (2003) suggestion as a foundation, the indicators used to assess the goodness of the model included the Chi-square to degree of freedom (Chi-square/df), goodness of fit index (GFI), comparative fit index (CFI), and root mean square error of approximation (RMSEA), and Tucker–Lewis index (TLI) was also included for further model fitness assurance (Prasetyo et al., 2020). All of the recorded indices appeared to comply with the cut-off criteria suggested by earlier researchers, which supported the measurement models' claim that they completely fit the acquired data based on the results listed in Table 3.4.

The skewness and kurtosis coefficients were investigated to determine if they met the CFA's normalcy requirements. More specifically, it was suggested that skewness values should range from −2 to +2 to show the characteristics of a normal distribution and that kurtosis values should vary from −7 to +7 to be appropriate for SEM study (Watkins, 2018). The skewness values for 32 items in the current study ranged from −0.918 to 2.756. The values of skewness for 32 items were simultaneously found to range between −1.351 and 3.233.

Internal consistency and item-total correlation were calculated to make sure the scale was reliable. The relationship between each scale item and the overall scale score was known as item-total correlation. The items were indicative of a comparable construct when there was a positive and high

Table 3.3 EFA results of the readiness of Industry 5.0 implementation

Factor/item		Factor loading
Institutional readiness. Eigen value: 3.324; Variance explained: 9.538%		
INR1	We closely adhere to industry 5.0 implementation regulations and standards.	0.780
INR2	We pay great attention to recommendations from professional associations and other government agencies on promising future directions.	0.770
INR3	We implement Industry 5.0 to get financial, tax, and other incentives from the government.	0.786
INR4	We implement Industry 5.0 for public image certifications and recognitions	0.898
Task readiness. Eigen value: 2.918; Variance explained: 8.451%		
TAR1	Our organization implements environmental solutions	0.823
TAR3	Our organization implements sustainability-focused business model	0.797
TAR4	Our organization implements strategic planning participation	0.819
TAR5	Our organization tracks sustainability metrics	0.802
Management readiness. Eigen value: 2.727; Variance explained: 7.850%		
MAR1	Our management has a plan to digitize the production process.	0.766
MAR2	The Industry 5.0 production process should be taken into consideration, according to our management.	0.864
MAR3	We have the right leadership in place to implement digitized production.	0.782
MAR5	Our management has commitment to upgrade the production process.	0.767
Human resource readiness. Eigen value: 2.699; Variance explained: 7.682%		
HURR1	Our human resource acquires technical knowledge	0.808
HURR3	Our human resource acquires technology management capability.	0.730
HURR4	Our human resource acquires relational knowledge.	0.779
HURR5	Our senior managers acquire managerial skills	0.744
Ecosystem actor readiness. Eigen value: 2.303; Variance explained: 6.589%		
ECAR1	The capacity of a human actor to select from a variety of actors and resources to collaborate with or use during the Industry 5.0 implementation process	0.803
ECAR2	The capacity of a human actor to detect and appropriately react to the good and negative effects of other actors during the Industry 5.0 implementation process	0.724
ECAR3	The capacity of a human actor, when dealing with other actors during the Industry 5.0 implementation process, to adapt to diverging institutions (i.e., various norms, regulations, or beliefs).	0.766
ECAR4	During the Industry 5.0 implementation process, a human actor's self-assurance in his or her capacity to master particular skills or jobs	0.729

(Continued)

Table 3.3 (Continued) EFA results of the readiness of Industry 5.0 implementation

Factor/item		Factor loading
Financial readiness. Eigen value: 2.228; Variance explained: 6.427%		
FIR1	Our organization is financially ready to switch to digital production.	0.697
FIR2	Industry 5.0-related financial incentives from agencies are known to our organization.	0.733
FIR3	Agency financing for Industry 5.0 is available to our organization.	0.725
FIR4	Our organization is aware of the agencies to whom our organization can apply for funding for Industry 5.0.	0.747
Technology readiness. Eigen value: 2.151; Variance explained: 6.232%		
TER1	Our information technology setup might be improved for the Industry 5.0 manufacturing process.	0.684
TER2	For Industry 5.0 processes, our main equipment could be networked.	0.672
TER3	The size of our factory floor allows for digital production.	0.767
TER4	Our infrastructure's adaptability fits Industry 5.0.	0.744
Cybersecurity readiness. Eigen value: 2.029; Variance explained: 5.831%		
CYR1	Our organization implements the identification and detection process for cybersecurity risk management	0.623
CYR2	Our organization implements the protection process for cybersecurity risk management	0.749
CYR3	Our organization implements the response process for cybersecurity risk management	0.603
CYR4	Our organization implements the recovery process for cybersecurity risk management	0.759

Table 3.4 Results of measurement model analysis

Fit indices	Measurement model	Suggested value	Global model fit	References
CMIN/DF	1.490	≤ 3	Yes	Hair et al. (2014)
GFI	0.927	≥ 0.9	Yes	Little (2013)
CFI	0.969	≥ 0.9	Yes	Tiglao et al. (2020); Dominguez et al. (2019)
TLI	0.966	≥ 0.9	Yes	Tiglao et al. (2020); Dominguez et al. (2019)
RMSEA	0.025	≤ 0.06	Yes	Xia and Yang (2018)

correlation between them (Koc & Barut, 2016). The correlation was considered to have a cut-off value of 0.30 (Pallant, 2007). Cronbach's alpha consistency coefficient was used to calculate the scale's internal consistency level. For internal consistency, an Cronbach's alpha of 0.70 or above was required (DeVellis & Thorpe, 2021).

Table 3.5 Results summary of measurement model assessment

Construct	Item abbreviation	AVE	Cronbach's alpha	Composite reliability	Inference
Technology factors					
Technology readiness	TER	0.514	0.808	0.809	Retained
Cybersecurity readiness	CYR	0.500	0.774	0.778	Retained
Task readiness	TAR	0.655	0.884	0.884	Retained
Organizational factors					
Financial readiness	FIR	0.524	0.814	0.815	Retained
Management readiness	MAR	0.631	0.870	0.872	Retained
Human resource readiness	HURR	0.585	0.849	0.849	Retained
Environmental factors					
Institutional readiness	INR	0.656	0.882	0.883	Retained
Ecosystem actor readiness	ECAR	0.570	0.840	0.841	Retained

Additionally, it was decided that the composite reliabilities of all the structures would surpass the specified internal consistency accomplishment level of 0.70. (Hair et al., 2018). For the purpose of demonstrating a measuring model with equivalent convergent validity, the AVE of all constructs had to be greater than the minimal threshold of 0.5 (Hair et al., 2018). The specifics of these outputs in Table 3.5 highlight the measurement model's validity, internal consistency, and dependability.

3.3.2.3 Model development and testing

The goal of the current study was to create a model that would explain how the elements of RII interacted with the benefits that SMEs may expect from adopting the technology. The distinct data set using the valid and trustworthy RII scale was captured to accomplish this goal. Figure 3.1 shows the research model with the hypotheses. Below are the hypotheses that, according to earlier literature, explain the links. Empirical observations were provided after information about participants and the data-gathering process.

Hypothesis development procedure. The technological component of the TOE framework denoted the pool of internal as well as external technologies to the enterprises (Chittipaka et al., 2022). In the current research, technological factors were characterized by the presence of technology readiness, cybersecurity readiness, and task readiness for Industry 5.0 implementation.

In order to adopt advanced technologies, businesses must be technologically prepared. Technology readiness is described as an individual inclination to adopt and apply new technologies to achieve goals in their personal and professional lives (Parasuraman, 2000). At the organizational level, technology readiness can be understood as the organizational inclination to

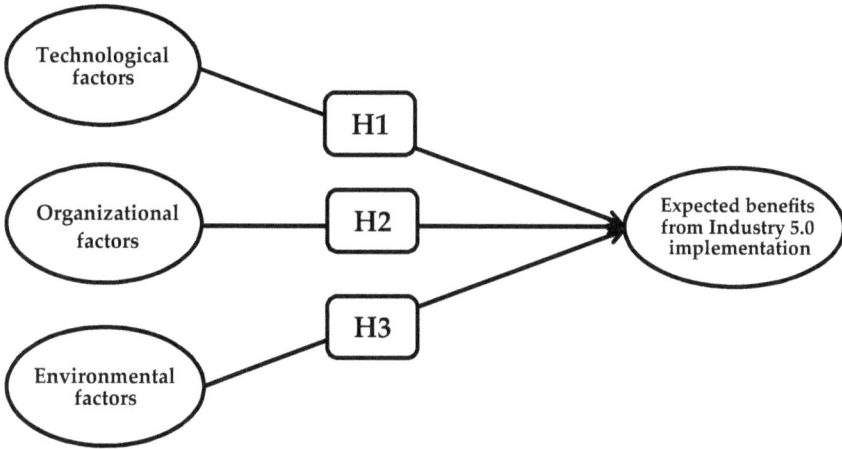

Figure 3.1 Hypothesized model.

adopt and apply new technologies to achieve the overall goals of the organization. How technologically proficient an organization is based on how ready it is to implement Industry 5.0 technology in its unique organization to meet organizational goals. This is a vital component since how well an organization manages these technologies will influence how successfully Industry 5.0 is implemented.

The harmonization between the development of advanced information technology and socio-economic issues has become the main concern of adequate assimilation of Industry 5.0 principles. It may lead to a significant risk toward the successful rollout of Industry 5.0 (Deguchi & Karasawa, 2020). In response to this, the deployment of environmental solutions and a business plan incorporating sustainability, participation in strategic planning, and keeping track of sustainability metrics will enable SMEs to be well-prepared for Industry 5.0 implementation. In light of the combined impact of the four subdimensions, the formation of task readiness mirrors the organizational tendency to approach and implement Industry 5.0 from a holistic level.

Industry 5.0 is anticipated to introduce a multitude of interconnected systems, processes, machines, devices, platforms, and nodes that will help to close the gap between the physical and digital worlds (Porambage et al., 2021). Although the aforementioned components constitute a close network, businesses must ensure there is no security breach without sacrificing service and operational standards (Maddikunta et al., 2022). To ensure the success of Industry 5.0, businesses must address security issues relating to authentication, integrity, access control, and audit (Liyanage et al., 2020). As a result, enterprises can use cybersecurity readiness as a roadmap to enhance the security of their information technology infrastructure and online services. In light of the issues raised above, the second hypothesis is conjectured accordingly.

Hypothesis 1(H1). *TEF instigates a significant effect on EBI in a positive manner.*

According to Gide and Sandu (2015), the internal qualities and resources of the company, including both tangible and intangible resources, are included in the TOE framework's organizational perspective (Chittipaka et al., 2022). In earlier investigations, different elements that include the organizational aspect have been identified (Alsetoohy et al., 2019; Gutierrez et al., 2015; Senyo et al., 2016). The organizational elements are heavily influenced by internal factors, management support and awareness, and employee involvement (Alraja et al., 2022). Additionally, the company needed to have access to specialists and financial resources in order to implement new technologies (Iacovou et al., 1995; Parasuraman, 2000; Chwelos et al., 2001). Therefore, businesses are more likely to succeed in adopting digital technology if they invest more money in doing so (Nikopoulou et al., 2023). Similar to this, numerous studies discovered that the availability of skilled personnel has an impact on an organization's readiness to adopt new technology (Ifinedo, 2011). On the other side, increasing expertise and knowledge enables technology users to utilize technology more effectively (Rogers, 2003; Gao et al., 2012). Building on the perspectives of the TOE framework, the presence of appropriate and enabling organizational elements enhances Industry 5.0 adoption by organizations. In the current research, organizational factors were characterized by the presence of financial readiness, management readiness, and human resource readiness for Industry 5.0 implementation. In light of the issues raised above, the second hypothesis is conjectured accordingly.

Hypothesis 2 (H2). *ORF instigates a significant effect on EBI in a positive manner.*

An organization's operating environment includes all relevant factors, such as laws and regulations (Tornatzky et al., 1990). The environmental dimension in the TOE framework refers to the external environment of an organization (Alraja et al., 2022; Chittipaka et al., 2022). Concretely, the environmental context represents the external social and technological support influencing the implementation of advanced technology in organizations (Tornatzky et al., 1990; Pan & Jang, 2008; Lin, 2014). In the current research, environmental factors were characterized by the presence of institutional readiness and ecosystem actor readiness for Industry 5.0 implementation. Building a successful digital economy depends on having effective information technology policies and regulatory frameworks (Othman et al., 2023). As proposed by Zhu et al. (2004), the impact of government is crucial within the TOE framework. Government laws are essential for the adoption and use of digital technology (Intan Salwani et al., 2009). Accordingly, government policies include incentives and prohibitions (Awa et al., 2017).

As stated by Baker (2012), when governments place restrictions on the market, the adoption of digital technology may be forced. On the other hand, manufacturing and product/service delivery nowadays are growingly realized among a wide range of actors outside of dyadic contacts, driven by technological advancements. The actor ecosystem readiness enables a customer or an employee to navigate the Industry 5.0 implementation process effectively. In light of the issues raised above, the third hypothesis is posited accordingly.

Hypothesis 3 (H3). *ENF instigates a significant effect on EBI in a positive manner.*

Operationalization and purification of constructs. The indicators for each construct of the current hypothesized model stemmed from precedent literature and the scale development phase in this research.

Technology context. The measurement scale for the technology context covered three ingredients, namely, technology readiness, cybersecurity readiness, and task readiness. More concretely, the criteria employed to evaluate technology readiness comprised four dimensions. The criteria employed to evaluate cybersecurity readiness comprised four dimensions. The criteria employed to evaluate task readiness comprised four dimensions.

Organizational context. The measurement scale for organizational context covered three ingredients, namely, financial readiness, management readiness, and human resource readiness. More concretely, the criteria employed to evaluate financial readiness comprised four dimensions. The criteria employed to evaluate management readiness comprised four dimensions. The criteria employed to evaluate human resource readiness comprised four dimensions.

Environmental context. The measurement scale for environmental context covered two ingredients, namely, institutional readiness and ecosystem actor readiness. More concretely, the criteria employed to evaluate institutional readiness comprised four dimensions. The criteria employed to evaluate ecosystem actor readiness comprised four dimensions.

Expected benefits from Industry 5.0 implementation. The criteria employed to evaluate EBI comprised five dimensions which were created from those proposed by Venkatesh et al. (2003); Chung and Park (2021), Ragu-Nathan et al. (2004), Khan et al. (2013), and Jung et al. (2020) and adjusted to be appropriate with the current study setting.

Participants and data collection procedure. This study gathered quantitative survey data from accountants in SMEs to further analyze the scale's validity and reliability. The SMEs in Vietnam made up the sample unit of analysis for this study, and the accountants in SMEs were selected as the study's target participants. According to Singh and Srivastava

(2019), there has not been a scholarly agreement on the precise sample size to use for CB-SEM deployment. In particular, a sample size that would yield an informant-to-informant ratio of between 1:4 and 1:10 was advised (Hinkin, 1995). While Hair et al. (2014) maintained that the sample size should fall within the range of 100–400 to be adequate to implement SEM, Wolf et al. (2013) suggested a sample size varied from 30 to 460, while Urbach and Ahlemann (2010) reasoned that the sample size should fall within the range of 200–800 responses for CB-SEM approach. From the begin of October 2022 to the mid of March 2023, the survey took place. After removing all invalid responses, a total of 1,283 legitimate surveys were obtained, representing an 83.31 percent response rate. Given that there were 1,283 responses, this number of samples met the prerequisites. A detailed description of demographic data collected from the survey is demonstrated in Table 3.6.

Results of the measurement model. Factor loadings of the items to their respective constructions were supported as convergent if they were higher than the criterion of 0.6 (Kline, 2015). In contrast, AVE values larger than 0.5 provide accepted proof of convergent validity measurements for the notion. (Hanus & Wu, 2016). Based on established loadings and measurement errors for each item, the CR was used to evaluate the reliability of each concept (Corsini et al., 2019). It was advised that the 0.70 literature cutoffs serve as the accepted threshold for the values of Cronbach's alpha and CR (Hair et al., 2017). The results in Table 3.7 highlight the fact that all of the model's constructs met the aforementioned requirements.

In order to determine if the constructions' components were uncorrelated with other constructs (Corsini et al., 2019) or obtained a low correlation with other constructs in the research, discriminant validity was used (Hair et al., 2010). For each latent variable, the square roots of the AVE correlations comparison and the inter-constructs comparison were used to test the discriminant validity (Alzahrani et al., 2017). The square root of the AVE was proposed to be more than the correlation between the constructs based on Fornell and Larcker's (1981) criterion. Additionally, each component should show a higher loading on its related factor than in the cross-loading (Gotz et al., 2010). When the correlations between any latent variables were noticeably less than 1.0, the discriminant validity of the constructs was also attained (Sekaran & Bougie, 2016). According to the statistical findings shown in Table 3.8, the hypothesized model reflected the aforementioned conditions, and as a result, the constructs gained discriminant validity.

Using Kaynak's (2003) suggestion as a foundation, the indicators used to assess the goodness of the model included the Chi-square to the degree of freedom (Chi-square/df), goodness of fit index (GFI), comparative fit index (CFI), root mean square error of approximation (RMSEA) and TLI was also included for further model fitness assurance (Prasetyo et al., 2020). All of

Table 3.6 Distribution of participants based on demographic characteristics

Item	Contents	Frequency	Valid (%)
Gender			
	Male	790	61.57
	Female	493	38.43
Age (years)			
	20 to Under 30	148	11.54
	30 to Under 40	719	56.03
	40 to Under 50	268	20.89
	50 to Under 60	148	11.54
Education			
	Undergraduate	1,248	97.27
	Postgraduate	35	2.73
Experience (years)			
	Under 10	148	11.54
	10 to Under 20	723	56.35
	20 to Under 30	269	20.96
	30 to Under 40	143	11.15
Business type			
	Production of food items	313	24.40
	Production of other machinery and equipment	53	4.13
	Production of fabricated metal products	34	2.65
	Production of electrical equipment	45	3.51
	Manufacture of electronic components, computers, visual, sounding, and communication equipment	29	2.26
	Production of goods made of rubber and plastic	84	6.55
	Production of medical, precision, and optical instruments, watches, and clocks	23	1.79
	Production of motor vehicles, trailers, and semitrailers	41	3.20
	Production of chemicals and chemical products, except pharmaceuticals and medicinal chemicals	32	2.49
	Production of other transport equipment	24	1.87
	Production of goods made of fur, clothes, and accessories	33	2.57
	Production of leather, luggage, and footwear	32	2.49
	Production of other non-metallic mineral products	45	3.51

(Continued)

Table 3.6 (Continued) Distribution of participants based on demographic characteristics

Item	Contents	Frequency	Valid (%)
	Production of textiles, except apparel	83	6.47
	Production of pulp, paper, and paper products	73	5.69
	Production of furniture	93	7.25
	Production of wood and products of wood and cork	42	3.27
	Production of fundamental metals	134	10.44
	Production of pharmaceuticals, medicinal chemicals	22	1.71
	Production of coke, briquettes, and refined petroleum products	48	3.75

Table 3.7 Results summary of measurement model assessment

		Convergent validity		Construct reliability		
Construct	Item abbreviation	Factor loadings ranges	AVE	Cronbach's alpha	Composite reliability	Inference
Technology factors						
Technology readiness	TER	0.707–0.769	0.547	0.828	0.828	Retained
Cybersecurity readiness	CYR	0.630–0.737	0.5	0.774	0.777	Retained
Task readiness	TAR	0.785–0.841	0.651	0.882	0.882	Retained
Organizational factors						
Financial readiness	FIR	0.688–0.767	0.521	0.812	0.813	Retained
Management readiness	MAR	0.770–0.862	0.636	0.872	0.874	Retained
Human resource readiness	HURR	0.733–0.811	0.584	0.848	0.848	Retained
Environmental factors						
Institutional readiness	INR	0.756–0.907	0.647	0.878	0.88	Retained
Ecosystem actor readiness	ECAR	0.726–0.782	0.557	0.834	0.834	Retained
Expected benefits from Industry 5.0	EBI	0.717–0.852	0.591	0.877	0.878	Retained

Table 3.8 Results of discriminant validity

	EBI	TAR	INR	MAR	HURR	ECAR	TER	FIR	CYR
EBI	I								
TAR	0.111	I							
INR	0.105	0.123	I						
MAR	0.03	0.145	0.148	I					
HURR	0.201	0.078	0.019	0.078	I				
ECAR	0.196	−0.003	0.116	0.072	0.146	I			
TER	0.114	0.178	0.139	0.085	0.049	0.07	I		
FIR	0.054	0.035	0.056	0.017	0.103	0.118	0.065	I	
CYR	0.134	0.055	0.027	0.075	0.102	0.124	0.153	0.076	I

the recorded indices appeared to comply with the cut-off criteria suggested by earlier researchers, which supported the measurement model and structural model in that they completely fit the acquired data based on the results listed in Table 3.9.

Results of the structural model. With reference to the parameter estimates and outcomes of the models' hypotheses detailed in Table 3.10, the CR for all variables exceeded the value of 1.96, and the p-value met the threshold, implying that the regression coefficients for the three direct effects were significant. Concretely, the positive and substantial path coefficients evinced that TEF demonstrated vigorous impacts on EFF ($\beta = 0.232$; p = 0.002) and ORF demonstrated vigorous impacts on EFF ($\beta = 0.156$; p = 0.008). On the other hand, the direct effect of ENF on EFF was positive and statistically significant ($\beta = 0.346$; p = 0.002). Thus, H1–H3 were accepted.

Table 3.9 The outcomes of measurement and structural model analysis

Model fitting index	Model fitting values	Standard values of the measurement model	Parameter estimates of the structural model	Result judgment	Recommended by
Chi-square/ df	< 3	1.838	1.996	Satisfied	Hair et al. (2014)
TLI	≥ 0.9	0.973	0.967	Satisfied	Little (2013)
CFI	≥ 0.9	0.976	0.970	Satisfied	Tiglao et al. (2020); Dominguez et al. (2019)
GFI	≥ 0.9	0.955	0.950	Satisfied	Tiglao et al. (2020); Dominguez et al. (2019)
RMSEA	< 0.06	0.026	0.028	Satisfied	Xia and Yang (2018)

Table 3.10 Structural coefficients (β) of the hypothesized model

Hypothesis No.	Hypothesized path			Standardized	SE	CR	p-value	Inference
H1	TEF	→	EBI	0.232	0.017	3.131	0.002	Buttressed
H2	ORF	→	EBI	0.156	0.086	2.672	0.008	Buttressed
H3	ENF	→	EBI	0.346	0.049	3.056	0.002	Buttressed

3.3.2.4 Discussion: Phase 2

Through a series of analyses, the interlinks and patterns within the codes were unearthed. Figure 3.2 presents the first-order codes, second-order theoretical sub-categories, and aggregate theoretical dimensions collectively. Following are three categories of themes that are revealed by the three aggregate dimensions.

The current research develops a scale for measuring the structural characteristics of RII within SMEs in emerging countries. The structural characteristics of RII include three core pillars: technological factors, organizational factors, and environmental factors. More concretely, the technological factors act as a high-order construct comprising three unidimensional constructs, namely, technology readiness, cybersecurity readiness, and task readiness. Particularly, by introducing numerous interconnected systems, processes, machines, devices, platforms, and nodes, Industry 5.0 is predicted to close the gap between the physical and digital worlds (Porambage et al., 2021). Although the aforementioned things are interconnected in a dense network, businesses must ensure that there is no security breach without sacrificing operating and service standards (Maddikunta et al., 2022). To support the success of Industry 5.0, businesses must address security issues relating to authentication, integrity, access control, and audit (Liyanage et al., 2020). On the other hand, Industry 5.0 is still in its infancy, but as time goes on, it will include a web of millions of interconnected equipment, devices, systems, and people that will collectively improve sustainability, productivity, and resilience. To scale up and connect to and create such a broad network, businesses must be sensitive to the demand and growing workload (Mukherjee et al., 2023).

Organizational factors served as a high-order construct comprising three unidimensional constructs, namely, financial readiness, management readiness, and human resource readiness. Building on the perspectives of Mukherjee et al. (2023), researchers believe that when Industry 5.0 emerges, robots may begin to perform routine daily tasks. The ability to handle unusual circumstances, inventiveness, and design are only a few of the specialized skill sets that humans may need at the same time. Given that Industry 5.0 is still in its very early stages, the government, society, and organizations may not yet be prepared to provide the necessary trainers and training programs (Sheridan, 2016). Additionally, Industry 5.0 would feature excessive

First-order codes	Second-order theoretical sub-categories codes	Aggregate theoretical dimensions

- Our information technology setup might be improved for the Industry 5.0 manufacturing process.
- For Industry 5.0 processes, our main equipment could be networked.
- The size of our factory floor allows for digital production.
- Our infrastructure's adaptability fits Industry 5.0.

Technology readiness

- Our organization implement the identification and detection process for cyber security risk management
- Our organization implement the protection process for cyber security risk management
- Our organization implement the response process for cyber security risk management
- Our organization implement the recovery process for cyber security risk management

Cybersecurity readiness

- Our organization implements environmental solutions
- Our organization implements sustainability-focused business model
- Our organization implements strategic planning participation
- Our organization tracks sustainability metrics

Task readiness

Technological factors

- Our organization is financially ready to switch to digital production.
- Industry 5.0-related financial incentives from agencies are known to our organization.
- Agency financing for Industry 5.0 is available to our organization.
- Our organization is aware of the agencies to whom our organization can apply for funding for Industry 5.0.

Financial readiness

- Our management has a plan to digitize the production process.
- The Industry 5.0 production process should be taken into consideration, according to our management.
- We have the right leadership in place to implement digitized production.
- Our management has commitment to upgrade production process.

Management readiness

- Our human resource acquires technical knowledge
- Our human resource acquires technology management capability.
- Our human resource acquires relational knowledge.
- Our senior managers acquire managerial skills

Human resource readiness

Organizational factors

- We closely adhere to industry 5.0 implementation regulations and standards.
- We pay great attention to recommendations from professional associations and other government agencies on promising future directions.
- We implement Industry 5.0 to get financial, tax, and other incentives from the government.
- We implement Industry 5.0 for public image certifications and recognitions

Institutional readiness

- The capacity of a human actor to select from a variety of actors and resources to collaborate with or use during the Industry 5.0 implementation process
- The capacity of a human actor to detect and appropriately react to the good and negative effects of other actors during the Industry 5.0 implementation process
- The capacity of a human actor, when dealing with other actors during the Industry 5.0 implementation process, to adapt to diverging institutions (i.e., various norms, regulations, or beliefs).
- During the Industry 5.0 implementation process, a human actor's self-assurance in his or her capacity to master particular skills or jobs

Ecosystem actor readiness

Environmental factors

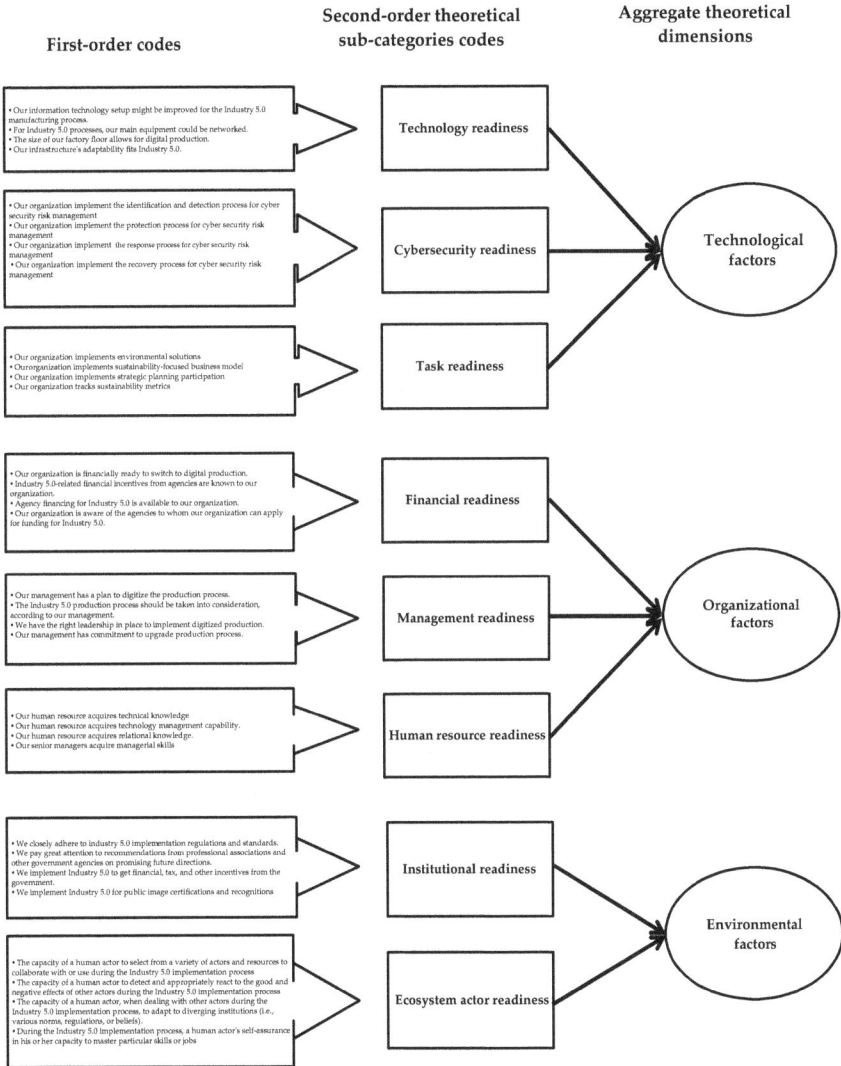

Figure 3.2 Data analysis and construct evolution.

technological exploitation, coworking between humans and robots, and mass customization through networked gadgets (Fukuda, 2020). These would necessitate large investments from governments as well as SMEs.

Environmental factors act as a high-order construct comprising two uni-dimensional constructs, namely, institutional readiness and ecosystem actor readiness. As argued by Maddikunta et al. (2022), human–robot coworking, 6G, digital twins, additive manufacturing, drone technology, networked, intelligent devices, and numerous systems are just a few of the newer, smarter ways of working that are anticipated to come with Industry 5.0. In order to manage each of these, new laws and regulations for overall ecosystem orientation, including laws for human–robot coworking, discriminating between different sorts of robots, as well as different types of machines such as drones and robots, need to be created (Demir et al., 2019). Additionally, it is necessary to develop clear regulations that include agility and flexibility in order to promote innovation in the ecosystem and in technology (Xu et al., 2021).

3.4 FINAL CONSIDERATION AND FUTURE EXTENSIONS

3.4.1 General discussion

In the past several years, significant attention has been made to human–robot collaboration in Industry 5.0, with a focus primarily on the role of the human in the technological transformation (Javaid & Haleem, 2020; Nahavandi, 2019; Pathak et al., 2019; Xu et al., 2021). Additionally, a number of studies examine the human role from various angles, such as technical, ethical, operational, societal, and safety, and this has emerged as one of the main research directions to shape this new industrial revolution (Demir et al., 2019; Callaghan, 2019). However, radical changes to Industry 5.0 will not be feasible in terms of the fundamental values without taking the SMEs into account (Madhavan et al., 2022). Even though the readiness of SMEs for digital transformation is critical, firms have a limited understanding of the required future skills and experience and their readiness to embark on the Industry 5.0 transformation. Thus, it is crucial to investigate SMEs' RII. The main aim of this research is to explore RII which is defined from the perspectives of SMEs and formulate a measurement scale utilizing mixed methods in two phases. In the first phase, hinged on the scale construction phase recommendations from DeVellis and Thorpe (2021) and Churchill (1979), Phase 1 in this research sought to produce a suitable item pool that accurately matched the construct of RII. The methods used in this study created a clear, logical, and systematic process that guaranteed the quality of the first item pool. In a nutshell, the initial item pool was improved in Phase 1 using several techniques, yielding 32 items for additional quantitative scale validation in Study 2. In the second phase, the current research develops a scale for measuring the structural characteristics of RII within SMEs in emerging countries. This study gathered quantitative survey data from a sample of 723 accountants in SMEs to further analyze the scale's validity and reliability as well as scale characteristics reexamination. The EFA results and CFA results of RII released the structural characteristics of RII. Subsequently, the SEM was employed in the model development and testing procedure to create a model that would explain how the elements of RII interacted with the benefits that SMEs may expect from adopting the technology by analyzing the statistical data from a sample of 1,283 accountants in SMEs. Consequently, the structural characteristics of RII comprised three core pillars: technological factors, organizational factors, and environmental factors. More concretely, technological factors act as a high-order construct comprising three unidimensional constructs, namely, technology readiness, cybersecurity readiness, and task readiness. While organizational factors serve as a high-order construct comprising three unidimensional constructs, namely, financial readiness, management readiness, and human resource readiness, the environmental factors act as a high-order construct comprising two unidimensional constructs, namely, institutional readiness and ecosystem actor readiness.

3.4.2 Practical implications

Policy makers and practitioners will greatly benefit from the research's empirical findings. In order to implement Industry 5.0 successfully, senior managers in SMEs should focus even more on the technological aspect, according to the acquired findings, which will help prioritize the practices. As a result, senior managers in SMEs should be aware of the benefits of contemporary information technology and look for practical, workable solutions to speed up the adoption of Industry 5.0 across all internal organizational functions. Senior managers in SMEs are also urged to make every effort to equally share the necessary resources for enhancing and strengthening the organization's information technology infrastructure and implementing good human resource management practices.

The findings of this study also increase the leaders of SMEs' understanding of the importance of human resources and management productivity. In this regard, it has also been mandated that senior managers in SMEs focus on enhancing organizational staff proficiency through appropriate training programs to keep them up to date with the Industry 5.0 model. Intriguingly, the current publication sheds light on the significance of environmental factors, and this finding can lead to a deeper understanding for individuals looking for fresh approaches to problems relating to RII. Simply defined, the most important aspects of implementing Industry 5.0 will be the legal and regulatory framework as well as the close cooperation amongst ecosystem actors.

It is also envisaged that the findings of this study will assist policymakers and other relevant organizations in developing standards for the implementation of Industry 5.0 that will encourage their widespread acceptance among SMEs. Hardware and software makers as well as merchants are requesting that efforts focus more intently on the benefits and drawbacks of Industry 5.0 in order to advance cutting-edge systems in a way that is suited for SMEs' unique needs. More concretely, technological initiatives and cybersecurity should be taken into consideration. As such, the hardware and software makers should pay more concern to the identification and detection process, protection process, response process, and recovery process for cybersecurity risk management. On the other hand, merchants should place much more attention on the organizational tasks of potential customers to provide the most beneficial product/service.

3.4.3 Conclusion and future lines of study

Industry 5.0 is far from being implemented by manufacturing firms in developing countries (Xu et al., 2021), and many industries in emerging economies have not yet adopted similar new technology (Sharma et al., 2022; Raj et al., 2020). It is also unclear how well-equipped sectors are to make this digital transformation; it is difficult to make all processes digital because it takes a lot of time, money, and manpower. The shift of SMEs from Industry 4.0 to Industry 5.0 is in fact fraught with challenges, such as a lack of resources (tools, expertise, and motivation), a lack of open innovation and strategy, and sustainability concerns. Despite the importance of SMEs' readiness for the digital transformation, firms' knowledge of the necessary future knowledge and experience and their readiness for the Industry 5.0 transition is low. The primary goal of this study is to investigate RII, as defined from the viewpoints of SMEs, and to develop a measurement scale using a combination of approaches over the course of two phases. Following Churchill's scale construction approach, the first phase determines three dimensions— the technological, organizational, and environmental aspects—with eight components that have suitable convergent and discriminant validity.

The nomological validity of the RII scale is assessed in the second phase by examining how it connects to other theoretical conceptions. In spite of highlighting notable results, the current research effort in a new field of study ran into a number of unavoidable limitations. Nevertheless, the limitations detailed could offer the starting points for new avenues generation for future works. First, the empirical sample in this study is limited to Vietnamese SMEs. Future research may also examine larger populations and various cultures. Additionally, future research examining the same characteristics might take into account businesses from other nations. The factors examined in this study could have varying degrees of value in other contextual circumstances. Second, this study is beneficial to SMEs and policymakers, but due to time restraints during data collection, the sample size is restricted. Follow-up researchers may be able to get a more comprehensive view of the issue regarding status and attitude toward Industry 5.0 adoption and preparedness with a larger sample size. To get more interesting results, more groups with different levels of preparation (ready vs. not ready) should be researched. To comprehend the time-oriented behavior of Industry 5.0 ready models, more complex research including a longitudinal study of the dimension prior to, during, and following the deployment of Industry 5.0 should be conducted.

Finally, this study depends on individual-level data's capacity to provide information regarding organizational-level constructs. Surveys are an effective way to gather data and offer a great place to start for additional activities, despite the fact that their explanatory power is restricted. However, it is crucial to recognize that survey data is gathered at the individual level and can be utilized to infer traits of the larger organization through equivalency testing. Nonetheless, the information in this research is based on an interview with a single respondent and the responses to a questionnaire survey. Building an analysis on one individual's perspective can be viewed as a constraint because one cannot be certain that this person has a comprehensive understanding of the subjects being studied. Future studies can thus investigate these phenomena by using several responders from each organization to reinforce the results.

REFERENCES

Abed, S. S. (2020). Social commerce adoption using TOE framework: An empirical investigation of Saudi Arabian SMEs. *International Journal of Information Management*, *53*, 102118. doi:10.1016/j.ijinfomgt.2020.102118

Alraja, M. N., Imran, R., Khashab, B. M., & Shah, M. (2022). Technological innovation, sustainable green practices and SMEs sustainable performance in times of crisis (COVID-19 pandemic). *Information Systems Frontiers*, *24*, 1081–1105. doi:10.1007/s10796-022-10250-z

Alsetoohy, O., Ayoun, B., Arous, S., Megahed, F., & Nabil, G. (2019). Intelligent agent technology: What affects its adoption in hotel food supply chain management? *Journal of Hospitality and Tourism Technology*, *10*(3), 317–341. doi:10.1108/JHTT-01-2018-0005

Alzahrani, A. I., Mahmud, I., Ramayah, T., Alfarraj, O., & Alalwan, N. (2017). Extending the theory of planned behavior (TPB) to explain online game playing among Malaysian undergraduate students. *Telematics and Informatics*, *34*(4), 239–251. doi:10.1016/j.tele.2016.07.001

Awa, H. O., Ukoha, O., & Emecheta, B. C. (2016). Using T-O-E theoretical framework to study the adoption of ERP solution. *Cogent Business & Management*, *3*(1), 1–23. doi:10.1080/23311975.2016.1196571

Awa, H. O., Ukoha, O., & Igwe, S. R. (2017). Revisiting technology-organization-environment (T-O-E) theory for enriched applicability. *The Bottom Line*, *30*(1), 2–22. doi:10.1108/bl-12-2016-0044

Baker, J. (2012). The technology–organization–environment framework. In: Dwivedi, Y., Wade, M., Schneberger, S. (Eds.), *Information Systems Theory. Integrated Series in Information Systems*, vol 28. Springer, New York, NY. doi:10.1007/978-1-4419-6108-2_12

Callaghan, C. W. (2019). Transcending the threshold limitation: A fifth industrial revolution? *Management Research Review*, *43*(4), 447–461. doi:10.1108/mrr-03-2019-0102

Chau, P. Y. K., & Tam, K. Y. (1997). Factors affecting the adoption of open systems: An exploratory study. *MIS Quarterly*, *21*(1), 1–24. doi:10.2307/249740

Chittipaka, V., Kumar, S., Sivarajah, U., Bowden, J. L.-H., & Manish Mohan Baral, M. M. (2022). Blockchain technology for supply chains operating in emerging markets: An empirical examination of technology-organization-environment (TOE) framework. *Annals of Operations Research*, 1–28. doi:10.1007/s10479-022-04801-5

Chung, S.-I., & Park, H.-S. (2021). The influencing factors of SME's acceptance intention to advance smart factory. *Journal of Digital Convergence*, 19(6), 199–211. doi:10.14400/JDC.2021.19.6.199

Churchill, G. (1979). A paradigm for developing better measures of marketing constructs. *Journal of Marketing Research*, 16, 64–73. doi:10.2307/3150876

Chwelos, P., Benbasat, I., & Dexter, A.S. (2001). Empirical test of an EDI adoption model. *Information Systems Research*, 12(3), 304–321.

Ciasullo, M. V., Orciuoli, F., Douglas, A., & Palumbo, R. (2022). Putting Health 4.0 at the service of Society 5.0: Exploratory insights from a pilot study. *Socio-Economic Planning Sciences*, 80, 1–12. doi:10.1016/j.seps.2021.101163

Corsini, F., Appio, F. P., & Frey, M. (2019). Exploring the antecedents and consequences of environmental performance in micro-enterprises: The case of the Italian craft beer industry. *Technological Forecasting and Social Change*, 138, 340–350. doi:10.1016/j.techfore.2018.10.018

Deguchi, A., & Karasawa, K. (2020). *Issues and Outlook*, vol. 155. Hitachi-UTokyo Laboratory (H-UTokyo Lab.).

DeJonckheere, M., & Vaughn, L. M. (2019). Semi-structured interviewing in primary care research: A balance of relationship and rigour. *Family Medicine and Community Health*, 7(2), 1–8. doi:10.1136/fmch-2018-000057

Demir, K. A., Döven, G., & Sezen, B. (2019). Industry 5.0 and human-robot co-working. *Procedia Computer Science*, 158, 688–695. doi:10.1016/j.procs.2019.09.104

DeVellis, R. F. (2017). *Scale Development: Theory and Applications* (4th ed.). Thousand Oaks, CA: Sage.

DeVellis, R. F., & Thorpe, C. T. (2021). *Scale Development: Theory and Applications*. London: Sage Publications.

Dominguez, L. C., Silkens, M.E.W.M., & Sanabria, A. (2019). The Dutch residency educational climate test: Construct and concurrent validation in Spanish language. *International Journal of Medical Education*, 10, 138–148. doi:10.5116/ijme.5d0c.bff7

Doyle-Kent, M., & Kopacek, P. (2019). Industry 5.0: Is the manufacturing industry on the cusp of a new revolution? In *Proceedings of the International Symposium for Production Research*. Springer: Berlin/Heidelberg, Germany.

Dunn, T. J., Baguley, T., & Brunsden, V. (2013). From alpha to omega: A practical solution to the pervasive problem of internal consistency estimation. *British Journal of Psychology*, 105(3), 399–412. doi:10.1111/bjop.12046

Espina-Romero, L., Guerrero-Alcedo, J., Avila, N. G., Sánchez, J. G. N., Hurtado, H. G., & Li, A. Q. (2023). Industry 5.0: Tracking scientific activity on the most influential industries, associated topics, and future research agenda. *Sustainability*, 15(6), 1–20. doi:10.3390/su15065554

Feng, X., & Behar-Horenstein, L. (2019). Maximizing NVivo utilities to analyze open-ended responses. *The Qualitative Report*, 24(3), 563–571. doi:10.46743/2160-3715/2019.3692

Fetscherin, M., & Stephano, R.-M. (2016). The medical tourism index: Scale development and validation. *Tourism Management*, 52, 539–556. doi:10.1016/j.tourman.2015.08.010

Fornell, C., & Larcker, D. F. (1981). Evaluating structural equation models with unobservable variables and measurement error. *Journal of Marketing Research*, *18*(1), 39–50. doi:10.1177/002224378101800104

Fukuda, K. (2020). Science, technology and innovation ecosystem transformation toward Society 5.0. *International Journal of Production Economics*, *220*, 1–14. doi:10.1016/j.ijpe.2019.07.033

Gao, T. (Tony), Leichter, G., & Wei, Y. (Susan). (2012). Countervailing effects of value and risk perceptions in manufacturers' adoption of expensive, discontinuous innovations. *Industrial Marketing Management*, *41*(4), 659–668. doi:10.1016/j.indmarman.2011.09

Ghobakhloo, M., Iranmanesh, M., Mubarak, M.F., Mubarik, M., Rejeb, A., Nilashi, M. (2022). Identifying Industry 5.0 contributions to sustainable development: A strategy roadmap for delivering sustainability values. *Sustainable Production and Consumption*, *33*, 716–737. doi:10.1016/j.spc.2022.08.003

Gide, E., & Sandu, R. (2015). A Study to Explore the Key Factors Impacting on Cloud Based Service Adoption in Indian SMEs. *Proceedings – 12th IEEE International Conference on E-Business Engineering, ICEBE 2015*, 387–392. doi:10.1109/ICEBE.2015.72

Gotz, O., Liehr-Gobbers, K., Krafft, M. (2010). Evaluation of structural equation models using the partial least squares (PLS) approach. In: Vinzi, V.E., Chin, W.W., Henseler, J., Wang, H. (Eds.), *Handbook of Partial Least Squares* (pp. 691–711). Heidelberg: Springer.

Gutierrez, A., Boukrami, E., & Lumsden, R. (2015). Technological, organisational and environmental factors influencing managers' decision to adopt cloud computing in the UK. *Journal of Enterprise Information Management*, *28*(6), 788–807. doi:10.1108/jeim-01-2015-0001

Ha, D. T., Le, T., Fisher, G., & Nguyen, T. T. (2022). Factors affecting the adoption extent of the balanced scorecard by Vietnamese small- and medium-sized enterprises. *Accounting Research Journal*, *35*(4), 543–560. doi:10.1108/ARJ-08-2020-0221

Hair, J. F., Black, W. C., Babin, B. J., & Anderson, R. E. (2010). *Multivariate Data Analysis* (7th ed.). Englewood Cliffs, NJ: Prentice Hall.

Hair, J. F., Black, W. C., Babin, B. J., & Anderson, R. E. (2014). *Multivariate Data Analysis* (7th ed.). Essex, UK: Pearson Education.

Hair, J. F., Black, W. C., Babin, B. J., & Anderson, R. E. (2018). *Multivariate Data Analysis* (8th ed.). United Kingdom: Cengage Learning.

Hair, J., Ringle, C. and Sarstedt, M. (2011). PLS-SEM: Indeed a silver bullet. *Journal of Marketing Theory and Practice*, *19*, 139–151. doi:10.2753/MTP1069-6679190202

Hair, J.F., Hult, G.T.M., Ringle, C.M., & Sarstedt, M. (2017). *A Primer on Partial Least Squares Structural Equation Modeling* (PLS-SEM) (2nd ed.). Thousand Oaks, CA: Sage Publications Inc.

Hanus, B., & Wu, Y. "Andy." (2016). Impact of users' security awareness on desktop security behavior: A protection motivation theory perspective. *Information Systems Management*, *33*(1), 2–16. doi:10.1080/10580530.2015.1117842

Harrington, D. (2009). *Confirmatory Factor Analysis*. New York, NY: Oxford University Press.

Hinkin, T. R. (1995). A review of scale development practices in the study of organizations. *Journal of Management*, *21*(5), 967–988. doi:10.1177/014920639502100509

Hoang, N.A., & Otake, T. (2014). Credit participation and credit source selection of Vietnam small and medium enterprises. *The South East Asian Journal of Management*, 8(2), 104–128. doi:10.21002/seam.v8i2.3929

Huang, S., Wang, B., Li, X., Zheng, P., Mourtzis, D., & Wang, L. (2022). Industry 5.0 and Society 5.0—Comparison, complementation and co-evolution. *Journal of Manufacturing Systems*, 64, 424–428. doi:10.1016/j.jmsy.2022.07.010

Iacovou, C. L., Benbasat, I., & Dexter, A. S. (1995). Electronic data interchange and small organizations: Adoption and impact of technology. *MIS Quarterly*, 19(4), 465–485.

Ifinedo, P. (2011). An empirical analysis of factors influencing internet/e-business technologies adoption by SMEs in Canada. *International Journal of Information Technology & Decision Making*, 10(04), 731–766. doi:10.1142/s0219622011004543

Intan Salwani, M., Marthandan, G., Daud Norzaidi, M., & Choy Chong, S. (2009). E-commerce usage and business performance in the Malaysian tourism sector: Empirical analysis. *Information Management & Computer Security*, 17(2), 166–185. doi:10.1108/09685220910964027

Jafari, N., Azarian, M., & Yu, H. (2022). Moving from Industry 4.0 to Industry 5.0: What are the implications for smart logistics? *Logistics*, 6(26), 1–7. doi:10.3390/logistics6020026

Javaid, M., & Haleem, A. (2020). Critical components of Industry 5.0 towards a successful adoption in the field of manufacturing. *Journal of Industrial Integration and Management*, 5, 327–348. doi:10.1142/s2424862220500141

Jung, W.-K., Kim, D.-R., Lee, H., Lee, T.-H., Yang, I., Youn, B. D., Zontar, D., Brockmann, M., Brecher, C., & Ahn, S.-H. (2020). Appropriate smart factory for SMEs: Concept, application and perspective. *International Journal of Precision Engineering and Manufacturing*, 22(1), 201–215. doi:10.1007/s12541-020-00445-2

Kaynak, H. (2003). The relationship between total quality management practices and their effects on firm performance. *Journal of Operations Management*, 21(4), 405–435. doi:10.1016/s0272-6963(03)00004-4

Khan, S. A., Lederer, A. L., & Mirchandani, D. A. (2013). Top management support, collective mindfulness, and information systems performance. *Journal of International Technology and Information Management*, 22(1), 95–122.

Kline, R. B. (2015). *Principles and Practice of Structural Equation Modeling*. New York, NY: Guilford publications.

Koc, M., & Barut, E. (2016). Development and validation of New Media Literacy Scale (NMLS) for university students. *Computers in Human Behavior*, 63, 834–843. doi:10.1016/j.chb.2016.06.035

Koch, P. J., van Amstel, M. K., Dębska, P., Thormann, M. A., Tetzlaff, A. J., Bøgh, S., & Chrysostomou, D. (2017). A skill-based robot co-worker for industrial maintenance tasks. *Procedia Manufacturing*, 11, 83–90. doi:10.1016/j.promfg.2017.07.141

Kumar, A., & Nayyar, A. (2020). si 3-Industry: A sustainable, intelligent, innovative, internet-of-things industry. *A roadmap to Industry 4.0: Smart production, sharp business and sustainable development*, 1–21.

Kummer, T.-F., Pelzl, S., & Bick, M. (2021). A conceptualisation of privacy risk and its influence on the disclosure of check-in services information. *International Journal of Information Management*, 57, 1–11. doi:10.1016/j.ijinfomgt.2020.102266

Leng, J., Sha, W., Wang, B., Zheng, P., Zhuang, C., Liu, Q., Wuest, T., Mourtzis, D., & Wang, L. (2022). Industry 5.0: Prospect and retrospect. *Journal of Manufacturing Systems*, 65, 279–295. doi:10.1016/j.jmsy.2022.09.017

Lin, H.-F. (2014). Understanding the determinants of electronic supply chain management system adoption: Using the technology–organization–environment framework. *Technological Forecasting and Social Change*, 86, 80–92. doi:10.1016/j.techfore.2013.09.001

Little, T. D. (2013). *Longitudinal Structural Equation Modeling.* New York, NY: The Guilford Press.

Liyanage, M., Braeken, A., Kumar, P., & Ylianttila, M. (Eds.). (2020). *IoT Security: Advances in Authentication* New Jersey, NJ: John Wiley and Sons.

Longo, F., Padovano, A., & Umbrello, S. (2020). Value-oriented and ethical technology engineering in Industry 5.0: A human-centric perspective for the design of the factory of the future. *Applied Sciences*, 10(12), 1–25. doi:10.3390/app10124182

Maddikunta, P. K. R., Pham, Q.-V., Prabadevi B., Deepa, N., Dev, K., Gadekallu, T. R., Ruby, R., Liyanage, M. (2022). Industry 5.0: A survey on enabling technologies and potential applications. *Journal of Industrial Information Integration*, 26, 1–19. doi:10.1016/j.jii.2021.100257

Madhavan, M.; Wangtueai, S.; Sharafuddin, M.A.; Chaichana, T. (2022). The precipitative effects of pandemic on open innovation of SMEs: A scientometrics and systematic review of Industry 4.0 and Industry 5.0. *Journal of Open Innovation: Technology, Market, and Complexity*, 8, 1–16. doi:10.3390/joitmc8030152

Masood, T., & Egger, J. (2019). Augmented reality in support of Industry 4.0— Implementation challenges and success factors. *Robotics and Computer-Integrated Manufacturing*, 58, 181–195. doi:10.1016/j.rcim.2019.02.003

Mehdiabadi, A., Shahabi, V., Shamsinejad, S., Amiri, M., Spulbar, C., & Birau, R. (2022). Investigating Industry 5.0 and its impact on the banking industry: Requirements, approaches and communications. *Applied Science*, 12, 1–25. doi:10.3390/app12105126

Muhamad, M. Q. B., Mohamad, S. J. A. N. S., & Nor, N. M. (2021). Technological-Organisational-Environmental (TOE) framework in Industry 4.0 adoption among SMEs in Malaysia: An early outlook. *ASEAN Entrepreneurship Journal*, 6(3), 13–19.

Mukherjee, A. A., Raj, A., & Aggarwal, S. (2023). Identification of barriers and their mitigation strategies for industry 5.0 implementation in emerging economies. *International Journal of Production Economics*, 257, 1–15. doi:10.1016/j.ijpe.2023.108770

Nahavandi, S. (2019). Industry 5.0—A human-centric solution. *Sustainability*, 11(16), 1–13. doi:10.3390/su11164371

Nayyar, A., & Kumar, A. (Eds.). (2020). *A Roadmap to Industry 4.0: Smart Production, Sharp Business and Sustainable Development* (pp. 1–21). Berlin: Springer.

Nguyen, T., & Wongsurawat, W. (2012). The impact of government policies on the development of small- and medium-sized enterprises: The case of Vietnam. *Journal for International Business and Entrepreneurship Development*, 6(2), 188–200. doi:10.1504/jibed.2012.048567

Nikopoulou, M., Kourouthanassis, P., Chasapi, G., Pateli, A., & Mylonas, N. (2023). Determinants of digital transformation in the hospitality industry: Technological, organizational, and environmental drivers. *Sustainability*, 15, 1–20. doi:10.3390/su15032736

Othman, A., Al Mutawaa, A., Al Tamimi, A., & Al Mansouri, M. (2023). Assessing the readiness of government and semi-government institutions in Qatar for inclusive and sustainable ICT accessibility: Introducing the MARSAD tool. *Sustainability*, *15*, 1–25. doi:10.3390/su15043853

Özdemir, V., & Hekim, N. (2018). Birth of Industry 5.0: Making sense of big data with artificial intelligence, "The Internet of Things" and next-generation technology policy. *OMICS: A Journal of Integrative Biology*, *22*(1), 65–76. doi:10.1089/omi.2017.0194

Pallant J. (2007). *SPSS Survivalmanual: A Step By Step Guide to data analysis using SPSS*. Sydney: Allen & Unwin, Sabon by Bookhouse.

Pan, M.-J., & Jang, W.-Y. (2008). Determinants of the adoption of enterprise resource planning within the technology-organization-environment framework: Taiwan's communications industry. *Journal of Computer Information Systems*, *48*, 94–102. doi:10.1080/08874417.2008.11646025

Parasuraman, A. (2000). Technology Readiness Index (TRI): A multiple-item scale to measure readiness to embrace new technologies. *Journal of Service Research*, *2*(4), 307–320. doi:10.1177/109467050024001

Paschek, D., Mocan, A., & Draghici, A. (2019). Industry 5.0-The expected impact of next Industrial Revolution. In *Proceedings of the Thriving on Future Education, Industry, Business, and Society, Proceedings of the Make Learn and TIIM International Conference*, Piran, Slovenia, 15–17 May 2019; pp. 15–17.

Pathak, P., Pal, P. R., Shrivastava, M., & Ora, M. S. (2019). Fifth revolution: Applied AI and human intelligence with cyber physical systems. *International Journal of Engineering and Advanced Technology*, *8*(3), 23–27.

Pham, Q. H., & Vu, K. P. (2022). Digitalization in small and medium enterprise: A parsimonious model of digitalization of accounting information for sustainable innovation ecosystem value generation. *Asia Pacific Journal of Innovation and Entrepreneurship*, *16*(1), 2–37. doi:10.1108/APJIE-02-2022-0013

Porambage, P., Gür, G., Osorio, D. P. M., Liyanage, M., Gurtov, A., & Ylianttila, M. (2021). The roadmap to 6G security and privacy. *IEEE Open Journal of the Communications Society*, *2*, 1094–1122.

Prasetyo, Y. T., Castillo, A. M., Salonga, L. J., Sia, J. A., & Seneta, J. A. (2020). Factors affecting perceived effectiveness of COVID-19 prevention measures among Filipino during enhanced community quarantine in Luzon, Philippines: Integrating protection motivation theory and extended theory of planned behavior. *International Journal of Infectious Diseases*, *99*, 312–323. doi:10.1016/j.ijid.2020.07.074

Ragu-Nathan, B. S., Apigian, C. H., Ragu-Nathan, T. S., & Tu, Q. (2004). A path analytic study of the effect of top management support for information systems performance. *Omega*, *32*(6), 459–471. doi:10.1016/j.omega.2004.03.001

Raj, A., Dwivedi, G., Sharma, A., Lopes de Sousa Jabbour, A. B., & Rajak, S. (2020). Barriers to the adoption of industry 4.0 technologies in the manufacturing sector: An inter-country comparative perspective. *International Journal of Production Economics*, *224*, 1–50. doi:10.1016/j.ijpe.2019.107546

Rajbhandari, S., Devkota, N., Khanal, G., Mahato, S., & Paudel, U. R. (2022). Assessing the industrial readiness for adoption of industry 4.0 in Nepal: A structural equation model analysis. *Heliyon*, *8*(2), 1–11. doi:10.1016/j.heliyon.2022.e08919

Robu, M. (2013). The dynamic and importance of SMEs in economy. *The USV Annals of Economics and Public Administration*, *13*, 84–89.

Rogers, E.M. (2003). *Diffusion of Innovations* (15th ed.). Tampa, FL: Free Press, Sch.

Sachsenmeier, P. (2016). Industry 5.0—The relevance and implications of bionics and synthetic biology. *Engineering, 2*(2), 225–229. doi:10.1016/j.eng.2016.02.015

Sekaran, U., & Bougie, R. (2016). *Research Methods for Business: A Skill Building Approach*. New Jersey, NJ: John Wiley & Sons.

Senyo, P. K., Effah, J., & Addae, E. (2016). Preliminary insight into cloud computing adoption in a developing country. *Journal of Enterprise Information Management, 29*(4), 505–524. doi:10.1108/jeim-09-2014-0094

Sharma, M., Joshi, S., Luthra, S., & Kumar, A. (2022). Managing disruptions and risks amidst COVID-19 outbreaks: Role of blockchain technology in developing resilient food supply chains. *Operations Management Research, 15*, 268–281. doi:10.1007/s12063-021-00198-9

Sheridan, T. B. (2016). Human–Robot interaction. *Human Factors: The Journal of the Human Factors and Ergonomics Society, 58*(4), 525–532. doi:10.1177/0018720 816644364

Singh, K. K., Nayyar, A., Tanwar, S., & Abouhawwash, M. (Eds.). (2021). *Emergence of Cyber Physical System and IoT in Smart Automation and Robotics*. Advances in Science, Technology & Innovation. Switzerland: Springer. doi:10.1007/978-3-030-66222-6

Singh, S., & Srivastava, S. (2019). Engaging consumers in multichannel online retail environment: A moderation study of platform type on interaction of e-commerce and m-commerce. *Journal of Modelling in Management, 14*(1), 49–76. doi:10.1108/ JM2-09-2017-0098

Stevens, J. (2002). *Applied Multivariate Statistics of the Social Sciences* (4th ed.). Mahwah, NJ: Erlbaum.

Tiglao, N. C. C., De Veyra, J. M., Tolentino, N. J. Y., & Tacderas, M. A. Y. (2020). The perception of service quality among paratransit users in Metro Manila using structural equations modelling (SEM) approach. *Research in Transportation Economics, 83*, 1–13. doi:10.1016/j.retrec.2020.100955

Tornatzky, L.G., Fleischer, M., & Chakrabarti, A.K. (1990). *Processes of Technological Innovation*. Washington, DC: Lexington Books.

Urbach, N., & Ahlemann, F. (2010). Structural equation modeling in information systems research using partial least squares. *Journal of Information Technology Theory and Application, 11*(2), 5–40.

Venaik, A., Jain, S., & Nayyar, A. (2023). Industry 4.0—Its Advancement and Effects on Security of Whistle-Blowers on Dark Web. In: Singh, G., Goel, R., Garg, V. (Eds.), *Industry 4.0 and the Digital Transformation of International Business*. Singapore: Springer. doi:10.1007/978-981-19-7880-7_6

Venkatesh, V., Morris, M. G., Gordon, B., & Davis, F. D. (2003). User acceptance of information technology: Toward a unified view. *MIS Quarterly, 27*(3), 425–478. doi:10.2307/30036540

Watkins, M. W. (2018). Exploratory factor analysis: A guide to best practice. *Journal of Black Psychology, 44*(3), 219–246. doi:10.1177/0095798418771807

Wolf, E. J., Harrington, K. M., Clark, S. L., & Miller, M. W. (2013). Sample size requirements for structural equation models. *Educational and Psychological Measurement, 73*(6), 913–934. doi:10.1177/0013164413495237

Xia, Y., & Yang, Y. (2018). RMSEA, CFI, and TLI in structural equation modeling with ordered categorical data: The story they tell depends on the estimation methods. *Behavior Research Methods, 51*, 409–428. doi:10.3758/s13428-018-1055-2

Xu, X., Lu, Y., Vogel-Heuser, B., & Wang, L. (2021). Industry 4.0 and Industry 5.0—Inception, conception and perception. *Journal of Manufacturing Systems, 61,* 530–535. doi:10.1016/j.jmsy.2021.10.006

Zhu, K., Kraemer, K. L., & Dedrick, J. (2004). Information technology payoff in E-Business environments: An international perspective on value creation of E-Business in the Financial Services Industry. *Journal of Management Information Systems, 21*(1), 17–54. doi:10.1080/07421222.2004.1104579

Zybery, I., & Rova, L. (2014). The role of the accountants in the framework of the modern technological developments and digital accounting systems. *European Scientific Journal, 24,* 30–48. doi:10.19044/esj.2011.v24n0p%25p

Chapter 4

Thriving in Industry 5.0

The evolving role of employees in the age of automation and innovation

V. Harish
PSG Institute of Management, India

Anand Nayyar
Duy Tan University, Vietnam

4.1 INTRODUCTION

In the last few decades, technology has made significant advancements and expansions, particularly with the advent of the Internet. People are witnessing a new era of rapid development of technologies such as the Internet of Things (IoT), Artificial Intelligence (AI), Big Data (BD), robotics, 3D printing, Cloud Computing (CC), and Mobile Devices (MD), are driving significant developments in business and society.

The evolution of new-age technologies has paved the transition to Industry 4.0 (I 4.0); as a result, technology has evolved as a vital part of daily life, impacting every aspect of people and society (Van Dijk, 2020). As a result, the use of ICT in industry distinguishes I 4.0 from previous industrial transformations, driving the development of Cyber-Physical Systems (CPS) (Nikolakis et al., 2019).

The emergence of I 4.0 has transformed the manufacturing sector as well as other sectors and has presented employees with new challenges and opportunities (Koh et al., 2019). However, technological and automation advancements have continued to revolutionise the industry, and we are now witnessing the beginning of Industry 5.0 (I 5.0). This era of production will necessitate a unique set of skills, competencies, and mentalities from the employees (Xu et al., 2021). Therefore, it is essential to investigate the impact of I 5.0 on employees and the strategies that can be implemented to successfully navigate these changes.

I 4.0 signifies the assimilation of new technologies such as artificial intelligence (AI), Internet of Things (IoT), and machine learning (ML) into the manufacturing sector (Sony & Naik, 2020). This integration has transformed the manufacturing of goods, resulting in augmented productivity and profitability.

However, the key focus of I 4.0 was on automation and the role of machines in manufacturing, which led to concerns about employment losses

DOI: 10.1201/9781003489269-4

and human obsolescence (Culot et al., 2020). I 5.0, with its emphasis on "human–machine collaboration" and the use of technology to augment human capabilities, emerged in response to these concerns (Kaasinen et al., 2022). I 5.0 seeks to balance the roles of technology and human expertise, emphasising the significance of machine-human collaboration (Xu et al., 2021). I 5.0, which emphasises collaboration between humans *and* machines and sustainable manufacturing, represents the evolution of I 4.0 and it aims to create a future in which people and technology work together to accomplish productivity and innovation while preserving the environment and promoting a socially responsible approach to manufacturing (Adel, 2022).

I 5.0 is characterised by decentralised production and service offerings, product and service customisation, and sustainable manufacturing (Yao et al., 2022) and decentralised production allows for the making of customised products and services closer to the consumer, thereby decreasing lead times and transportation costs. I 5.0 places an emphasis on reducing waste and carbon emissions by the adoption of renewable energy and ecological manufacturing practices (ElFar et al., 2021). I 5.0 will also result in a more decentralised approach to manufacturing, with an emphasis on customisation and smaller-batch production. This will necessitate a workforce that is more agile and adaptable to changing customer requirements and demands (Tiwari et al., 2022). By establishing new occupations and requiring new skills and capabilities, I 5.0 will radically alter employment options (Battini et al., 2022).

I 5.0 will open up new job roles such as robot coordinators, experts in digital clones, and augmented reality technicians (Mashelkar, 2018). Coordinators of robots will supervise the incorporation of robots into the production process and ensure their proper operation. Experts in digital twins will utilise digital representations of products, equipment, and processes to optimise production and reduce costs. Technicians specialising in augmented reality will use AR technology to enhance the user experience, facilitate remote maintenance and training, and enhance quality control (Nagaraj, 2020).

Employees will be required to adapt to new-age skills and abilities such as digital literacy, data analytics, and problem-solving as a result of I 5.0. Because the incorporation of advanced technologies will require employees to utilise software applications, operate machinery, and analyse data, digital literacy is crucial (Mitchell & Guile, 2022). As companies collect and analyse enormous quantities of data to optimise production and identify opportunities for improvement, data analytics will become indispensable. As the complexity of the production process increases, employees will be required to make fast decisions and troubleshoot issues, which will necessitate problem-solving skills.

I 5.0 will generate new employment opportunities requiring new skills and competencies. It will also change the nature of current job responsibilities, requiring employees to adapt to new technology and collaborate with machines (Nahavandi, 2019). For firms to stay competitive in this new age of manufacturing, individuals and organisations must accept the changes and invest in training programmes.

Apart from the new job roles and talents required, I 5.0 would alter the nature and environment of the workspace. Employees will work alongside robots and other machines with a focus on maximising efficiency (Kaasinen et al., 2022), allowing for increased productivity, efficiency, and safety. This will result in a shift in employment roles, with an increased emphasis on problem-solving, decision-making, and creative thinking.

In addition, I 5.0 will have a substantial effect on the education and training industry. Individuals may not be adequately prepared for the new job roles and skills demanded by I 5.0 through the use of traditional education models. Consequently, education and training programmes will need to be planned to integrate novel technologies and skills, such as "Big data and machine learning" (Broo et al., 2022).

I 5.0 will alter the employment landscape by generating innovative job categories and necessitating varied skills and competencies (Jerman et al., 2020) and it will also result in a workforce that is more flexible and adaptable to altering market demands. The focus of this chapter is to delve into the topic of "An Employee in the Era of I 5.0" and offer perspectives into the encounters that employees will face in this new era. I 5.0 will be defined and its distinguishing characteristics relative to I 4.0 will be highlighted, and the chapter will then explore the influence of I 5.0 on employees and the skills and competencies required to flourish in this new era. In addition, the chapter will discuss the training and development programmes required to prepare employees for I 5.0 and the challenges organisations might face when facing them, and finally, it will provide organisations and employees with strategies to surmount the challenges and capitalise on the chances offered by I 5.0.

4.1.1 Objectives of the chapter

The objectives of the chapter are

- To understand the evolution of Industry 1.0 to I 5.0;
- To understand the effect of I 5.0 on employees;
- To identify the key competencies for career development for people in I 5.0;
- To analyse the challenges faced by employees in I 5.0;
- And, to understand the strategies that stakeholders can take to overcome the challenges faced by employees in I 5.0.

4.1.2 Organisation of the chapter

The chapter is organised as follows: Section 4.2 proceeds with a detailed literature study that investigates the development of industry 1.0 to 5.0, focusing on the distinctive qualities of each generation and specifically comparing I 4.0 and I 5.0. Section 4.3 looks into the characteristics of I 5.0 as well as the impact that it has on personnel. Section 4.4 discusses the abilities that workers in I 5.0 must possess as well as the difficulties that workers in I 5.0 must overcome. Section 4.5 discusses the impact of I 5.0 on employment across various industries. Section 4.6 discusses the competencies for career development and growth for people in I 5.0. Section 4.7 highlights the skills required for employees in I 5.0. Section 4.8 enlightens the challenges faced by employees in I 5.0. Section 4.9 extends the discussion on strategies to overcome challenges in I 5.0. Section 4.10 elaborates on the human–robot collaboration (HRC) in I 5.0. And, finally, Section 4.11 concludes the chapter.

4.2 LITERATURE REVIEW

I 5.0 will drastically alter employment opportunities by creating new positions and requiring new skills and capabilities (Doyle-Kent & Kopacek, 2020). I 5.0 emphasises human–machine collaboration, which will create new job responsibilities that require specialised skills and knowledge.

4.2.1 Definitions of Industry 5.0

I 5.0 refers to the amalgamation of advanced digital technologies, such as the IoT, AI, and robotics, with human intelligence to create a new level of manufacturing productivity and efficiency. The chapter provides a comprehensive analysis of I 5.0, discussing its historical context, defining characteristics, potential benefits, and obstacles. It provides researchers, practitioners, and policymakers with vital insight into the future prospects of this emerging paradigm (Leng et al., 2022).

I 5.0 signifies the integration of advanced digital technologies with human expertise and creativity to advance flexible smart factories that are autonomous and self-optimising. The chapter presents a paradigm for transitioning from I 4.0 to I 5.0, emphasising the need to move from "digital manufacturing to digital society".

The I 5.0 evolution strategy is innovative. The report concludes that I 5.0 will require a holistic and collaborative strategy for technological and social digital change (Skobelev & Borovik, 2017).

"I 5.0 is characterized by the seamless integration of digital and physical technologies, enabling machines and humans to work together in a flexible and adaptive way". The chapter discusses a macro-perspective on I 5.0, highlighting the necessity for a holistic and collaborative framework that reflects social, environmental, and economic factors. The I 5.0 transition framework is innovative and the chapter concludes that I 5.0's full potential requires a human-centric strategy (Tiwari et al., 2022). "I 5.0 is the next generation of manufacturing that brings together humans, machines, and data in a way that maximises the strengths of each" (Nahavandi, 2019).

I 5.0 is the fusion of smart technologies with conventional processes, resulting in the conception of intelligent factories that are proficient in acclimatising to changes in the environment and production requirements. The chapter discusses EU companies' I 4.0 to 5.0 transformation. Analysing indicators and models to assess companies' I 5.0 readiness is a novelty factor of the chapter. To move to I 5.0, organisations must invest in people, technology, and sustainability (Alexa et al., 2022).

I 5.0 based on the various definitions can be defined as follows:

I 5.0 is a paradigm for manufacturing that integrates sophisticated digital technologies, such as the Internet of Things, artificial intelligence, and robotics, with human expertise and creativity to create smart factories. It is characterised by the seamless integration of digital and physical technologies, which enables humans and machines to work together in a flexible and adaptable manner to maximise manufacturing productivity and efficiency. I 5.0 also combines humans, machinery, and data to create intelligent factories that can adapt to environmental and production demands.

I 5.0 will generate new occupations such as robot coordinators, experts in digital clones, and augmented reality technicians. Coordinators of robots will supervise the incorporation of robots into the production process and ensure their proper operation (Elangovan, 2021). Experts in digital twins will utilise digital representations of products, equipment, and processes to optimise production and reduce costs (Nahavandi, 2019). Technicians specialising in augmented reality will use AR technology to enhance the user experience, facilitate remote maintenance and training, and enhance quality control.

I 5.0 will also necessitate new skills and competencies, such as digital literacy, data analytics, and problem-solving, from employees. Because the incorporation of advanced technologies will require employees to utilise software applications, operate machinery, and analyse data, digital literacy is crucial (Maddikunta et al., 2022). As companies collect and analyse enormous quantities of data to optimise production and identify opportunities for improvement, data analytics will become indispensable. As the complexity of the production process increases, employees will be required to make fast decisions and troubleshoot issues, which will necessitate problem-solving skills (Mitchell & Guile, 2022).

I 5.0 will generate new employment opportunities requiring new skills and competencies. It will also change the nature of current job responsibilities, requiring employees to adapt to new technology and collaborate with machines (Kolade & Owoseni, 2022). In addition to the new job roles and talents required, I 5.0 would alter the nature and environment of work. Employees will work alongside robots and other machines with a concentration on "human–machine collaboration", allowing for augmented productivity, efficiency, and safety (Demir et al., 2019).

The increasing complexity of manufacturing processes, which necessitates more advanced technologies and greater association between humans and machinery, is one of the primary drivers of I 5.0 (Özdemir & Hekim, 2018) and this complexity is driven by the need for increased customisation and manufacturing flexibility, as well as the expanding demand for sustainability and energy efficiency. I 5.0 is viewed as a remedy to these issues, offering a new framework for collaboration and innovation in industry and manufacturing (Rodríguez-Garcia et al., 2023).

I 5.0 research has focused on a variety of topics, such as the influence of emerging technologies on the manufacturing process, the role of human factors, and the potential to generate innovation and economic growth; and the potential for I 5.0 to improve the competence and sustainability of manufacturing methods has been the subject of some research. Skobelev and Borovik (2017), in their work, examined the usage of AI and IoT in smart production, highlighting the potential for these technologies to improve manufacturing processes' productivity, waste reduction, and sustainability.

Alves et al. (2023) examined the influence of I 5.0 on the workplace, highlighting the potential for new technologies to improve job satisfaction and the authors emphasised the significance of devising user-friendly technologies and work environments that promote employee well-being.

The potential for I 5.0 to spur innovation and economic growth has also been investigated. Sharma et al. (2022), for instance, investigated the effect of I 5.0 on the German engineering sector, emphasising the potential for new technologies to drive productivity gains and enhance competitiveness. Authors highlighted the significance of establishing a policy environment that encourages investment in new technologies and skill development. Concerns exist regarding the influence of automation on jobs and the position of humans in the traditional process, despite the potential advantages of I 5.0. For instance, Ghobakhloo et al. (2022) examined the impact of automation on job displacement and emphasised the significance of reskilling and upskilling programmes to facilitate the changeover to I 5.0.

4.2.2 Evolution from Industry 1.0 to 5.0

Figure 4.1 illustrates the transition/evolution of Industry 1.0 to 5.0.

I 1.0, which was characterised by manual labour, employed unskilled labour with poor working conditions and minimal job security. The majority of the workforce was engaged in physically demanding and repetitive duties, with minimal training and development opportunities. The compensation structure was founded on hourly wages and lacked benefits and job security (Vinitha et al., 2020).

INDUSTRY

1.0	2.0	3.0	4.0	5.0

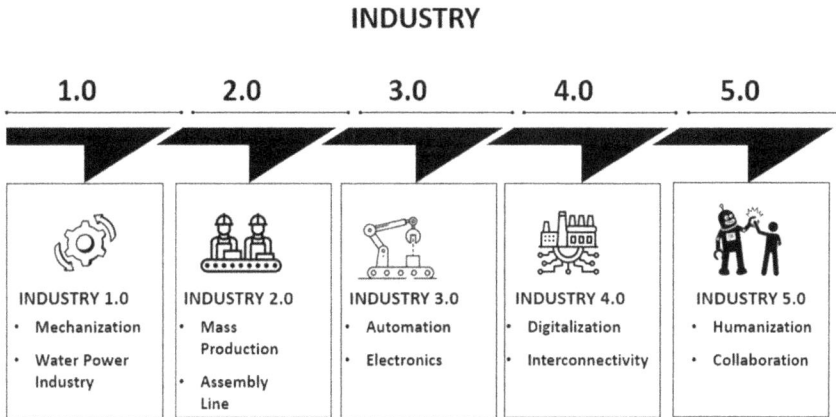

INDUSTRY 1.0	INDUSTRY 2.0	INDUSTRY 3.0	INDUSTRY 4.0	INDUSTRY 5.0
• Mechanization • Water Power Industry	• Mass Production • Assembly Line	• Automation • Electronics	• Digitalization • Interconnectivity	• Humanization • Collaboration

Figure 4.1 Transition from Industry 1.0 to 5.0.

Source: Authors' own source.

The onset of mechanisation in production processes characterised Industry 1.0, started in the 18th century and sustained till the middle of the 19th century. The invention of steam-powered machinery, which increased productivity leading to the progress of factory systems, was the most significant technological advancement of this time period. Among the earliest examples of this type of apparatus were the cotton gin and the spinning jenny. During this era, the production of iron and coal also advanced significantly.

The start of Industry 2.0 (I 2.0) led to the employment of skilled workers with enhanced working conditions and job security (Yavari & Pilevari, 2020). The introduction of machines increased output and decreased labour hours. The wage structure shifted to incorporate hourly wages with benefits such as health insurance and retirement plans.

Industry 2.0 commenced in the late 19th century and endured until the beginning of the 20th century. The start of industrial manufacturing through the use of electrical power, assembly lines, and conveyor belts marked this phase. This caused an increase in output and efficiency, allowing manufacturers to produce more goods at a reduced price and the automobile sector was one of the most significant sectors to emerge during this time.

I 3.0 introduced automation, which led to the employment of highly skilled employees with improved working conditions, job security, and flexible hours (Thangaraj & Narayanan, 2018). The workplace has become increasingly automated, necessitating both technical and soft skills in order to operate and maintain the machinery. The wage structure shifted from hourly pay to salaried compensation with benefits such as paid leave and overtime (Zakoldaev et al., 2020).

I 3.0 commenced in the second half of the 20th century and endured until the beginning of the 21st century. This era was characterised by the launch of computers and the manufacturing process. Manufacturers were able to produce products with greater precision and accuracy thanks to computer-controlled machines, resulting in quality improvements, and, in addition, this time period witnessed the emergence of international supply chains and the implementation of lean management techniques to boost productivity.

I 4.0, which is characterised by digitalisation, has resulted in the hiring of multiskilled employees with flexible compensation and on-demand work hours (Lasi et al., 2014) and in order to function in a fast-paced, digital space, the workforce must possess technical, social, and digital skills, and the workplace has evolved to include remote work, and the wage structure is more flexible, with the possibility of individualised compensation arrangements.

I 4.0 began and is ongoing at the start of the 21st century and this era is distinguished by the incorporation of cutting-edge technologies such as the Internet of Things (IoT), artificial intelligence (AI), and big data analytics into production processes and this has led to the progress of intelligent factories that can promote autonomous communication and decision-making. I 4.0 has the potential to revolutionise the manufacturing industry by increasing efficiency, decreasing costs, and facilitating mass customisation.

Recent years have seen the advent of the notion of I 5.0, which builds on the achievements of I 4.0 and it is characterised by the incorporation of cutting-edge technologies with human skills in order to produce personalised goods and services that meet societal requirements. Adopting technologies such as AI, robotics, and augmented reality can augment the capabilities of human employees in I 5.0. Human-centred manufacturing, which emphasises the value of human input and the creation of meaningful work, is the focus of I 5.0. I 5.0 has the potential to stimulate economic expansion while addressing social and environmental issues.

With each industrial revolution, workplace safety has increased (Agarwal et al., 2023). In Industry 1.0, production was prioritised at the expense of worker safety. Industry 2.0 was characterised by the introduction of machinery and an emphasis on enhancing workplace safety. With the implementation of safety protocols and training programmes, Industry 3.0 further enhanced workplace safety.

The interface between people and gadgets has also evolved with each industrial revolution. In the First Industrial Revolution, manual labour was the norm. Industry 2.0 introduced assistance from machines, with humans operating machines to execute tasks. Industry 3.0 introduced automated processes that are controlled by humans (Papcun et al., 2018) Human-augmented intelligence is the norm in I 4.0, with workers using sophisticated technologies to enhance their productivity and decision-making.

In conclusion, the evolution from I 1.0 to I 4.0 has resulted in substantial enhancements to working conditions, job security, and workplace safety. The workforce has shifted from unskilled labour in Industry 1.0 to workers with multiple technical, soft, and digital abilities in I 4.0. Continuous training and development are now required for workers to remain pertinent in a rapidly evolving technological landscape.

4.2.3 Comparison of Industry 1.0, 2.0, 3.0, 4.0, and 5.0

Table 4.1 highlights an exhaustive comparison of Industry 1.0, 2.0, 3.0, 4.0, and 5.0 across various aspects (Yin et al., 2018, Koh et al., 2019, Xu et al., 2021, Konstantinidis et al., 2022, Mathur et al., 2022, Joseph, 2023).

Table 4.1 outlines the major characteristics of industrial development during various time periods. The first period, Industry 1.0, was characterised by the mechanisation of manual labour predominantly through the steam engine. The personnel consisted of artisans and craftsmen with specialised training. With a semi-skilled labour force, Industry 3.0 witnessed the automation of routine duties via electronics and automated production. "cyber-physical systems" and automation, driven by data and connectivity, automated production, and the amalgamation of progressive technologies into the employees increased productivity and efficiency in I 4.0, and knowledgeable workers and qualified technicians became the standard. The emphasis on personalisation, customisation, and innovative products and services has led to exceptionally high levels of efficiency, adaptability, and customisation. I 5.0 is characterised by advanced and real-time data collection and ubiquitous connectivity, with minimal environmental impact. The change from Industry 1.0 to I 5.0 has been accompanied by significant workforce challenges, including long work hours, employment displacement, a skills gap, rapid technological change, and mental health issues.

4.2.4 Comparison between Industry 4.0 and 5.0

While the progression from I 1.0 to 5.0 has been significant, researchers have focussed more on I 4.0 to I 5.0 as it is from the present to the future. Some of the significant differences between I 4.0 and 5.0 are presented in Table 4.2.

Table 4.2 indicates that I 4.0 and I 5.0 are two distinct phases of the manufacturing industry's evolution, each characterised by distinct technological advancements and focuses. "Cyber-physical systems, the IoT, AI, big data, and cloud computing" are the primary technologies that drive I 4.0 (Ammar et al., 2021). These technologies facilitate real-time "machine-to-machine and machine-to-human communication", allowing for greater control and optimisation of production processes. I 4.0 aims to construct a

Table 4.1 Comparison of Industry 1.0 to 5.0

Aspect	Industry 1.0	Industry 2.0	Industry 3.0	Industry 4.0	5.0
Time Period	Late 18th century to mid-19th century	Late 19th century to mid-20th century	Mid-20th century to late 20th century	Late 20th century to present	Present and future
Main technology	Mechanical power	Mass production and assembly line	Electronics and automated production	Cyber-physical systems and automation	Integration of advanced technologies with workforce
Main driver	Steam engine	Electricity	Computers and automation	Data and connectivity	Artificial intelligence and the Internet of Things
Workforce	Skilled artisans and craftsmen	Unskilled labour force	Semi-skilled labour force	Knowledgeable workers and skilled technicians	Highly skilled and adaptive workforce
Focus	Mechanisation of manual tasks	Efficiency and mass production	Automation of routine tasks	Personalisation and customisation	Integration of humans and machines
Products	Textiles, iron, coal, steam engines	Cars, appliances, consumer goods	Electronics, computers, aerospace	Software, e-commerce, online services	Advanced products and services
Efficiency	Low	Medium	High	Very high	Extremely high
Flexibility	Low	Low to medium	High	Very high	Extremely high
Customisation	Low	Low	Medium	High	Very high
Data Collection	Minimal	Minimal	Basic	Extensive	Advanced and real-time
Connectivity	Minimal	Minimal	Basic	Advanced	Advanced and pervasive
Environmental impact	High	Medium	Medium	Low to medium	Low
Workforce challenges	Long working hours, dangerous working conditions	Repetitive tasks, low job satisfaction	Job displacement, need for retraining	Skills gap, need for lifelong learning	Rapid technological change, job insecurity, privacy concerns, mental health issues

Table 4.2 Comparison between I 4.0 and I 5.0

Aspect	Industry 4.0	Industry 5.0	Citation
Focus	Automation and electronic data interchange	Inclusion of AI, IoT, and robotics	Massaro (2021), Tiwari et al. (2022), Akundi et al. (2022)
Technology	"Cyber-physical systems, IoT, AI, cloud computing and big data"	AI, IoT, robotics, additive manufacturing, and advanced materials	Ammar et al. (2021), Khorasani et al. (2022)
Goal	Efficiency and productivity gains through automation	Humans and machines Collaboration	Ozkeser (2018), Leng et al. (2022)
Human– Technology Interaction	Workers monitor and control machines	Machines and humans work collaboratively	Huang et al. (2022), Maddikunta et al. (2022)
Impact on Workforce	A few jobs are substituted by automation	Emerging new jobs and roles	Nahavandi (2019), Battini et al. (2022)
Education and Skills	Emphasis on STEM education and digital skills	Greater emphasis on creativity, problem-solving, and adaptability skills	Saniuk and Grabowska (2022), Rathod and Agrawal (2023), Shukla and Singh (2023)

production system that is highly automated, interconnected, and able to respond rapidly to shifting market demands.

I 4.0 is characterised by human–machinery collaboration where employees monitor and control machines to ensure their correct and efficient operation. Apprehensions have been high about the possible effect of the technologies on the workforce due to the introduction of automation, which results in the replacement of some positions by machines (Battini et al., 2022). The drive of I 5.0 is to progress a more personalised and human-centred production system capable of producing goods that are tailored to the requirements and choices (Ozkeser, 2018). In I 5.0, humans and algorithms collaborate to develop innovative solutions to complex problems such as sustainability and social responsibility. I 5.0 is characterised in terms of human–technology interaction by machines and humans cooperating to accomplish a common goal. This collaboration necessitates a workforce with both technical and emotional skills, including creativity, problem-solving, and adaptability.

4.2.5 Employment in I 5.0

Table 4.3 enlightens 5Ws and 1 H regarding employment in Industry 5.0.

Table 4.3 5W and H analysis table for employment in I 5.0

Aspect	Question	Answer	Citation
When	When will I 5.0 impact employment?	I 5.0 is already having an effect on employment, with the integration of "cutting-edge technologies" such as IoT, AI and automation robots creating new job positions and requiring new skills from employees. In the coming years, the full impact on employment is anticipated to persist.	• Huang et al. (2022) • Maddikunta et al. (2022) • Leng et al. (2022)
What	What is I 5.0?	I 5.0 is the incorporation of advanced technologies such that there is a sense of human–machine collaboration in various processes. This integration is generating new employment opportunities and altering the nature of labour.	• Østergaard (2018) • Xu et al. (2021) • Nahavandi (2019)
Why	Why is I 5.0 important for employment?	I 5.0 is significant for employment because it generates new job openings in fields like data analysis, robotics maintenance, and AI. It is also varying the work, requiring the expansion of new jobs in the market.	• Aheleroff et al. (2022) • Konstantinidis et al. (2022) • Raja Santhi and Muthuswamy (2023)
Whom	Whom does I 5.0 impact?	I 5.0 has consequences for both employers and employees. To remain competitive, employers must acclimatise to new technologies and think about emerging business models, while employees must acquire new skills to remain employable in a changing job market.	• Jafari et al. (2022) • Paschek et al. (2019) • Tiwari et al. (2022)
Where	Where is I 5.0 impacting employment?	I 5.0 has an effect on employment in almost all industries. Countries and regions invest in the adoption of I 5.0 technologies, to reap the benefits of the various technologies.	• Maddikunta et al. (2022) • Javaid et al. (2020) • Huang et al. (2022)
How	How is I 5.0 changing employment?	I 5.0 is transforming employment by automating routine and repetitive tasks, creating new job roles in areas such as data analysis and automation maintenance, and requiring workers to acquire new skills such as creativity, problem-solving, and adaptability.	• Mitchell and Guile (2022) • Hussain et al. (2023) • Broo et al. (2022)

4.3 CHARACTERISTICS OF I 5.0

As I 5.0 is a comparatively new notion, its characteristics are still being defined and refined. However, based on current discussions and research, the following are some of the most prominent characteristics are as follows (Figure 4.2).

- **Human–machine collaboration:** The alliance between humans and machinery is one of the defining characteristics of I 5.0. Instead of viewing machines as a replacement for human labour, to accomplish shared goals (Pizoń & Gola, 2023), this strategy acknowledges the assets of both humans and machines and aims to create a synergistic effect that improves performance and outcomes. Humans and machines collaborate to attain higher productivity and innovation in I 5.0. "The vision of I 5.0 is to achieve symbiosis between humans and machines" (Leng et al., 2022).
- **Customisation and flexibility:** I 5.0 emphasises customisation and flexibility. Instead of relying on mass production techniques, I 5.0 aims to produce smaller batches of customised products that are customised to the requirements and preferences of specific customers (Destouet et al., 2023). This strategy necessitates a workforce that is more adaptable, agile and able to rapidly adapt to shifting customer demands and production requirements. I 5.0 emphasises the objective of satisfying the varying requirements of customers. "I 5.0 aims to deliver individualised products and services to customers" (Adel, 2022).
- **Sustainability:** Sustainability is an additional important characteristic of I 5.0 wherein it aims to accomplish sustainable production by optimising the use of resources, minimising waste, and minimising the manufacturing process' carbon footprint. This strategy emphasises

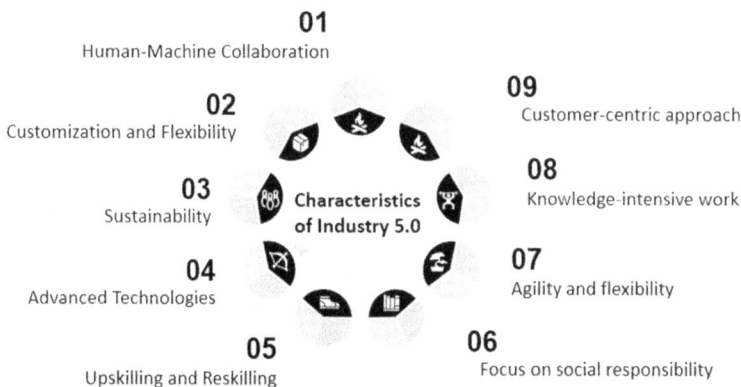

01
Human-Machine Collaboration

02
Customization and Flexibility

03
Sustainability

04
Advanced Technologies

05
Upskilling and Reskilling

Characteristics of Industry 5.0

09
Customer-centric approach

08
Knowledge-intensive work

07
Agility and flexibility

06
Focus on social responsibility

Figure 4.2 Characteristics of I 5.0.

social and environmental responsibility and drives to generate a more impartial distribution of wealth by ensuring that the potential benefits of technological adoption are shared among all stakeholders (Ghobakhloo et al., 2022). "I 5.0 represents the convergence of advanced technologies with the physical world". "I 5.0 puts a strong emphasis on sustainability and green production" (Zeb et al., 2022). To achieve sustainability in I 5.0, however, all stakeholders, including manufacturers, policymakers, and consumers must collaborate (Kasinathan et al., 2022).

- **Advanced technologies:** I 5.0 relies on sophisticated technologies to accomplish its objectives which facilitate increased automation, enhanced efficiency, and enhanced human–machine interaction and collaboration. In the literature, the significance of advanced technologies in I 5.0 has been extensively discussed. According to Adel (2022), the goals of I 5.0 significantly rely on "AI and machine learning". Utilising these technologies permits manufacturers to acquire and analyse vast quantities of data in real time, thereby improving the processes and enhancing quality control. Similarly, Tiwari et al. (2022) assert that I 5.0 necessitates the integration of robotics and IoT in order to create highly connected and automatic manufacturing.

- **Upskilling and reskilling:** I 5.0 necessitates a highly competent staff that can work with advanced technologies, analyse the data obtained, and make quick decisions. Consequently, one of the key aspects of I 5.0 is the emphasis on upskilling workers to acclimatise to the evolving nature of work (Gagnidze, 2022). This approach acknowledges that the skills required for I 5.0 may differ from those required for previous industrial revolutions and aims to prepare workers for future employment. Upskilling and reskilling programmes will play a dynamic role in facilitating the transition of workers into the new occupations and roles created by I 5.0. These programmes must emphasise the development of skills in areas such as data analytics, digital literacy, and programming.

- **Focus on social responsibility:** I 5.0 emphasises the social responsibility of businesses and the necessity of generating a positive social impact (Ivanov, 2023). "I 5.0 requires businesses to prioritise their social responsibilities and have a positive impact on society". I 5.0 places a heavy emphasis on corporate social responsibility (CSR) environmental concerns (Akundi et al., 2022). Companies implementing I 5.0 are expected to consider the impact of their operations on their stakeholders, and the community at large. The emphasis on social responsibility in I 5.0 is also expected to improve companies' reputations, attract customers, and increase their competitive advantage (Cillo et al., 2022).

- **Agility and flexibility:** I 5.0 requires businesses to be agile to acclimatise to rapid technological change and fluctuating market conditions. "I 5.0 requires companies to be agile and flexible in order to remain competitive in a rapidly changing market" (Akkaya & Ahmed, 2022).

I 5.0 emphasises the significance of agility and flexibility in organisations through the incorporation of cutting-edge technologies and systems. These systems enable real-time data analysis, which expedites decision-making and improves process efficiency. I 5.0's increased flexibility and agility can reduce lead times, lessen expenses, and enhance customer satisfaction (Huang et al., 2022). In addition, the ability to quickly and effectively adapt to varying market factors is becoming an increasingly important success factor for businesses.

- **Knowledge-intensive work**: I 5.0 places greater emphasis on knowledge-intensive work and requires highly trained and adaptable employees. "I 5.0 requires a workforce that is highly skilled and capable of adapting to new technologies and processes" (Mitchell & Guile, 2022). I 5.0 emphasises knowledge-intensive work, which necessitates highly competent employees who can manage complex systems and processes, and knowledge-intensive work requires technical proficiency and problem-solving skills as emerging technologies and processes are implemented in the workplace; I 5.0 also emphasises the significance of continuous learning and upskilling (Rodríguez-Garcia et al., 2023). This emphasis on knowledge-intensive work provides new opportunities for highly skilled workers while increasing manufacturing sector productivity and efficiency; nonetheless, it poses a challenge for businesses to attract and retain skilled employees and may necessitate investments in training and development programmes.
- **Customer-centric approach**: I 5.0 involves a customer-first approach to production and operations, with an emphasis on delivering customer value and "I 5.0 requires companies to adopt a customer-centric approach and deliver products and services that provide value for customers" (Barata & Kayser, 2023). I 5.0 emphasises a customer-centric approach in which products are designed and manufactured based on customer requirements and preferences and this is accomplished by employing cutting-edge technologies such as artificial intelligence and the IoT, which enable real-time data collection and analysis to better comprehend consumer behaviour and preferences. I 5.0 also enables greater product customisation and personalisation, allowing businesses to meet the specific needs and preferences of each consumer (Alexa et al., 2022), and by emphasising a customer-centric approach, I 5.0 can assist businesses in enhancing customer satisfaction, fostering brand loyalty, and gaining a competitive advantage in the market.

The available literature on the I 5.0 characteristics is expanding. A study by Pizoń and Gola (2023), for instance, highlights the significance of "human–machine" collaboration and customisation in I 5.0 and the study emphasises that I 5.0 aims to create a manufacturing process that is more adaptable to fluctuating consumer demands and production requirements. I 5.0 aims to accomplish sustainable production by minimising waste and reducing the environmental impact of firms.

Carayannis and Morawska-Jancelewicz (2022) highlighted the significance of sustainability and social responsibility in I 5.0. I 5.0 seeks to establish a more equitable distribution of wealth by ensuring that the goodness of technological adoption is distributed to all stakeholders; additionally, the study emphasises the significance of upskilling and reskilling workers in order to change the nature of work in I 5.0.

Moreover, a study by Ma et al. (2022) emphasised the significance of innovative technologies and these technologies facilitate increased automation, enhanced efficiency, and enhanced human–machine collaboration. Additionally, the study emphasises the need for a highly competent labour force capable of operating and maintaining these technologies.

Literature suggests that I 5.0 emphasises collaboration, customisation, sustainability, advanced technologies, and upskilling and reskilling. These characteristics represent a shift from viewing manufacturing as a solely technological process to one that emphasises the importance of human labour in achieving maximum productivity, efficiency, and sustainability.

4.4 INDUSTRY 5.0'S EFFECT ON EMPLOYEES

The rise of I 5.0 is likely to have a substantial influence on manufacturing employees. Some of the prospective employee impacts of I 5.0 are discussed along with supporting literature.

- **Upskilling and reskilling:** As previously stated, I 5.0 requires a highly skilled staff capable of working with advanced technologies. This necessitates skill upgrades and retraining for many workers to remain employable. According to a study by Zizic et al. (2022), employees in I 5.0 will need an assortment of skills, including technical knowledge, data analysis, problem-solving abilities, and life-long learning.
- **Changes in job roles:** I 5.0 will likely result in changes to job duties, with some tasks being automated and others requiring new skills. According to a study by Pacher et al. (2023), employees will need to be more flexible and adaptable in I 5.0. The research indicates that employees will need to be able to switch between duties and collaborate with machines.
- **Improved working conditions:** The potential exists for I 5.0 to enhance the workplace atmosphere for employees. For instance, advanced technologies can lessen the risk of workplace stress and injuries and allow employees to complete tasks more effectively. According to Zizic et al. (2022), I 5.0 can also generate new employment opportunities in maintenance, programming, and data analysis as well as provide better working conditions for employees.
- **Job displacement:** Despite the possible benefits of I 5.0, there is a danger that some employees will be replaced by automation. I 5.0 could contribute to the displacement of low-skilled workers who perform

repetitive tasks (Ghobakhloo et al., 2022). The research focuses on the need for policies and programmes to assist displaced workers in finding new employment.

I 5.0 is likely to have complex and varied effects on employees. It has the scope to improve working conditions and establish innovative employment opportunities, but it also poses a threat of job displacement for some workers. Therefore, employees impacted by I 5.0 require policies and programmes to support upskilling and reskilling, job transition, and social protection.
   Table 4.4 specifies Industry 5.0's effect on employees.

Table 4.4 I 5.0's effect on employees

Point	Description	References
I	Upskilling and reskilling become essential to remain employable in I 5.0.	Zizic et al. (2022), Mukherjee et al. (2023)
2	The implementation of novel technologies leads to the displacement of low-skill jobs, which may negatively impact certain segments of the workforce.	Ghobakhloo et al. (2022), Battini et al. (2022)
3	I 5.0 enables the creation of new job roles that require skills in AI, robotics, and data analytics.	Kasinathan et al. (2022), Huang et al. (2022)
4	Association between humans and machines becomes more prevalent, which may require new ways of working and interacting with technology.	Tiwari et al. (2022), Maddikunta et al. (2022)
5	The use of smart technologies may improve working conditions, such as reducing physical strain and minimising exposure to hazardous materials.	Battini et al. (2022), De Giovanni (2023)
6	I 5.0 emphasises a customer-centric approach, which may require employees to develop new skills in various domains such as design thinking and customer service.	Mansurali et al. (2023), Joglekar et al. (2023)
7	Flexibility and adaptability become critical skills for employees in I 5.0, as companies are expected to be responsive to changes in market demand.	Modgil et al. (2023), Raja Santhi and Muthuswamy (2023)
8	The usage of data analytics and AI may lead to increased surveillance of employees in the workplace.	Rožanec et al. (2022), Zizic et al. (2022)
9	The development of a sustainable process is a key dimension of I 5.0, which may require employees to work with a greater emphasis on sustainability and environmental impact.	Ghobakhloo et al. (2022), Saniuk and Grabowska (2022)
10	The implementation of I 5.0 may need significant modifications in organisational culture and management practices, which may impact employee engagement and job satisfaction.	Maddikunta et al. (2022), Panday and Kaur (2023)

Source: Authors' own work.

4.5 IMPACT OF INDUSTRY 5.0 ON EMPLOYMENT ACROSS VARIOUS INDUSTRIES

Table 4.5 summarises the impact of I 5.0 on employment in ten prominent industries, along with citations for the supporting literature.

It is anticipated that the introduction of novel technologies will automate repetitive and monotonous activities, allowing workers to concentrate on more creative and complex tasks. It is also anticipated that the technology

Table 4.5 Impact of I 5.0 on employment in prominent industries

Industry	Impact on employment	Authors
Automotive	Higher demand for skilled workers to operate and maintain advanced machinery, but lower demand for assembly-line workers due to automation	Patel et al. (2023), Rastogi et al. (2022)
Aerospace	Increased need for engineers and technicians to design and maintain advanced aircraft and space vehicles, but lower demand for manual labour in manufacturing and assembly	Massaro (2021), Sharma and Arya (2022)
Electronics	Higher demand for engineers and technicians to develop and maintain advanced electronics, but lower demand for assembly-line workers due to automation	Qahtan et al. (2022)
Pharmaceuticals	Increased need for researchers and engineers to develop advanced medicines and therapies, but lower demand for manual labour in manufacturing and packaging	Sharma et al. (2022)
Food and Beverage	Higher demand for engineers and technicians to develop and maintain advanced production and packaging equipment, but lower demand for manual labour in processing and assembly	Saptaningtyas and Rahayu (2020)
Construction	Increased use of automation and robotics to streamline building processes and increase efficiency, leading to lower demand for manual labour	Kasinathan et al. (2022)
Healthcare	Higher demand for skilled workers to operate and maintain advanced medical equipment and robots, but lower demand for manual labour in certain aspects of patient care	Javaid et al. (2020)
Energy	Increased need for engineers and technicians to develop and maintain advanced renewable energy technologies, but lower demand for manual labour in traditional energy production	Carayannis et al. (2021)

integration will create new employment roles, such as data analysts, robot technicians, and AI engineers.

4.6 KEY COMPETENCIES FOR CAREER DEVELOPMENT AND GROWTH FOR PEOPLE IN INDUSTRY 5.0

As previously discussed, I 5.0 necessitates a highly trained labour force capable of operating and maintaining progressive technologies and some of the skills required for personnel in I 5.0, supported by relevant literature (Figure 4.3):

1. **Technical expertise**: Employees must have technical expertise in areas such as automation, AI and the IoT to work in I 5.0. According to a study by Mitchell and Guile (2022), employees in I 5.0 will need to be able to program and operate advanced machines, as well as analyse and interpret the data produced by these machines.
2. **Problem-solving skills**: I 5.0 requires employees to be able to solve complex manufacturing process-related problems. According to a study by Neubert et al. (2015), I 5.0 necessitates a manufacturing process that is more adaptable to altering customer demands and production requirements, and in order to maximise production efficiency, staff should analyse data and make well-informed choices.
3. **Collaborative skills**: I 5.0 necessitates that employees collaborate with machinery and other workers and Leng et al. (2022) report that I 5.0 seeks to establish a more equitable distribution of wealth by ensuring that the benefits of technological advancements are shared with all stakeholders and this necessitates that employees be able to effectively work in teams and communicate with machinery and other employees.

Technical Expertise	Problem-Solving Skills	Collaborative Skills
Soft Skills	Leadership Skills	Project Management Abilities
Continuous Learning	Customer Service	Critical Thinking
	Ethics and Social Responsibility	

Figure 4.3 Key competencies for career development and growth for people in I 5.0.

Source: Authors' own source.

4. **Soft skills:** Employees in I 5.0 must possess a variety of soft skills, including communication, creativity, and adaptability. I 5.0, according to a study by Chin (2021), requires workers to be more flexible and adaptable in their job responsibilities and this necessitates that employees be able to switch between duties and collaborate with machines and other employees.

5. **Leadership skills:** I 5.0 requires employees to have leadership skills in order to manage teams, drive innovation, and effectively implement changes, and a study by Bakir and Dahlan (2022) indicated that I 5.0 necessitates an innovative culture in which employees are empowered to make decisions and take risks and this implies that employees must be capable of inspiring and motivating others, communicating a vision, and setting an example.

6. **Project management abilities:** In order to manage complex initiatives involving multiple stakeholders and technologies, I 5.0 employees must have solid project management abilities, and a study by Cakmakci (2019) indicated that I 5.0 necessitates a more collaborative approach to project management, in which employees must collaborate closely with machinery and other workers to complete projects on time and within budget.

7. **Continuous learning:** I 5.0 requires employees to be committed to continuous learning in order to keep up with swiftly evolving technologies and processes, and, according to a study by (Carayannis et al., 2021), I 5.0 necessitates a culture of learning and development in which employees are encouraged to acquire new skills and knowledge and this implies that employees must be proactive in their pursuit of learning opportunities and skill upgrades.

8. **Customer service:** I 5.0 requires employees to have a customer-centric mindset to satisfy the changing needs and demands of customers, and, according to a study by Al Faruqi (2019), I 5.0 necessitates a customer-centric approach in which employees must be able to anticipate customer requirements and provide customised solutions, therefore, employees must possess strong communication and interpersonal skills in order to establish and maintain relationships with consumers.

9. **Critical thinking:** I 5.0 requires employees to possess critical thinking skills in order to analyse data, identify problems, and create solutions and a study by Javaid and Haleem (2020) indicates that I 5.0 requires employees to make decisions in a more analytical and data-driven manner and this necessitates that employees be able to analyse data, recognise patterns and trends, and devise strategies to maximise production efficiency.

10. **Ethics and social responsibility:** Employees in I 5.0 must have an inclination towards ethics and social responsibility and according to a study by Zhang et al. (2021), I 5.0 necessitates a more responsible

and sustainable approach to manufacturing, in which workers must consider the social and environmental impact of their work and this implies that employees must have a strong ethical conscience and be able to make decisions that are consistent with the company's values and mission.

4.7 SKILLS REQUIRED FOR EMPLOYEES IN INDUSTRY 5.0

Employees in I 5.0 require a range of technical, problem-solving, collaborative, and soft skills and as all sectors continue to progress continuously, it becomes imperative for employees to continuously upskill and reskill to remain employable.

To flourish, employees must possess a wide variety of skills and these abilities consist of critical thinking, problem-solving, communication, leadership, adaptability, creativity, time management, collaboration, emotional intelligence, and decision-making.

Critical thinking is a fundamental skill that enables employees to objectively analyse information and make sensible decisions based on this analysis and this skill is essential for identifying problems, creating solutions, and evaluating results and complex technological systems in I 5.0 require continuous attention and problem-solving to optimise performance and prevent breakdowns, making problem-solving an important skill.

Collaboration between humans and machinery, as well as managing teams with disparate skill sets and backgrounds, requires effective communication, and effective communication skills are crucial in I 5.0, where the emphasis is on developing and implementing new and emerging technologies.

In I 5.0, leadership is also an essential talent and it is necessary to motivate and manage teams towards a common objective, to manage diverse skill sets and backgrounds, to drive innovation, and to respond to industry-specific changes.

In I 5.0, where technology is perpetually evolving, adaptability is a crucial talent. Employees must be able to rapidly adopt new tools and procedures in order to maintain competitiveness. Another essential talent for driving innovation and developing new technologies and products is creativity.

Emotional intelligence is essential for team management and conflict resolution in the workplace. It allows employees to recognise and control their own emotions as well as those of others. In I 5.0, decision-making skills are essential for evaluating new technologies and processes and identifying areas for enhancement in existing systems.

The skills can be classified into three basic skills, namely, personal, interpersonal, and business skills as shown in Figure 4.4.

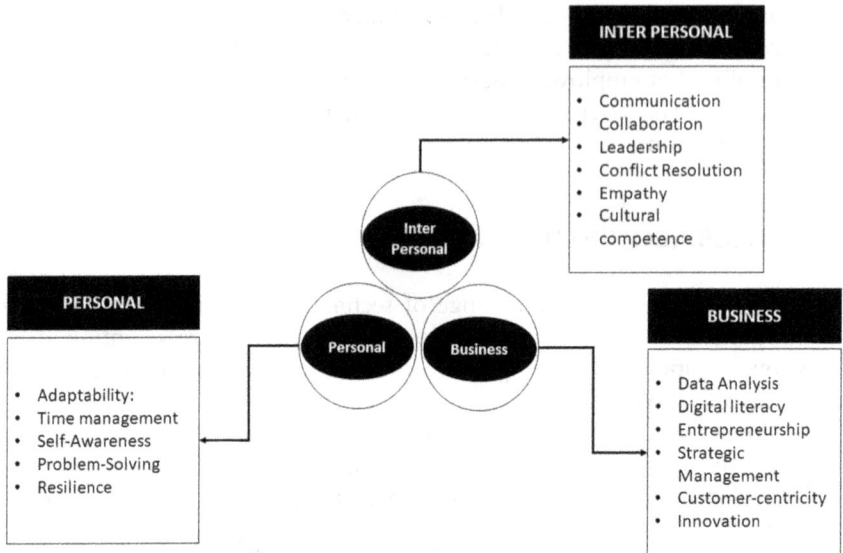

Figure 4.4 Skills required for employees in I 5.0.

Source: Authors' own source.

4.7.1 Personal skills

Personal skills refer to an individual's capacity to effectively manage themselves. Adaptability, time management, self-awareness, problem-solving, and resiliency are among these skills. Adaptability refers to the characteristics of being receptive to change, learning rapidly, and being resourceful and time management involves establishing objectives, establishing priorities, and meeting deadlines. Self-awareness is the ability to recognise one's own strengths and weaknesses, to seek feedback, and to use that feedback to develop while problem-solving requires data analysis, critical reasoning, and the development of inventive solutions and resilience involves managing stress, recovering from setbacks, and maintaining optimism (McKinsey Global Institute, 2017, Mitchell & Guile, 2022, Neubert et al., 2015, Saniuk & Grabowska, 2022).

- **Adaptability:** I 5.0 requires individuals to be open to change and fast learners, so adaptability is a crucial personal skill, and success requires the ability to adapt to new technologies and processes, as well as diverse work environments, and also essential are resourcefulness and the ability to devise original solutions to issues (Ivanov, 2023).
- **Time management:** Time management is another essential skill for the I 5.0 era and setting objectives, prioritising duties, and meeting deadlines are essential for productive and efficient work, and time

management also requires the ability to juggle numerous duties and responsibilities (Tabuenca et al., 2022).

- **Self-awareness:** Self-awareness is the capacity to recognise one's own assets and weaknesses, as well as to solicit feedback from others, and this competency is essential for both personal and professional development, and individuals are also able to identify areas for improvement and take measures to enhance their skills (London et al., 2023).
- **Problem-solving:** Problem-solving is an essential skill for I 5.0, as it requires data analysis, critical reasoning, and innovative solutions, and this ability is essential for identifying and resolving complex problems in the workplace, as well as creating innovative solutions to meet business requirements (Hollenstein et al., 2022).
- **Resilience:** Resilience is the capacity to manage stress, recover from setbacks, and maintain optimism in the face of adversity, and managing the high-pressure demands of I 5.0 work environments, as well as adapting to change and overcoming obstacles, requires this skill while developing resilience also requires self-care practises, such as exercise, mindfulness, and reaching out to colleagues and mentors for assistance (Mangalaraj et al., 2023).

4.7.2 Interpersonal skills

Interpersonal skills refer to a person's capacity to interact effectively with others and these competencies consist of communication, collaboration, leadership, conflict resolution, empathy, and cultural competence. Communication requires precise expression and attentive listening, and the provision and reception of feedback and collaboration requires effective teamwork, respect for diversity, and relationship building while leadership requires inspiring and motivating others, making decisions, and delegating responsibilities. Resolving a conflict entails addressing differences of opinion, locating common ground, and negotiating, and compassionate empathy involves comprehending and responding to the perspectives of others. Understanding diverse cultures and customs, working across borders, and valuing diversity are all components of cultural competence (McKinsey Global Institute, 2017, Mitchell & Guile, 2022, Neubert et al., 2015, Saniuk & Grabowska, 2022).

- **Communication:** This skill requires the ability to communicate ideas plainly and effectively, to actively listen to others, and to provide constructive feedback; individuals must develop effective communication skills to avoid misunderstandings and ensure that ideas are conveyed accurately (Cho & Lee, 2022).
- **Collaboration:** Collaboration entails working effectively with others to achieve a common objective and this requires valuing diversity,

fostering relationships, and recognising the various perspectives and experiences of team members (Olson & Johnson, 2023).

- **Leadership:** Leadership entails inspiring and motivating others, making intelligent decisions, and delegating tasks efficiently. Leaders must possess effective communication skills, the capacity to influence others, and a strategic perspective (Akundi et al., 2022).
- **Conflict resolution:** Conflict resolution requires the ability to address disagreements constructively, identify common ground, and negotiate a solution that works for all parties and this skill is essential in business for maintaining positive relationships and preventing conflicts from escalating (Rathod & Agrawal, 2023).
- **Empathy:** The capacity to comprehend and respond to the perspectives and emotions of others and it is essential for establishing relationships and collaborating effectively with diverse groups (Mehdiabadi et al., 2022).
- **Cultural competence:** Cultural competence refers to the ability to work effectively with people of various cultures and backgrounds and this includes comprehending diverse customs and norms, valuing diversity, and communicating effectively across cultural barriers (Whewell et al., 2022).

4.7.3 Business skills

Business skills refer to an individual's capacity to manage and function effectively in a business setting and these abilities include data analysis, digital literacy, entrepreneurship, strategic management, and a focus on the consumer while data analysis entails accumulating and interpreting information and making decisions based on that analysis the components of digital literacy are data analysis, computing, and cybersecurity. Entrepreneurship involves recognising opportunities, taking risks, and being innovative, and the components of strategic management are planning, budgeting, and project management. User experience, customer service, and relationship management are all components of customer-centricity (McKinsey Global Institute, 2017, Mitchell & Guile, 2022, Neubert et al., 2015, Saniuk & Grabowska, 2022).

- **Data analysis:** The ability to collect, analyse, and interpret data is crucial for making sound business decisions and utilising statistical and computational tools to identify trends, patterns, and insights that can guide strategic decision-making (Hollenstein et al., 2022).
- **Digital literacy:** Employees must have a fundamental understanding of technology, such as computing and cybersecurity, as businesses become increasingly digital and digital literacy entails the ability to use and comprehend digital tools and technologies, as well as to recognise and mitigate cybersecurity risks (Mitchell & Guile, 2022).

- **Entrepreneurship:** Entrepreneurship entails recognising opportunities, taking calculated risks, and innovating to develop new products, services, or enterprises and it requires creativity, strategic thought, business acumen, and risk management skills (Paschek et al., 2019).
- **Strategic management:** Strategic management entails the capacity to effectively plan, budget, and manage projects to attain business objectives and it requires a comprehensive comprehension of the business environment and the ability to identify and prioritise key objectives, allocate resources, and manage risks (Kasinathan et al., 2022).
- **Customer-centricity:** Consumer-centricity entails placing the consumer at the centre of all business decisions and it requires an in-depth knowledge of customer requirements and preferences, as well as the ability to design and deliver products and services that satisfy those needs and this ability also entails managing customer relationships to increase customer loyalty and improve the overall customer experience (Wulf et al., 2017).
- **Innovation:** Innovation is the process of identifying new concepts, products, or services that have the potential to add value to a business and it requires a combination of originality, market savvy, and strategic thought, as well as the ability to manage risk and uncertainty while skills in innovation can be used to create new products and services, enhance business processes, and generate new revenue streams (Aslam et al., 2020).

In conclusion, I 5.0 requires its employees to possess a wide variety of skills. For success in the industry, it is essential to be able to think critically, solve problems creatively, communicate effectively, lead teams, adapt to change, manage time, collaborate with others, possess emotional intelligence, and make informed decisions.

4.8 CHALLENGES FACED BY EMPLOYEES IN INDUSTRY 5.0

I 5.0 is characterised by the incorporation of advanced technologies to coexist with the workforce posing some challenges to the employees. Here are ten challenges encountered by I 5.0 employees:

1. **Rapid technological changes:** Keeping up with the continuous changes and advancements in technology is one of the greatest tests faced by employees in I 5.0 and a significant majority of the core skills required for current employment will change in the future, indicating the need for continuous upskilling and reskilling (Leng et al., 2022).
2. **Upgrading skills:** With the increasing use of advanced and new/ updated technologies in I 5.0, employees are required to perpetually

upgrade their skills and according to a report by Deloitte, the workforce must acquire cognitive abilities, social and affective skills, and advanced technological skills to support technological advancements (Deloitte, 2017).

3. **Job insecurity:** Automation and robotics are progressively replacing human labourers, resulting in job losses and creating job insecurity for workers and a McKinsey Global Institute report states that automation in the manufacturing and service sector could displace up to 800 million positions worldwide by 2030 (McKinsey Global Institute, 2017).

4. **Privacy and security concerns:** With the expanding use of technology, employees run the risk of their personal information being compromised or misappropriated and cybersecurity threats are on the rise, and organisations must take precautions to protect sensitive data (Raja Santhi & Muthuswamy, 2023).

5. **Mental health concerns:** The integration of technology in the workplace can increase employee stress, anxiety, and exhaustion and according to the National Institute for Occupational Safety and Health, excessive technology use in the workplace was associated with greater psychological distress (Padma et al., 2015).

6. **Resistance to change:** Staff and employees may be resistant to the changes introduced by new technologies or work processes, resulting in decreased productivity and efficiency and organisations must take steps to effectively manage change, including providing employees with training and support (Moraru and Popa, 2021).

7. **Intergenerational divide:** The use of advanced technologies may create a chasm between older and younger employees with varying degrees of technological proficiency (Elangovan, 2021) and organisations must take steps to close the digital divide and guarantee that all employees have access to training and support.

8. **Collaboration and communication:** The increased use of technology may reduce face-to-face communication, resulting in communication gaps and the possibility of misinterpretation and organisations must promote effective communication and collaboration among employees, utilising technology as necessary (Ivanov, 2023).

9. **Ethical concerns:** The use of advanced technologies may raise ethical issues regarding their impact on society, employment, and the environment and organisations must be cognisant of these issues and ensure that technology is utilised ethically and responsibly (Longo et al., 2020).

10. **Skills gap:** The fast-evolving technological advancements may create a skills gap between the workforce and the requirements of I 5.0, making it challenging to locate qualified workers, and organisations must take steps to close this divide, including providing employees with training and support (Doyle-Kent & Kopacek, 2020).

4.9 STRATEGIES TO OVERCOME CHALLENGES IN INDUSTRY 5.0

While research indicates that I 5.0 is likely to pose many challenges to employees, there are certain things that can be done by organisations, governments, and individuals (employees) to overcome these challenges which are presented below.

Strategies that can be taken by companies to negate the challenges faced by employees in I 5.0 include the following.

1. **Rapid technological changes**: Companies can invest in regular training and development programs to help employees stay up-to-date with the latest technologies and industry trends and one of the most effective strategies is to constantly relook into the needs of the employees and have a continuous learning program in the organisation to help employees stay up to date with the latest technologies and industry trends and this ensures that employees have the necessary skills to adapt to the evolving technological landscape.

2. **Upgrading skills**: Companies can provide incentives to encourage employees to acquire new skills and offer opportunities for them to attend workshops, conferences, and other training programs, and firms can come out with innovative incentive mechanisms and create a robust eco system in order to facilitate skill upgradation by employees and this encourages continuous learning and development, enabling employees to remain relevant in a rapidly changing technological environment.

3. **Job insecurity**: Companies can focus on retraining employees whose jobs are at risk of being automated, and invest in the development of new roles that complement technology rather than replacing it and companies by doing this can improve the morale of the employees by eliminating the employees job insecurity by upskilling them, and this ensures that employees are equipped with skills that are in high demand and have a promising future.

4. **Privacy and security concerns**: Companies can ensure that they have adequate measures in place to protect employee data, such as implementing strong data encryption, access controls, and regular security audits, and this helps to safeguard employee data and minimise the risk of data breaches and cyber-attacks.

5. **Mental health issues**: Companies can prioritise employee well-being by providing flexible work arrangements, offering mental health support programs, and encouraging a healthy work–life balance, and this helps to reduce stress and improve employee morale and productivity.

6. **Resistance to change**: Companies can involve employees in the decision-making process when implementing new technologies, provide adequate training and support, and communicate the benefits of

change, and this helps to overcome resistance to change and facilitate a smooth transition to new technologies.

7. **Intergenerational divide:** Companies can provide training programs that cater to different levels of technological proficiency, encourage cross-generational mentorship and collaboration, and create a culture that values diversity and companies can involve employees in the decision-making process, provide adequate training and support, and communicate the benefits of change and this helps to overcome resistance to change and facilitate a smooth transition to new technologies.

8. **Collaboration and communication:** Companies can provide employees with tools and platforms that facilitate communication and collaboration, encourage regular face-to-face interactions, and prioritise open and transparent communication, and the openness in communication can certainly assist in reducing conflicts and minimise communication gaps and this helps to foster a collaborative culture and improve team dynamics.

9. **Ethical concerns:** Companies can establish ethical frameworks and guidelines for the use of new technologies and engage in open and transparent dialogues with employees, customers, and other stakeholders about the impact of technology on society, and this helps to ensure that technology is used responsibly and for the benefit of society as a whole.

10. **Skills gap:** Companies can work with educational institutions to bridge the skills gap by providing internships, apprenticeships, and other forms of experiential learning, and invest in developing talent pipelines to attract and retain skilled workers.

4.9.1 Strategies that governments can take to overcome the challenges faced by employees in Industry 5.0

1. Governments can
   - invest in education and training programmes that empower employees with the necessary skills to keep up with the fast advancements in various technologies.
   - provide funding and support for vocational training programmes and encourage initiatives that promote lifelong learning.
   - implement policies and programmes that assist displaced employees, such as unemployment insurance, retraining programmes, and job placement services.
   - create regulations and standards to ensure the preservation of employee data and privacy, as well as penalties for noncompliance.
   - implement policies that inspire a healthy work–life balance, flexibility in the work arrangements, and mental health support programmes.
   - collaborate with businesses to create change management programmes that involve employees and promote innovation across the various segments.

- promote diversity and inclusion in the workforce by encouraging the development of education and training programmes that appeal to various levels of technological proficiency.
- promote the development of communication and collaboration platforms that facilitate dialogue between employers and employees, as well as the use of digital tools for remote work.
- establish ethical guidelines and frameworks that govern the use of new technologies and promote public awareness and engagement on the social and environmental effects of technology.
- collaborate with industry stakeholders to identify skills deficits, develop policies and programmes that support the development of talent pipelines, and invest in various research initiatives.

4.9.2 Individual strategies for overcoming the challenges encountered by employees in I 5.0

1. Individuals should embrace change, be receptive to acquiring new skills, and view obstacles as opportunities to develop and improve.
2. Employees must continually develop their skills and update their knowledge in order to remain pertinent in the industry, given the rapid pace of technological advancement.
3. Embrace technology: Employees must learn to utilise technology to enhance their productivity and work processes.
4. As I 5.0 requires a human touch, employees must concentrate on enhancing their creativity and problem-solving abilities.
5. Collaboration and cross-functional collaboration will become increasingly crucial in I 5.0. Employees in co-existence with machines must be able to work effectively in diverse organisations and possess strong communication and interpersonal skills.
6. Prioritise mental and physical health, as the adoption of technology and automation may increase stress and exhaustion. It is crucial to prioritise mental and physical well-being by taking breaks, engaging in physical activity, and seeking support when necessary.
7. Employees should cultivate an entrepreneurial mindset and search for opportunities to innovate and create new value within their organisations.
8. Seek out mentors: Learning from seasoned colleagues and mentors can help individuals navigate the challenges of I 5.0 by providing valuable insights.
9. Establish a professional network: Establishing and maintaining a network of industry contacts can provide opportunities for professional advancement and development.
10. Keeping abreast of the latest industry trends, best practices, and technological developments can help employees remain competitive in the employment market.

Table 4.6 Strategies to be adopted by different stakeholders to overcome the challenges of I 5.0 towards employment

Organisations	Government	Individuals
Invest in regular training programs	Invest in education	Develop skills and knowledge
Provide incentives for employees to acquire new skills	Support vocational training	Embrace technology
Retrain employees whose jobs are at risk of being automated	Assist displaced employees	Enhance creativity and problem-solving abilities
Ensure adequate measures are in place to protect employee data	Ensure privacy & security	Work effectively in diverse organisations
Prioritise employee well-being	Promote mental health	Prioritise mental and physical health
Involve employees in decision-making	Create change management	Cultivate an entrepreneurial mindset
Provide training programs for technological proficiency	Encourage diversity & inclusion	Seek out mentors
Provide tools that facilitate communication and collaboration	Promote collaboration & communication	Establish a professional network
Establish ethical frameworks and guidelines	Establish ethical guidelines & frameworks	Keep abreast of the latest industry trends and developments
Work with educational institutions to bridge the skills gap	Identify skills deficits	Foster awareness of industry trends

Individuals can overcome the challenges of I 5.0 and position themselves for success in the new era of employment by employing these strategies.

Strategies to be undertaken by different stakeholders to overcome the challenges of I 5.0 towards employment are illustrated in Table 4.6.

4.10 MANAGING HUMAN–ROBOT COLLABORATION IN INDUSTRY 5.0

In the era of I 5.0, robots and AI play an increasingly crucial role in reshaping our lives and workplaces. Workers are now expected to collaborate with robots and other autonomous systems due to the rise of intelligent automation. This co-existence between humans and machines is a key driver of I 5.0, and it has the possibility to change the workplace by enhancing human

1

2

3

Mitigating risks and ensuring safety in human -robot collaboration

Designing effective human-robot teams

Strategies for overcoming communication and coordination challenges

Figure 4.5 Managing human–robot collaboration in I 5.0.

capabilities, boosting productivity, and increasing efficiency (Adel, 2022). However, there are obstacles to the effective collaboration between humans and automata. Humans and robots, for example, have distinct assets and weaknesses. Humans are superior at decision-making, problem-solving, and creative thinking, whereas robots excel at repetitive and routine duties. Therefore, the challenge is to design collaboration systems that capitalise on the strengths of both humans and robots while compensating for their deficiencies. Figure 4.5 highlights managing HRC in Industry 5.0.

A multidisciplinary approach encompassing robotics, artificial intelligence, human factors engineering, and organisational psychology is necessary for the fruitful integration of HRC (Lu et al., 2022). Researchers have witnessed the transformative effects of HRC on industries such as manufacturing, service industry, healthcare, logistics, and agriculture. Humans are still required to perform tasks that require manual dexterity, creativity, and decision-making because robots cannot perform all duties (Skobelev & Borovik, 2017).

4.10.1 Designing effective human–robot teams

In recent years, there has been an increasing interest in designing effective human–robot teams for applications as diverse as manufacturing, healthcare, and search-and-rescue missions. A comprehensive comprehension of the cognitive, social, and emotional factors that influence human–robot interaction and collaboration is necessary for the development of such teams.

Assuring effective communication and coordination between team members is one of the primary challenges of designing human–robot teams (Kaasinen et al., 2022) and this requires the development of algorithms and interfaces that allow robots to interpret and respond to human gestures, facial expressions, and verbal cues and, for instance, scientists have developed algorithms that enable robots to recognise and respond to hand gestures and facial expressions, allowing them to communicate with humans in a more natural and intuitive manner and similarly, researchers have created interfaces that allow humans to control robots using voice commands or hand gestures, facilitating team coordination (Green et al., 2008).

Ensuring that robots can adapt to shifting environments and tasks is a significant challenge in the design of effective human–robot teams and this requires the development of algorithms and control systems that enable robots to modify their behaviour based on their interactions with humans and their surroundings (Oliff et al., 2020) and, for instance, researchers have created algorithms that enable robots to learn from human demonstrations, allowing them to execute complex tasks such as assembly or manipulation with greater precision and efficiency.

Designing effective human–robot teams requires addressing social and affective factors that influence human–robot interaction and collaboration, in addition to technical obstacles, and, for instance, researchers have discovered that humans are more likely to trust and collaborate with robots that demonstrate human-like social behaviours such as empathy, humour, and civility (Demir et al., 2019) and therefore, designing robots capable of exhibiting such behaviours can improve their ability to collaborate with humans.

Finally, it is essential to consider the ethical and legal implications of designing human–robot partnerships, especially in terms of privacy, security, and responsibility, and designing robots that can collect and retain sensitive information about humans, for instance, may raise privacy and security concerns and similarly, the development of automata capable of autonomous decision-making may raise concerns regarding accountability and responsibility.

4.10.2 Strategies for overcoming communication and coordination challenges

Communication and coordination are indispensable to the success of any team, including human–robot partnerships; however, there are a number of obstacles that can hinder effective communication and coordination in human–robot teams, such as language, cultural, and technical differences, and developing lucid communication protocols and interfaces is one of the primary strategies for overcoming communication and coordination challenges (Akhilesh, 2020). This requires the development of algorithms and interfaces that enable humans and robots to exchange information and instructions in a plain and concise manner, and, for instance, researchers have created interfaces that allow humans to control robots using voice commands or hand gestures, facilitating seamless interaction between humans and robots, and researchers have also developed algorithms that enable robots to interpret and respond to human gestures, facial expressions, and verbal signals, allowing them to communicate with humans in a more natural and intuitive manner.

Establishing a common language and culture is an additional method for fostering effective communication and coordination in human–robot partnerships and this involves establishing a shared understanding of the team's guiding objectives, values, and norms, for instance, researchers have created frameworks for establishing a shared culture between humans and robots, such as the concept of "joint attention", in which humans and robots focus their attention on the same object or task (Siposova & Carpenter, 2019).

Establishing distinct roles and responsibilities for each team member is another technique for fostering effective communication and coordination in human–robot teams (Dahiya et al., 2023) and this involves identifying the tasks and responsibilities assigned to each team member and establishing explicit communication channels and decision-making authority, For instance, researchers have created algorithms that enable robots to learn from human demonstrations, allowing them to execute complex tasks such as assembly or manipulation with greater precision and efficiency.

In order to promote effective communication and coordination in human–robot partnerships, ongoing training and evaluation is necessary and this includes providing team members with regular feedback and training, as well as monitoring and evaluating the team's performance as a whole, for instance, researchers have created instruments for monitoring and assessing the performance of human–robot teams, such as the Robot Activity Support Scale, which measures the level of assistance provided by robots during daily activities (Prati et al., 2021).

In conclusion, a combination of technical, social, and affective strategies is necessary to promote effective communication and coordination in human–robot teams and this requires the development of clear communication protocols and interfaces, the establishment of a common language and culture, the consideration of social and emotional factors, the establishment of clear roles and responsibilities, and the provision of ongoing training and evaluation; by implementing these strategies, we can surmount communication and coordination obstacles and create effective, efficient, and socially responsible human–robot teams.

4.10.3 Mitigating risks and ensuring safety in human–robot collaboration

HRC has emerged as a result of the increasing use of robots in various industries and in HRC, robots work alongside humans to complete various duties. HRC has a number of potential advantages but it also poses some hazards and safety concerns (Parra et al., 2020) and in order to assure the safety of humans and robots working together, it is essential to identify and mitigate these risks and this chapter will discuss strategies for mitigating risks and guaranteeing safety in HRC.Before deploying robots in the workplace, conducting a risk assessment is one of the most important methods for ensuring safety in HRC and this involves identifying potential risks and hazards associated with robot use, such as collisions, entrapment, and electrical hazards (Huck et al., 2021). Once the risks have been identified, appropriate countermeasures, such as implementing safety sensors, protective barriers, and emergency stop buttons, can be implemented.

Designing robots with safety features is another technique for mitigating hazards in HRC and this involves the design of robots endowed with sensors and software to detect and respond to potential safety hazards, for instance, robots can be programmed to automatically halt or slow down when they sense an obstacle in their path or a human in their workspace (Zacharaki et al., 2020) and in addition to technical measures, promoting safety in HRC necessitates the consideration of social and cultural factors.

Establishing defined roles and responsibilities for every team member is another method for promoting safety within HRC and this involves identifying the tasks and responsibilities assigned to each team member and establishing explicit communication channels and decision-making authority (Caruana & Francalanza, 2023) and this can aid in preventing confusion and conflicts that could result in safety incidents.

Finally, promoting HRC safety requires ongoing training and evaluation and this includes training team members on how to work safely with automata and evaluating the team's performance as a whole and this can aid in identifying areas for development and addressing safety concerns prior to the occurrence of incidents.

4.11 CONCLUSION

I 5.0 is a paradigm shift that emphasises the need for businesses to embrace automation and innovation while empowering their employees and this chapter examines the changing position of employees in the age of automation and innovation, highlighting the challenges they face and the skills they must possess to thrive in this new era of business.

The first section of the chapter introduces the concept of I 5.0 and its essential characteristics and the section then examines the relevant literature, tracing the development of the industry from its earliest origins to the present day and the comparison with I 4.0 emphasises the significant differences between the two paradigms, especially I 5.0's increased emphasis on collaboration and human–robot interaction.

The chapter then discusses the impact of I 5.0 on employees, outlining the essential skills they must acquire to flourish in this new industrial era and the challenges they encounter, such as the need to continuously acquire new skills and adapt to new technologies, are also examined.

Strategies to surmount these obstacles, including those that companies, governments, and individuals, can be implemented to support employees in I 5.0, are then discussed and emphasis is also placed on the significance of managing HRC, as this will be crucial to the success of I 5.0.

Finally, the chapter emphasises the potential for future research in this discipline and there will be new challenges and opportunities for employees, businesses, and governments as I 5.0 continues to evolve and understanding how to navigate these changes will call for ongoing research and creativity.

REFERENCES

Adel, A. (2022). Future of Industry 5.0 in society: Human-centric solutions, challenges and prospective research areas. *Journal of Cloud Computing*, *11*(1), 1–15.

Agarwal, A., Mathur, P., & Walia, S. (2023). Journey of HR From Industry 1.0 to 5.0 and the Road Ahead. In *Opportunities and Challenges of Business 5.0 in Emerging Markets* (pp. 172–184). IGI Global.

Aheleroff, S., Huang, H. Y., Xu, X., & Zhong, R. R. (2022). Toward sustainability and resilience with Industry 4.0 and INDUSTRY 5.0. *Frontiers in Manufacturing Technology*, *2*, 951643. https://doi.org/10.3389/fmtec.2022.951643

Akhilesh, K. B. (2020). *Smart Technologies—Scope and Applications* (pp. 1–16). Springer Singapore.

Akkaya, B., & Ahmed, J. (2022). VUCA-RR Toward INDUSTRY 5.0. In *Agile Management and VUCA-RR: Opportunities and Threats in Industry 4.0 towards Society 5.0* (pp. 1–11). Emerald Publishing Limited.

Akundi, A., Euresti, D., Luna, S., Ankobiah, W., Lopes, A., & Edinbarough, I. (2022). State of INDUSTRY 5.0—Analysis and identification of current research trends. *Applied System Innovation*, *5*(1), 27.

Al Faruqi, U. (2019). Future service in INDUSTRY 5.0. *Jurnal Sistem Cerdas*, *2*(1), 67–79.

Alexa, L., Pîslaru, M., & Avasilcăi, S. (2022). From Industry 4.0 to INDUSTRY 5.0—an overview of European Union enterprises. *Sustainability and Innovation in Manufacturing Enterprises: Indicators, Models and Assessment for INDUSTRY 5.0*, 221–231.

Alves, J., Lima, T. M., & Gaspar, P. D. (2023). Is INDUSTRY 5.0 a Human-Centred Approach? A Systematic Review. *Processes*, *11*(1), 193.

Ammar, M., Haleem, A., Javaid, M., Walia, R., & Bahl, S. (2021). Improving material quality management and manufacturing organizations system through Industry 4.0 technologies. *Materials Today: Proceedings*, *45*, 5089–5096.

Aslam, F., Aimin, W., Li, M., & Ur Rehman, K. (2020). Innovation in the era of IoT and INDUSTRY 5.0: Absolute innovation management (AIM) framework. Information, *11*(2), 124.

Bakir, A., & Dahlan, M. (2022). Higher education leadership and curricular design in INDUSTRY 5.0 environment: A cursory glance. *Development and Learning in Organizations: An International Journal*, (ahead-of-print).

Barata, J., & Kayser, I. (2023). INDUSTRY 5.0–Past, present, and near future. *Procedia Computer Science*, *219*, 778–788.

Battini, D., Berti, N., Finco, S., Zennaro, I., & Das, A. (2022). Towards INDUSTRY 5.0: A multi-objective job rotation model for an inclusive workforce. *International Journal of Production Economics, 250,* 108619.

Broo, D. G., Kaynak, O., & Sait, S. M. (2022). Rethinking engineering education at the age of INDUSTRY 5.0. *Journal of Industrial Information Integration, 25,* 100311.

Cakmakci, M. (2019). Interaction in project management approach within industry 4.0. In: Trojanowska, J., Ciszak, O., Machado, J., Pavlenko, I. (eds) *Advances in Manufacturing II: Volume 1-Solutions for Industry 4.0* (pp. 176–189). Springer International Publishing, Cham. https://doi.org/10.1007/978-3-030-18715-6_15

Carayannis, E. G., Draper, J., & Bhaneja, B. (2021). Towards fusion energy in the INDUSTRY 5.0 and Society 5.0 context: Call for a global commission for urgent action on fusion energy. *Journal of the Knowledge Economy, 12*(4), 1891–1904.

Carayannis, E. G., & Morawska-Jancelewicz, J. (2022). The futures of Europe: Society 5.0 and INDUSTRY 5.0 as driving forces of future universities. *Journal of the Knowledge Economy,* 1–27.

Caruana, L., & Francalanza, E. (2023). A Safety 4.0 Approach for Collaborative Robotics in the Factories of the Future. *Procedia Computer Science, 217,* 1784–1793.

Chin, S. T. S. (2021). Influence of emotional intelligence on the workforce for INDUSTRY 5.0. *Journal of Human Resources Management Research, 2021*(2021), 882278.

Cho, H. Y., & Lee, H. J. (2022). Digital transformation for efficient communication in the workplace: Analyzing the flow coworking tool. *Business Communication Research and Practice, 5*(1), 20–28.

Cillo, V., Gregori, G. L., Daniele, L. M., Caputo, F., & Bitbol-Saba, N. (2022). Rethinking companies' culture through knowledge management lens during INDUSTRY 5.0 transition. *Journal of Knowledge Management, 26*(10), 2485–2498.

Culot, G., Nassimbeni, G., Orzes, G., & Sartor, M. (2020). Behind the definition of Industry 4.0: Analysis and open questions. *International Journal of Production Economics, 226,* 107617.

Dahiya, A., Aroyo, A. M., Dautenhahn, K., & Smith, S. L. (2023). A survey of multi-agent Human–Robot Interaction systems. *Robotics and Autonomous Systems, 161,* 104335.

De Giovanni, P. (2023). Sustainability of the Metaverse: A Transition to INDUSTRY 5.0. *Sustainability, 15*(7), 6079.

Deloitte. (2017). Future-ready workforce: Adapting to the new world of work. Retrieved from https://www2.deloitte.com/content/dam/Deloitte/ca/Documents/consulting/ca-future-ready-workforce-en-aoda.pdf

Demir, K. A., Döven, G., & Sezen, B. (2019). INDUSTRY 5.0 and human-robot co-working. *Procedia Computer Science, 158,* 688–695.

Destouet, C., Tlahig, H., Bettayeb, B., & Mazari, B. (2023). Flexible job shop scheduling problem under INDUSTRY 5.0: A survey on human reintegration, environmental consideration and resilience improvement. *Journal of Manufacturing Systems, 67,* 155–173.

Doyle-Kent, M., & Kopacek, P. (2020). INDUSTRY 5.0: Is the manufacturing industry on the cusp of a new revolution?. In *Proceedings of the International Symposium for Production Research 2019* (pp. 432–441). Springer International Publishing.

Elangovan, U. (2021). *INDUSTRY 5.0: The Future of the Industrial Economy*. CRC Press.

ElFar, O. A., Chang, C. K., Leong, H. Y., Peter, A. P., Chew, K. W., & Show, P. L. (2021). Prospects of INDUSTRY 5.0 in algae: Customization of production and new advance technology for clean bioenergy generation. *Energy Conversion and Management: X, 10*, 100048.

Gagnidze, I. (2022). Industry 4.0 and INDUSTRY 5.0: Can clusters deal with the challenges? (A systemic approach). Kybernetes.

Ghobakhloo, M., Iranmanesh, M., Mubarak, M. F., Mubarik, M., Rejeb, A., & Nilashi, M. (2022). Identifying INDUSTRY 5.0 contributions to sustainable development: A strategy roadmap for delivering sustainability values. *Sustainable Production and Consumption, 33*, 716–737.

Green, S. A., Billinghurst, M., Chen, X., & Chase, J. G. (2008). Human-robot collaboration: A literature review and augmented reality approach in design. *International Journal of Advanced Robotic Systems, 5*(1), 1.

Hollenstein, L., Thurnheer, S., & Vogt, F. (2022). Problem solving and digital transformation: Acquiring skills through pretend play in kindergarten. *Education Sciences, 12*(2), 92.

Huang, Sihan, Baicun Wang, Xingyu Li, Pai Zheng, Dimitris Mourtzis, and Lihui Wang. (2022) "INDUSTRY 5.0 and Society 5.0—Comparison, complementation and co-evolution." *Journal of Manufacturing Systems, 64*(2022), 424–428.

Huck, T. P., Münch, N., Hornung, L., Ledermann, C., & Wurll, C. (2021). Risk assessment tools for industrial human-robot collaboration: Novel approaches and practical needs. *Safety Science, 141*, 105288.

Hussain, S., Singh, A. M., Mohanty, P., & Gavinolla, M. R. (2023). Next generation employability and career sustainability in the hospitality INDUSTRY 5.0. *Worldwide Hospitality and Tourism Themes, 15*(3), 308–321.

Ivanov, D. (2023). The INDUSTRY 5.0 framework: Viability-based integration of the resilience, sustainability, and human-centricity perspectives. *International Journal of Production Research, 61*(5), 1683–1695.

Jafari, N., Azarian, M., & Yu, H. (2022). Moving from Industry 4.0 to INDUSTRY 5.0: What are the implications for smart logistics? *Logistics, 6*(2), 26.

Javaid, M., & Haleem, A. (2020). Critical components of INDUSTRY 5.0 towards a successful adoption in the field of manufacturing. *Journal of Industrial Integration and Management, 5*(3), 327–348.

Javaid, M., Haleem, A., Singh, R. P., Haq, M. I. U., Raina, A., & Suman, R. (2020). INDUSTRY 5.0: Potential applications in COVID-19. *Journal of Industrial Integration and Management, 5*(4), 507–530.

Jerman, A., Pejić Bach, M., & Aleksić, A. (2020). Transformation towards smart factory system: Examining new job profiles and competencies. *Systems Research and Behavioral Science, 37*(2), 388–402.

Joglekar, S., Kadam, S., Director, I. C. T., & Dharmadhikari, S. (2023). INDUSTRY 5.0: Analysis, applications and prognosis. *The Online Journal of Distance Education and e-Learning, 11*(1), 1.

Joseph, J. (2023). Work-Life Balance and Its Socio-cultural Inclination from Industry 1.0 to Industry 4.0. In Singh, G., Goel, R., Garg, V. (eds) *Industry 4.0 and the Digital Transformation of International Business* (pp. 287–304). Springer Nature, Singapore. https://doi.org/10.1007/978-981-19-7880-7_17

Kaasinen, E., Anttila, A. H., Heikkilä, P., Laarni, J., Koskinen, H., & Väätänen, A. (2022). Smooth and resilient human–machine teamwork as an INDUSTRY 5.0 design challenge. *Sustainability, 14*(5), 2773.

Kasinathan, P., Pugazhendhi, R., Elavarasan, R. M., Ramachandaramurthy, V. K., Ramanathan, V., Subramanian, S., ... Alsharif, M. H. (2022). Realization of sustainable development goals with disruptive technologies by integrating INDUSTRY 5.0, Society 5.0, smart cities and villages. *Sustainability, 14*(22), 15258.

Khorasani, M., Loy, J., Ghasemi, A. H., Sharabian, E., Leary, M., Mirafzal, H., ... Gibson, I. (2022). A review of Industry 4.0 and additive manufacturing synergy. *Rapid Prototyping Journal,* (ahead-of-print).

Koh, L., Orzes, G., & Jia, F. J. (2019). The fourth industrial revolution (Industry 4.0): Technologies disruption on operations and supply chain management. *International Journal of Operations & Production Management, 39*(6/7/8), 817–828.

Kolade, O., & Owoseni, A. (2022). Employment 5.0: The work of the future and the future of work. *Technology in Society, 71,* 102086.

Konstantinidis, F. K., Myrillas, N., Mouroutsos, S. G., Koulouriotis, D., & Gasteratos, A. (2022). Assessment of industry 4.0 for modern manufacturing ecosystem: A systematic survey of surveys. *Machines, 10*(9), 746.

Lasi, H., Fettke, P., Kemper, H. G., Feld, T., & Hoffmann, M. (2014). Industry 4.0. *Business & Information Systems Engineering, 6,* 239–242.

Leng, J., Sha, W., Wang, B., Zheng, P., Zhuang, C., Liu, Q., ... Wang, L. (2022). INDUSTRY 5.0: Prospect and retrospect. *Journal of Manufacturing Systems, 65,* 279–295.

London, M., Sessa, V. I., & Shelley, L. A. (2023). Developing self-awareness: Learning processes for self-and interpersonal growth. *Annual Review of Organizational Psychology and Organizational Behavior, 10,* 261–288.

Longo, F., Padovano, A., & Umbrello, S. (2020). Value-oriented and ethical technology engineering in INDUSTRY 5.0: A human-centric perspective for the design of the factory of the future. *Applied Sciences, 10*(12), 4182.

Lu, Y., Zheng, H., Chand, S., Xia, W., Liu, Z., Xu, X., ... Bao, J. (2022). Outlook on human-centric manufacturing towards INDUSTRY 5.0. *Journal of Manufacturing Systems, 62,* 612–627.

Ma, X., Mao, C., & Liu, G. (2022). Can robots replace human beings?—assessment on the developmental potential of construction robot. *Journal of Building Engineering, 56,* 104727.

Maddikunta, P. K. R., Pham, Q. V., Prabadevi, B., Deepa, N., Dev, K., Gadekallu, T. R., ... Liyanage, M. (2022). INDUSTRY 5.0: A survey on enabling technologies and potential applications. *Journal of Industrial Information Integration, 26,* 100257.

Mangalaraj, G., Nerur, S., & Dwivedi, R. (2023). Digital transformation for agility and resilience: An exploratory study. *Journal of Computer Information Systems, 63*(1), 11–23.

Mansurali, A., Harish, V., & Ramakrishnan, S. (2023). INDUSTRY 5.0–The Co-creator in Marketing. In Saini, A. and Garg, V. (ed.) *Transformation for Sustainable Business and Management Practices: Exploring the Spectrum of INDUSTRY 5.0* (pp. 5–15). Emerald Publishing Limited. https://doi.org/10.1108/978-1-80262-277-520231002

Mashelkar, R. A. (2018). Exponential technology, industry 4.0 and future of jobs in India. *Review of Market Integration, 10*(2), 138–157.

Massaro, A. (2021). *Electronics in Advanced Research Industries: Industry 4.0 to INDUSTRY 5.0 Advances*. John Wiley & Sons.

Mathur, A., Dabas, A., & Sharma, N. (2022, December). Evolution From Industry 1.0 to INDUSTRY 5.0. In *2022 4th International Conference on Advances in Computing, Communication Control and Networking (ICAC3N)* (pp. 1390–1394). IEEE.

McKinsey Global Institute. (2017). Jobs lost, jobs gained: What the future of work will mean for jobs, skills, and wages. Retrieved from https://www.mckinsey.com/featured-insights/future-of-work/jobs-lost-jobs-gained-what-the-future-of-work-will-mean-for-jobs-skills-and-wages

Mehdiabadi, A., Shahabi, V., Shamsinejad, S., Amiri, M., Spulbar, C., & Birau, R. (2022). Investigating INDUSTRY 5.0 and its impact on the banking industry: Requirements, approaches and communications. *Applied Sciences*, *12*(10), 5126.

Mitchell, J., & Guile, D. (2022). Fusion skills and INDUSTRY 5.0: Conceptions and challenges. *Insights Into Global Engineering Education After the Birth of INDUSTRY 5.0*, 53.

Modgil, S., Singh, R. K., & Agrawal, S. (2023). Developing human capabilities for supply chains: An INDUSTRY 5.0 perspective. *Annals of Operations Research*, *2023*, 1–31. https://doi.org/10.1007/s10479-023-05245-1

Moraru, G. M., & Popa, D. (2021). Potential Resistance of Employees to Change in the Transition to INDUSTRY 5.0. In *MATEC Web of Conferences* (Vol. 343, p. 07005). EDP Sciences.

Mukherjee, A. A., Raj, A., & Aggarwal, S. (2023). Identification of barriers and their mitigation strategies for INDUSTRY 5.0 implementation in emerging economies. *International Journal of Production Economics*, 108770.

Nagaraj, S. V. (2020). Disruptive technologies that are likely to shape future jobs. *Procedia Computer Science*, *172*, 502–504.

Nahavandi, S. (2019) INDUSTRY 5.0—A human-centric solution. *Sustainability*, *11*(16), 4371.

Neubert, J. C., Mainert, J., Kretzschmar, A., & Greiff, S. (2015). The assessment of 21st century skills in industrial and organizational psychology: Complex and collaborative problem solving. *Industrial and Organizational Psychology*, *8*(2), 238–268.

Nikolakis, N., Maratos, V., & Makris, S. (2019). A cyber physical system (CPS) approach for safe human-robot collaboration in a shared workplace. *Robotics and Computer-Integrated Manufacturing*, *56*, 233–243.

Oliff, H., Liu, Y., Kumar, M., Williams, M., & Ryan, M. (2020). Reinforcement learning for facilitating human-robot-interaction in manufacturing. *Journal of Manufacturing Systems*, *56*, 326–340.

Olson, T., & Johnson, P. (2023). Case study in digital transformation collaboration. In *INTED2023 Proceedings* (pp. 533–536). IATED.

Østergaard, E. H. (2018). Welcome to INDUSTRY 5.0. *Retrieved Febr, 5*, 2020.

Özdemir, V., & Hekim, N. (2018). Birth of INDUSTRY 5.0: Making sense of big data with artificial intelligence, "the internet of things" and next-generation technology policy. *Omics: A Journal of Integrative Biology*, *22*(1), 65–76.

Ozkeser, B. (2018). Lean innovation approach in INDUSTRY 5.0. *The Eurasia Proceedings of Science Technology Engineering and Mathematics*, *2*, 422–428.

Padma, V., Anand, N. N., Gurukul, S. S., Javid, S. S. M., Prasad, A., & Arun, S. (2015). Health problems and stress in Information Technology and Business Process Outsourcing employees. *Journal of Pharmacy & Bioallied Sciences*, *7*(Suppl 1), S9.

Pacher, C., Woschank, M., & Zunk, B. M. (2023). The role of competence profiles in INDUSTRY 5.0-related vocational education and training: Exemplary development of a competence Profile for industrial logistics engineering education. *Applied Sciences*, 13(5), 3280.

Panday, P., & Kaur, G. (2023). Talent management and employee outlook on INDUSTRY 5.0. In *Handbook of Research on Education Institutions, Skills, and Jobs in the Digital Era* (pp. 299–306). IGI Global.

Papcun, P., Kajáti, E., & Koziorek, J. (2018, August). Human machine interface in concept of industry 4.0. In *2018 World Symposium on Digital Intelligence for Systems and Machines (DISA)* (pp. 289–296). IEEE.

Parra, P. S., Calleros, O. L., & Ramirez-Serrano, A. (2020). Human-robot collaboration systems: Components and applications. In *International Conference of Control, Dynamic Systems, and Robotics*.

Paschek, D., Mocan, A., & Draghici, A. (2019, May). INDUSTRY 5.0—The expected impact of next industrial revolution. In *Thriving on future education, industry, business, and Society, Proceedings of the MakeLearn and TIIM International Conference*, Piran, Slovenia (pp. 15–17).

Patel, P. H., Angrish, A. K., & Nadda, V. (2023). A Cross-Sector Comparison of INDUSTRY 5.0: Digital Technologies in Supply Chain Management of FMCG and the Automotive Sector. In *Opportunities and Challenges of Business 5.0 in Emerging Markets* (pp. 99–123). IGI Global.

Pizoń, J., & Gola, A. (2023). Human–Machine relationship—perspective and future roadmap for INDUSTRY 5.0 solutions. *Machines*, 11(2), 203.

Prati, E., Peruzzini, M., Pellicciari, M., & Raffaeli, R. (2021). How to include User eXperience in the design of Human-Robot Interaction. *Robotics and Computer-Integrated Manufacturing*, 68, 102072.

Qahtan, S., Alsattar, H. A., Zaidan, A. A., Pamucar, D., & Deveci, M. (2022). Integrated sustainable transportation modelling approaches for electronic passenger vehicle in the context of INDUSTRY 5.0. *Journal of Innovation & Knowledge*, 7(4), 100277.

Raja Santhi, A., & Muthuswamy, P. (2023). INDUSTRY 5.0 or industry 4.0 S? Introduction to industry 4.0 and a peek into the prospective INDUSTRY 5.0 technologies. *International Journal on Interactive Design and Manufacturing (IJIDeM)*, 1–33.

Rastogi, R., Sharma, B. P., & Gupta, M. (2022). Registration of vehicles with validation and obvious manner through Blockchain: Smart City approach in INDUSTRY 5.0. *Smart City Infrastructure: The Blockchain Perspective*, 127–161.

Rathod, S., & Agrawal, P. (2023). Emerging challenges and opportunities for leaders in INDUSTRY 5.0: A roadmap through literature review. *Managing Technology Integration for Human Resources in INDUSTRY 5.0*, 2, 20–30.

Rodríguez-Garcia, C., León-Mateos, F., López-Manuel, L., & Sartal, A. (2023). Assessing the drivers behind innovative and creative companies. The importance of knowledge transfer in the field of INDUSTRY 5.0. In *INDUSTRY 5.0: Creative and Innovative Organizations* (pp. 91–114). Cham: Springer International Publishing.

Rožanec, J. M., Novalija, I., Zajec, P., Kenda, K., Tavakoli Ghinani, H., Suh, S., ... Soldatos, J. (2022). Human-centric artificial intelligence architecture for INDUSTRY 5.0 applications. *International Journal of Production Research*, 61(20), 6847–6872.

Saniuk, S., & Grabowska, S. (2022). Development of knowledge and skills of engineers and managers in the era of INDUSTRY 5.0 in the light of expert research. *Zeszyty Naukowe. Organizacja i Zarządzanie/Politechnika Śląska*.

Saptaningtyas, W. W. E., & Rahayu, D. K. (2020). A proposed model for food manufacturing in SMEs: Facing INDUSTRY 5.0. In *Proceedings of the International Conference on Industrial Engineering and Operations Management* (pp. 1653–1661).

Sharma, R., & Arya, R. (2022). UAV based long range environment monitoring system with INDUSTRY 5.0 perspectives for smart city infrastructure. *Computers & Industrial Engineering, 168*, 108066.

Sharma, M., Sehrawat, R., Luthra, S., Daim, T., & Bakry, D. (2022). Moving towards INDUSTRY 5.0 in the pharmaceutical manufacturing sector: Challenges and solutions for Germany. *IEEE Transactions on Engineering Management*, 1–18. https://doi.org/10.1109/TEM.2022.3143466

Shukla, P., & Singh, S. (2023). INDUSTRY 5.0 and Digital Innovations: Antecedents to Sustainable Business Model. In *Transformation for Sustainable Business and Management Practices: Exploring the Spectrum of INDUSTRY 5.0* (pp. 17–29). Emerald Publishing Limited.

Siposova, B., & Carpenter, M. (2019). A new look at joint attention and common knowledge. *Cognition, 189*, 260–274.

Skobelev, P. O., & Borovik, S. Y. (2017). On the way from Industry 4.0 to INDUSTRY 5.0: From digital manufacturing to digital society. *Industry 4.0, 2*(6), 307–311.

Sony, M., & Naik, S. (2020). Key ingredients for evaluating Industry 4.0 readiness for organizations: A literature review. *Benchmarking: An International Journal, 27*(7), 2213–2232.

Tabuenca, B., Greller, W., & Verpoorten, D. (2022). Mind the gap: Smoothing the transition to higher education fostering time management skills. *Universal Access in the Information Society, 21*(2), 367–379.

Thangaraj, J., & Narayanan, R. L. (2018). *Industry 1.0 to 4.0: The Evolution of Smart Factories*. Chicago, IL: APICS.

Tiwari, S., Bahuguna, P. C., & Walker, J. (2022). INDUSTRY 5.0: A macroperspective approach. In *Handbook of Research on Innovative Management Using AI in INDUSTRY 5.0* (pp. 59–73). IGI Global.

Van Dijk, J. (2020). *The Digital Divide*. John Wiley & Sons.

Vinitha, K., Prabhu, R. A., Bhaskar, R., & Hariharan, R. (2020). Review on industrial mathematics and materials at Industry 1.0 to Industry 4.0. *Materials Today: Proceedings, 33*, 3956–3960.

Whewell, E., Caldwell, H., Frydenberg, M., & Andone, D. (2022). Changemakers as digital makers: Connecting and co-creating. *Education and Information Technologies, 27*(5), 6691–6713.

Wulf, J., Mettler, T., & Brenner, W. (2017). Using a digital services capability model to assess readiness for the digital consumer. *MIS Quarterly Executive, 16*(3), 171–195.

Xu, X., Lu, Y., Vogel-Heuser, B., & Wang, L. (2021). Industry 4.0 and INDUSTRY 5.0—Inception, conception and perception. *Journal of Manufacturing Systems, 61*, 530–535.

Yao, X., Ma, N., Zhang, J., Wang, K., Yang, E., & Faccio, M. (2022). Enhancing wisdom manufacturing as industrial metaverse for industry and society 5.0. *Journal of Intelligent Manufacturing, 35*, 235–255. https://doi.org/10.1007/s10845-022-02027-7

Yavari, F., & Pilevari, N. (2020). Industry revolutions development from Industry 1.0 to INDUSTRY 5.0 in manufacturing. *Journal of Industrial Strategic Management*, *5*(2), 44–63.

Yin, Y., Stecke, K. E., & Li, D. (2018). The evolution of production systems from Industry 2.0 through Industry 4.0. *International Journal of Production Research*, *56*(1–2), 848–861.

Zacharaki, A., Kostavelis, I., Gasteratos, A., & Dokas, I. (2020). Safety bounds in human robot interaction: A survey. *Safety Science*, *127*, 104667.

Zakoldaev, D. A., Korobeynikov, A. G., Shukalov, A. V., Zharinov, I. O., & Zharinov, O. O. (2020). Industry 4.0 vs Industry 3.0: The role of personnel in production. In *IOP Conference Series: Materials Science and Engineering* (Vol. 734, No. 1, p. 012048). IOP Publishing.

Zeb, S., Mahmood, A., Khowaja, S. A., Dev, K., Hassan, S. A., Qureshi, N. M. F., … Bellavista, P. (2022). INDUSTRY 5.0 is coming: A survey on intelligent nextG wireless networks as technological enablers. *arXiv preprint arXiv:2205.09084*.

Zhang, Q., Chen, Y., Lin, W., & Chen, Y. (2021). Optimizing Medical Enterprise's Operations Management considering Corporate Social Responsibility under INDUSTRY 5.0. *Discrete Dynamics in Nature and Society*, *2021*, 1–13.

Zizic, M. C., Mladineo, M., Gjeldum, N., & Celent, L. (2022). From industry 4.0 towards INDUSTRY 5.0: A review and analysis of paradigm shift for the people, organization and technology. *Energies*, *15*(14), 5221.

Chapter 5

Exploring the impact of collaborative robots on human–machine cooperation in the era of Industry 5.0

R. Damaševičius
Vytautas Magnus University, Kaunas, Lithuania

M. Vasiljevas, L. Narbutaitė, and T. Blažauskas
Kaunas University of Technology, Kaunas, Lithuania

5.1 INTRODUCTION

Industry 5.0 is a new paradigm in the field of industrial automation that emphasizes the integration of human skills and expertise with advanced technologies (Maddikunta et al., 2022). It represents a significant shift from the previous industrial revolutions where the focus was on replacing human labor with machines to improve efficiency and productivity. Industry 5.0 seeks to leverage the strengths of both humans and machines in order to create a more flexible, adaptable, and sustainable industrial ecosystem.

Cobots are a key technology that supports the vision of Industry 5.0 as an evolution of the concept of Industry 4.0 (Nayyar & Kumar, 2020; Weiss et al., 2021). Cobots are designed to work alongside human workers in a collaborative manner, performing repetitive, dangerous, or physically demanding tasks while freeing up human workers to focus on more complex or creative tasks that require human intelligence and problem-solving skills (Liu et al., 2022). Unlike traditional industrial robots that are designed to work independently in a cage or a fixed workspace, cobots are designed to be flexible and adaptable, capable of interacting with human workers in a safe and intuitive manner (Faccio et al., 2023).

The development of cobots has been driven by a number of factors, including advances in artificial intelligence, sensors, and other enabling technologies (Batth et al., 2018) such as the Internet of Things (Kumar & Nayyar, 2019), as well as a growing recognition of the importance of human-centric design (Rožanec et al., 2022) and the need for more flexible and adaptable production systems. Cobots are being used in a wide range of industries, from automotive and aerospace manufacturing (Javaid et al., 2022) to healthcare (Pauliková et al., 2021) and logistics.

DOI: 10.1201/9781003489269-5

Despite the potential benefits of cobots, there are also challenges to their adoption and integration into existing production systems (Dobra & Dhir, 2020). These challenges include issues related to safety (Boschetti et al., 2023), training, and job displacement, as well as questions about the ethical and legal implications of using robots in the workplace. Addressing these challenges will require a collaborative effort between industry, academia, and government, as well as a better understanding of the social and cultural factors that influence the adoption and acceptance of new technologies (Leesakul et al., 2022).

The aim of this chapter is to provide an overview of the background of Industry 5.0 and cobots and their potential impact on the industrial ecosystem. The chapter also aims to identify the challenges and opportunities associated with the adoption of cobots and to highlight the need for collaborative efforts to address these challenges.

The novelty of this chapter lies in its focus on the integration of cobots within the context of Industry 5.0, and its exploration of the challenges and opportunities associated with this integration. The chapter also provides a comprehensive overview of the background and development of Industry 5.0 and cobots and identifies key trends and drivers in the field.

This chapter contributes to the understanding of the potential impact of cobots on the industrial ecosystem and identifies key challenges and opportunities associated with their adoption. The chapter also highlights the need for a collaborative approach to addressing these challenges and provides insights into the social and cultural factors that influence the adoption and acceptance of new technologies. The chapter is intended to be a useful resource for researchers, practitioners, and policymakers interested in the integration of cobots within the context of Industry 5.0.

5.1.1 Organization of chapter

The remaining parts of this chapter are organized as follows. Section 5.2 discusses the development and implementation of cobots. Section 5.3 discusses the integration of cobots into industrial manufacturing. Section 5.4 presents a case study of integrating cobots in a manufacturing company. Section 5.5 summarizes the role of cobots in Industry 5.0 and future research directions. And, finally, Section 5.6 concludes the chapter.

5.2 DEVELOPMENT AND IMPLEMENTATION OF COBOTS

5.2.1 Evolution of collaborative robots

Cobots are a type of industrial robot designed to work alongside human operators in a collaborative and safe manner (Vicentini, 2021). The evolution of cobots can be traced through three generations (Figure 5.1).

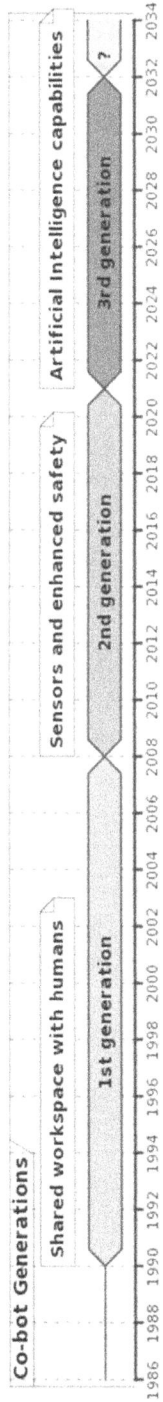

Figure 5.1 Timeline of cobot generations.

The first generation of cobots emerged in the early 1990s and were designed to operate in a shared workspace with humans. These cobots were characterized by their low payload capacity, limited range of motion, and low speed. They were primarily used for tasks such as assembly, packaging, and inspection.

The second generation of cobots emerged in the late 2000s and were designed to be more flexible and adaptable to a wider range of tasks. These cobots had improved safety features, such as sensors and software that allowed them to detect and avoid collisions with humans. They also had improved payloads and range of motion, which made them suitable for tasks such as welding, painting, and material handling.

The third generation of cobots is currently emerging and is characterized by advanced sensing and artificial intelligence capabilities. These cobots are designed to be more autonomous and are capable of learning from their environment and adapting to new tasks. They also have improved safety features, such as advanced vision systems and force sensors, which allow them to work safely and efficiently alongside humans. These cobots are being used for a wide range of tasks, including logistics, healthcare, and education.

The evolution of cobots is driven by advances in technology, including improvements in sensors, software, and artificial intelligence (Borboni et al., 2023). As cobots continue to evolve, they are expected to become more capable, adaptable, and autonomous, which will enable them to play an increasingly important role in the industrial ecosystem (Rodriguez-Guerra et al., 2021).

5.2.2 Different types of cobots and their applications in various industries

There are several types of cobots, each designed for specific applications in different industries. Some of the most common types of cobots and their applications are summarized in Table 5.1. Assembly cobots are designed to perform assembly tasks alongside human operators in industries such as automotive, electronics, and consumer goods. They are typically equipped with grippers, vision systems, and force sensors to detect and manipulate parts. Inspection cobots are designed to perform quality control tasks, such as inspecting products for defects or identifying discrepancies in manufacturing processes. They are used in industries such as aerospace, pharmaceuticals, and food and beverage. Packaging cobots are designed to package products in industries such as food and beverage, pharmaceuticals, and consumer goods. They are typically equipped with end-of-arm tooling, conveyors, and vision systems to package products efficiently and accurately. Material-handling cobots are designed to move materials and products from one location to another in industries such as logistics, warehousing,

Table 5.1 Table of different types of cobots, their application areas, tasks, and parts

Cobot type	Area	Tasks	Parts
Assembly cobots (Andersson et al., 2021; Calvo & Gil, 2022; Navas-Reascos et al., 2022; Rossi et al., 2020; Yin & Li, 2023)	Automotive, electronics, consumer goods	Perform assembly tasks alongside human operators	Grippers, vision systems, and force sensors to detect and manipulate parts
Inspection cobots (Hameed et al., 2023)	Aerospace, pharmaceuticals, food and beverage	Perform quality control tasks such as inspecting products for defects or identifying discrepancies in manufacturing processes	2D and 3D cameras for detecting defects or inconsistencies in products; laser scanners for measuring dimensions and checking for deviations; force sensors for detecting variations in pressure or weight; thermal cameras for detecting heat or temperature variations; X-ray or ultrasound systems for inspecting internal structures of products; spectrometers for analyzing chemical composition of products
Packaging cobots (Romanov et al., 2022)	Food and beverage, pharmaceuticals, consumer goods	Package products	End-of-arm tooling, conveyors, and vision systems to package products efficiently and accurately
Material-handling cobots (Fager et al., 2020; Javaid et al., 2022)	Logistics, warehousing, e-commerce	Move materials and products from one location to another	Conveyors, sensors, and vision systems to navigate their environment and pick up and move objects
Welding cobots (Jones et al., 2015)	Automotive, aerospace, construction	Perform welding tasks	Equipped with welding torches, vision systems, and sensors to perform precise and accurate welding

(Continued)

Table 5.1 (Continued) Table of different types of cobots, their application areas, tasks, and parts

Cobot type	Area	Tasks	Parts
Healthcare cobots (Chiriatti et al., 2021)	Hospitals, clinics, long-term care facilities	Assist healthcare professionals with tasks such as patient monitoring, medication delivery, and sterilization	Sensors such as cameras, microphones, and infrared sensors to detect and respond to changes in their environment and monitor patient health; end effectors such as grippers, syringe injectors, and sterilization equipment
Education cobots (Damaševičius et al., 2017; Plauska & Damaševičius, 2014; Ponsa & Tornil-Sin, 2022)	Classrooms, research labs	Teach and assist students in educational settings	Grippers to pick up and manipulate objects during the learning process; sensors to detect the environment and interact with objects and humans safely; cameras for vision-based tasks such as object recognition, tracking, and detection; an interactive display screen is used to display information and engage with students; speakers for audio interaction.

and e-commerce. They are typically equipped with conveyors, sensors, and vision systems to navigate their environment and pick up and move objects. Welding cobots are designed to perform welding tasks in industries such as automotive, aerospace, and construction. They are typically equipped with welding torches, vision systems, and sensors to perform precise and accurate welding. Healthcare cobots are designed to assist healthcare professionals with tasks such as patient monitoring, medication delivery, and sterilization. They are used in hospitals, clinics, and long-term care facilities. Education cobots are designed to teach and assist students in educational settings, such as classrooms and research labs. They are used to teach coding, robotics, and other STEM (Science, Technology, Engineering, Mathematics) subjects.

Cobots have several advantages over traditional industrial robots, including lower costs, greater flexibility, and improved safety features. As a result, they are becoming increasingly popular in industries such as manufacturing, logistics, and healthcare. As technology continues to advance, cobots are expected to become even more capable and adaptable, enabling them to perform a wider range of tasks in a variety of industries.

5.2.3 Characteristics of cobots

Cobots are designed to work safely alongside human workers and have several characteristics that differentiate them from traditional industrial robots, which are summarized in Table 5.2.

Overall, cobots are designed to be more flexible, user-friendly, and collaborative than traditional industrial robots, with a focus on working safely and effectively alongside human workers.

5.3 INTEGRATION OF COBOTS INTO INDUSTRIAL MANUFACTURING

5.3.1 Defects of human work and machine work

Complementarity is the concept that cobots can work alongside humans to complement and enhance their skills, rather than replacing them (Sowa et al., 2021). Cobots can address the defects of both human work and machine work by combining the strengths of both.

Humans have the ability to learn and adapt quickly, make decisions based on context and intuition, and perform delicate or complex tasks that require dexterity and creativity. However, humans are prone to fatigue, error, and injury when performing repetitive, monotonous, or physically demanding tasks. In contrast, machines excel at performing repetitive, precise, and consistent tasks without tiring, but lack the ability to learn, adapt, or make context-based decisions.

Table 5.2 Features of cobots

Features	Description
Safety (Adriaensen et al., 2022; Bi et al., 2022; Boschetti et al., 2023; Villani et al., 2018)	Cobots are equipped with sensors and other safety features that prevent collisions and other accidents. They can detect the presence of humans and automatically slow down or stop when a human enters their workspace.
Flexibility (Buerkle et al., 2023)	Cobots are designed to be easily reprogrammed or reconfigured to perform different tasks or to adapt to changes in the manufacturing process, making them more flexible than traditional industrial robots.
Ease of use (Buerkle et al., 2023; Fournier et al., 2022)	Cobots are designed to be easy to program and operate, with user-friendly interfaces that allow workers to quickly teach them new tasks.
Compact size	Cobots are typically smaller and lighter than traditional industrial robots, making them easier to integrate into existing workspaces and to move around as needed.
Lower cost	Cobots are generally less expensive to purchase and maintain than traditional industrial robots, making them more accessible to small- and medium-sized enterprises.
Human-like movement (Papanagiotou et al., 2021; Sauer et al., 2021)	Cobots are designed to mimic human movement and behavior, making them well-suited for tasks that require delicate or precise movements.
Collaboration with human workers (Amin et al., 2020; Liang et al., 2021; Villani et al., 2018)	Cobots are designed to work collaboratively with human workers, performing tasks that are too repetitive, dangerous, or physically demanding for humans to do alone.

Cobots can address these defects of human work and machine work by working collaboratively with humans to perform tasks that require both cognitive and physical abilities. For example, a cobot can work alongside a human operator in an assembly line, handling heavy or repetitive tasks while the human performs more delicate or complex tasks. This can help to reduce worker fatigue and injury, while also increasing efficiency and productivity.

Furthermore, cobots can help to improve the quality of work by reducing errors and variability. For example, an inspection cobot can detect defects in a product that a human operator may miss, while a welding cobot can perform precise and consistent welds that are difficult for a human to achieve consistently.

5.3.2 Challenges in integrating cobots with human workers

Integrating cobots with human workers is a challenging task that requires careful planning, design, and implementation (Calvo & Gil, 2022; Rossi et al., 2020). The following are some of the challenges involved in integrating cobots with human workers:

- **Safety:** Safety is the primary concern when integrating cobots with human workers. Cobots need to be designed and programmed to avoid collisions, reduce the risk of injury, and ensure safe operations in the presence of human workers.
- **Training:** Human workers need to be trained to work with cobots effectively (Qiu et al., 2021). This includes training in operating the cobot, understanding its capabilities and limitations, and learning how to collaborate with it safely.
- **Communication:** Effective communication between human workers and cobots is critical for successful collaboration. Cobots need to be designed to communicate their actions and status effectively, and human workers need to be able to communicate their intentions, needs, and feedback to the cobot (Salehzadeh et al., 2022).
- **Compatibility:** Cobots need to be compatible with existing workflows and processes in the industry. This requires careful planning and design to ensure that cobots can integrate seamlessly with existing systems and processes.
- **Cost:** The cost of integrating cobots with human workers can be significant, including the cost of purchasing, programming, and training. This can be a barrier to adoption, particularly for small- and medium-sized enterprises.
- **Resistance to change:** Resistance to change is a common challenge when introducing new technologies, including cobots, into the workplace. Human workers may be resistant to the idea of working with cobots, particularly if they perceive them as a threat to their jobs or livelihoods.

Addressing these challenges requires a collaborative effort between designers, engineers, human factors experts, and industry stakeholders. Effective integration of cobots with human workers can improve productivity, reduce errors, and enhance worker safety and satisfaction. However, to achieve these benefits, it is essential to address the challenges involved in integrating cobots with human workers effectively.

5.3.3 Human-centered design of cobots

Human-centered design is a design approach that emphasizes the needs, preferences, and limitations of users throughout the design process (Nguyen Ngoc et al., 2022). The human-centered design of cobots requires a deep understanding of human needs and limitations and a willingness to adapt the design of the robot to meet those needs (Boschetti et al., 2023; Chiriatti et al., 2022). When applied to the development of cobots, human-centered design principles (Table 5.3) ensure that the robots are designed with the goal of improving human performance and well-being in mind.

One principle of the human-centered design of cobots is to ensure that they are intuitive to use (Villani et al., 2018). The robots should be designed with a simple and clear interface that can be easily understood and operated by humans. This will allow workers to use the robots effectively without requiring extensive training or specialized knowledge.

Table 5.3 Human-centered design of cobots

Principle	Human requirements	Cobot requirements
Safety	Ensuring human safety and well-being in cobot workspaces by using safety sensors, providing safety training, and designing cobots that minimize the risk of physical harm	Equipping cobots with safety sensors and features, minimizing sharp edges or dangerous moving parts, and ensuring that cobots can safely collaborate with humans
Intuitiveness	Designing cobots that are intuitive and easy to use for human operators, with clear instructions and minimal training required	Providing cobots with clear and simple user interfaces, easily understood instructions, and clear indicators of cobot status and performance
Adaptability	Cobots that can adapt to different human work styles, preferences, and needs, and that can work with a range of human abilities	Cobots that can easily be reprogrammed or reconfigured to adapt to different tasks or work environments, and that can adjust their movements or behavior to suit different human needs
Collaboration	Cobots that can collaborate with human operators in a way that maximizes productivity, efficiency, and well-being for both humans and cobots	Cobots that can communicate and collaborate effectively with human operators, that can perform tasks that complement human abilities, and that can adapt their behavior to suit the needs and preferences of human operators
Usability	Designing cobots that are usable and accessible for a wide range of human operators, including those with physical or cognitive disabilities, and that can be integrated easily into existing work environments	Providing cobots with features such as adjustable height, easy-to-reach controls, and voice-activated commands, and ensuring that cobots can be easily integrated with existing equipment and workspaces

Another principle is to ensure that cobots are designed to support collaboration between humans and robots (Li et al., 2023). This requires designing the robot to be aware of human movements and intentions and to be able to adapt its behavior to support the human (Jahanmahin et al., 2022). For example, the robot should be able to anticipate when the human needs assistance and provide that assistance in a timely and appropriate manner.

The human-centered design of cobots should also consider the physical and cognitive limitations of the human user (Fournier et al., 2022). The robot should be designed to minimize physical strain and fatigue on the human user and to support their cognitive abilities. This may involve designing the robot to take over repetitive or physically demanding tasks, leaving the worker free to focus on more complex and creative tasks.

Finally, the human-centered design of cobots should consider the social and emotional aspects of human–robot interaction (Cusano, 2022; Eyam et al., 2021). The robot should be designed to foster a sense of trust and collaboration with the human user and to provide feedback and recognition that supports the worker's motivation and well-being (Kumar et al., 2021).

5.3.4 User interface (UI) design of cobots

The user experience (UX) of cobot interfaces is an essential aspect of designing effective and efficient human–robot collaboration in industrial environments (Duarte et al., 2022; Frijns & Schmidbauer, 2021). A well-designed UX can increase worker satisfaction, productivity, and safety while reducing errors and the need for extensive training.

Effective UX design of cobot interfaces can significantly improve human–robot collaboration, increase productivity and safety, and enhance worker satisfaction. By considering the principles of simplicity, clarity, flexibility, feedback, compatibility, and usability testing, designers can create effective and efficient interfaces that optimize the UX and support human performance. The key considerations for designing cobot interfaces that optimize the UX are summarized in Table 5.4.

5.3.5 Cobot as a digital twin

Digital twin technology is a virtual representation of a physical system, product, or process that can be used for simulation, monitoring, and optimization purposes (Mazumder et al., 2023). It can be implemented using virtual reality technology (Bermúdez I Badia et al., 2022). In the context of cobots, digital twin technology can play a critical role in improving the performance and safety of these collaborative systems (Park et al., 2022).

Table 5.4 UX considerations of cobot design

Consideration	Explanation
Simplicity	The interface of the cobot should be easy to use, intuitive, and require minimal training. Users should be able to quickly understand the functions of the robot and operate it efficiently.
Clarity	The interface should provide clear and concise information about the cobot's status, actions, and feedback. This information should be presented in a way that is easy to understand and visually appealing.
Flexibility	The interface should allow for customization and flexibility to adapt to different tasks and user preferences. This can include the ability to change the robot's movements, speed, and actions.
Feedback	The interface should provide real-time feedback to the user, such as visual and auditory signals, to ensure that the user is aware of the robot's actions and status. This can help to prevent errors and increase safety.
Compatibility	The interface should be compatible with different types of devices and platforms, such as smartphones, tablets, and PCs. This can increase the flexibility and accessibility of the cobot system.
Usability testing	The interface should be tested with real users in a simulated environment to identify usability issues and improve the UX. This can involve user testing, observation, and feedback.

One of the key benefits of digital twin technology is that it allows for the creation of a virtual environment where the behavior of the cobot can be tested and optimized before it is deployed in the real world (Wang et al., 2021). This can help to identify and address any potential issues or safety concerns before they arise, reducing the risk of accidents and improving the overall efficiency of the cobot. Another important aspect of digital twin technology is its ability to enable real-time monitoring and control of the cobot.

By collecting and analyzing data from sensors and other sources, the digital twin can provide insights into the performance of the cobot, identifying areas for improvement and enabling operators to make more informed decisions about how to optimize its operation.

In addition, digital twin technology can also facilitate collaboration between humans and cobots. By creating a virtual environment that accurately reflects the real-world conditions in which the cobot will operate, it is possible to train human workers on how to interact safely and effectively with the cobot, reducing the risk of accidents and improving overall productivity.

To summarize, the role of digital twin technology in cobots is to enhance their performance, safety, and overall efficiency. As cobots become increasingly common in industrial settings, digital twin technology is likely to play an increasingly important role in ensuring that these collaborative systems operate safely and effectively.

5.3.6 The use of AI methods

Artificial Intelligence (AI) methods are increasingly being used in cobot technology to enhance human–robot collaboration and ensure safety in the workplace (Borboni et al., 2023). Some of the AI methods used by cobots include the following.

- **Machine learning:** This involves training a cobot to recognize patterns in data and make predictions based on that data. Machine learning algorithms can be used to improve cobots' ability to detect and respond to objects in their environment, to learn from human behavior and adjust their movements accordingly, and to predict and prevent errors or accidents (Semeraro et al., 2022).
- **Computer vision:** Cobots use computer vision algorithms to detect, recognize, and classify objects in their environment, which allows them to interact with objects and people more accurately and safely. Computer vision algorithms can also be used to identify defects or quality issues in manufacturing processes and to monitor workers' activities for safety and productivity (Mart´ınez-Franco & Alvarez Mart´ınez, 2021).
- **Natural language processing:** This involves teaching cobots to understand and respond to natural language commands and queries. Cobots with natural language processing capabilities can communicate more effectively with human workers, enabling more collaborative and efficient workflows (Angleraud et al., 2021; Tirmizi et al., 2019).
- **Reinforcement learning:** This involves training a cobot to learn from its own experiences and adjust its behavior accordingly. Reinforcement learning algorithms can be used to optimize cobots' movements and actions and to learn from the feedback provided by human workers (Alessio et al., 2022; Gomes et al., 2021; Li, 2021; Winter et al., 2019).
- **Deep learning:** This involves training cobots to recognize complex patterns in data using deep neural networks. Deep learning algorithms can be used to improve cobots' accuracy and efficiency in tasks such as object recognition and motion planning (Luneckas et al., 2021).

The AI methods play a critical role in enabling cobots to operate intelligently, autonomously, and collaboratively with human workers, improving efficiency, safety, and productivity in manufacturing and other industries. One of the key applications of AI in cobot technology is in human behavior prediction, safe zone detection, task planning and optimization, and quality control.

- Human behavior prediction involves using machine learning algorithms to analyze human behavior and predict future actions. By analyzing the patterns and movements of human workers, cobots can anticipate their behavior and adjust their movements accordingly (Male & Martinez-Hernandez, 2021). For example, if a human worker is about to walk into the cobot's work area, the cobot can slow down or stop its movements to avoid any potential collisions.

- Safe zone detection involves using sensors and AI algorithms to detect the presence of humans in the cobot's work area (Hammer & Jung, 2020). The cobot can then adjust its movements to ensure that it stays within its designated work area and avoids any potential collisions with humans. For example, if a human worker enters the cobot's work area, the cobot can stop its movements until the worker has left the area.
- Task planning and optimization involves using algorithms to plan and optimize the cobot's movements and actions based on the specific task it is performing (El Zaatari et al., 2019). By analyzing factors such as the location and orientation of objects, the cobot can determine the most efficient path to complete the task. This can lead to increased productivity and efficiency in manufacturing and other industries that use cobots.
- Quality control in manufacturing processes involves using machine vision and AI algorithms that can detect defects in products as they are being manufactured. By analyzing images of the products, the cobot can identify defects and either remove the defective products from the assembly line or alert human workers to make necessary adjustments.

Therefore, the AI methods enable cobots to perform tasks more efficiently and effectively, while also enhancing safety and reducing the risk of errors and accidents.

5.4 CASE STUDY: INTEGRATING COBOTS INTO A MANUFACTURING COMPANY

5.4.1 Generic architecture of cobots

The generic architecture of cobots can be divided into the mechanical structure, control system, sensor and perception system, and human–machine interface (HMI) (Segura et al., 2021).

The mechanical structure of a cobot includes the robot arm, grippers, and end-effectors (Devaraja et al., 2021). The robot arm is the core of the mechanical structure, which provides the cobot with the capability of movement and flexibility (Devaraja et al., 2020). The gripper and end-effectors are used to manipulate and interact with the objects in the environment.

The control system of a cobot is responsible for controlling the movement and operation of the robot. It includes the motion control system, power supply system, and communication system. The motion control system is used to control the speed and direction of the robot. The power supply system provides the necessary power to the robot, and the communication system allows the robot to communicate with other machines or humans.

The sensor and perception system of a cobot includes various sensors such as cameras, LiDAR, and tactile sensors. These sensors are used to capture the environmental data and provide the robot with the necessary information to perform the task. The perception system uses the data from the sensors to recognize and understand the objects and environment.

The HMI is used to enable communication and interaction between the cobot and human workers (Coronado et al., 2022). This interface can take various forms, including buttons, touchscreens, voice recognition, and gestures. The interface is designed to be intuitive and easy to use, allowing the human worker to interact with the robot effectively.

The generic architecture of cobot (Figure 5.2) is designed to enable effective human–machine cooperation and support human workers in various industries. There are five main components: the Human, the Robot, the Sensor, the Actuator, and the Controller. The Human interacts with the Controller by sending commands, while the Robot receives commands from the Controller. The Sensor sends data to the Controller, which collects data from multiple sensors. The Actuator receives data from the Controller and sends commands to the physical system. Finally, the Controller interacts with the Human by receiving feedback and sends commands to the Robot. The architecture is flexible and can be customized based on the specific requirements of the application.

5.4.2 Integration of cobots in the production process

The integration of cobots in the production process requires careful consideration and planning to ensure that they can work effectively with human

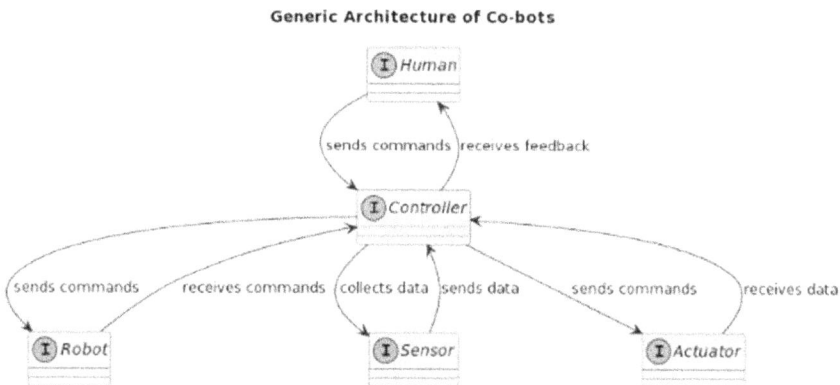

Figure 5.2 Generic architecture of cobots.

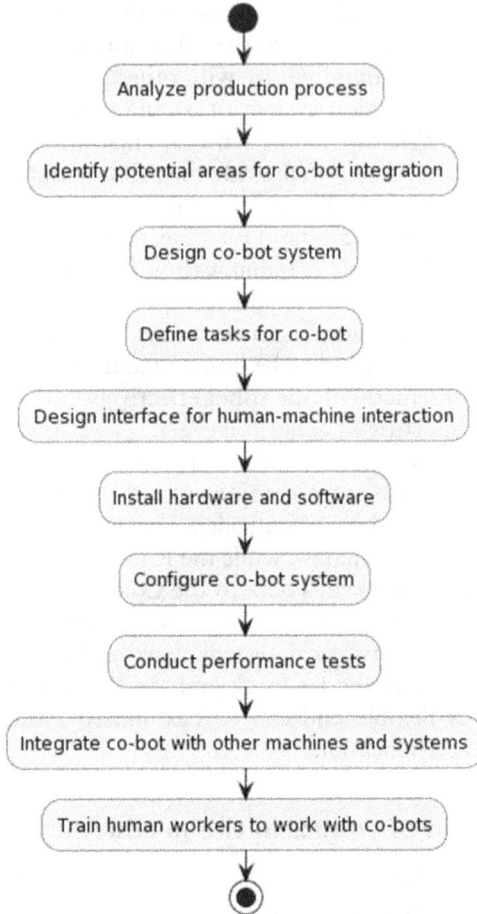

Figure 5.3 Integration of cobots in the production process.

workers and other machines in the production line. This integration involves several steps (Figure 5.3):

- **Analyzing the production process:** The first step is to analyze the production process and identify areas where cobots can be integrated. This analysis should consider factors such as the type of task, the level of complexity, the skill required, and the potential benefits of using cobots.
- **Designing the cobot system:** Once the potential areas for cobot integration are identified, the next step is to design the cobot system. This includes selecting the appropriate cobot type, defining the tasks it will perform, and designing the interface for human–machine interaction (Hentout et al., 2019).

- **Installation and testing:** After designing the cobot system, it needs to be installed and tested to ensure that it operates smoothly and efficiently. This involves installing the hardware and software, configuring the system, and conducting tests to verify its performance.
- **Integration with other machines and systems:** Cobots must be integrated with other machines and systems in the production line to ensure that they can operate in coordination with them. This integration may involve modifying existing systems or designing new ones that can work with the cobot system.
- **Training and support:** Human workers must be trained to work with cobots and to understand their capabilities and limitations. They should also be provided with support and assistance to ensure that they can use the cobot effectively.

The integration of cobots in the production process can improve productivity, reduce costs, and enhance the quality of products (Cohen et al., 2022). However, it requires careful planning, design, and implementation to ensure that cobots can work effectively with human workers and other machines in the production line.

5.4.3 Real-life scenarios of human and cobot operation

5.4.3.1 Generic scenario of cobot operation

Here is a typical operation scenario of a cobot in industrial manufacturing.

(1) **Initialization:** The cobot is initialized by the operator, who inputs the necessary parameters and task details, such as the type of product to be manufactured, the required quality specifications, and the quantity.
(2) **Material handling:** The cobot is programmed to pick up the raw materials required for the production process and transport them to the production line.
(3) **Assembly and machining:** Once the raw materials are transported to the production line, the cobot is programmed to assemble and machine the parts according to the product specifications. This may include cutting, drilling, welding, and other operations.
(4) **Quality control:** As the product is being assembled, the cobot is programmed to perform quality control checks to ensure that the product meets the required specifications.
(5) **Packaging and shipping:** Once the product is assembled and inspected, the cobot is programmed to package the product and prepare it for shipping.
(6) **Maintenance:** The cobot is programmed to perform routine maintenance tasks such as cleaning, oiling, and changing worn parts to ensure the smooth operation of the cobot and the production process.

Throughout the process, the cobot communicates with the operator to receive instructions, provide feedback, and alert the operator if there are any issues that need to be addressed. The cobot is also equipped with sensors and safety mechanisms to ensure that it operates safely and efficiently.

5.4.3.2 Collaboration scenario for the assembly task

Further, we discuss the example scenario in which a cobot and a human worker collaborate to assemble a product as follows (Figure 5.4). This sequence diagram represents the collaboration between a human worker and a cobot in an assembly process. The human worker initiates the process by requesting the cobot to start the assembly process. The cobot then sends a signal to the manufacturing system to start the process and the conveyor belt is started. As objects are detected on the conveyor belt, the manufacturing system sends a signal to the cobot to pick up the object. The cobot then requests confirmation of the object from the human worker, who confirms it. The cobot then sends a signal to the manufacturing system to continue the process and move to the next station. The conveyor belt is stopped and the cobot moves to the next station to continue the assembly process.

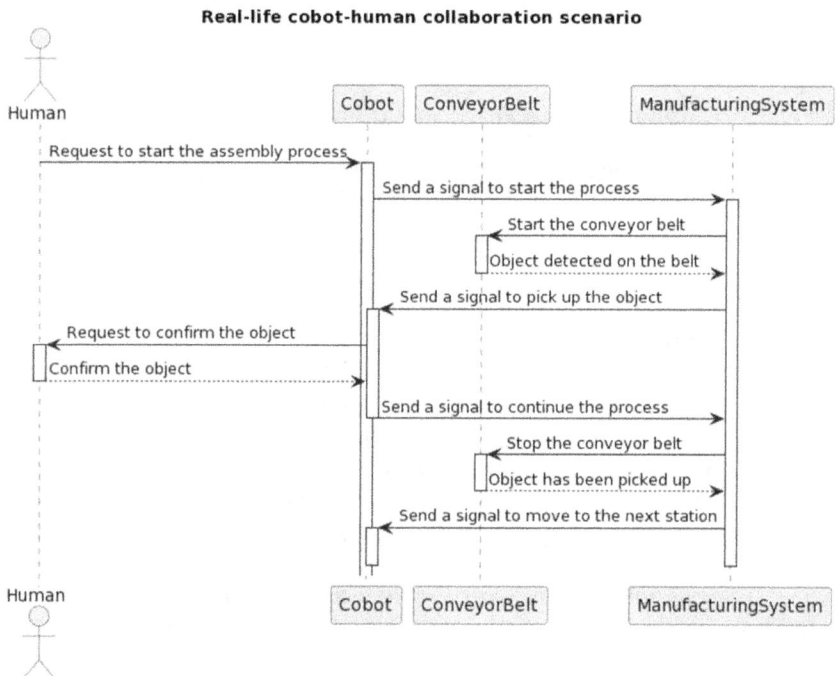

Figure 5.4 Example of a real-life scenario of assembly.

5.4.3.3 Collaboration scenario for the painting–coating task

In another example, we examine the industrial painting–coating task. The sequence diagram (Figure 5.5) illustrates a real-life scenario of painting and coating by a cobot, with the participation of a human worker and the paint application process. The first step is performed by the human worker, who prepares the paint and the object to be painted. Then, the human worker positions the object, and the cobot starts scanning it to generate a 3D model. The cobot adjusts the position and orientation of the object based on the model and then applies the base coat. The human worker assists the cobot with masking and positioning during the application process. After applying the base coat, the cobot waits for it to dry before proceeding to apply the second coat. The human worker inspects and touches up the paint as needed.

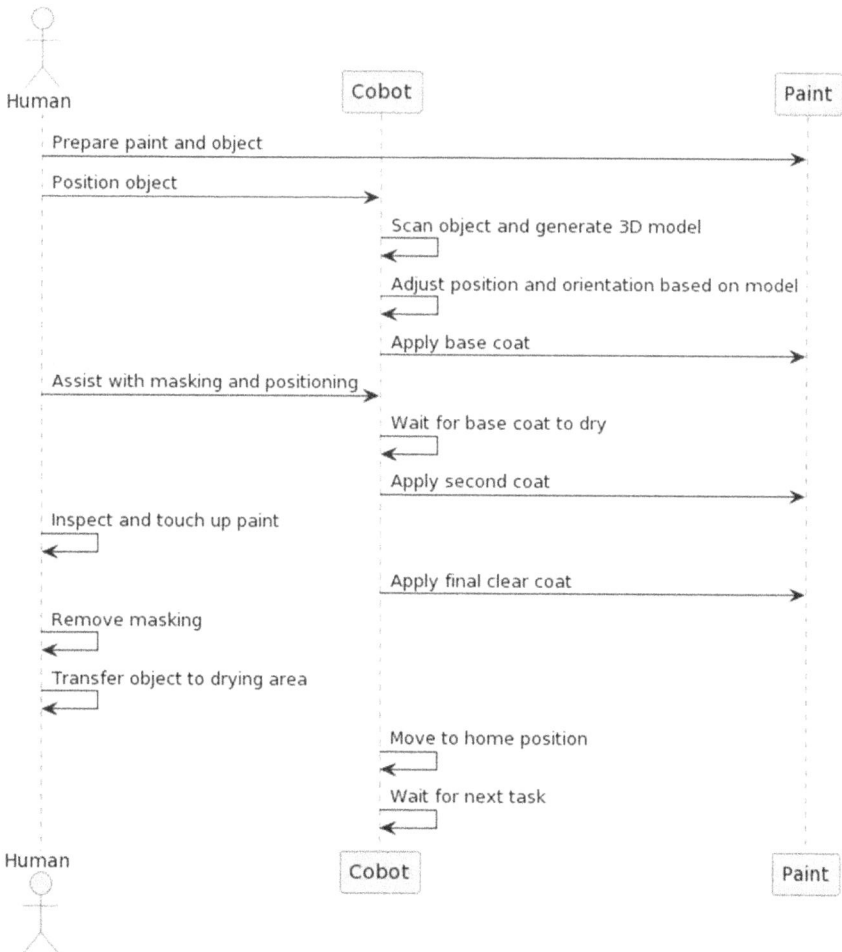

Figure 5.5 Example of a real-life scenario of painting and coating.

Then, the cobot applies the final clear coat, and the human worker removes the masking. Finally, the human worker transfers the object to the drying area, while the cobot moves to its home position and waits for the next task.

5.5 THE ROLE OF COBOTS IN INDUSTRY 5.0 AND FUTURE RESEARCH DIRECTIONS

5.5.1 Benefits and drawbacks of cobots in Industry

Cobots offer several benefits to industries, including increased productivity, improved quality of work, greater flexibility, lower costs, and improved safety (Giberti et al., 2022). Cobots can work alongside humans and perform repetitive or dangerous tasks, resulting in higher productivity. They can also reduce errors and defects in the manufacturing process by performing tasks with consistent accuracy and precision. Cobots can be reprogrammed or reconfigured easily, making them more flexible than traditional robots. They are also less expensive to purchase and maintain, making them more accessible to small- and medium-sized businesses. Cobots are designed with sensors and other safety features to work safely alongside human workers, preventing accidents and collisions.

The drawbacks of cobots in industry include limited capabilities, integration challenges, maintenance and repair, lack of human touch, and the need for specialized skills. Cobots may not be suitable for tasks that require a high degree of force or speed and may require significant changes to the manufacturing process to integrate with existing equipment and systems. They also require regular maintenance and specialized skills to program and operate and may lack the human touch necessary for some tasks.

The advantages and drawbacks of cobots in industry are summarized in Table 5.5. The benefits of cobots in industry outweigh the drawbacks, making them a promising technology for improving productivity, quality, and safety in manufacturing and other industries. However, careful consideration and planning are necessary to ensure successful integration and effective use of cobots in the manufacturing process.

5.5.2 The potential of cobots in improving work support and productivity

Cobots have the potential to greatly improve work support and productivity in a variety of industries. Some potential benefits of cobots in this regard include the following.

- **Reducing physical strain on workers:** Cobots can be used to perform physically demanding tasks, such as lifting heavy objects, that can cause strain or injury to human workers. This can help reduce the risk of workplace injuries and improve worker safety.

Table 5.5 Advantages and drawbacks of cobots

Advantages of cobots	Drawbacks of cobots
Increased productivity: Cobots can work alongside human workers, performing repetitive or dangerous tasks, freeing up human workers to focus on more complex tasks, and increasing overall productivity.	**Limited capabilities**: Cobots are generally less powerful than traditional industrial robots and may not be able to perform tasks that require a high degree of force or speed.
Improved quality: Cobots can perform tasks with consistent accuracy and precision, reducing errors and defects in the manufacturing process.	**Integration challenges**: Cobots may require significant changes to the manufacturing process and may be difficult to integrate with existing equipment and systems.
Greater flexibility: Cobots can be easily reprogrammed or reconfigured to perform different tasks or to adapt to changes in the manufacturing process, making them more flexible than traditional industrial robots.	**Maintenance and repair**: Cobots require regular maintenance and may be more difficult to repair than traditional industrial robots due to their complex and interconnected systems.
Lower costs: Cobots are typically less expensive to purchase and maintain than traditional industrial robots, making them more accessible to small- and medium-sized enterprises.	**Lack of human touch**: Cobots may lack the human touch in some tasks where human intuition is necessary to make decisions.
Improved safety: Cobots are designed to work safely alongside human workers, with sensors and other safety features that prevent collisions and other accidents.	**Need for specialized skills**: Cobots require specialized skills to program and operate, which may require additional training or hiring of specialized personnel.

- **Improving efficiency and productivity**: Cobots can work alongside human workers to complete tasks more quickly and efficiently, reducing production times and increasing output.
- **Improving quality and accuracy**: Cobots are able to perform tasks with a high level of precision and accuracy, helping to improve the overall quality of products and reducing the number of errors or defects.
- **Enabling more flexible production processes**: Cobots can be quickly reprogrammed or reconfigured to perform different tasks or to adapt to changes in the manufacturing process, making production processes more flexible and adaptable.
- **Providing real-time data and insights**: Cobots can collect and analyze data in real-time, providing valuable insights into production processes and helping to identify areas for improvement.
- **Enhancing worker skills and training**: Cobots can be used to train workers on new skills and tasks, providing opportunities for skill development and career growth.

The potential of cobots to improve work support and productivity is significant. By working alongside human workers to perform tasks more quickly, accurately, and safely, cobots have the potential to revolutionize production processes and improve outcomes for workers and companies alike.

5.5.3 Real-time operation of cobots

Real-time operation is critical for cobots in the discussed case study and in most robot operation scenarios. The cobots are designed to work alongside human workers, and their movements need to be precise and coordinated with the movements of the human workers. Any delays or lags in the cobot's operation can lead to accidents and injuries. For example, in the painting and coating scenario discussed earlier, the cobot needs to apply the paint or coating to the object precisely and in a timely manner to avoid drips and ensure a smooth finish. If there is a delay in the cobot's operation, the paint or coating may dry out or be applied unevenly, leading to defects in the finished product.

Real-time operation is also important for cobots' safety features, such as sensors that detect the presence of humans in their work area. If the cobot's sensors have a delay in detecting the presence of a human worker, it may not be able to adjust its movements in time to avoid a collision. Therefore, real-time operation is critical for cobots to perform their tasks efficiently and safely. This requires advanced control systems and algorithms that can ensure that the cobot's movements are coordinated with the movements of human workers and other machines in the production line and that any safety risks are quickly detected and addressed.

5.5.4 Ethical and legal issues of cobots in industry

The use of cobots in industry raises ethical and legal issues that must be addressed (Chromjakova et al., 2021). Some of the key issues include the following.

- **Job displacement:** The use of cobots has the potential to displace human workers, raising questions about the impact on employment and the potential for social and economic disruption.
- **Safety and liability:** Cobots must be designed and programmed to operate safely in close proximity to human workers. Any failure to ensure safety could lead to injuries or other accidents, raising questions of liability and legal responsibility.
- **Privacy and data security:** Cobots collect and transmit data, raising concerns about data privacy and security. Companies must take steps to protect sensitive data and ensure that cobots are not used to violate the privacy rights of workers or customers (Venaik et al., 2023).

- **Autonomy and control**: As cobots become more advanced, they may have the ability to make decisions and act autonomously. This raises questions about who is responsible for decisions made by cobots and how their actions can be controlled or regulated.
- **Fairness and bias**: Cobots are programmed to make decisions based on data and algorithms, but these algorithms may contain biases or perpetuate unfair practices. This raises questions about the potential for discrimination and the need to ensure fairness in cobot decision-making.
- **Intellectual property**: Cobots may be used to automate processes or tasks that were previously performed by human workers (Damaševičius et al., 2020; Lefranc et al., 2022). This raises questions about the ownership of intellectual property and the rights of workers who may have contributed to the development of these processes.

To address these ethical and legal issues, companies must take a proactive approach to designing and implementing cobot systems. This includes engaging with workers and other stakeholders to address concerns about job displacement and safety, implementing strong data privacy and security measures, and ensuring that cobot decision-making is transparent and fair. Policymakers and regulators must work to develop guidelines and regulations that promote the responsible use of cobots in industry, balancing the potential benefits with the need to protect workers and other stakeholders.

5.5.5 Future research directions in human–machine cooperation and cobots

The field of human–machine cooperation and cobots is rapidly evolving, with new technologies and applications emerging at a rapid pace. Future research in this area is likely to focus on a number of key areas, including the following.

(1) **Human–robot interaction**: Researchers will continue to explore ways to improve the interaction between humans and cobots, focusing on areas such as natural language processing, speech recognition, and object recognition.
(2) **Artificial intelligence**: As cobots become more advanced, researchers will explore new ways to integrate artificial intelligence into cobot systems, enabling them to make more complex decisions and operate autonomously.
(3) **Ethics and social impact**: The use of cobots raises a number of ethical and social impact questions, such as job displacement, privacy, and data security. Future research will focus on developing frameworks and guidelines for the responsible use of cobots and exploring the potential social and economic impact of these technologies.

(4) **Industry-specific applications**: Cobots have the potential to revolutionize a wide range of industries, from manufacturing and logistics to healthcare and retail. Future research will focus on developing cobot applications tailored to the specific needs and challenges of different industries.

(5) **Training and education**: As cobots become more prevalent in the workforce, there will be a growing need for workers to be trained in their use and operation. Future research will focus on developing training and education programs to ensure that workers have the necessary skills to work effectively with cobots.

(6) **Human-centered design**: As cobots become more integrated into the workforce, there will be a growing need for human-centered design approaches that take into account the needs and preferences of human workers. Future research will focus on developing cobot systems that are designed to work seamlessly with human workers, improving productivity and safety in the workplace.

To summarize, the future of cobots and human–machine cooperation is bright, with significant potential to transform a wide range of industries and improve the lives of workers and consumers alike. By focusing on these key research areas, researchers and practitioners can ensure that cobot technology is developed and implemented in a responsible and effective manner.

5.6 CONCLUSION

Cobots have the potential to significantly increase productivity and flexibility in industrial manufacturing by providing human–machine collaboration and support. Different types of cobots have unique applications and benefits in various industries and manufacturing processes. Cobots can improve workplace safety by taking over dangerous or repetitive tasks, reducing the risk of injuries for human workers. Human-centered design principles and well-designed user interfaces are crucial for the successful integration and adoption of cobots in the production process.

The research article provides important implications for the adoption of cobots in Industry 5.0. The findings suggest that cobots have significant potential to improve work support, productivity, and safety in various industries. Therefore, companies should consider integrating cobots into their production processes to improve efficiency and quality while reducing costs. However, there are also challenges associated with cobot adoption that need to be addressed. These include ethical and legal issues, integration with human workers, and the need for human-centered design principles to ensure the usability and effectiveness of cobot interfaces. To fully realize the potential of cobots, companies need to prioritize employee training and education to ensure that workers are comfortable and knowledgeable about working alongside these machines. Additionally, companies need to establish clear ethical and legal guidelines for the use of cobots and implement appropriate safeguards to prevent potential harm or misuse.

Future research should aim to address these limitations by conducting more comprehensive studies that explore the adoption and use of cobots across a range of industries and contexts. This could involve longitudinal studies that track the evolution of cobot use over time, as well as studies that incorporate user feedback and perspectives. Another important area for future research is the development of more advanced cobots that are capable of more complex tasks and interactions with human workers. This could involve the integration of advanced technologies such as AI, machine learning, and computer vision, as well as the development of more sophisticated interfaces and control systems. Currently, most cobots use simple machine learning algorithms to analyze human behavior and predict their movements. However, with more advanced AI algorithms, cobots could become even more efficient and effective in working alongside human workers. Another area of future research could be the development of more advanced sensors and safety features for cobots. While current cobots are designed to work safely alongside human workers, there is always room for improvement. By developing more advanced sensors and safety features, cobots could become even safer and more reliable. In addition, there is a need for more research into the social and ethical implications of cobots in the workplace. As cobots become more prevalent in industrial production, there may be concerns about the impact on employment, worker safety, and privacy. More research is needed to address these concerns and develop strategies for mitigating any negative effects. There is a lot of potential for future research and development in the field of cobots in industrial production. By continuing to improve and refine cobot technology, we can create a safer, more efficient, and more productive workplace for human workers.

REFERENCES

Adriaensen, A., Costantino, F., Di Gravio, G., & Patriarca, R. (2022). Teaming with industrial cobots: A socio-technical perspective on safety analysis. *Human Factors and Ergonomics in Manufacturing*, 32 (2), 173–198.

Alessio, A., Aliev, K., & Antonelli, D. (2022). Robust adversarial reinforcement learning for optimal assembly sequence definition in a cobot workcell.

Amin, F. M., Rezayati, M., van de Venn, H. W., & Karimpour, H. (2020). A mixed-perception approach for safe human–robot collaboration in industrial automation. *Sensors*, 20 (21), 1–20.

Andersson, S. K. L., Granlund, A., Bruch, J., & Hedelind, M. (2021). Experienced challenges when implementing collaborative robot applications in assembly operations. *International Journal of Automation Technology*, 15 (5), 678–688.

Angleraud, A., Mehman Sefat, A., Netzev, M., & Pieters, R. (2021). Coordinating shared tasks in human-robot collaboration by commands. *Frontiers in Robotics and AI*, 8.

Batth, R. S., Nayyar, A., & Nagpal, A. (2018). Internet of robotic things: Driving intelligent robotics of future - concept, architecture, applications and technologies. In *2018 4th International Conference on Computing Sciences (ICCS)* (pp. 151–160).

Bermúdez I Badia, S., Silva, P. A., Branco, D., Pinto, A., Carvalho, C., Menezes, P., ...Pilacinski, A. (2022). Virtual reality for safe testing and development in collaborative robotics: Challenges and perspectives. *Electronics*, 11 (11).

Bi, Z. M., Chen, B., Xu, L., Wu, C., Malott, C., Chamberlin, M., & Enterline, T. (2022). Security and safety assurance of collaborative manufacturing in industry 4.0. *Enterprise Information Systems*, 16 (12).

Borboni, A., Reddy, K. V. V., Elamvazuthi, I., Al-Quraishi, M. S., Natarajan, E., & Azhar Ali, S. S. (2023). The expanding role of artificial intelligence in collaborative robots for industrial applications: A systematic review of recent works. *Machines*, 11 (1).

Boschetti, G., Faccio, M., & Granata, I. (2023). Human-centered design for productivity and safety in collaborative robots cells: A new methodological approach. *Electronics*, 12 (1).

Buerkle, A., Eaton, W., Al-Yacoub, A., Zimmer, M., Kinnell, P., Henshaw, M., ...Lohse, (2023). Towards industrial robots as a service (IRaaS): Flexibility, usability, safety and business models. *Robotics and Computer-Integrated Manufacturing*, 81.

Calvo, R., & Gil, P. (2022). Evaluation of collaborative robot sustainable integration in manufacturing assembly by using process time savings. *Materials*, 15 (2).

Chiriatti, G., Ciccarelli, M., Forlini, M., Franchini, M., Palmieri, G., Papetti, A., & Germani, (2022). Human-centered design of a collaborative robotic system for the shoe-polishing process. *Machines*, 10 (11).

Chiriatti, G., Palmieri, G., & Palpacelli, M. C. (2021). *Collaborative robotics for rehabilitation: A multibody model for kinematic and dynamic analysis* (Vol. 91).

Chromjakova, F., Trentesaux, D., & Kwarteng, M. A. (2021). Human and cobot cooperation ethics: The process management concept of the production workplace. *Journal of Competitiveness*, 13 (3), 21–38.

Cohen, Y., Shoval, S., Faccio, M., & Minto, R. (2022). Deploying cobots in collaborative systems: major considerations and productivity analysis. *International Journal of Production Research*, 60 (6), 1815–1831.

Coronado, E., Kiyokawa, T., Ricardez, G. A. G., Ramirez-Alpizar, I. G., Venture, G., & Yamanobe, N. (2022). Evaluating quality in human-robot interaction: A systematic search and classification of performance and human-centered factors, measures and metrics towards an industry 5.0. *Journal of Manufacturing Systems*, 63, 392–410.

Cusano, N. (2022). *Cobot and sobot: For a new ontology of collaborative and social robots*. Foundations of Science.

Damaševičius, R., Maskeliūnas, R., Narvydas, G., Narbutaitė, R., Polap, D., & Woźniak, M. (2020). Intelligent automation of dental material analysis using robotic arm with jerk optimized trajectory. *Journal of Ambient Intelligence and Humanized Computing*, 11 (12), 6223–6234.

Damaševičius, R., Narbutaite, L., Plauska, I., & Blažauskas, T. (2017). Advances in the use of educational robots in project-based teaching. *TEM Journal*, 6 (2), 342–348.

Devaraja, R. R., Maskeliūnas, R., & Damaševičius, R. (2020). *Aisra: Anthropomorphic robotic hand for small-scale industrial applications* (Vol. 12249 LNCS).

Devaraja, R. R., Maskeliūnas, R., & Damaševičius, R. (2021). Design and evaluation of anthropomorphic robotic hand for object grasping and shape recognition. *Computers*, 10 (1), 1–14.

Dobra, Z., & Dhir, K. S. (2020). Technology jump in the industry: human–robot cooperation in production. *Industrial Robot*, 47 (5), 757–775.

Duarte, I. M., Pinto, A., Carvalho, C., Zornoza, A., & Santos, J. (2022). The contribution of the user experiences goals for designing better cobots: A systematic literature review. *Applied System Innovation*, 5 (6).

El Zaatari, S., Marei, M., Li, W., & Usman, Z. (2019). Cobot programming for collaborative industrial tasks: An overview. *Robotics and Autonomous Systems*, 116, 162–180.

Eyam, A. T., Mohammed, W. M., & Martinez Lastra, J. L. (2021). Emotion-driven analysis and control of human-robot interactions in collaborative applications. *Sensors*, 21 (14).

Faccio, M., Granata, I., Menini, A., Milanese, M., Rossato, C., Bottin, M., ... Rosati, G. (2023). Human factors in cobot era: a review of modern production systems features. *Journal of Intelligent Manufacturing*, 34 (1), 85–106.

Fager, P., Sgarbossa, F., & Calzavara, M. (2020, December). Cost modelling of onboard cobot-supported item sorting in a picking system. *International Journal of Production Research*, 59 (11), 3269–3284.

Fournier, E., Kilgus, D., Landry, A., Hmedan, B., Pellier, D., Fiorino, H., & Jeoffrion, C. (2022). The impacts of human-cobot collaboration on perceived cognitive load and usability during an industrial task: An exploratory experiment. *IISE Transactions on Occupational Ergonomics and Human Factors*, 10 (2), 83–90.

Frijns, H. A., & Schmidbauer, C. (2021). Design guidelines for collaborative industrial robot user interfaces (Vol. 12934 LNCS).

Giberti, H., Abbattista, T., Carnevale, M., Giagu, L., & Cristini, F. (2022). A methodology for flexible implementation of collaborative robots in smart manufacturing systems. *Robotics*, (1).

Gomes, N. M., Martins, F. N., Lima, J., & Wörtche, H. (2021). Deep reinforcement learning applied to a robotic pick-and-place application (Vol. 1488 CCIS).

Hameed, A., Ordys, A., Możaryn, J., & Sibilska-Mroziewicz, A. (2023). Control system design and methods for collaborative robots: Review. *Applied Sciences*, 13 (1).

Hammer, C., & Jung, N. (2020). Research project beyondSPAI – the safe and reliable monitoring of adaptive safety zones in the proximity of collaborating industrial robots using an intelligent InGaAs camera system (Vol. 12198 LNCS).

Hentout, A., Aouache, M., Maoudj, A., & Akli, I. (2019). Human–robot interaction in industrial collaborative robotics: a literature review of the decade 2008–2017. *Advanced Robotics*, 33 (15–16), 764–799.

Jahanmahin, R., Masoud, S., Rickli, J., & Djuric, A. (2022). Human-robot interactions in manufacturing: A survey of human behavior modeling. *Robotics and Computer-Integrated Manufacturing*, 78.

Javaid, M., Haleem, A., Singh, R. P., Rab, S., & Suman, R. (2022). Significant applications of cobots in the field of manufacturing. *Cognitive Robotics*, 2, 222–233.

Jones, J. E., Dydo, J. R., Rhoades, V. L., Fast, K., Beard, J., Bryant, A., ... Gaffney, J. H. (2015). Development of a collaborative robot (COBOT) for increased welding productivity and quality in the shipyard. In *SNAME 5th World Maritime Technology Conference*, WMTC 2015.

Kumar, A., & Nayyar, A. (2019, November). si3-industry: A sustainable, intelligent, innovative, internet-of-things industry. In *A roadmap to industry 4.0: Smart production, sharp business and sustainable development* (pp. 1–21). Springer International Publishing.

Kumar, S., Savur, C., & Sahin, F. (2021). Survey of human-robot collaboration in industrial settings: Awareness, intelligence, and compliance. *IEEE Transactions on Systems, Man, and Cybernetics: Systems*, 51 (1), 280–297.

Leesakul, N., Oostveen, A., Eimontaite, I., Wilson, M. L., & Hyde, R. (2022). Workplace 4.0: Exploring the implications of technology adoption in digital manufacturing on a sustainable workforce. *Sustainability*, 14 (6), 3311.

Lefranc, G., Lopez-Juarez, I., Osorio-Comparán, R., & Peña-Cabrera, M. (2022). Impact of cobots on automation. *Procedia Computer Science*, 214, 71–78.

Li, S., Zheng, P., Liu, S., Wang, Z., Wang, X. V., Zheng, L., & Wang, L. (2023). Proactive human–robot collaboration: Mutual-cognitive, predictable, and self-organising perspectives. *Robotics and Computer-Integrated Manufacturing*, 81.

Li, W. D. (2021). Reinforcement learning-based learning from demonstrations for collaborative robots. In *IEEE international conference on automation science and engineering* (Vol. 2021, August, pp. 1642–1647).

Liang, C., Wang, X., Kamat, V. R., & Menassa, C. C. (2021). Human-robot collaboration in construction: Classification and research trends. *Journal of Construction Engineering and Management*, 147 (10).

Liu, L., Guo, F., Zou, Z., & Duffy, V. G. (2022). Application, development and future opportunities of collaborative robots (cobots) in manufacturing: A literature review. *International Journal of Human-Computer Interaction*.

Luneckas, M., Luneckas, T., Kriaučiūnas, J., Udris, D., Plonis, D., Damaševičius, R., & Maskeliūnas, R. (2021). Hexapod robot gait switching for energy consumption and cost of transport management using heuristic algorithms. *Applied Sciences*, 11 (3), 1–13.

Maddikunta, P. K. R., Pham, Q., Prabadevi, P., Deepa, N., Dev, K., Gadekallu, T. R., ... Liyanage, M. (2022). Industry 5.0: A survey on enabling technologies and potential applications. *Journal of Industrial Information Integration*, 26.

Male, J., & Martinez-Hernandez, U. (2021). Collaborative architecture for human-robot assembly tasks using multimodal sensors. In *2021 20th International Conference on Advanced Robotics, ICAR 2021* (pp. 1024–1029).

Martínez-Franco, J. C., & Alvarez Martínez, D. (2021). Machine vision for collaborative robotics using synthetic data-driven learning (Vol. 987).

Mazumder, A., Sahed, M. F., Tasneem, Z., Das, P., Badal, F. R., Ali, M. F., ...Islam, M. R. (2023). Towards next generation digital twin in robotics: Trends, scopes, challenges, and future. *Heliyon*, 9 (2).

Navas-Reascos, G. E., Romero, D., Rodriguez, C. A., Guedea, F., & Stahre, J. (2022). Wire harness assembly process supported by a collaborative robot: A case study focus on ergonomics. *Robotics*, 11 (6).

Nayyar, A., & Kumar, A. (Eds.). (2020). *A roadmap to industry 4.0: Smart production, sharp business and sustainable development*. Springer International Publishing.

Nguyen Ngoc, H., Lasa, G., & Iriarte, I. (2022). Human-centred design in industry 4.0: case study review and opportunities for future research. *Journal of Intelligent Manufacturing*, 33 (1), 35–76.

Papanagiotou, D., Senteri, G., & Manitsaris, S. (2021). Egocentric gesture recognition using 3d convolutional neural networks for the spatiotemporal adaptation of collaborative robots. *Frontiers in Neurorobotics*, 15.

Park, J., Caroe Sorensen, L., Faarvang Mathiesen, S., & Schlette, C. (2022). A digital twin-based workspace monitoring system for safe human-robot collaboration. In *2022 10th International Conference on Control, Mechatronics and Automation, ICCMA 2022* (p. 24–30).

Pauliková, A., Babeľová, Z. G., & Ubárová, M. (2021). Analysis of the impact of human–cobot collaborative manufacturing implementation on the occupational health and safety and the quality requirements. *International Journal of Environmental Research and Public Health*, 18 (4), 1–15.

Plauska, I., & Damaševičius, R. (2014). Educational robots for internet-of-things supported collaborative learning (Vol. 465).

Ponsa, P., & Tornil-Sin, S. (2022). Exploring the practical use of a collaborative robot for academic purposes. In *Sefi 2022 - 50th Annual Conference of the European Society for Engineering Education, Proceedings* (p. 1445–1453).

Qiu, S., Natarajarathinam, M., Johnson, M. D., & Roumell, E. A. (2021). The future of work: Identifying future-ready capabilities for the industrial distribution workforce. In *ASEE Annual Conference and Exposition, Conference Proceedings*.

Rodriguez-Guerra, D., Sorrosal, G., Cabanes, I., & Calleja, C. (2021). Human-robot interaction review: Challenges and solutions for modern industrial environments. *IEEE Access*, 9, 108557–108578.

Romanov, D., Korostynska, O., Lekang, O. I., & Mason, A. (2022). Towards human-robot collaboration in meat processing: Challenges and possibilities. *Journal of Food Engineering*, 331, 111117.

Rossi, F., Pini, F., Carlesimo, A., Dalpadulo, E., Blumetti, F., Gherardini, F., & Leali, F. (2020). Effective integration of cobots and additive manufacturing for reconfigurable assembly solutions of biomedical products. *International Journal on Interactive Design and Manufacturing*, 14 (3), 1085–1089.

Rožanec, J. M., Novalija, I., Zajec, P., Kenda, K., Tavakoli Ghinani, H., Suh, S., … Soldatos, (2022). Human-centric artificial intelligence architecture for industry 5.0 applications. *International Journal of Production Research*. https://doi.org/10.1080/00207543.2022.2138611

Salehzadeh, R., Gong, J., & Jalili, N. (2022). Purposeful communication in human-robot collaboration: A review of modern approaches in manufacturing. *IEEE Access*, 10, 129344–129361.

Sauer, V., Sauer, A., & Mertens, A. (2021). Zoomorphic gestures for communicating cobot states. *IEEE Robotics and Automation Letters*, 6 (2), 2179–2185.

Segura, P., Lobato-Calleros, O., Ramírez-Serrano, A., & Soria, I. (2021). Human-robot collaborative systems: Structural components for current manufacturing applications. *Advances in Industrial and Manufacturing Engineering*, 3.

Semeraro, F., Griffiths, A., & Cangelosi, A. (2022). Human–robot collaboration and machine learning: A systematic review of recent research. *Robotics and Computer-Integrated Manufacturing*, 79.

Sowa, K., Przegalinska, A., & Ciechanowski, L. (2021, March). Cobots in knowledge work. *Journal of Business Research*, 125, 135–142.

Tirmizi, A., Cat, B. D., Janssen, K., Pane, Y., Leconte, P., & Witters, M. (2019). User-friendly programming of flexible assembly applications with collaborative robots. In *Proceedings of the 2019 20th International Conference on Research and Education in Mechatronics, REM 2019.*

Venaik, A., Jain, S., & Nayyar, A. (2023). Industry 4.0—its advancement and effects on security of whistle-blowers on dark web. In *Industry 4.0 and the digital transformation of international business* (pp. 103–121). Springer Nature Singapore.

Vicentini, F. (2021). Collaborative robotics: A survey. *Journal of Mechanical Design, Transactions of the ASME,* 143 (4).

Villani, V., Pini, F., Leali, F., & Secchi, C. (2018). Survey on human–robot collaboration in industrial settings: Safety, intuitive interfaces and applications. *Mechatronics,* 55, 248–266.

Wang, X., Liang, C., Menassa, C. C., & Kamat, V. R. (2021). Interactive and immersive process-level digital twin for collaborative human-robot construction work. *Journal of Computing in Civil Engineering,* 35 (6). https://doi.org/10.1061/(ASCE) CP.1943-5487.0000988

Weiss, A., Wortmeier, A., & Kubicek, B. (2021). Cobots in Industry 4.0: A roadmap for future practice studies on human-robot collaboration. *IEEE Transactions on Human-Machine Systems,* 51 (4), 335–345.

Winter, J. D., Beir, A. D., Makrini, I. E., de Perre, G. V., Nowé, A., & Vanderborght, B. (2019). Accelerating interactive reinforcement learning by human advice for an assembly task by a cobot. *Robotics,* 8 (4).

Yin, M., & Li, J. (2023). A systematic review on digital human models in assembly process planning. *International Journal of Advanced Manufacturing Technology.*

Chapter 6

Harmonising human and robotic workforces in Industry 5.0

Creating a resilient, human-centric, and sustainable organisational ecosystem

Kanchan Pranay Patil and Mugdha Shailendra Kulkarni
Symbiosis International (Deemed University), Pune, India

6.1 INTRODUCTION

Extending the concepts of "Industry 4.0," which prioritises automation, connectivity, and digitalisation, "Industry 5.0" refers to the next phase of industrialisation (Skobelev & Borovik, 2017). While Industry 4.0 placed a strong emphasis on the application of cutting-edge technologies to boost output and efficiency, Industry 5.0 goes a step further by integrating resilience, sustainability, and human-centricity into the industrial processes, especially for micro-, small-, and medium-sized businesses (Organisations) (Mourtzis et al., 2022; Leng et al., 2022). The EU defines Industry 4.0 as "a vision of an industry that aims beyond efficiency and productivity as the sole goals and reinforces the role and the contribution of industry to society." It is the new 4.0 Industrial Revolution which focused on humanism and ethics (Berrah et al., 2021) and is more phased towards advanced technology. Artificial intelligence (AI)-based robots, the Internet of Things (IoT), blockchain, smart contracts, etc., add to the efficiency as well as productivity of the organisation. The human-centric motivation, as well as augmented resilience, emphasises sustainability. Even though CSR activities were always prevailing, there was a need to put the idea of prioritising people and the environment over profits to shift the industry's emphasis (Spence & Rutherford, 2001). The idea of Industry 5.0 widens the organisation, becoming more human-centric, resilient, and sustainable as the main foundations of Industry 5.0 to develop a more comprehensive vision than Industry 4.0 (Saniuk et al., 2022).

The combination of man and machinery progresses the means and efficiency of production is known as the "Industry 5.0" revolution (Nahavandi, 2019). The automotive industry is witnessing a paradigm shift with the integration of human brains and mechanical muscles. This interaction has stimulated the need for a new set of managerial competencies. Certain managers' competencies must be identified and examined in this study for human brains and mechanical muscles to work together successfully in the automotive industry. While interpersonal skills like leadership, teamwork, and communication are crucial for efficient team management, technical skills like knowledge of technology, engineering, and mechanics are critical for making informed decisions (Ngwenyama & Lee, 1997). Several definitions of a manager from the time of Fayol (2016) said that a manager's duties should include planning, directing, and controlling. Management tasks are gradually performed with bigger stakeholder groups and intelligent agents. To develop and implement management activities in the future, a more contemporary idea that refers to managers as "architects of context" (Wrzesniewski & Dutton, 2001, pp. 179–201) may be beneficial and more flexible.

To encourage innovation throughout the lifespan of the organisation, Industry 5.0 is required. According to Hung et al. (2010), managers play a crucial role in providing resources and empowering employees to experiment and work together to develop innovative ideas. Creativity involves a certain amount of risk, and managers can give those who might otherwise feel susceptible to criticism the justification they need to take risks. Many concepts can be developed, but not all will result in valuable advancements. To recognise individuals who show promise and require further examination, leaders and managers may have a role to play. The Industrial Revolution, which brought about significant changes in technology and society, has resulted in many industries becoming mechanised, factories expanding, and replacing labour incentives with machine incentives. These adjustments significantly impacted family life and helped create a new working class (Mokyr, 2001). In contrast to Industry 4.0, which concentrated on digitising and automating production processes, Industry 5.0 is based on prior industrial revolutions and strives to combine cutting-edge technology with a human-centric approach. This strategy particularly emphasises the importance of social responsibility, adaptability, and sustainability.

Implementing Industry 5.0 requires a socio-technical viewpoint that carefully takes into account inter-human, inter-system, and contextual relationships. Industry 5.0 identifies the industry's capability to contribute to societal goals beyond employment and economic expansion and to develop into a dependable wealth source by ensuring organisation production. This study will look at how managers might notice Industry 5.0's potential to advance social objectives beyond profit in the context of the car sector.

From being technology-driven in the Fourth Industrial Revolution to becoming value-driven in the Fifth Industrial Revolution, managerial competencies must be transformed. Instead of being fearful about losing traditional employment prospects, managers in the automobile industry can concentrate on utilising cutting-edge technical breakthroughs, facilitating communication, business intelligence, and production/service delivery (Schröder, 2016). However, it is crucial to look into how, in an Industry 5.0 environment, human agents will get the competencies needed to act autonomously and advance the techno-social revolution (Bednar & Welch, 2020).

To reflect new thinking and provide performance, changes are expected to follow technological changes in relationships among organisational stakeholders. Bednar and Welch (2020) defined "smart" working as the "changes in approaches to work, work cultures, business architectures, premises, decision making, communications, and collaboration." Organisational challenges, like a lack of agility and cultural complacency, were revealed as even more of a significant issue of worry than technical or security difficulties in a poll of managers' views towards the use of technology (Li, 2015). The base of industrial conditions is determined to be equipment, and embedded intelligent equipment enhancement is a crucial prerequisite for digital twins. Fault identification and rectification can be accelerated simultaneously by the automatic real-time process analysis that digital twins can provide between connected machines and data sources. Digital twins can also dramatically reduce expenses and boost productivity in industrial operations. Digital twins indicate its prospective application value and ensuing potential value in Industry 5.0 through the prospect. This methodical review will work as a technical road map for the intelligent development of industrial manufacture and the improvement of the efficacy of the complete business process in the Industrial 5.0 era.

Therefore, the growth of Industry 5.0 is examined in this study from the viewpoint of how each management perceives their job roles and skills to be human-centric, resilient, and sustainable (European Commission, 2021). The study concludes that for the automotive industry to prosper in the future, it is essential to have specific managerial competencies. Software for manufacturing cost estimation can be used to estimate the price of a new product. It automates the costing processes and shortens the time to market new products (Hammarlund & Trakanavicius, 2023). Combining humans and machines can boost production and efficiency across several industries, including the auto sector (Raja Santhi & Muthuswamy, 2023). Hence, organisations must possess a distinct set of skills than traditional managerial abilities to integrate humans and robots successfully (Agolla, 2018).

6.1.1 Objectives of the chapter

The objectives of the chapter are as follows:

- To evaluate each manager's competence in an automotive organisation embracing Industry 5.0;
- To understand human-centric values managers perceive crucial to involvement in Industry 5.0;
- Understanding managers' role in encouraging team members' resilience is crucial to involvement in Industry 5.0;
- And, to understand how sustainability principles in the workplace are perceived as crucial to involvement in Industry 5.0.

6.1.2 Organisation of chapter

The rest of the chapter is organised as follows: Section 6.2 elaborates a review of the literature on Industry 5.0, emphasising its evolution and relevance. Section 6.3 focuses on research methodology. Section 6.4 elucidates data analysis and findings with regard to manager competencies to promote human-centricity, resilience, and sustainability. Section 6.5 highlights discussion and recommendations with regard to the factors playing a dominant role in Industry 5.0 and human-centricity, resilience, and sustainability. And, finally, Section 6.6 concludes the chapter with implications.

6.2 REVIEW OF LITERATURE

Engineers, roboticists, and management researchers have all shown an interest in the interaction of human-sectored mechanical muscles. According to the literature on this subject, combining humans and machines can boost production and efficiency across several industries, including the auto sector. However, managers must possess a distinct set of skills than traditional managerial abilities to integrate humans and robots successfully. According to the literature, managers in the automotive sector must be thoroughly aware of both the capabilities of humans and machines and how they may work together to produce the necessary results. The review is categorised into three parts: first, explaining the evolution of Industry 5.0, second, mentioning the impact on the automotive sector, and, further, understanding of manager competencies to promote human-centricity values, strategic resilience, and sustainability.

6.2.1 Evolution of the Industrial Revolution

The First Industrial Revolution took place between the 1760s and around 1840. In this region, the Second Industrial Revolution began. This was known as "the technological revolution," primarily in Germany and America, according to historians (Verhave & van Hoorn, 2014). It began in the early days of computing, where initially computers were hefty and disproportionately bulky, the time taken to compute was long, and the power they could produce was less. These challenges set a stage for a modern society that is only feasible thanks to computer technology (Krippendorff, 2018).

As part of the Third Industrial Revolution, factory automation increased around 1970, using electronics and IT (information technology). Manufacturing and automation have substantially advanced, thanks to Internet availability, connectivity, and renewable energy (Bloem et al., 2014).

Industry 3.0 replaced humans with automated systems on the production line using Programmable Logic Controllers (PLCs). Even if mechanical systems existed, they could not operate without human input. The Industrial

IoT (IIoT), as it is currently known, enables this information exchange. The following are some of the significant elements of Industry 4.0: A "cyber-physical system" (CPS) is a mechanical apparatus managed by computer-based algorithms (Sharma & Singh, n.d.). The IoT is a system of interconnected devices, vehicles, and other items that enables computers to sense, detect, and keep an eye on them from a distance. Visionaries are already forecasting the arrival of Industry 5.0 less than ten years after the initial discussions on Industry 4.0 in the manufacturing sector (Wójcicki et al., 2022). Industry 5.0 is expected to place a greater emphasis on the rein-tegration of human hands and brains into the industrial framework than the current revolution, which is focused on transforming factories into IoT-enabled innovative facilities that use cognitive computing and link via cloud servers (George & George, 2020; Maddikunta et al., 2022). The "Industry 5.0" revolution combines man and machine to increase production methods and efficiency (Nahavandi, 2019, p. 4371), even when firms begin utilising cutting-edge technologies. The automotive industry is witnessing a para-digm shift with the integration of human brains and mechanical muscles (Kolarevic & Parlac, 2015).

As the Industrial Revolution continued, the expansion of Industry 4.0 continued through digital transformation and beyond and can be traced back to the Industrial Revolution. Each stage emphasises how the Industrial Revolution has increased output, which has changed how we view and do our work in the industry (Morrar et al., 2017; Vinitha et al., 2020).

The most recent business and manufacturing buzzword is "Industry 5.0." Industry 5.0 drives further to Industry 4.0 by fusing digital & phys-ical technology to improve human–machine collaboration (Carayannis & Morawska-Jancelewicz, 2022). Industry 4.0 was all about the integra-tion of digital technologies. Industry 5.0 seeks to develop a more human-centric manufacturing and industrial approach to enhance productivity and efficiency. Each following stage emphasises how the Industrial Revolution has increased production and changed how we view and work in the industry (Longo et al., 2020). Industry 5.0 seeks to develop a more human-centric manufacturing and industrial approach to enhance productivity and efficiency (Zizic et al., 2022). The textile industry and transportation underwent a tremendous transformation due to industri-alisation. Fuel sources like steam and coal made machine use more prac-ticable, and the idea of producing with machines quickly gained traction (Wrigley, 1962)—machines allowed for the quick and straightforward production of items and the development of several innovations and technology.

As a result of this interaction, an entirely new set of managerial skills is now necessary. A literature review on practical organisational skills com-bines personal, relational and technical skills. While technical skills like information technology, manufacturing, and mechanics are vital for making methodological conclusions, relational skills like leadership, cooperation, and communication are vital for actual team management (Giunipero et al., 2006). The literature suggests that managers in the organisation need to deeply consider both the capabilities of humans and machines and how they could collaborate to generate the desired results. It combines the methodological and interpersonal skills which are required for management. While interpersonal skills like leadership, teamwork, and communication are crucial for efficient team management, technical skills like knowledge of technology, engineering, and mechanics are essential, critical, and informed judgements (Robles, 2012). Specific managerial skills need to be recognised and researched for mechanical muscles and human brains to function effectively in the automotive industry. Human skills like leadership, cooperation, and communication are essential for effective team management, but technical abilities like computer literacy and making informed decisions require an understanding of engineering and mechanics. Human characteristics like cooperation, leadership, and communication are crucial for efficient team management, but technical skills like computer literacy and the ability to make sound decisions also come into play.

6.2.2 Impact on the automotive sector and technologies used in Industry 5.0

Due to the fusion of mechanical muscles and human minds, the automotive sector has undergone a considerable revolution. The sector industry automates its processes to improve productivity and efficiency, resulting in collaboration between humans and robots (George & George, 2020). A new set of managerial skills is now required to successfully integrate humans and machines as a result of this interaction. Managers in the automotive business require a specific skill set to manage interactions between people and machines (Carver & Turoff, 2007). To ensure everyone is on the same page with the goals and objectives, managers must effectively communicate with their teams. Successful team management depends on collaboration, and managers must promote a collaborative environment to ensure everyone is working towards the same objective (Bedwell et al., 2012). The emphasis of Industry 5.0 is on both technology and people. It blends cutting-edge technologies, such as robotics, the IoT, and AI, with a human-centred viewpoint. Managers and staff members can manage complex problems requiring judgement and expertise. According to Industry 5.0, the production process values human abilities and innovation. It also emphasises how important it is to use it morally and sensibly.

IIoT and Industry 5.0 technologies can improve the resilience of systems in their physical, digital and social dimensions. A dynamic and self-adapting system supported by AI/ML and real-time intelligence for predictive cyber risk analytics. This knowledge was applied to design a step-by-step approach for the SME's automotive supply chain integration with IIoT technologies in Industry 5.0.

6.2.3 Manager competencies to promote human-centricity values

The breadth and depth of automation made possible by Industry 4.0 should not make us lose sight of the value of excellent organisational analysis, especially given the political and social implications.

It will be necessary to look into the organisational analysis that forms the framework of Industry 5.0. The need for rapid technology exploitation will increase the strain on corporate and business planning procedures. Building a platform using tools to facilitate reciprocal exploration of divergent employees and group viewpoints should help all stakeholder groups communicate more effectively (Bednar, 2000). Strategic planning, decision-making, and knowledge management will all receive a new boost from managing stakeholder connections and interactions.

Employees must be ethically aligned to experience positive outcomes regarding job satisfaction, meaningfulness of work, organisational commitment, and lower intentions to quit. To be ethically aligned, employees must perceive that their manager acts ethically and feel that their values align with their organisations (Longo et al., 2020).

Organisations can acquire specialised capabilities that prioritise the welfare and empowerment of individuals to adopt human-centric values in Industry 5.0. The human-centric approach "promotes talents, diversity, and empowerment" (Carayannis & Morawska-Jancelewicz, 2022). Humans, as a resource, need to focus. Focusing on human resource talent will help to transition into the digital era across many industries, including education, by placing the interpersonal and human components at the centre of our activities (Swanson, 2022; Roepke et al., 2000). However, the modern digital world emphasises "human-centricity" as a crucial element of the future workplace (Kolade & Owoseni, 2022; Jafari et al., 2022). Integrating CPS into industrial processes, Industry 5.0 and Industry 4.0 aims to merge the physical and digital worlds and establish a self-managing network linking people, machines, products, and other connected items. The emphasis on human-centricity, which enables a paradigm change from independent automated and human operations towards a human–automation symbiosis, is a crucial component of the present evolution to Industry 4.0 (Zarte et al., 2020).

Industrial and commercial uses of virtual reality, artificial intelligence, and highly integrated manufacturing systems are now available, thanks to technological advancements. As work activities and markets have used information and communication technologies to operate remotely, it has also liberated corporate training from a focus on location. There are discussions on developing Industry 5.0, combining social and technological systems to provide mass-customized goods and services.

The phrase "smart working" has been used to refer to an evolutionary transition occurring across a variety of work-related characteristics. According to

Iannotta et al., (2020), "changes in approaches to work, work cultures, business architectures, premises, decision making, communications, and collaboration" have occurred in recent years. The significance of location in professional activities has decreased, and there is no more room for collaboration, employee autonomy, talent management, and an emphasis on innovation). Flexibility is a crucial component of modern, innovative working practices. Flexibility and spatial displacement are only one aspect of intelligent work, however.

While Industry 4.0 has increased production, it has also raised issues among stakeholders and the general public. Many businesses are still hesitant to take advantage of the possibilities of the cyber-physical output because they worry that doing so will make them overly dependent on systems susceptible to embedded errors, cyberattacks, and power failures. Smaller organisations may need more skills to change to Industry 4.0, or current managers may fear losing control over systems they have a limited understanding of. Concerns from workers include but are not limited to, the possibility that they would lose their jobs due to the use of robotics in manufacturing or providing services.

Smart working demands the right mix of abilities, engagement, and enabling technologies, as well as professional training and dedication from workers. However, whether or not they do so critically depends on the perceptions and perspectives of the involved players and the extent to which they can explore and express them. Social networks can be seen as intertwined elements of cultural behaviour. Systemically desirable and culturally viable possibilities must be sought when thinking about purposeful change in a system. It will be challenging to undertake change that is not culturally feasible within specific socio-technical ecosystems.

Innovative ideas and analytical thinking are essential for success in business and life. When we look at the history of technological advancement, it becomes clear that scientific breakthroughs were formerly, and frequently now are, made by lone researchers or small teams working with meagre resources. In this context, innovation refers to a way of thought that improves processes, systems, goods, and services rather than producing new items.

Professionals who must make judgements based on information must be able to think critically and analyse instead of just pulling things apart; it is a process of separating the facts that matter, their implications (both positive and negative), and choosing the best course of action.

For professionals, emotional intelligence is becoming more and more crucial. It takes self-control to manage situations for the most outstanding results to communicate effectively with individuals at work, on task teams, and with peers. Empathetic recognition and understanding of emotions and their effects are also necessary. Employees must develop their emotional intelligence to succeed as a team; leaders and supervisors are not responsible for coaching staff in this area.

6.2.4 Manager competencies to promote strategic resilience

While making sense of complex experiences has always been a part of human life, the speed of technological advancement has increased in the twenty-first century to the point where people can now avoid encountering uncertainty and unfamiliar obstacles regularly. Managers are the designers of context and can shape internal behaviour by fostering dialogue and offering the right resources (Bednar, 2000).

Their capacity to gather contextual knowledge from experts performing their duties, interacting with clients, and engaging in cooperative experimentation and innovation may play a crucial role. The ability to recognise subtle changes as they happen is a trait of the most resilient organisations, according to (Weick & Sutcliffe, 2015), Industry 5.0 also emphasises the importance of resilience in disruptions such as natural disasters and pandemics. This involves developing flexible and adaptable manufacturing processes that quickly respond to changing conditions. It also consists of implementing risk-management strategies that can mitigate the impact of disruptions (Leng et al., 2022; Grzybowska & Stachowiak, 2022).

According to the European Commission, a resilient approach is where companies are agile and flexible and change by adapting to new technology. However, organisations high on flexibility and agility are only sometimes resilient (Carvalho et al., 2011). Effectiveness and profit enlargement are significant factors of the business. The addition of flexibility and adaptability with respect to the "lean" version is specifically motivated by the importance of competence (Kochan et al., 2018). Some organisations are anti-fragile organisations that learn to foresee, respond, and earn systematically from the crisis to guarantee stable and long-lasting performance. A robust industry that can quickly adjust to the approaching flood of new technology is required by the ever-changing times; nonetheless, a solid social environment composed of organic social values is necessary for a functioning human civilisation. Due to the speed with which its technologies are being adopted and their sound effects across all industries, Industry 4.0 has recently become the standard for applications. These advancements have neglected the environment by giving machines precedence over people, but they still need to produce the intended results. As a result, Industry 5.0 (I5.0) revolution is a call to action for putting sustainability concepts into practice, fusing technology and human values, and is seen as a step towards achieving sustainable development goals. In light of this, the current study suggests a paradigm for examining I5.0 enablers for reaching.

6.2.5 Manager competencies to promote sustainability

An organisation focused on sustainability fully understands its responsibilities to various stakeholder groups (Perrini & Tencati, 2006). Such a company deliberately enhances its social and ecological performance while considering socio-technical concerns. Any creative development carried out in business will provide the organisation with advantages. However, these benefits will have varying values for various stakeholder groups. Within an evolution towards more sustainable business practices, it might take time to reconcile divergent and competing requirements (Bednar & Welch, 2020). Industry 5.0 technology enables manufacturing industries to be sustainable through circular processes. Organisations must prioritise the reuse, repurposing, and recycling of natural resources and reduce waste and environmental impact to eventually create a circular economy with improved resource efficiency and effectiveness (Xu et al., 2021). Industry 5.0 places a strong emphasis on sustainability. Using sustainable production techniques and renewable energy sources, the industry seeks to minimise waste and lower its carbon footprint. This includes applying circular economy ideas and using eco-friendly materials (Fraga-Lamas et al., 2021). All 17 of the Sustainable Development Goals, as well as the Triple Bottom Line's three Ps, should be prioritised by organisations. Instead of minimising negative impact, a sustainable company strategy emphasises positive impact. People and companies with these abilities can successfully lead their organisations.

6.3 RESEARCH METHODOLOGY

6.3.1 Qualitative study

This research has used the grounded theory technique (Glaser & Strauss, 1967) better to understand managerial competencies in the context of Industry 5.0; the research also examines the ideas and insights of managers directly involved in providing goods and services related to the automotive sectors in emerging economies.

The following steps enlist the qualitative study conducted.

To find the responses' themes, patterns, and trends, content analysis (Krippendorff, 2018) is used to examine and evaluate qualitative data, such as interview transcripts. Researchers might acquire important insights into particular themes or research objectives by thoroughly examining the interviewees' responses while employing the semi-structured interview method. After all interviews have been coded, group the codes into more general themes or groups. To develop a complete and cogent representation of the data, look for repeating patterns and connections between codes.

a) Examine the categorised data to find recurring themes, significant revelations, and new trends. To comprehend the relevance of each theme, pay attention to the topic's frequency and depth.
b) Considering the research objectives while interpreting the content analysis's findings. Discuss and connect the results' implications to the main study issues.

c) Starting with the semi-structured interviews, transcribe the data. In order to have a thorough record of the participant's responses, ensure that every interview is wholly and accurately transcribed.

d) Clearly describe the study goals or subjects you wish to investigate through the content analysis. This will direct your investigation and assist in keeping your attention on pertinent themes and patterns.

e) Create some code by creating a coding framework depending on the study's goals. In coding, labels or codes are given to text passages that connect to particular topics, concepts, or ideas. Ensuring the codes are unambiguous, mutually exclusive, and all-inclusive is essential.

f) To begin coding, read the transcripts line by line and assign the appropriate codes to the text passages corresponding to the predetermined themes. If a segment covers more than one theme, it could get more than one code.

g) Creating a coding framework before writing any code depends on the research objectives. Labs or codes are assigned to text sections that relate to specific subjects, ideas, or concepts. Making sure the codes are clear, mutually exclusive, and inclusive is crucial.

h) To start coding, read the transcripts line by line and give the text passages that relate to the present topics the relevant codes. A segment may receive more than one code if it covers more than one theme.

i) Consider the content analysis findings in light of the study's goals. Discuss and connect the results' implications to the main study issues.

j) Consider using several coders for intercoder reliability checks to guarantee the accuracy of the content analysis. In order to do this, a portion of the data must be independently coded by several researchers, and their level of agreement must then be evaluated.

k) Finally, present the results clearly and cogently. To illustrate the themes and patterns that emerged from the investigation, utilise tables, charts, and narratives. To thoroughly explain the research process, be open and honest about your approach and limits.

These steps can be used to combine content analysis with the semi-structured interview method.

6.3.2 Data collection

After contacting 25 companies, we conducted semi-structured interviews with ten managers from ten automotive industries with headquarters in Pune and the Pimpri-Chinchwad automotive industry belt. For three months, nine in-person interviews and one telephone interview were done. On average, the managers we spoke with had over five years of experience in a leadership position within the automotive organisation and were knowledgeable about standard business procedures. The Mahratta Chamber of Commerce, Industries and Agriculture (MCCIA) database, a network and advocacy organisation for Indian businesses in Pune, contained information on the sample of managers. The following selection criteria is used to decide which managers are qualified for this study's interviews: (1) deliver products and services in their respective organisations; (2) have the power to reallocate company resources as necessary; and (3) be in charge of boosting organisational performance.

Following the recording of each interview, the transcripts were read to draw conclusions and do content analysis. With the available data, drawing reproducible and reliable implications from text or other related materials in the contexts of their use is the goal of the research technique known as content analysis. When used as a research technique, content analysis presents new viewpoints, increases the researcher's understanding of certain occurrences, or directs helpful action. Due to the sensitive material presented in these interviews, organisations and individuals had to maintain their anonymity. Identifier codes were given to each organisation and its members to achieve this anonymity. Table 6.1 displays these identifiers and the accompanying case information.

In total, ten individuals across ten organisations were interviewed – Table 6.1 presents the summary.

6.3.3 Interview guide

We conducted interviews with managers from several firms in the automotive sector during the research phase to get valuable insights into their thoughts on Industry 5.0. These interviews were done in two stages lasting 60–75 minutes on average. We asked participants to provide an overview of their various businesses, including their major line of business, production processes, product offers, target markets, and key competitors, during the first stage of the interviews. This first stage set the groundwork for comprehending the

Table 6.1 Anonymised organisations and participant breakdown summary

Manufacturing industry code	Individual code	Position	Industry
O1	I1	Managers	Electronics/electrical
O2	I2	Assistant Manager	Engineering
O3	I3	Managing Director	Aerospace
O4	I4	CEO	Chemical
O5	I5	Technology Consultant	Automotive
O6	I6	Managing Director	Engineering
O7	I7	Assistant Manager	Automotive
O8	I8	Management Consultant	Electronics/electrical
O9	I9	Managing Director	Automotive
O10	I10	Assistant Manager	Automotive

framework in which these managers worked and the specific difficulties and opportunities they encountered in the Industry 5.0 era.

Moving on to the second level, we explored further the expected roles and skills of managers in the fast-expanding Industry 5.0 environment. We developed open discussions, allowing managers to freely communicate their opinions and experience the issues they thought were significant in the context of Industry 5.0, which provided us with excellent insight into the real-world challenges automotive industry leaders face.

We used an iterative data collection and analysis approach, adjusting our coding approach as new insights emerged from the interviews. This iterative procedure confirmed that we thoroughly collected and comprehended the workings of the managers' comments. We wanted to discover how to embrace human-centricity, devise resilient strategies, and promote sustainability efforts within their firms. These three essential categories piqued our interest since they represented critical criteria for success and flexibility in Industry 5.0.

6.4 DATA ANALYSIS AND FINDINGS

All of the interviews were recorded and then transcribed verbatim. The coding strategy recommended by Strauss & Corbin (1997) was used independently by the two researchers who participated in this study to analyse the data. The independent assessments' results helped build the human-centricity, resilience, and sustainability factors. Thematic analysis by NVivo 12 Pro software was used to analyse the transcribed interviews; significant themes related to sustainable development functions like human-centricity, socio-environment sustainability, and resilience and their association with Industry 5.0 competencies were identified for further pattern analysis. It is for this reason that thematic analysis became the preferred method to analyse qualitative data (Skovdal & Cornish, 2015). Tables 6.2a, b, 6.3a, b, and 6.4a, b provide semi-structured interview questions, respective themes, and further interview coding.

6.4.1 Manager competencies to promote human-centricity

Industry 5.0 facilitates the human-centric approach, which places fundamental human wants and interests at the centre of the manufacturing process. Management must consider industry workers not as "cost" but as "investment." Industry 5.0 technology used in manufacturing is adaptive to the needs and diversity of industry workers (Xu et al., 2021). Management has to safeguard workers' physical and mental well-being, making the working environment safe, maintaining the work–life balance, and being conducive to growth (Xu et al., 2021). Managers have

Table 6.2a Semi-structured questions for manager competencies to promote human-centricity

Sr. No.	Semi-structured questions	Theme from the questions
	How do managers comprehend the significance of values that put employees first in Industry 5.0?	
1	How do you articulate a clear vision of Industry 5.0 and how it aligns with human-centric values while inspiring and motivating employees to work towards that vision?	Employee empowerment
2	Does empowering employees by providing autonomy, trust, and opportunities for growth allow them to contribute their unique skills to Industry 5.0 initiatives?	Diversity and inclusion
3	Can Industry 5.0 initiatives show empathy for workers' work–life balance needs and implement policies that promote mental health and well-being in place?	Work–life balance
4	Do you provide resources and opportunities for employees to enhance their skills and knowledge?	Skill development and lifelong learning
5	They understand the importance of diverse perspectives and experiences in driving Industry 5.0 innovation and fostering an inclusive environment.	Ethical and transparent practices
6	How do you ensure prioritising of customer needs and expectations when designing Industry 5.0 projects and solutions?	Customer-centric
7	Do you think upholding a robust ethical code and promoting transparent practices within the organisation is required while implementing Industry 5.0?	Social responsibility

Table 6.2b Human-centricity factors in Industry 5.0 context

Manager competencies to promote human-centricity	I1	I2	I3	I4	I5	I6	I7	I8	I9	I10	Total
Employee empowerment	√		√		√	√		√		√	6
Diversity and inclusion		√			√		√			√	4
Work–life balance	√		√		√	√	√	√	√	√	8
Skill development and lifelong learning	√		√		√	√	√	√	√	√	8
Ethical and transparent practices	√		√			√		√			4
Customer-centric	√			√	√	√	√	√		√	7
Social responsibility	√		√		√		√		√		5

Table 6.3a Semi-structured questions for manager competencies to promote resilience

Sr. No.	Semi-structured questions	Theme from the questions
	What role does resilience play for managers in their teams and organisations?	
1	How do you define Industry 5.0, and what are the critical strategic implications for businesses in this era of technological advancement?	Strategic thinking
2	Describe a situation where you had to adapt your management style or approach to address unexpected challenges or changes in the industry or market.	Adaptability and flexibility
3	How do you identify and assess potential risks associated with Industry 5.0 initiatives and technological advancements?	Management of risk
4	Describe a complex situation where you had to make a critical decision with limited information and in the face of uncertainty. How did you approach it?	Decision-making in an unpredictable environment
5	How do you demonstrate emotional intelligence as a leader, especially when guiding your team through challenging projects or periods of change in Industry 5.0?	Leadership and emotional intelligence
6	How do you approach building and maintaining solid relationships with stakeholders, both internal and external, in the context of Industry 5.0 projects?	Building strong relationships with stakeholders
7	How do you promote a culture of continuous learning and personal development within your team to keep up with Industry 5.0 advancements?	Continuous learning and personal development
8	Describe a critical crisis or unexpected event that you had to manage in your previous role, and how you handled the situation to minimise its impact on the organisation.	Crisis management

to safeguard the autonomy, human dignity, and privacy of the employees (Huang et al., 2022). The following are semi-structured interview questions and respective themes in Table 6.2a.

According to Table 6.2b, out of ten respondents, six agree that employee empowerment is a significant aspect of human-centricity. Industry 5.0 is all about enhancing the digital transformation of enterprises through the effective fusion of people and technology so that they may work more productively together in a secure setting. It fosters mechanised efficiency while promoting human creativity through emerging technologies such as 3D printing, collaborative robots (cobots), AI/ML, digital twins, and blockchain. Human-centricity includes diversity in terms of varied resources and inclusive cultural outcome results, which boosts productivity and a sense of belonging at work. Even though less than 50% of the respondents agree with diversity and inclusion, a company like Accenture asserts that organisations use a human-centric approach. The strategy may grow adaptable

Table 6.3b Resilience factors in Industry 5.0 context

Manager competencies for resiliency	I1	I2	I3	I4	I5	I6	I7	I8	I9	I10	Total
Strategic thinking	√	√		√	√	√		√	√	√	8
Adaptability and flexibility		√	√	√	√	√		√		√	7
Management of risk	√	√	√			√		√	√	√	7
Decision-making in an unpredictable environment		√	√			√		√	√	√	6
Leadership and emotional intelligence	√	√	√		√			√	√	√	7
Building strong relationships with stakeholders	√	√		√	√	√		√	√	√	8
Continuous learning and personal development	√	√	√	√	√	√			√	√	8
Crisis management		√	√		√	√		√	√		6

and resilient in challenging and unpredictable social and economic settings (Cillo et al., 2021); 80% of survey participants acknowledge the challenges in work–life balance, highlighting Industry 5.0 as a solution with its growing prevalence of flexible working arrangements, particularly in industrialized nations. However, it is unclear how they will affect their health and happiness. Lifelong learning is an important aspect where companies have been obliged to evolve, embracing continuous learning (lifelong learning) to provide their employees with the training and skills necessary for the new professional responsibilities to reduce this training gap. As shown in Table 6.2b, 70% of the respondents agree on the ethics and transparent factors. Industry 5.0 asserts that it will move away from technical productivity methods and towards a more human-centred strategy. Socio-technical interoperability between people and AI systems is a necessary first step. The globe has undergone extensive digitisation, and the new modern workplace is now primarily virtual. If we want people to focus on innovation while being more creative, we must develop and use AI systems and technologies deemed trustworthy and ethical. This is only possible if humans use the proper technological governance while considering productivity and a sustainable economy and environment.

Thus, Industry 5.0 emphasises the value of human workers, including their talents, creativity, and capacity for problem-solving. Industry 5.0 intends to empower employees by integrating technology to improve their

Table 6.4a Semi-structured questions for manager competencies to promote resilience

Sr. No.	Semi-structured questions	Theme from the questions

How do managers perceive the influence of integrating sustainability principles in the workplace?

1	How do you perceive the role of Industry 5.0 in addressing environmental challenges and promoting sustainability?	Environmental awareness
2	Describe how you integrate sustainability goals into strategic planning to ensure the long-term environmental viability of projects and initiatives.	Strategic planning for sustainability
3	How do you encourage and support a sustainable innovation culture within your team or organisation?	Sustainable innovation
4	a) How do you measure and track the environmental performance of projects or initiatives related to Industry 5.0?b) Can you provide examples of key performance indicators (KPIs) that you have used to evaluate the success of sustainability efforts within your team or organisation?	Performance measurement and reporting
5	Describe how you build collaborative relationships with external stakeholders, such as suppliers, partners, or industry peers, to advance environmental sustainability goals in Industry 5.0 projects.	Collaboration and partnerships
6	How do you integrate circular economy principles into designing and implementing Industry 5.0 processes and products?	Circular process
7	How do you instil a culture of sustainability within your team or organisation, making it an integral part of the company's values and identity?	Sustainability values

Table 6.4b Sustainability factors in Industry 5.0 context

Manager competencies for sustainability	I1	I2	I3	I4	I5	I6	I7	I8	I9	I10	Total
Environmental awareness	√		√		√	√		√	√	√	7
Strategic planning for sustainability		√			√		√	√		√	5
Sustainable innovation:	√		√		√		√	√	√		6
Performance measurement and reporting	√		√		√	√		√	√	√	7
Collaboration and partnerships	√	√	√			√	√	√	√		7
Circular process	√			√	√	√	√	√		√	7
Sustainability values		√	√		√	√	√				5

productivity and well-being instead of automating and replacing human jobs with machines. For instance, collaborative robots (cobots) can support people in physically taxing or repetitive jobs, resulting in safer and more effective workplaces. Industry 5.0 promotes a productive workplace culture that embraces human potential by giving human needs and talents priority.

6.4.2 Manager competencies to promote resilience

Industry 5.0 also emphasises the importance of resilience in disruptions such as natural disasters and pandemics. This involves developing flexible and adaptable manufacturing processes that quickly respond to changing conditions. It also consists of implementing risk-management strategies that can mitigate the impact of disruptions. Table 6.3a illustrates the semi-structured interview questions and respective themes.

According to Table 6.3b, resilient organisations can maintain their enthusiasm under stress, handle disruptive changes, and adapt. They recover quickly after failures. They also conquer significant obstacles without acting dysfunctional or hurting other people. Hence, more than 70% of the respondents agree that strategic thinking, adaptability and flexibility are significant factors in Industry 5.0. Resilient leaders can maintain enthusiasm under stress, handle disruptive changes, and adapt. They recover quickly after failures. They also conquer significant obstacles without acting dysfunctional or hurting other people. Also, 60% more than half of the respondents agree that decision-making in an unpredictable environment faces challenging decisions that must be made quickly and effectively. However, risk and uncertainty can significantly affect how decisions bother and worry organisations. This examines the challenges of choosing wisely in risky and ambiguous circumstances and provides insightful guidance, including scenario planning and prototyping, communication, and risk management. According to psychological research, organisations with emotional intelligence may also be crucial in high-stress battle circumstances where cognitive performance is limited. This lack of cognitive function frequently occurs in stressful and high-pressure professional contexts. This is noticed as 80% of the organisations agree that EI is important; 60% agree that crisis management and the capacity to thrive despite disaster risk. Also, managers can prepare for, deal with, withstand, and recover from disasters, for example, COVID crisis management with work from home, especially regarding the sector.

6.4.3 Manager competencies to promote sustainability

Sustainability is a critical element of Industry 5.0. The industry aims to reduce its carbon footprint and minimise waste through renewable energy sources and sustainable manufacturing practices. This includes the use of eco-friendly materials and the implementation of circular economy principles. The survey questionnaire and theme are given in Table 6.4a.

According to Table 6.4b, seven out of ten respondents agree that organisations must emphasise environmentally friendly practices using cutting-edge technology. This covers the application of renewable energy sources—like solar or wind energy—in manufacturing operations. In addition, Industry 5.0 encourages the efficient use of resources and waste reduction by applying cutting-edge sensors, data analytics, and intelligent systems. Such strategic planning for sustainability is agreed upon by five out of ten. Six respondents emphasized the idea of sustainable innovation in Industry 5.0, emphasizing its dedication to environmental sustainability and resilience, focusing on human-centred design. Given the increasing regulatory and financial attention on lowering carbon emissions and environmental impact, businesses must assess the resource footprint that supports their manufacturing process. Analysing the source of raw materials, the amount of waste produced, the environmental impact, the energy efficiency of operations, and the energy sources may all be part of this. Performance measurement and reporting, where seven out of the ten respondents agreed, mentions that organisations can assure compliance with environmental laws, industry standards, and sustainability frameworks by measuring and reporting on their sustainability performance. Organisations can identify gaps or areas of non-compliance and take corrective measures to align with applicable legislation and standards by routinely monitoring and reporting their sustainability performance. A circular process where Industry 5.0 technology can make manufacturing processes more resource-efficient by increasing product lifespan, decreasing waste, and optimising output. For instance, IoT sensors and data analytics can optimise energy and material to ensure that resources are used effectively throughout the production cycle. Sustainable values out of ten, seven respondents agreed. Organisations should actively seek collaborations and partnerships with external organisations, including NGOs, research institutions, and industry associations, to share information and best practices values in sustainability. Organisations can learn about new sustainable values, exchange experiences, and use group efforts to address sustainability concerns by joining sustainability networks and projects.

Organisations should uphold the highest ethical standards and encourage openness throughout their business operations. This entails acting honestly and responsibly, ensuring workers are treated fairly, and engaging in ethical business practices. Organisations gain the trust of their staff, clients, and stakeholders by acting with integrity and ethics. Organisations should focus on their consumers' demands and preferences when designing products and services so that they meet their expectations. This entails undertaking market research, actively listening to customer feedback, and constantly updating products based on feedback. Organisations may establish lasting bonds and client loyalty by strongly emphasising customer satisfaction.

6.5 DISCUSSION AND RECOMMENDATIONS

Industry 5.0, which emphasises human-centricity, strives to establish a harmonious partnership between people and machines in which technology enhances rather than replaces human talents. This method produces superior results in innovation and problem-solving and improves job satisfaction, engagement, and productivity. The following sub-sections discuss diverse factors that play a dominant role in Industry 5.0 and human-centricity, resilience, and sustainability.

6.5.1 Factors playing a dominant role in Industry 5.0 and human-centricity

The following points highlight diverse factors that play a dominant role in Industry 5.0 and human-centricity.

- **Employee empowerment:** Organisations in Industry 5.0 understand the importance of including workers in decision-making. Organisations may tap into their collective intelligence and make better decisions by allowing employees to share their views and knowledge. Employee ownership, accountability, and dedication are all fostered by this involvement. Organisations should promote an atmosphere that gives employees agency and promotes active engagement (Murillo & Lozano, 2006). This entails encouraging open communication, involving workers in decision-making processes, and offering chances for career advancement (Bednar et al., 2016). Employees who feel empowered are more driven, engaged, and willing to contribute to the organisation's success (Taneja et al., 2015; Yukl & Becker, 2006).
- **Diversity and inclusion:** Harnessing the potential of diversity and fostering an inclusive culture become even more critical for small- and medium-sized enterprises to grow and remain competitive in Industry 5.0, where cooperation between people and cutting-edge technologies is required. Small- and medium-sized businesses (organisations) should appreciate diversity and foster inclusive workplaces that value and respect individual differences (Boehm et al., 2014). This entails encouraging diversity in hiring procedures, ensuring every employee has an equal opportunity to succeed, and building an inclusive and respectful workplace atmosphere (Pless & Maak, 2004). Organisations may stimulate innovation and creativity by utilising a diverse workforce's distinctive perspectives and skills (Jayne, 2004).
- **Work–life balance:** Organisations in Industry 5.0 prioritise maintaining workers' physical and emotional health. This fosters a culture that values self-care, opposes excessive working hours, and supports work–life balance. Organisations may put rules restricting communication after hours, promoting vacation use, or providing staff members time off

to unwind. Organisations should prioritise their employees' well-being and understand the value of work–life balance. This entails providing flexible work schedules, encouraging healthy work habits, supplying tools for stress management and preserving one's mental and physical well-being. Supporting work–life balance increases productivity, job satisfaction, and employee satisfaction (Shipman et al., 2023).

- **Skill development and lifelong learning**: Organisations should support lifelong learning and invest in their personnel's ongoing skill development (Schuetze, 2006; Nevis et al., 2009). This can be done by encouraging staff members to learn new skills and information through mentorship programmes, workshops, and training sessions. Organisations can help their employees adapt to shifting market conditions and technological demands by building a learning culture which promotes personal development and career advancement.

6.5.2 Factors playing a dominant role in Industry 5.0 & resilience

Industry 5.0 competencies for resilience are adept at organising and leading teams in stressful situations while remaining composed and making wise choices (Morgan et al., 1998). They know the significance of effective communication, resource allocation, and quick problem-solving in crises.

The following points highlight diverse factors that play a dominant role in Industry 5.0 and resilience.

- **Strategic thinking**: Resilient organisations can strategies and foresee upcoming difficulties and disruptions. They know the broader market dynamics, technical developments, and new trends that could impact the sector. They can direct the organisation towards resilience by considering various circumstances and creating proactive measures (Jarvenpaa & Ives, 1994).
- **Adaptability and flexibility**: Resilient organisations are adaptive and flexible in the face of change. To respond to unforeseen occurrences and changes in the market, they can swiftly modify their plans, procedures, and business practices. They welcome innovation and are constantly looking for new chances to advance (Seville et al., 2015; He et al., 2023).
- **Management of risk**: Resilient organisations are adept at recognising, evaluating, and controlling risks (Waters, 2011). They are thoroughly aware of the business's possible hazards and take proactive steps to reduce them (Luo & Tung, 2007). This entails putting in place risk-management frameworks, creating backup plans, and ensuring operations continue during an interruption.
- **Making decisions in an uncertain world**: Resilient organisations are skilled at making defensible choices in complex and challenging

circumstances (Gregory et al., 2012). Before making strategic decisions, they acquire and analyse pertinent data, communicate with stakeholders, and consider various viewpoints. They can make prompt judgements to guide the organisation towards resilience because they can easily take calculated risks (Chang et al., 2014).

- **Leadership and emotional intelligence:** Resilient organisations have excellent leadership and emotional intelligence. Especially in trying circumstances, they can motivate their colleagues and effectively communicate. They are sympathetic to the worries of their employees since they are aware of how emotions affect people. Resilient organisations build a supportive workplace environment that encourages teamwork, trust, and resilience (Goleman, 2001).
- **Building solid relationships with stakeholders:** Collaborating effectively are skills that resilient organisations possess. They actively interact with vendors, clients, business partners, and other relevant entities to promote cooperation and learning from one another. They use these connections to strengthen their resilience and adjust to shifting market conditions (Zack, 2003).
- **Continuous learning and personal development:** Resilient organisations are dedicated to their personal growth. They stay current on business trends, technological developments, and resilient best practices. They actively support their team members' progress as professionals by looking for possibilities for professional development (Bande et al., 2015).

6.5.3 Factors playing a dominant role in Industry 5.0 & sustainability

Organisations should take ownership of their social duty by actively supporting the communities in which they operate. This includes participating in CSR programmes, lending a hand to neighbourhood projects, and considering their activities' social and environmental effects. Organisations can contribute to community development, participate in sustainability projects, and behave ethically as businesses (Jenkins, 2009). By fostering these competencies, organisations can establish a workplace emphasising individuals' well-being, well-being, development and empowerment. As a result, the organisation experiences an uptick in employee happiness, productivity, and overall performance (Markos & Sridevi, 2010).

The following points highlight diverse factors that play a dominant role in Industry 5.0 and sustainability.

- **Environmental awareness:** Organisations must be well-versed in sustainability concepts and environmental challenges. They should be current on ecological laws, new fashions, and sustainable living guidelines. Because of this awareness, they can include environmental factors in decision-making processes, encouraging sustainable practices (Rani & Mishra, 2014).

- **Strategic planning for sustainability**: Organisations must be able to create and practice sustainability strategies consistent with the organisation's objectives and core values. They should consider sustainability's environmental, social, and economic facets when creating a long-term strategy. Organisations can direct the organisation towards sustainable practises by establishing precise sustainability objectives and incorporating them into the overall company plan (Burke & Gaughran, 2007; Clarke and Roome, 1999).

- **Environmental awareness**: Organisations must understand sustainability concepts and environmental challenges. They should be current on ecological laws, new fashions, and sustainable living guidelines. Because of this awareness, they can include environmental factors in decision-making processes, encouraging sustainable practices. Effective organisations know the significance of engaging with stakeholders to promote sustainability. They should foster partnerships with suppliers, consumers, employees, communities, and regulatory agencies to identify sustainability goals and work together on sustainability projects (Eckersley, 2003). To foster trust and a sense of shared commitment to sustainability, organisations should engage stakeholders in meaningful discussions and publicly convey the organisation's sustainability activities. Energy, water, materials, and waste management are all areas in which organisations should be well-versed. They should locate areas where the organisation can improve resource efficiency and waste reduction operations. The organisation can reduce their influence on the environment and improve resource efficiency by implementing sustainable procurement procedures, making the most of its energy use, and embracing circular economy ideas (Milios, 2018; Tukker, 2015).

- **Sustainable innovation**: Organisations should encourage a culture of innovation that fosters the development of sustainable products and the improvement of processes. Employee idea generation and implementation that minimises adverse environmental effects and develops sustainable solutions should be encouraged. Organisations may promote innovation for sustainability by creating a culture of creativity and supporting the investigation of sustainable technology and practices (Milios, 2018; Jones & Wynn, 2019).

- **Performance measurement and reporting**: Organisations should set up systems to track and evaluate the company's sustainability performance (Epstein & Buhovac, 2010). They should monitor important indicators like carbon emissions, water usage, waste generation, and social effects to gauge progress towards sustainability objectives. The organisation's dedication to accountability and openness is demonstrated by effective reporting of sustainability performance, both internally and internationally (Krajnc & Glavič, 2003; McDonald & Young, 2012).

- **Change management and employee engagement:** To successfully lead sustainability programmes, organisations must have excellent change management abilities (Cameron & Green, 2019). Employees should be informed of the value of sustainability, given the opportunity to participate in decision-making, and given training and information on sustainable practices (Renwick et al., 2013). Organisations may help the organisation adopt sustainable behaviours by encouraging employee involvement and developing a sense of purpose.

6.6 CONCLUSION AND IMPLICATIONS OF INDUSTRY 5.0

Industry 5.0 technology synergy promotes human-centricity, sustainability, and resilience by fusing human and technological strengths. It recognises the value of humans in the production process, develops their abilities through interaction with robots, and encourages a positive work atmosphere. It promotes sustainable practices by minimising resource usage and relying on renewable energy sources. Finally, it improves resilience by enabling agile and adaptable production systems and utilising data-driven decision-making. Industry 5.0 provides the way for a more inclusive, sustainable, and resilient future by upholding these ideals.

Resolution is achieved by a paradigm shift brought about by Industry 5.0, which will place less emphasis on technology and assume that cooperation between humans and robots is the foundation for progress. The Industrial Revolution is increasing customer satisfaction by deploying personalised products. Industry 5.0 is required for contemporary enterprises to obtain competitive advantages and foster economic growth for production (Alves et al., 2023; Pang et al., 2023).

With the viewpoint of keeping the human being at the focal point of industrial production processes, Industry 5.0 trails Industry 4.0. As a result, there will be a shift from production systems centred on human beings and high customisation to production systems centred on technology advancement and high productivity. As a result, technology no longer controls human beings; instead, they control technology and use it for their benefit. Researchers are advancing the idea of Industry 5.0 even though enterprises and industries have yet to embrace it fully. This is because the industrial reality of today still faces issues related to Industry 4.0 and the digitalisation era. So far, the concentration of humans in industrial systems has been the objective behind Industry 5.0 (Yitmen et al., 2023; Ghobakhloo et al., 2022).

Industry 5.0 helps generate a more ecologically friendly and sustainable industrial ecosystem by exploiting energy and material utilisation. Industry 5.0 identifies resilient systems needing to regulate and bounce back from setbacks. Industry 5.0 makes available real-time data collection, analysis, and further decision-making based on fusing technologies like the IoT and AI. This makes preventive maintenance, predictive analytics, and quick response to unplanned events possible, reducing downtime and enhancing industrial processes' overall resilience. Additionally, Industry 5.0 promotes flexible manufacturing and decentralised production, allowing businesses to respond quickly to shifting consumer needs and automotive supply chain disruptions.

Thus, concluding that automation and AI are combined to enhance rather than completely replace human abilities. This strategy ensures that the production process relies heavily on human creativity, problem-solving abilities, and emotional intelligence. This finding suggests that Industry 5.0 not only assists employees in adjusting to new tasks but also helps acquire the skills required to run and maintain sophisticated machinery and systems; businesses and governments may concentrate on upskilling and reskilling programmes. Another significant finding is that a circular economy model, which encourages the reuse, renovation, and recycling of materials and goods, may be supported by Industry 5.0. The sector may lessen its environmental and waste production impact by designing products with recyclability in mind and streamlining supply networks.

REFERENCES

Agolla, J. E. (2018). Human capital in the intelligent manufacturing and Industry 4.0 revolution. *Digital transformation in Smart Manufacturing*, 2, 41–58.

Alves, J., Lima, T. M., & Gaspar, P. D. (2023). Is Industry 5.0 a human-centred approach? *A Systematic Review. Processes*, 11(1), 193.

Bande, B., Fernández-Ferrín, P., Varela, J. A., & Jaramillo, F. (2015). Emotions and salesperson propensity to leave: The effects of emotional intelligence and resilience. *Industrial Marketing Management*, 44, 142–153.

Bednar, P. M. (2000). A contextual integration of individual and organisational learning perspectives as part of IS analysis. *Informing Science: The International Journal of an Emerging Transdiscipline*, 3, 145–156.

Bednar, P. M., & Welch, C. (2020). Socio-technical perspectives on smart working: Creating meaningful and sustainable systems. *Information Systems Frontiers*, 22(2), 281–298.

Bednar, P. M., Welch, C., & Milner, C. (2016). Excellence in practice through a socio-technical, open systems approach to process analysis and design. *International Journal of Systems and Society (IJSS)*, 3(1), 110–118.

Bedwell, W. L., Wildman, J. L., DiazGranados, D., Salazar, M., Kramer, W. S., & Salas, E. (2012). Collaboration at work: An integrative multilevel conceptualisation. *Human Resource Management Review*, 22(2), 128–145.

Berrah, L., Clivillé, V., Trentesaux, D., & Chapel, C. (2021). Industrial performance: An evolution incorporating ethics in the context of Industry 4.0. *Sustainability*, 13(16), 9209.

Bloem, J., Van Doorn, M., Duivestein, S., Excoffier, D., Maas, R., & Van Ommeren, E. (2014). The fourth industrial revolution. *Things Tighten*, 8(1), 11–15.

Boehm, S. A., Kunze, F., & Bruch, H. (2014). Spotlight on age-diversity climate: The impact of age-inclusive HR practices on firm-level outcomes. *Personnel Psychology*, 67(3), 667–704.

Burke, S., & Gaughran, W. F. (2007). Developing a framework for sustainability management in engineering SMEs. *Robotics and Computer-Integrated Manufacturing*, *23*(6), 696–703.

Cameron, E., & Green, M. (2019). *Making sense of change management: A complete guide to the models, tools and techniques of organisational change*. Kogan Page Publishers.

Carayannis, E. G., & Morawska-Jancelewicz, J. (2022). The futures of Europe: Society 5.0 and Industry 5.0 as driving forces of future universities. *Journal of the Knowledge Economy*, *13*(4), 3445–3471.

Carvalho, H., Duarte, S., & Machado, V. C. (2011). Lean, agile, resilient and green: divergencies and synergies. *International Journal of Lean Six Sigma*, *2*(2), 151–179.

Carver, L., & Turoff, M. (2007). Human-computer interaction: The human and computer teams in emergency management information systems. *Communications of the ACM*, *50*(3), 33–38.

Chang, S. E., McDaniels, T., Fox, J., Dhariwal, R., & Longstaff, H. (2014). Toward disaster-resilient cities: Characterising resilience of infrastructure systems with expert judgments. *Risk Analysis*, *34*(3), 416–434.

Cillo, V., Gregori, G. L., Daniele, L. M., Caputo, F., & Bitbol-Saba, N. (2021). Rethinking companies' culture through knowledge management lens during Industry 5.0 transition. *Journal of Knowledge Management*, *26*(10), 2485–2498.

Clarke, S., & Roome, N. (1999). Sustainable business: Learning–action networks as organisational assets. *Business Strategy and the Environment*, *8*(5), 296–310.

Eckersley, R. (2003). *Deliberative democracy, ecological representation and risk: Towards a democracy of the affected. In Democratic Innovation*. Routledge.

Epstein, M. J., & Buhovac, A. R. (2010). Solving the sustainability implementation challenge. *Organizational Dynamics*, *39*(4), 306.

European Commission. (2021). *Industry 5.0: Towards a more sustainable, resilient and human-centric industry*. Retrieved from https://ec.europa.eu/info/news/industry-50-towards-more-sustainable-resilient-and-human-centric-industry

Fayol, H. (2016). *General and industrial management*. Ravenio Books.

Fraga-Lamas, P., Lopes, S. I., & Fernández-Caramés, T. M. (2021). Green IoT and edge AI as key technological enablers for a sustainable digital transition towards a smart circular economy: An Industry 5.0 use case. *Sensors*, *21*(17), 5745.

George, A. S., & George, A. H. (2020). Industrial Revolution 5.0: The transformation of the modern manufacturing process to enable man and machine to work hand in hand. *Journal of Seybold Report, Report ISSN NO, 1533*, 9211.

Ghobakhloo, M., Iranmanesh, M., Mubarak, M. F., Mubarik, M., Rejeb, A., & Nilashi, M. (2022). Identifying Industry 5.0 contributions to sustainable development: A strategy roadmap for delivering sustainability values. *Sustainable Production and Consumption*, *33*, 716–737.

Giunipero, L., Handfield, R. B., & Eltantawy, R. (2006). Supply management's evolution: Key skill sets for the supply manager of the future. *International Journal of Operations & Production Management*, *26*(7) 822–844.

Glaser, B. G., & Strauss, A. L. (1967). *The discovery of grounded theory: Strategies for qualitative research*. Chicago: Aldine Publishers.

Goleman, D. (2001). The emotionally intelligent workplace: How to select for, measure, and improve emotional intelligence in individuals, groups, and organisations. In E. B. Cherniss, & Daniel, *The emotionally intelligent workplace* (Vol. 1, pp. 27–44).

Gregory, R., Failing, L., Harstone, M., Long, G., McDaniels, T., & Ohlson, D. (2012). *Structured decision making: A practical guide to environmental management choices.* John Wiley & Sons.

Grzybowska, K., & Stachowiak, A. (2022). Global changes and disruptions in supply chains—Preliminary research to sustainable resilience of supply chains. *Energies, 15*(13), 4579.

Hammarlund, C., & Trakanavicius, M. (2023). Digital transformation and new product development at incumbent manufacturing firms a case study on the effects of digital transformation on new product development.

He, Z., Huang, H., Choi, H., & Bilgihan, A. (2023). Building organisational resilience with digital transformation. *Journal of Service Management, 34*(1), 147–171.

Huang, Y., Zhou, Y., Wong, T. K., Luo, D., Zhang, G., Chen, J., & Smith, G. D. (2022). Inpatient Dignity Scale: Mandarin translation and psychometric characteristics evaluation. *Nursing Open, 9*(1), 500–512.

Hung, R. Y.Y., Lien, B. Y., Fang, S. C., & McLean, G. N. (2010). Knowledge as a facilitator for enhancing innovation performance through total quality management. *Total Quality Management, 21*(4), 425–438.

Iannotta, M., Meret, C., & Marchetti, G. (2020). Defining leadership in smart working contexts: A concept synthesis. *Frontiers in Psychology, 11*, 556933.

Jafari, N., Azarian, M., & Yu, H. (2022). Moving from Industry 4.0 to Industry 5.0: What are the implications for smart logistics? *Logistics, 6*(2), 26.

Jarvenpaa, S. L., & Ives. (1994). The global network organisation of the future: Information management opportunities and challenges. *Journal of Management Information Systems, 10*(4), 25–57.

Jayne, M. E. (2004). Leveraging diversity to improve business performance: Research findings and recommendations for organisations. *Human Resource Management: Published in Cooperation with the School of Business Administration, The University of Michigan and in alliance with the Society of Human Resources Management*, pp. 409–424.

Jenkins, H. (2009). A 'business opportunity model of corporate social responsibility for small and medium-sized enterprises. *Business Ethics: A European Review, 18*(1), 21–36.

Jones, P., & Wynn, M. G. (2019). The circular economy, natural capital and resilience in tourism and hospitality. *International Journal of Contemporary Hospitality Management, 31*(6), 2544–2563.

Kochan, T. A., Lansbury, R. D., & MacDuffie, J. P. (2018). *After lean production: Evolving employment practices in the world auto industry.* Cornell University Press.

Kolade, O., & Owoseni, A. (2022). Employment 5.0: The work of the future and the future of work. *Technology in Society*, 102086.

Kolarevic, B., & Parlac, V. (2015). *Building dynamics: Exploring the architecture of change.* Routledge.

Krajnc, D., & Glavič, P. (2003). Indicators of sustainable production. *Clean Technologies and Environmental Policy, 5*, 279–288.

Krippendorff, K. (2018). *Content analysis: An introduction to its methodology.* Sage Publications. ISBN: 9781506395661

Leng, J., Sha, W., Wang, B., Zheng, P., Zhuang, C., Liu, Q., & Wang, L. (2022). Industry 5.0: Prospect and retrospect. *Journal of Manufacturing Systems, 65*, 279–295.

Li, S. X. (2015). The internet of things: A survey. *Information Systems Frontiers*, *17*(2), 243–259.

Longo, F., Padovano, A., & Umbrello, S. (2020). Value-oriented and ethical technology engineering in Industry 5.0: A human-centric perspective for the design of the factory of the future. *Applied Sciences*, *10*(12), 4182.

Luo, Y., & Tung, R. L. (2007). International expansion of emerging market enterprises: A springboard perspective. *Journal of International Business Studies*, pp. 38, 481–498.

Maddikunta, P. K., Pham, Q. V., Prabadevi, B., Deepa, N., Dev, K., Gadekallu, T. R., & Liyanage, M. (2022). Industry 5.0: A survey on enabling technologies and potential applications. *Journal of Industrial Information Integration*, *26*, 100257.

Markos, S., & Sridevi, M. S. (2010). Employee engagement: The key to improving performance. *International Journal of Business and Management*, *5*(12), 89.

McDonald, S., & Young, S. (2012). Cross-sector collaboration shaping corporate social responsibility best practice within the mining industry. *Journal of Cleaner Production*, pp. 37, 54–67.

Milios, L. (2018). Advancing to a Circular Economy: three essential ingredients for a comprehensive policy mix. *Sustainability Science*, *13*(3), 861–878.

Mokyr, J. (2001). The rise and fall of the factory system: technology, firms, and households since the industrial revolution. *In Carnegie-Rochester Conference Series on Public Policy*, *55*, pp. 1–45. North-Holland.

Morgan, D. L., Krueger, R. A., & King, J. A. (1998). *Planning focus groups*. Sage.

Morrar, R., Arman, H., & Mousa, S. (2017). The fourth industrial revolution (Industry 4.0): A social innovation perspective. *Technology Innovation Management Review*, *7*(11), 12–20.

Mourtzis, D., Angelopoulos, J., & Panopoulos, N. (2022). A literature review of the challenges and opportunities of the transition from Industry 4.0 to Society 5.0. *Energies*, *15*(17), 6276.

Murillo, D., & Lozano, J. M. (2006). SMEs and CSR: An approach to CSR in their own words. *Journal of Business Ethics*, *67*, 227–240.

Nahavandi, S. (2019). Industry 5.0—A human-centric solution. *Sustainability*, *11*(16), 4371.

Nevis, E. C., DiBella, A. J., & Gould, J. M. (2009). Understanding organisations as learning systems. *In knowledge, Groupware and the Internet*, pp. 43–63.

Ngwenyama, O. K., & Lee, A. S. (1997). Communication richness in electronic mail: Critical social theory and the contextuality of meaning. *MIS Q*, *21*(2), 145–167.

Pang, T. Y., Lee, T. K., & Murshed, M. (2023). Towards a new paradigm for digital health training and education in Australia: Exploring the implication of the fifth industrial revolution. *Applied Sciences*, *13*(11), 6854.

Perrini, F., & Tencati, A. (2006). Sustainability and stakeholder management: The need for new corporate performance evaluation and reporting systems. *Business Strategy and the Environment*, *15*(5), 296–308.

Pless, N., & Maak, T. (2004). Building an inclusive diversity culture: Principles, processes and practice. *Journal of Business Ethics*, *54*, 129–147.

Raja Santhi, A., & Muthuswamy, P. (2023). Industry 5.0 or Industry 4.0 S? Introduction to Industry 4.0 and a peek into the prospective Industry 5.0 technologies. *International Journal on Interactive Design and Manufacturing (IJIDeM)*, *17*(2), 947–979.

Rani, S., & Mishra, K. (2014). Green HRM: Practices and strategic implementation in the organisations. *International Journal on Recent and Innovation Trends in Computing and Communication*, 2(11), pp. 3633–3639.

Renwick, D. W., Redman, T., & Maguire, S. (2013). Green human resource management: A review and research agenda.*International Journal of Management Reviews*, 15(1), 1–14.

Robles, M. M. (2012). Executive perceptions of the top 10 soft skills needed in today's workplace. *Business Communication Quarterly*, 75(4), 453–465.

Roepke, R., Agarwal, R., & Ferratt, T. W. (2000). Aligning the IT human resource with business vision: The leadership initiative at 3M. *Mis Quarterly*, 327–353.

Saniuk, S., Grabowska, S., & Straka, M. (2022). Identification of social and economic expectations: Contextual reasons for the transformation process of Industry 4.0 into the Industry 5.0 concept. *Sustainability 4, 14*(3), 1391.

Schröder, C. (2016). *The challenges of Industry 4.0 for small and medium-sized enterprises*. Friedrich-Ebert-Stiftung: Bonn, Germany, 1–28.

Schuetze, H. G. (2006). International concepts and agendas of lifelong learning. *Compare*, 36(3), 289–306.

Seville, E., Van Opstal, D., & Vargo, J. (2015). A primer in resiliency: seven principles for managing the unexpected. *Global Business and Organizational Excellence*, *34*(3), 6–18.

Sharma, A., & Singh, B. J. (n.d.). Evolution of industrial revolutions: A review. *International Journal of Innovative Technology and Exploring Engineering*, 9(11), 66–73.

Shipman, K., Burrell, D. N., & Huff Mac Pherson, A. (2023). An organisational analysis of how managers must understand the mental health impact of teleworking during COVID-19 on employees. *International Journal of Organizational Analysis*, 31(4), 1081–1104.

Skobelev, P. O., & Borovik, S. Y. (2017). *On the way from Industry 4.0 to Industry 5.0: From digital manufacturing to digital society*. (2nd ed.). 6, 307–311.

Skovdal, M., & Cornish, F. (2015). Qualitative research for development: A guide for practitioners. Practical Action (Organization). ISBN 9781853398544

Spence, L. J., & Rutherford, R. (2001). Social responsibility, profit maximisation and the small firm owner-manager. *Journal of Small Business and Enterprise Development*, 8(2), 126–139.

Strauss, A., & Corbin, J. M. (1997). *Grounded theory in practice*. Sage.

Swanson, R. A. (2022). *Foundations of human resource development*. Berrett.

Taneja, S., Sewell, S. S., & Odom, R. Y. (2015). A culture of employee engagement: A strategic perspective for global managers. *Journal of Business Strategy*, 36(3), 46–56.

Tukker, A. (2015). Product services for a resource-efficient and circular economy–a review. *Journal of Cleaner Production*, pp. 97, 76–91.

Verhave, T., & van Hoorn, W. (2014). The temporalisation of the self. *Historical Social Psychology (Psychology Revivals)*, 325.

Vinitha, K., Prabhu, R. A., Bhaskar, R., & Hariharan, R. (2020). Review on industrial mathematics and materials at Industry 1.0 to Industry 4.0. *Materials Today: Proceedings*, 33, pp. 3956–3960.

Waters, D. (2011). *Supply chain risk management: vulnerability and resilience in logistics*. Kogan Page Publishers.

Weick, K. E., & Sutcliffe, K. M. (2015). *Managing the unexpected: Sustained performance in a complex world*. John Wiley & Sons.

Wójcicki, K., Biegańska, M., Paliwoda, B., & Górna, J. (2022). Internet of Things in industry: Research profiling, application, challenges and opportunities. *A Review. Energies, 15*(5), 1806.

Wrigley, E. A. (1962). The supply of raw materials in the industrial revolution. *The Economic History Review, 15*(1), 1–16.

Wrzesniewski, A., & Dutton, J. E. (2001). Crafting a job: Revisioning employees as active crafters of their work. *Academy of Management Review, 26*(2), 179–201.

Xu, X., Lu, Y., Vogel-Heuser, B., & Wang, L. (2021). Industry 4.0 and Industry 5.0—Inception, conception and perception. *Journal of Manufacturing Systems, 61*, 530–535.

Yitmen, I., Almusaed, A., & Alizadehsalehi, S. (2023). Investigating the causal relationships among enablers of the construction 5.0 Paradigm: Integration of Operator 5.0 and society 5.0 with human-centricity, sustainability, and resilience. *Sustainability, 15*(11), 9105.

Yukl, G. A., & Becker, W. S. (2006). Effective empowerment in organisations. *Organization Management Journal, 3*(3), 210–231.

Zack, M. H. (2003). Rethinking the knowledge-based organisation. *MIT Sloan Management Review*.

Zarte, M., Pechmann, A., & Nunes, I. L. (2020). Principles for human-centred system design in Industry 4.0-a systematic literature review. In *Advances in Human Factors and Systems Interaction: Proceedings of the AHFE 2020 Virtual Conference on Human Facts*.

Zizic, M. C., Mladineo, M., Gjeldum, N., & Celent, L. (2022). From Industry 4.0 towards Industry 5.0: A review and analysis of paradigm shift for the people, organisation and technology. *Energies, 15*(14), 5221.

Chapter 7

Impact of Industry 5.0 on sustainable development with machine learning and the role of internal audit

Literature review

Ali Rehman
A'Sharqiyah University, Ibra, Oman

Tariq Umar
University of the West England, Bristol, UK

Fathyah Hashim
Universiti Sains Malaysia, Penang, Malaysia

7.1 INTRODUCTION

A nation's economy cannot survive without the existence of industries (Milberg & Winkler, 2013). The economy is composed of various industries that can be further categorized into interrelated goods and services (Morrar et al., 2017). Dividing businesses into distinct industrial segments can provide investors and economists with insight into the diverse levels of economic activity (Rodrik, 2018; Audretsch et al., 2015). Familiarity with this organizational framework can enhance one's comprehension of the economy's overall structure. Industries refer to a group of businesses engaged in commercial activities (Müller et al., 2021) and can be divided into different sectors such as financial, services, and manufacturing. While many people associate industries with large-scale manufacturing, it can also include other commercial activities such as agriculture, transportation, and hospitality.

The different levels of industries include primary, secondary, tertiary, quaternary, and quinary industries (Dinu & Nosca, 2020). Primary industries extract or produce raw materials, such as mining, forestry, and fishing, while secondary industries transform raw materials into usable products through processing and manufacturing (Larchenko et al., 2020). Tertiary industries provide essential services and support to allow other levels of industry to function (Ge et al., 2022). Quaternary industries focus on the creation and transfer of information (Aithal, 2019), and quinary industries are responsible for controlling the industrial and government decision-making processes (Dinu et al., 2022). Understanding these levels can provide insight into the different activities that contribute to the economy and how they work together to create value.

Industries play a crucial role in the economy of a nation (Milberg & Winkler, 2013). They provide employment opportunities, generate income, and contribute to economic growth and development. Industries provide jobs to people, both skilled and unskilled, which contributes to reducing unemployment rates. Employment opportunities provided by industries also enable individuals to earn income and contribute to the economy (Manzoor et al., 2019). Industries generate income for individuals, businesses, and governments through taxes, fees, and other revenues. This income is then used to finance government expenditures, pay salaries, and support social welfare programs, which contribute to economic growth and development.

Industries often produce goods and services that can be exported, which helps to earn foreign exchange and improve the balance of payments (Manzoor et al., 2019). This can lead to increased economic growth and development. Industries are often at the forefront of innovation and technological advancement, which can lead to the transfer of knowledge and technology to other sectors of the economy, leading to the creation of new products and services, and increased productivity. Industries usually require extensive infrastructure such as transportation networks, power plants, and communication systems, which can lead to the development of the necessary infrastructure and create opportunities for other businesses to grow and develop, thus these industries are the cornerstone for the sustainable development of any nation (Liu et al., 2018).

Industry 5.0 is a term that is currently being used to describe the future of manufacturing and production. It is a concept that builds upon the previous industrial revolutions, namely, the mechanization of production (Industry 1.0), mass production (Industry 2.0), automation (Industry 3.0), and digitalization (Industry 4.0). Industry 5.0 envisions a new era of manufacturing that combines advanced technology with the creativity and ingenuity of human beings. In Industry 5.0, machines and humans will work together in a more collaborative and integrated way than ever before (Xu et al., 2021). Rather than simply automating repetitive tasks, Industry 5.0 seeks to empower workers with new tools and technologies that allow them to take on more complex and creative tasks (Nayyar & Kumar, 2020). This could include things like using virtual reality to design products or working alongside robots to assemble them. Industry 5.0 emphasizes sustainability and social responsibility and encourages manufacturing processes to be designed with the goal of minimizing environmental impact, reducing waste, and promoting ethical practices throughout the supply chain. While Industry 5.0 is still in the early stages of development, it has the potential to transform the way we think about manufacturing and production, creating new opportunities for innovation, collaboration, and social responsibility (Breque et al., 2021; Committee, 2021; Longo et al., 2020; Bednar & Welch, 2020).

Industry 5.0 envisions a future where humans and machines work together (Leng et al., 2022) in harmony to enhance their collaboration, leading to the creation of more successful and efficient businesses enabling to achieve sustainable development (Dwivedi et al., 2023). Such an interaction possesses many risk factors including fraud risk and can raise many red flags. Given that Industry 5.0 is a relatively new concept, it is possible that numerous risks associated with it have not yet been identified. According to the Association of Certified Fraud Examiners (ACFE), the median duration of fraud – which refers to the average time between when fraud occurs and when it is discovered – is 12 months (ACFE, 2022). Moreover, Figure 7.1 illustrates that the financial loss resulting from fraud increases with the duration of its detection.

Machine learning (ML) with the assistance of internal audit (IA) can identify such red flags and can ensure the path of sustainability. It is worth noting that fraud is one of the major barriers to the achievement of sustainable development (Rehman, 2022a). Industry 5.0 can be utilized to prevent fraud (Nurafia, 2023). Table 7.1 demonstrates how Industry 5.0 can be utilized to prevent fraud in some industries.

Sustainable development is required due to the urgent need to address climate change and its impact on the environment, society, and governance (Sinha et al., 2020). Sustainable development refers to a development model that meets the needs of the present without compromising the ability of future generations to meet their own needs. The United Nations has identified 17 Sustainable Development Goals (SDGs) to be

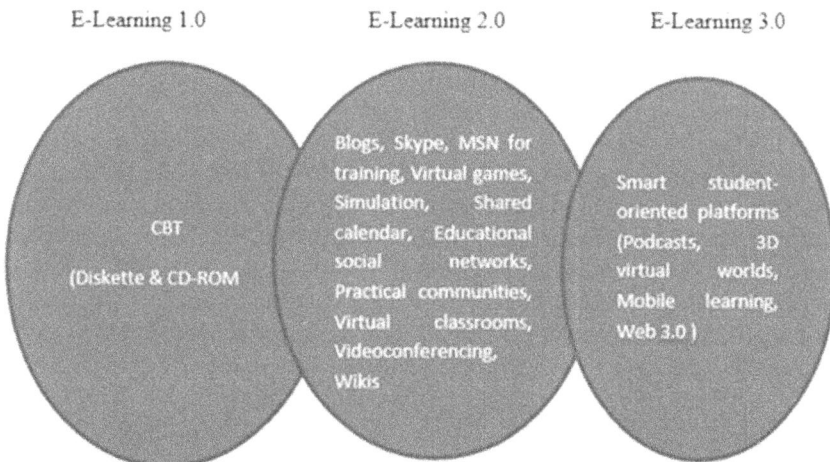

E-Learning 1.0   E-Learning 2.0   E-Learning 3.0

CBT

(Diskette & CD-ROM

Blogs, Skype, MSN for training, Virtual games, Simulation, Shared calendar, Educational social networks, Practical communities, Virtual classrooms, Videoconferencing, Wikis

Smart student-oriented platforms (Podcasts, 3D virtual worlds, Mobile learning, Web 3.0)

Figure 7.1 Increase in financial loss due to increase in time of its detection.

Source: ACFE (2022).

Table 7.1 Utilization of Industry 5.0 to prevent frauds

Industry	Technology	How it prevents fraud
Financial Services	Machine Learning	Analyzes transactional data to detect anomalies indicative of fraudulent activity.
Manufacturing	Machine Learning, IoT and Blockchain	Increases supply chain transparency and traceability to identify fraudulent activities such as counterfeit goods, theft, and tampering.
Retail	Machine Learning	Analyzes customer data to detect suspicious patterns of behavior that may indicate fraudulent activity.
Healthcare	Big Data Analytics and Machine Learning	Analyzes claims data to identify patterns and anomalies that may indicate fraudulent activity, such as duplicate claims or claims for services that were never provided.

achieved by 2030, covering areas such as poverty reduction, climate action, and responsible consumption and production (Assembly, 2015; Sinha et al., 2020).

Given that technological advancements have the power to transform the way value is created, exchanged, and distributed, it is imperative that these technologies are designed to support societal values of the future. As these changes continue to unfold, questions surrounding technological innovation have arisen, necessitating that the industry re-examine its position and role in society (Breque et al., 2021). The political priorities of the world have played a significant role in shaping this discourse. "Go Green" initiatives require a transition to a more circular economy and greater reliance on sustainable resources. Moreover, the COVID-19 crisis has underscored the importance of rethinking existing working methods and approaches (Rehman et al., 2023), particularly regarding the vulnerability of global supply chains, with a focus on building industries that are more resilient, sustainable, and human-centric in the future.

Industries contribute toward the sustainability of the nations (Morrar et al., 2017). Industries focus on creating reliable, sustainable, and resilient infrastructure, including cross-border and regional infrastructure, to support economic growth and enhance human well-being (Breque et al., 2021). Industries 5.0 can assure equitable and affordable access for all, promote inclusive and sustainable industrialization, increasing industry's contribution to employment and gross domestic product, particularly in developing countries (Kynčlová et al., 2020). Industries 5.0 can ensure and facilitate access to financial services, such as affordable credit, for small-scale industries and integrate them into markets and value chains. Industry 5.0 will utilize upgraded infrastructure and upgrade industries to adopt clean and environmentally sound technologies and processes, while also improving resource-use efficiency. It is expected that Industry 5.0 encourages scientific research and innovation, particularly in developing countries, and increases research and development (Gera et al., 2023).

There are several risks associated with Industry 5.0 and sustainable development which require mitigation. These risks create hindrances in the achievement of sustainability. Such risks are identified by the United Nations (UN) and include (UN, 2022) the following:

1. Lack of access to financial support for small manufacturers. Only 33% are benefiting from loans or lines of credit.
2. Industries of least developed countries are still suffering from post-pandemic impact. The gap between the recoverability of manufacturing growth between developed nations and least developed nations is approximately three times.
3. One-third of jobs in industries are experiencing negative impacts.
4. Lower tech-oriented countries are suffering lower production as compared to higher technology industries.

It is evident that technology plays a crucial role in achieving sustainability (Sarkodie, 2022; Waqas et al., 2022); however, it is important to acknowledge that not everyone may have access to it due to limited and expensive availability. Moreover, the mere availability of technology does not guarantee its proper implementation and desired outcomes (Mergel et al., 2019). To address these risks, machine learning can offer several promising solutions (Wang et al., 2022b). Machine learning is a rapidly evolving field of artificial intelligence (AI) that enables computers to learn from data and improve their performance over time without being explicitly programmed (Soori et al., 2023). The use of machine learning has the potential to contribute to sustainable development by enabling more efficient and effective decision-making, reducing waste and emissions, and improving resource management.

The adoption of machine learning also brings new challenges and risks, such as data privacy and biased decision-making due to limited information (Rahman et al., 2022). The successful implementation of machine learning technology in organizations is often hindered by various factors (Dwivedi, et al., 2019). In addition to potential algorithmic issues arising from improper training sets, other challenges may include a lack of strategic planning and limited experience with the technology, which can result in wasted time and resources. Security vulnerabilities may also arise from the use of outdated data sources, and regulatory challenges can arise from a lack of understanding of how the algorithm makes decisions (Shaw et al., 2019). Furthermore, third-party risks can pose a significant threat to organizations, as a provider may fail to properly govern the machine learning solution, leading to a data breach. Thus, it is important to identify and address these challenges to ensure the successful adoption and use of machine learning technology in organizations (Koptelov, 2022; Cohen, 2020; Myradov, 2022; Rahman et al., 2022).

According to a recent survey administered to over 2,000 individuals across diverse industries, the most significant barriers to the adoption of machine learning technology were identified as a lack of clear strategy and a shortage of talent with appropriate skill sets. Specifically, 43% of respondents cited a lack of a clear strategy as a primary obstacle, while 42% identified a dearth of skilled personnel as a challenge. These results underscore the importance of addressing these issues in order to effectively integrate machine learning technology into organizational operations (Diligent, 2021). The identified risks can be effectively addressed through the utilization of the internal audit function, which is an integral component of organizational governance management. By leveraging their expertise, internal auditors can identify and propose meaningful recommendations to mitigate the identified risks (Rehman, 2022a).

The role of internal audit is to ensure that the effective and ethical use of technologies mitigates risks associated with machine learning and implementation of Industry 5.0 (Verma et al., 2022). Internal audits can play a critical role in ensuring the alignment of machine learning initiatives with the organization's sustainability objectives, identifying, and mitigating risks, and promoting transparency and accountability (Rehman, 2022a). Internal audits play a crucial role in ensuring that the organization's sustainability goals are being met (Amoako et al., 2023) and that the adoption of Industry 5.0 technologies is aligned with these goals.

Internal audits can conduct audits to assess the effectiveness of the organization's sustainability initiatives, including the adoption of machine learning and Industry 5.0 (Verma, et al., 2022). These audits can help identify areas of improvement and ensure that the organization is effectively utilizing these technologies to meet its sustainability objectives (Assiri & Humayun, 2023). Internal audits can also help identify potential risks associated with the adoption of Industry 5.0 technologies, including data privacy and security risks. As machine learning requires the use of large amounts of data, it is essential that the organization has effective controls in place to protect this data from misuse or unauthorized access. Internal audits can conduct assessments of the organization's data privacy and security controls (Slapni et al., 2022) to ensure that they are effective and meet industry best practices.

The advent of digitization and digital transformation has brought significant changes to various industries. It is important for the internal audit to stay abreast of these changes, as it plays a critical role in collecting primary data from various information systems within a company (Kristensen, 2022), processing it, and transforming it into new information (Iliev, 2022). As the industry moves toward Industry 5.0, controls such as financial control, tax control, and evaluation control are all facing challenges due to digitization (Adel, 2022; Espina-Romero, et al., 2023).

It is crucial for internal audit functions to digitize and embrace the use of technology to remain effective (Adel, 2022). In some cases, the digitization of control or audit may even need to precede the digitization of processes in society and the economy. To this end, control processes have been digitized in recent years through the use of accounting programs, electronic invoicing, remote accounting, electronic signature, and submission of various declarations electronically to regulatory agencies (Coman, et al., 2022). Overall, the digitization of control functions is essential to keep pace with Industry 5.0 and to ensure the continued effectiveness of control processes.

Internal audits can also benefit from machine learning. Integrating machine learning into internal audit processes can offer several benefits (Shinde, 2021). The use of machine learning can automate audit procedures, making them more efficient and faster. This automation can also enable internal auditors to tackle more complex tasks and improve their skill set, particularly in the identification of business processes, associated risks, and fraud detection. Additionally, real-time anomaly detection and monitoring can help internal auditors address potential issues before they become major problems (Bruno & Pedrosa Isabel, 2020).

Stakeholder Theory suggests that organizations and/or industries have a responsibility to consider the interests of all stakeholders, including the environment and society, in addition to their shareholders (Freeman & Dmytriyev, 2017; Bridoux & Stoelhorst, 2022). The application of machine learning for sustainable development in Industry 5.0 can be seen as a way for industries to fulfill this responsibility, by reducing their environmental impact and promoting social responsibility (Freeman et al., 2021). Internal audits can play a critical role in ensuring that the organization is considering the interests of all stakeholders in its machine learning initiatives, including potential environmental and social impacts. This can help the organization build stronger relationships with its stakeholders and enhance its reputation as a responsible and sustainable business (Guo et al., 2023).

Resource-Based View (RBV) Theory of the industry suggests that the strategic resources and capabilities of an industry can be a source of sustainable competitive advantage (Freeman et al., 2021). The application of machine learning for sustainable development can be seen as a strategic resource that can enable industries to achieve their sustainability objectives and gain a competitive advantage in the market. Internal audits can help ensure that the industry is effectively utilizing its machine learning capabilities to achieve its sustainability objectives and gain a competitive advantage (Chen et al., 2022).

Ethical Decision-Making (EDM) Theory can be applied to the topic of machine learning for sustainable development, as the adoption of these technologies presents new ethical challenges and risks (Meier et al., 2022). This theory suggests that EDM involves considering the potential consequences of actions and ensuring that they align with ethical principles and values (Abel et al., 2016). Internal audits can help ensure that the organization's machine learning initiatives are aligned with its ethical principles and values and that potential ethical risks are identified and addressed.

To summarize, for the above-mentioned three theories, the Stakeholder Theory suggests that the adoption of machine learning for Industry 5.0 and sustainable development can help organizations fulfill their responsibility toward their stakeholders, while the RBV suggests that these initiatives can provide a strategic resource that enables the industries to achieve its sustainability objectives and gain a competitive advantage. The EDM Theory emphasizes the importance of considering the potential consequences and ethical implications of machine learning for sustainable development and ensuring that these initiatives are aligned with ethical principles and values for Industry 5.0. Internal audits can play a critical role in each of these theories by ensuring that the industry's machine learning initiatives are effectively aligned with its sustainability objectives, promote transparency and accountability, and are implemented in an ethical and responsible manner.

The purpose of this chapter is to analyze the effects of Industry 5.0 on sustainable development by using machine learning, and the role of internal audit in facilitating this transformation. This chapter aims to provide a comprehensive analysis and literature review by applying three theoretical perspectives: Stakeholder Theory, RBV, and EDM. To achieve this purpose, a literature review of relevant scholarly articles, reports, and case studies that explore the adoption of Industry 5.0 technologies and their implications for sustainability, machine learning, and the role of internal audit is conducted.

The success of Industry 5.0 adoption is dependent on industries adopting a holistic approach that considers the interests of all stakeholders, invests in critical resources such as data infrastructure and skilled personnel, and ensures ethical principles are upheld. By adopting these practices, organizations can effectively leverage Industry 5.0 technologies to promote sustainable development while mitigating risks and ensuring compliance with ethical standards. This chapter may encourage regulators and policymakers to revise their rules and policies and incorporate Industry 5.0 for sustainable development, while including internal audit and machine learning as compulsory resources. To the best of the researchers' knowledge, there is no study available that identifies the impact of Industry 5.0 on sustainability and the use of internal audit and machine learning.

7.1.1 Objectives of the chapter

The objectives of the chapter are:

- To identify that industries are vital to a nation's economy, contributing to employment, income generation, and economic growth;

- To identify Industry 5.0 contribution to sustainability and social responsibility;
- To analyze that ML can help with sustainable development and resource management;
- And, to analyze the role of IA in identifying and mitigating risks associated with Industry 5.0 and ML, and how IA must embrace technology to remain effective.

7.1.2 Organization of the chapter

The rest of the chapter is organized as follows: Section 7.2 illustrates the literature review and Section 7.3 enlightens associated theories. And, finally, Section 7.4 concludes the chapter with future scope.

7.2 LITERATURE REVIEW

This section aims to provide an in-depth analysis of Industry 5.0, which is the latest stage of industrial evolution, and explore how Industry 5.0 is set to impact SDGs and the potential risks associated with its implementation. Moreover, this section will also delve into two crucial aspects that can promote sustainability in Industry 5.0 – internal audit and machine learning. These activities have the potential to make a positive impact on the implementation of Industry 5.0 by ensuring proper control mechanisms are in place to enhance sustainability. Internal audit is a process of assessing the effectiveness of an organization's internal control systems, including its governance, risk management, and internal operations. It can help companies identify weaknesses and vulnerabilities that may hinder their efforts to achieve sustainability goals. Additionally, it can enable businesses to establish accountability and transparency in their operations, which is crucial for sustainable development. Similarly, machine learning is another critical tool for achieving sustainability in Industry 5.0. By analyzing large datasets, machine learning can help companies identify patterns and trends, making it easier to anticipate future challenges and risks. This knowledge can empower businesses to make informed decisions that positively impact sustainability. This section will also discuss in detail the associated theories, namely, the Stakeholder Theory, RBV of the industry, and EDM Theory.

7.2.1 Industry 5.0

7.2.1.1 Background of sustainable development

There has been a continuous effort since humans started their lives on the earth to make the earth a better place to live. History indicates that humans went through several stages, including the Stone Age, Bronze Age, and Iron

Age, before reaching the current modern lifestyle where life is more facilitated and comfortable (Jørgensen, 1989).

This development has, however, changed and affected the natural system of the earth, resulting in the issues of global warming and climate change (Weart, 2003). Humans have witnessed some greater natural disasters that are viewed as a result of human activities (Guo, 2010). The main reason for these disasters in many instances was the rise of the earth's temperature. There have been many studies and analyses that demonstrate the rise in earth's temperature, which has been very sharp since 1980 – the era also known as the Third Industrial Revolution (Figure 7.2) is defined by The New York Times (NYT) and informed the rise in temperature for a whole century (NYT, 2023; NASA, 2022; Cohen, 2018). These consequences of human activities on earth have forced us to think about the negative implications of these developments and how to avoid or reduce these consequences.

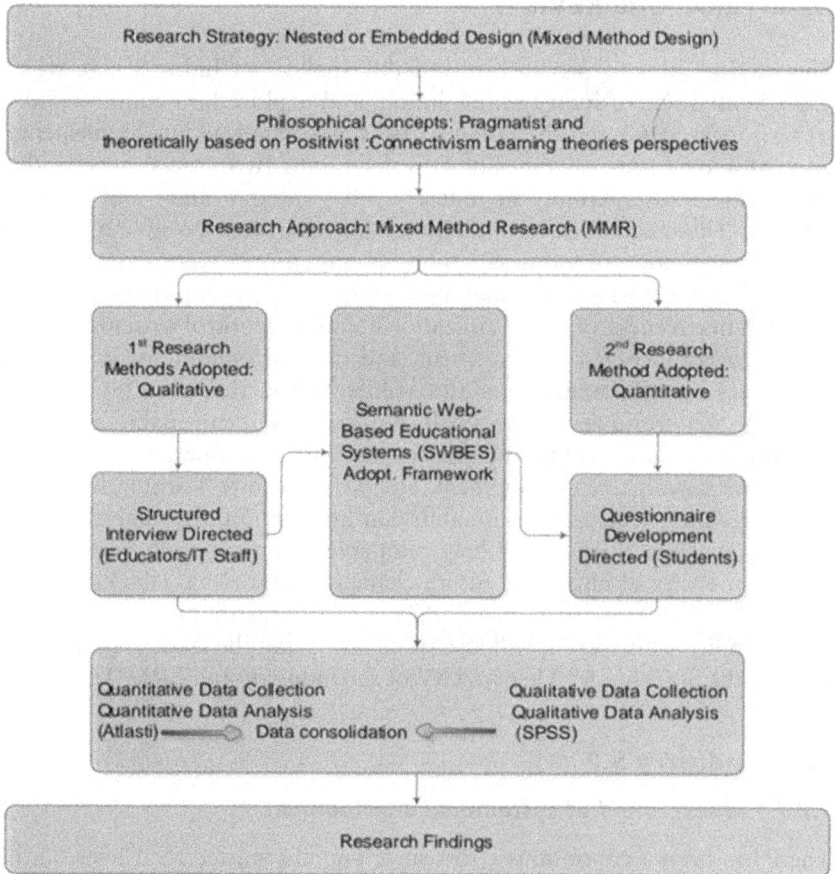

Figure 7.2 Rise in the earth's temperature from 1880 to 2022.

The whole idea is this is quite simple. We have only one earth to live on, and it is not only us but also the upcoming generation who will have only this earth to live; therefore, we need to protect and use the earth's resources wisely so that it can remain a better place for the future generation to live. Currently, there are scientific investigations for the settlement on other planets, but this appears to take a significant time; thus, the only available option so far is the earth (Arnhof, 2016). While there has been concern about the sustainability of the earth for a long time, the concept of sustainability first truly appeared in the Brundtland Commission Report, published in 1987. This report was submitted to the United Nations' 42nd General Assembly session. This report truly aimed to warn the countries about the negative environmental impact caused by economic development and globalization. The report further aimed to provide solutions to the problems arising from industrialization, urbanization, and population growth (Brundtland Commission, 1987).

The idea of sustainability developed in the early 1980s, as reported in the International Geosphere-Biosphere Programme, can be defined as "meeting fundamental human needs while preserving the earth natural environment" (IGBP, 1999). Since the earth's population is increasing, it is putting pressure on the earth's resources. According to the World Economic Forum, it is estimated that food production will need to double by 2050 to feed 10 billion people on the earth (WEF, 2018). Today, sustainability has three essential pillars – environmental protection, social development, and economic growth – and sustainable development can be defined as a development that meets the needs of the present without compromising the ability of future generations to meet their own needs (Sachs, 2015). This is something that cannot be achieved alone; therefore, collective commitment and efforts are needed, which has been evidenced at the global level in the form of the Paris Agreement and the United Nations' 17 SDGs (Paris Agreement, 2015; UN SDGs, 2015).

In the past, there were eight Millennium Development Goals (MDGs) – which range from halving extreme poverty to halting the spread of HIV/AIDS and providing universal primary education (MDGs, 2000). These MDGs are now replaced by the 17 SDGs. The principle of SDGs is very simple. The overall aim is to reduce the negative impacts of human activities on the environment without compromising socio-economic development. If we look at the list of the main pillars of sustainability, then environmental sustainability includes biodiversity conversation, efficient land use, and physical planning.

Social sustainability focuses on decent work, quality education, good health, and ensuring the rule of law and human rights. Finally, economic sustainability involves the reduction of the negative impact of human activities on the environment (Opoku, 2022; Mensah, 2019). Hopwood et al. (2005), viewed sustainable development as human-centered where sustainable development allows for balancing environmental and social dimensions provided that there is a strong commitment to social issues, for instance, ensuring good health for all. In other words, society depends on the environment, while the economy depends on society.

On one side, there is a great realization of sustainability and sustainable development, but on the other side, there is evidence that demonstrates the world is not meeting its commitment and promises. The earth has more challenges than before, therefore, to transfer the earth to the coming generation in a good liveable condition, all efforts at individual, societal, organizational, governmental, and international levels should be undertaken to realize the key elements of sustainability and sustainable development. If the world is not progressing the way it should progress toward sustainable development, this means that the efforts for that are derailed at one or another level. Keeping the importance of sustainable development and the world's commitment to make the earth a better place to live, there has been a great contribution of technology and industries. On the one hand, it has made our lives comfortable but on the other, side, there have been some consequences of these. The next section sheds light on the United Nations' SDGs which are set to be achieved by 2030.

7.2.1.2 Sustainable development goals

The United Nations' current SDGs were prepared in the United Nations Sustainable Development Conference that was organized in the city of Rio de Janeiro, Brazil, in 2012. The purpose of this conference was to arrive at a set of universal goals, connecting the environmental, political, and economic challenges that are faced by the world. Later in 2015, the United Nations officially adopted the 17 SDGs, replacing the eight MDGs and setting 2030 as a target of achievement for them (UN SDGs, 2015). Most of the countries around the world have adopted these goals as their national development goals and trying to achieve them through a collaborative way of working. These goals are guided by the principle of universality, which means that all countries and citizens have a role to play in their achievement (Fei et al., 2021). Each goal is further supported with several number of target and indicators as noted in Table 7.2.

Statistics indicate that on overage there has been good progress toward sustainability, as shown in Figure 7.3, but clearly, the line goes flat, meaning there has been no improvement since 2019 (Sachs et al., 2022). The current (2022) SDGs report provides a global overview of progress on the implementation of the 2030 Agenda for Sustainable Development, using the latest available data and estimates. The report clearly indicates that many countries are not on track toward the achievement of these goals by 2030. While there are many factors that are affecting the progress of SDGs, COVID-19 is the leading factor that has wiped out more than four years of progress (SDGR, 2022). The situation over the progress is not satisfactory which is well observed by the United Nations Secretary-General António Guterres in

Table 7.2 The United Nations' SDGs, its descriptions targets, and indicators

Goal number	Goal name	Goal description	Goal targets	Goal indicator
1	No Poverty	No Poverty: End poverty in all its forms everywhere	7	13
2	Zero Hunger	Zero Hunger: End hunger, achieve food security and improved nutrition, and promote sustainable agriculture	8	14
3	Good Health and Well-Being	Good Health and Well-Being: Ensure healthy lives and promote well-being for all at all ages	13	28
4	Quality Education	Quality Education: Ensure inclusive and equitable quality education and promote life-long learning opportunities for all	10	12
5	Gender Equality	Gender Equality: Achieve gender equality and empower all women and girls	9	14
6	Clean Water and Sanitation	Clean Water and Sanitation: Ensure availability and sustainable management of water and sanitation for all	8	11
7	Affordable and Clean Energy	Affordable and Clean Energy: Ensure access to affordable, reliable, sustainable, and modern energy for all	5	6
8	Decent Work and Economic Growth	Decent Work and Economic Growth: Promote sustained, inclusive, and sustainable economic growth, full and productive employment and decent work for all	12	16
9	Industry Innovation and Infrastructure	Industry, Innovation, and Infrastructure: Build resilient infrastructure, promote inclusive and sustainable industrialization, and foster innovation	08	12
10	Reduced Inequalities	Reduced Inequalities: Reduce inequality within and among countries	10	14
11	Sustainable Cities and Communities	Sustainable Cities and Communities: Make cities and human settlements inclusive, safe, resilient and sustainable	10	14
12	Responsible Consumption and Production	Responsible Consumption and Production: Ensure sustainable consumption and production patterns	11	13

(Continued)

Table 7.2 (Continued) The United Nations' SDGs, its descriptions targets, and indicators

Goal number	Goal name	Goal description	Goal targets	Goal indicator
13	Climate Action	Climate Action: Take urgent action to combat climate change and its impacts	5	8
14	Life below Water	Life below Water (Oceans): Conserve and sustainably use the oceans, seas, and marine resources for sustainable development	10	10
15	Life on land	Life on Land (Biodiversity): Protect, restore, and promote sustainable use of terrestrial ecosystems, sustainably manage forests, combat desertification, and halt and reverse land degradation and halt biodiversity loss	12	14
16	Peace, Justice, and Strong Institutions	Peace, Justice, and Strong Institution: Promote peaceful and inclusive societies for sustainable development, provide access to justice for all, and build effective, accountable, and inclusive institutions at all levels	12	24
17	Partnerships of Goals	Partnership for the Goals: Strengthen the means of implementation and revitalize the global partnership for sustainable development	19	24

his remarks stating, "We must rise higher to rescue the SDGs – and stay true to our promise of a world of peace, dignity and prosperity on a healthy planet".

It is a fact that there was a major interruption in the progress of SDGs for the last three years, starting from 2019, due to COVID, but statistics prior to COVID indicate that most of the countries were not on track to achieve these goals (Umar & Umeokafor, 2022). The successful achievement of all the goals is only possible when all the elements of society and all government organizations embed these goals in their relevant regional, national, and local policies and make an effective system of monitoring and cooperation among the organizations because, in some instances, it is possible that the effort of one organization can derail on efforts of the other, affecting one or many goals. This is also because the SDGs and targets are interlinked to each other where one can affect the other in a positive or negative way (Nilsson et al., 2016; Griggs et al., 2013; Fei et al., 2021; UNSC, 2019; Roy & Pramanick, 2019) – Figure 7.4.

The collaborative way of working to achieve these goals is well appreciated by many researchers. For instance, Adams (2017) that the fulfillment of

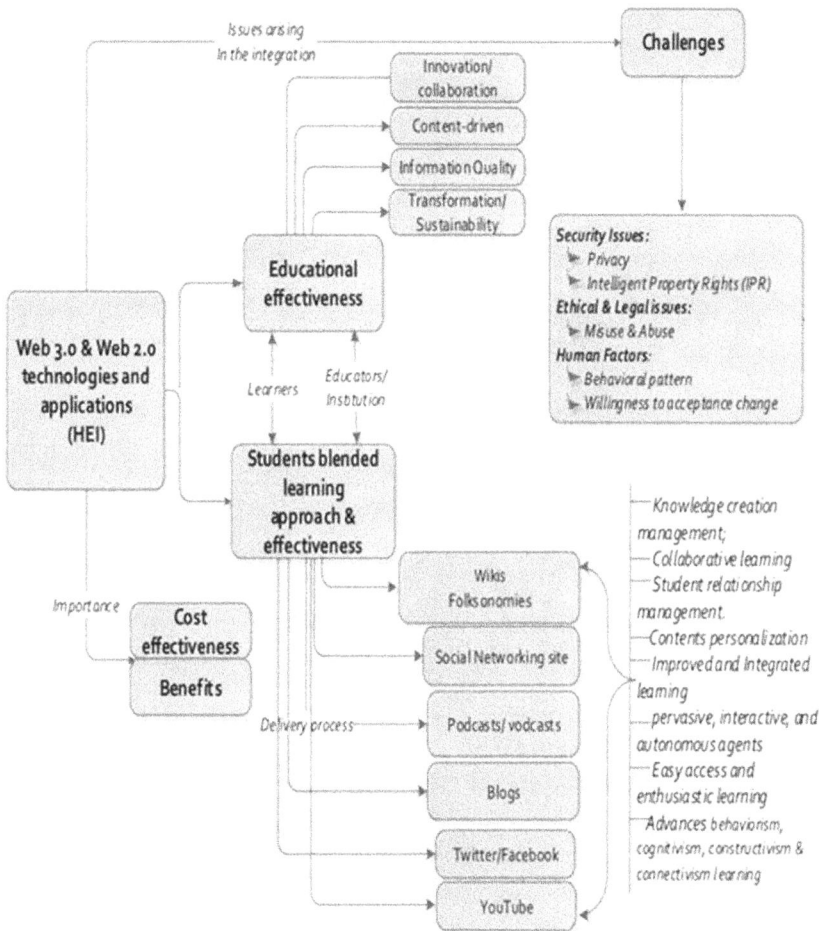

Figure 7.3 SDGs' index score over time.

Source: Sachs et al. (2022).

the SDGs can be difficult without collaboration between governments, private and public sectors, and civil society organizations. Fei et al. 2021) argued that the 17 SDGs can be divided into five "P"s – People, Planet, Prosperity, Peace, and Partnership – as shown in Table 7.3. All these five "P"s of the SDGs need the effective involvement of all the stakeholders, such as governments, institutions, and businesses. The built environment is also an important stakeholder; therefore, it needs to develop approaches and align its activities to SDGs. Likewise, SDGs require the transformations of different societal elements, including education and skills; health and well-being; clean energy and industry; sustainable land use; sustainable

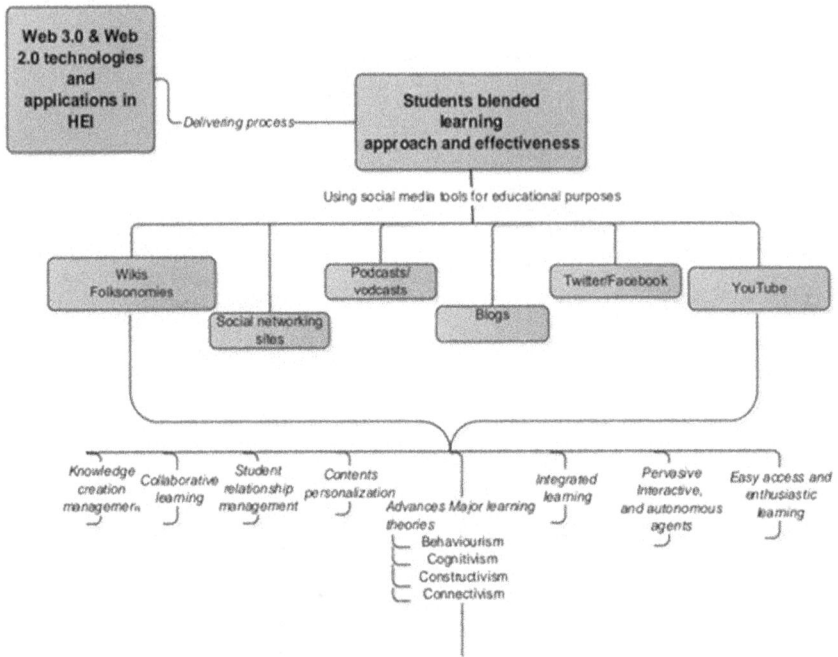

Figure 7.4 Interlinking of SDGs.

Source: Nilsson et al. (2016); Griggs et al. (2013); Fei et al. (2021); UNSC (2019); Roy and Pramanick (2019).

cities; and digital technologies (Sachs et al., 2019). The next section covers the industrial revolutions and their implications on the delivery of SDGs.

7.2.1.3 Industry 5.0 and other industrial revolutions

There have been several industrial revolutions throughout history, each marked by significant changes in the way goods are produced and services are delivered (Xu et al., 2018). The First Industrial Revolution started in the 18th century in Britain and was characterized by the mechanization of production using steam power (Deane, 1979). It led to the development of factories and the mass production of goods. The Second Industrial Revolution took place in the late 19th and early 20th centuries and was characterized by the widespread use of electricity, the assembly line, and mass production (Mokyr and Strotz, 1998). It also saw the emergence of the automobile and the development of new materials like steel. The Third Industrial Revolution started in the 1960s and is also known as the Digital Revolution (Janicke and Jacob, 2013). It was characterized by the use of electronics and computers in production, and the emergence of the internet,

Table 7.3 Five "P"s of SDGs

P1	P2	P3	P4	P5
People	*Planet*	*Prosperity*	*Peace*	*Partnership*
SDG 1: No Poverty	SDG 6: Clean Water and Sanitation	SDG 7: Clean and Affordable Energy	SDG 16: Peace Justice and Strong Institutions	SDG 17: Partnership for Goals
SDG 2: No Hunger	SDG 12: Responsible Consumption and Production	SDG 8: Decent Work and Economic Growth		
SDG 3: Good Health and Well-being	SDG 13: Climate Action	SDG 9: Industry, Innovation, and Infrastructure		
SDG 4: Quality Education	SDG 14: Life below Water	SDG 10: Reduced Inequalities		
SDG 5: Gender Equality	SDG 15: Life on land	SDG 11: Sustainable Cities and Communities		

which led to the widespread use of e-commerce and digital communications. The Fourth Industrial Revolution is currently underway and is characterized by the integration of physical, digital, and biological systems (Philbeck and Davis, 2018). It includes the use of technologies such as artificial intelligence, the Internet of Things (IoT), and 3D printing (Bloem et al., 2014). Each of these industrial revolutions has had a profound impact on society and the economy, leading to significant changes in the way we work, communicate, and live our lives.

The concept of a "Fifth Industrial Revolution" is still emerging and not yet fully defined. However, it is generally believed that the Fifth Industrial Revolution will build upon the technologies and trends of the Fourth Industrial Revolution, particularly in the areas of automation, connectivity, and artificial intelligence (Mourtzis, 2021). Some researchers predict that the Fifth Industrial Revolution will be marked by the widespread use of autonomous machines and robots, which will have the ability to work alongside humans and even replace them in certain tasks (Paschek et al., 2019). This could lead to significant changes in the way we work, as well as the way products and services are produced and delivered. Other potential trends that could define the Fifth Industrial Revolution include the use of blockchain technology for secure transactions and data sharing, the development of advanced materials and nanotechnology, and the expansion of renewable energy sources to meet the growing demand for sustainable solutions (Pathak et al., 2019). While the exact contours of the Fifth Industrial Revolution are still uncertain, however, it is clear that it will continue to transform the economy and society in profound ways, creating new opportunities and challenges for individuals and businesses alike (Coelho et al., 2023; Mourtzis, 2021). Some potential key elements of the Fifth Industrial Revolution will include the following.

7.2.1.3.1 Artificial Intelligence

AI is considered one of the main key elements of the Fifth Industrial Revolution and is already transforming many aspects of our lives, and it is likely to play an even greater role in the Fifth Industrial Revolution. AI technologies such as machine learning and natural language processing will enable machines to perform more complex tasks, make decisions, and interact with humans in more natural ways (Syam and Sharma, 2018).

7.2.1.3.2 Robotics and automation

Robotics and automation are expected to become even more advanced and sophisticated in the Fifth Industrial Revolution. Autonomous machines and robots will be able to work alongside humans and perform tasks that are too dangerous, repetitive, or difficult for humans (Noble et al., 2022).

The Fifth Industrial Revolution, or Industry 5.0, envisions a manufacturing process that is more collaborative and integrated, combining advanced technologies like machine learning, artificial intelligence, and robotics with human creativity and ingenuity. Robotics and automation play a critical role in this vision by making it possible for autonomous machines and robots to work alongside humans in a range of industrial and manufacturing settings.

One of the primary advantages of robotics and automation is their ability to perform tasks that are too dangerous, repetitive, or difficult for humans. This includes tasks like welding, assembly, and material handling, which can be done with greater precision and efficiency by robots. This, in turn, can lead to improved safety, reduced errors and waste, and increased productivity.

Robotic systems are also becoming increasingly sophisticated, with advancements in machine learning and artificial intelligence enabling them to perform more complex tasks. For example, robots can now learn from experience and adapt to changing environments, making them suitable for a wider range of applications. Additionally, robots can work collaboratively with humans, allowing for more flexible and efficient production processes.

However, as with any new technology, there are potential risks associated with the adoption of robotics and automation. These risks include job displacement, as robots and automation systems take over tasks traditionally performed by humans. Additionally, there may be concerns about the ethical and societal implications of increased reliance on automation, such as the potential for bias in decision-making processes.

7.2.1.3.3 The Internet of Things

The IoT will continue to connect more devices and objects, enabling greater communication and collaboration between machines, as well as between machines and humans. This will lead to new opportunities for data analysis and optimization, as well as the development of new products and services (Greengard, 2021).

IoT refers to the interconnection of various smart devices and objects, such as sensors, appliances, and vehicles, with the internet. This interconnectivity allows for the exchange of data between these devices, enabling them to communicate and interact with each other. As technology continues to advance, the IoT is expected to become even more prevalent in the Fifth Industrial Revolution.

With the increasing use of the IoT, there will be a significant increase in the amount of data generated. This data can be used for data analysis and optimization, providing valuable insights for businesses to improve their operations and decision-making processes. For example, in manufacturing, sensors on machines can provide real-time data on their performance, enabling businesses to identify and address any issues before they lead to downtime or equipment failure.

In addition to data analysis, the IoT also creates new opportunities for product and service development. For example, smart homes with interconnected devices can provide greater convenience and automation for homeowners, such as automatically adjusting temperature settings or turning on lights when someone enters a room. In the healthcare industry, IoT devices can be used to monitor patients remotely, providing doctors with real-time information on their health and enabling quicker response times to emergencies.

7.2.1.3.4 Blockchain technology

Blockchain technology has the potential to transform many industries, from finance to healthcare to supply chain management. The decentralized nature of blockchain makes it more secure, transparent, and efficient, and it could be a key element of the Fifth Industrial Revolution (Kimani et al., 2020).

Blockchain technology is a distributed ledger technology that allows for secure and transparent transactions and data storage. In the Fifth Industrial Revolution, blockchain technology could have a significant impact on various industries by enabling secure and efficient data exchange and collaboration between parties. For example, in the finance industry, blockchain technology could be used to streamline and secure financial transactions, reducing the need for intermediaries and decreasing transaction costs. In healthcare, blockchain technology could be used to secure and streamline medical records, improving patient care and reducing costs.

Moreover, blockchain technology could also have a significant impact on supply chain management by enabling greater transparency and accountability throughout the supply chain. By using blockchain technology, companies can track and verify the authenticity and integrity of goods and products, from raw materials to finished products, ensuring that they meet certain standards and regulations.

Additionally, blockchain technology could promote greater social responsibility by enabling greater transparency and accountability in the supply chain, which could help companies reduce their environmental footprint and improve labor conditions. Blockchain technology could also promote greater sustainability by enabling the tracking and monitoring of carbon emissions, waste, and resource usage.

7.2.1.3.5 Advanced materials and nanotechnology

The development of new materials and nanotechnology could enable the creation of new products and applications with unprecedented properties and capabilities (Ghosal and Chakraborty, 2021). This could include everything from stronger and lighter materials to more efficient batteries and sensors.

Advanced materials and nanotechnology have the potential to revolutionize the way we produce and use materials in the Fifth Industrial Revolution. The development of new materials with advanced properties, such as increased strength, durability, and flexibility, could lead to the creation of new products and applications in various industries such as aerospace, automotive, and construction. For example, the use of lightweight and strong materials such as graphene and carbon nanotubes could revolutionize the construction industry by enabling the development of more resilient and earthquake-resistant buildings.

In addition, nanotechnology could lead to the creation of new materials with unique properties that could have a significant impact on various applications. For instance, the development of nanoscale materials and devices could enable the creation of more efficient batteries and energy storage systems. Nanoscale sensors could also provide highly sensitive and selective detection of various chemicals, biomolecules, and other substances, leading to new applications in healthcare and environmental monitoring.

Moreover, the use of advanced materials and nanotechnology could also lead to significant improvements in manufacturing processes, such as enhanced precision and control, reduced waste and energy consumption, and increased automation. This could result in higher production efficiency, lower costs, and reduced environmental impact. Overall, the development of advanced materials and nanotechnology has the potential to transform various industries, enabling new products and applications with unprecedented properties and capabilities.

7.2.1.3.6 Machine learning

Machine learning is a subfield of AI that involves developing algorithms and statistical models that enable computers to learn from data and make predictions or decisions without being explicitly programmed (Mitchell, 2007). In other words, machine learning is a way of teaching computers to recognize patterns and relationships in data and use this information to make informed decisions or predictions about new data (Zhou, 2021). This is accomplished through the use of algorithms that are trained on large amounts of data, and which are then able to automatically identify patterns and make predictions based on new data inputs. There are many different types of machine learning, including supervised learning, unsupervised learning, and reinforcement learning (Ayodele, 2010).

In supervised learning, the algorithm is trained on labeled data, meaning that the input data is accompanied by the correct output, allowing the algorithm to learn from these examples (Burkart and Huber, 2021). In unsupervised learning, the algorithm is trained on unlabeled data, meaning that there is no correct output, and the algorithm must identify patterns and relationships on its own (Usama et al., 2019). In reinforcement learning, the algorithm learns through trial and error, receiving rewards or penalties for certain actions, and adjusting its behavior accordingly (Kaelbling et al., 1996).

7.2.2 The role of industrial revolutions and machine learning in SGDs

The industrial revolutions have had a significant impact on the achievement of the SDGs, which are a universal call to action to end poverty, protect the planet, and ensure that all people enjoy peace and prosperity (Popović, 2020). The First Industrial Revolution played a major role in driving economic growth and lifting millions of people out of poverty, but it also had negative impacts on the environment and public health (Prisecaru, 2016). The SDGs related to poverty reduction and economic growth are closely tied to the impacts of the First Industrial Revolution. However, the negative environmental impacts, such as air and water pollution, continue to be a major challenge today (Jacobson, 2009). The Second Industrial Revolution had a significant impact on industrialization and economic development, but it also contributed to environmental degradation and the depletion of natural resources (Ali, 2021). The SDGs related to sustainable cities and communities, responsible consumption and production, and climate action are closely tied to the impacts of the Second Industrial Revolution. The Third Industrial

Revolution, or Digital Revolution, has enabled greater connectivity and access to information, but it has also raised concerns about data privacy and security (Solangi et al., 2018). The SDGs related to quality education, innovation and infrastructure, and sustainable cities and communities are closely tied to the impacts of the Third Industrial Revolution. The Fourth Industrial Revolution has the potential to drive economic growth and innovation, but it also raises concerns about job displacement, income inequality, and environmental sustainability (Xu et al., 2018). The SDGs related to decent work and economic growth, industry, innovation, and infrastructure, and sustainable cities and communities are closely tied to the impacts of the Fourth Industrial Revolution. Overall, each industrial revolution has had a significant impact on sustainable development, both positive and negative. It is important to ensure that future industrial revolutions are designed and implemented in a way that supports the achievement of the SDGs and contributes to a more sustainable and equitable world.

Machine learning can also play an important role in supporting the achievement of the SDGs by enabling more efficient and effective data analysis, decision-making, and resource allocation (Asadikia et al., 2021). Machine learning can be used in environmental monitoring to analyze data from sensors and other monitoring devices to detect patterns and trends in environmental conditions (Kvamsdal et al., 2021). This can help to identify areas of concern and inform decision-making on issues such as air and water quality, deforestation, and climate change. It can also be used in precision agriculture to help farmers optimize crop yields, reduce water use, and minimize the use of fertilizers and pesticides.

By analyzing data from sensors and other sources, machine learning algorithms can provide insights into soil conditions, weather patterns, and plant health, enabling farmers to make more informed decisions about planting, harvesting, and resource management. Koolen et al. (2017) suggested that machine learning can be useful in energy management to optimize energy use and reduce waste in buildings, factories, and other facilities. By analyzing data from sensors, machine learning algorithms can identify opportunities for energy efficiency improvements, such as adjusting temperature settings or lighting levels, and provide recommendations for reducing energy consumption. Likewise, machine learning can be used in disaster response to analyze data from social media, satellite imagery, and other sources to identify areas of need and allocate resources more effectively in the event of a natural disaster (Ofli et al., 2016). This can help to improve response times and minimize the impact of disasters on communities. Similarly, Tizghadam et al. (2019) argued that machine learning can help optimize transportation routes and reduce emissions by analyzing data on traffic patterns, weather conditions, and other factors. By providing real-time insights and recommendations, machine learning algorithms can help reduce congestion, improve safety, and minimize the environmental impact of transportation. By enabling more efficient and effective data analysis and decision-making, machine learning can help accelerate progress toward the SDGs and create a more sustainable and equitable future.

7.2.2.1 Potential risks of machine learning to sustainable development

Machine learning has the potential to significantly impact sustainable development by enabling more efficient and effective use of resources, better decision-making, and improved monitoring and evaluation. However, there are also potential risks associated with the use of machine learning that could undermine SDGs (Ray, 2019). Machine learning algorithms are only as good as the data they are trained on. If the data used to train an algorithm is biased, the algorithm will also be biased, potentially leading to discrimination, and perpetuating existing inequalities (Mahesh, 2020). To reduce bias and discrimination, it is important to ensure that the data used to train machine learning models is diverse and inclusive (Mehrabi et al., 2021). This can be achieved by collecting data from a wide range of sources, including underrepresented communities, and by using tools to detect and correct bias in the data.

Likewise, overreliance on machine learning and other technological solutions may divert attention and resources away from addressing underlying social and economic issues that contribute to sustainable development challenges (Froomkin et al., 2019). To reduce overreliance on technology, it is important to encourage critical thinking and skepticism when it comes to machine learning algorithms (Chiang and Yin, 2021). This can be achieved by educating individuals on the limitations of machine learning, and by promoting the use of multiple sources of information and expertise. It is also important to implement human oversight and decision-making in critical areas. This can be achieved by using machine learning models to support human decision-making, rather than replacing it entirely (Campbell et al., 2007). Using multiple approaches to problem-solving involves employing both machine learning algorithms and traditional analytical methods, as well as integrating machine learning with human judgment when necessary.

Similarly, machine learning algorithms require significant computing power and energy, which could contribute to increased energy consumption and greenhouse gas emissions if not properly managed (Ghoddusi et al., 2019). To reduce the energy consumption of machine learning algorithms, it is important to optimize their efficiency (García-Martín et al., 2019). This can be achieved by using hardware that is specifically designed for machine learning, such as graphics processing units, and by implementing algorithms that are designed to minimize energy consumption. The use of machine learning for monitoring and surveillance raises concerns about privacy and security, particularly in the context of vulnerable populations. To protect data privacy and security, it is important to implement strong security protocols, including encryption and access controls. It is also important to ensure that individuals have control over their data and are able to give informed consent for its use in machine learning models (Zhu et al., 2020).

It is established that machine learning algorithms can be difficult to understand and interpret, making it challenging to hold organizations accountable for their decisions and actions (Van der Sommen et al., 2020). To increase transparency and explainability, it is important to use machine learning models that are designed to be interpretable. This can be achieved by using models that are based on decision trees, rule-based systems, or linear regression, which are easier to understand than more complex models like neural networks (Ahmad et al., 2018). Since machine learning algorithms become more sophisticated, there is a risk that they will replace human labor in certain industries (Wang et al., 2022a). This could lead to job losses and economic instability, particularly in low-skilled sectors. To mitigate the risk of job displacement, it is important to invest in reskilling and upskilling programs to help workers adapt to the changing job market. This can include programs that focus on developing skills in areas such as data analysis, machine learning, and other emerging technologies (Stearns, 2020).

Overall, machine learning needs to be used in a way that supports SDGs, it is important to prioritize the responsible and ethical use of these technologies. This includes ensuring that data used to train machine learning algorithms is diverse and representative, developing transparent and explainable algorithms, and establishing clear guidelines and regulations for the use of machine learning in sensitive contexts.

7.2.3 Internal audit

IA can be defined as an activity or function available within an organization that assists in the achievement of vision, mission, and goals (Rehman, 2022a). IA enhances the governance, risk, and control of industries which are required to safeguard the assets and protect stakeholders' rights (Dat et al., 2020). IA is an independent activity, and its independence is ensured through its reporting lines. In accordance with the International Professional Practicing Framework (IPPF) issued by the Institute of Internal Auditors (IIA) (Eulerich & Eulerich, 2020), IA functionally reports to the audit and risk committee of the board and administratively reports to the CEO of the industry. As IPPF is principal-based and not rule-based; therefore, several regulators in many countries made an amendment to the reporting line of IA. Based on this amendment, IA functionally and administratively reports to the audit and risk committee, eliminating the reporting to the CEO. Such practice provides the true meaning of independence to the IA enabling it to achieve sustainable goals for the organizations.

Fraud can be defined as an intentional act that is conducted to deceive people or organizations with the intention of obtaining personal benefits (Sithic & Balasubramanian, 2013). Fraud is the prime obstacle in the path

Table 7.4 Fraud cases in which organizations utilized technology

Name of company	Fraud amount (USD)	Reason	Source
Wirecard	2.1 billion	Fraud was carried out through a complex network of shell companies and involved the use of technology such as fake payment processing systems and doctored financial statements.	Taub (2023)
Theranos	121 million	The fraud was carried out by manipulating the results of blood tests using technology that did not work as advertised.	Williams (2022)
Volkswagen	19.1 billion	Volkswagen was exposed for cheating emissions tests by installing software in its diesel vehicles that could detect when they were being tested and adjust the performance of the engines to produce lower emissions. The fraud affected millions of vehicles and resulted in billions of dollars in fines and legal settlements.	Barth et al., 2022

of sustainability (Rehman, 2022b). With the inclusion of technologies, frauds and fraudsters are also evolving (Dorminey et al., 2012). Table 7.4 defines the few fraud cases that occurred in the last five years in which technology was utilized to the best extent possible.

It is worth noting that the role of external auditors for fraud prevention and detection is not visible. Opinions provided by external auditors are more prone to saving their reputation than to safeguard the company's assets or to inform the true and fair view (Rehman, 2022b). Table 7.5 identifies the few recent cases where external auditors were fined by the regulators for not preventing and/or detecting fraud or were part of the fraud.

IA has emerged from just being a compliance function to a trusted advisor. Failure of external auditors to provide the required satisfaction to shareholders and stakeholders highlighted the importance of IA.

IA performs that task and tries to provide overall control opinion for the entire organization (Eulerich & Eulerich, 2020); moreover, the constant presence of IA within the organization along with the major role in risk

Table 7.5 Fines on external auditors by regulators

Name of company	Year	Penalty (USD)	Source
EY	2020	11 million	Jones and Hobson (2020)
KPMG	2023	231 million	Louch et al., (2023)
PWC	2022	2.1 million	Bloomberg (2022)

assessment, makes IA the paramount source of fraud prevention, fraud detection who augmentations the governance, risk, and control. At this point, it would be necessary to briefly define the association of governance, risk, and control with the roles of IA.

- *Governance and role of internal audit:* Governance refers to the policies and procedures developed for the achievement of organizational value, mission, vision, and objectives. Governance can be classified as corporate governance and organizational governance (Hakimah et al., 2019). Corporate governance deals with the policies, objectives, and strategies of an organization that are approved by the board of directors (Singh et al., 2018). The board of directors is nominated and appointed by the shareholders (Ntim, 2018) and in some countries by the stakeholders as well (such as banks and major creditors), whereas organizational governance refers to the procedures, implementation guidelines, and organizational hierarchy which are developed by the executive/senior management to run day-to-day business (Klein et al., 2019).

IA plays a crucial role in the effectiveness of corporate governance. Due to IA's independence, objectivity, and knowledge, IA audits the entire organization and provides assurance and consulting activities. IA's opinions are based on the completed task and professional judgment. IA provides and safeguards the interest of shareholders and at the same time enhances organizational function through its meaningful recommendations (Vadasi et al., 2020).

- *Risk and role of internal audit:* Risk can be defined as any event that can prevent an organization from achieving its mission and vision (Rasborg, 2021). Although IA is not directly responsible for the identification of risk and development of a risk register (Slapni et al., 2022); however, IA develops its audit plan and conducts its work based on the risk of the organization (Esmail & Haque, 2022). With risk-based audits, IA ensures that the identified risks are properly mitigated and action plans are properly implemented.

In a few of the organizations, risk function also comes under the jurisdiction of IA. In such scenarios, it becomes very argumentative how IA can mitigate the conflict of interest, ensure segregation of duties, and audit the risks identified by the IA itself. For the best performance, the risk function should be under the organizational governance which can be audited by IA.

- *Controls and role of internal audit:* Controls are the policies and procedures designed to achieve organizational objectives (Kamaruddin & Ramli, 2018). In the current business environment, many controls are automated, making it obligatory for IA to update its function and adopt the automation in the process. Control is developed to ensure consistency and equality in operation (Mohamad Nasri et al., 2020). Controls are designed and updated to ensure that the current market trends and regulatory environment are catered for. The role of IA is to ensure that the developed controls are achieving the desired results or not (Bubilek, 2017). IA during the course of audit can identify many redundant and overlapping controls. Moreover, IA with robotic process automation (RPA) and machine learning, can conduct continuous control monitoring on a daily basis and identify control deficiencies immediately.

7.2.3.1 Three lines model

The Institute of Internal Auditors (IIA) defines a "three lines" model in which it is identified which authority in an organization is responsible for what (IIA, 2020). Previously, this model was identified as a "three lines of defense" model; however, it is updated as a three lines model where all the lines are responsible for defense, prevention, and act. Figure 7.5 defines the three lines model: arrows define accountability, reporting, delegation, direction sources, oversight, alignment, communication, and coordination.

These three lines can be defined and summarized in Table 7.6.

7.2.3.2 Internal audit and Industry 5.0

Industry 5.0, also known as the "human-centered industry", represents the latest stage in the evolution of industry and manufacturing (Longo, & Padovano, 2020). With a focus on the human experience and technology playing a supportive role, organizations can benefit from IA in supporting their transition to Industry 5.0. In accordance with the recent report by the

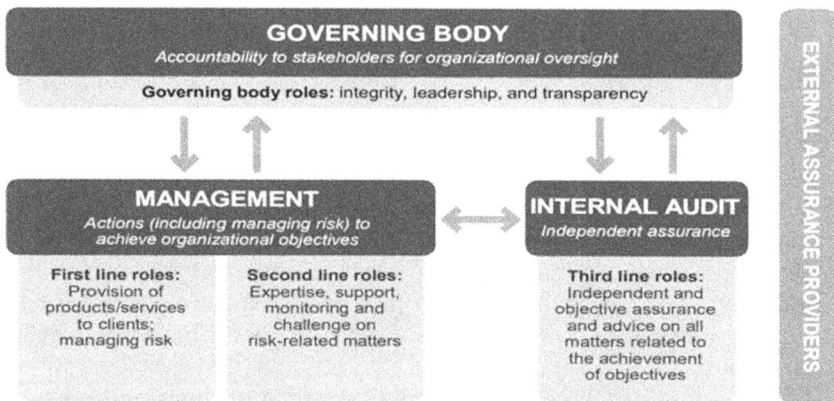

Figure 7.5 Three lines model.

Source: IIA (2020).

Table 7.6 Summary of the roles as per the three lines model

Authority	Role
Governing body	Governance bodies are the board of directors which are appointed by the shareholders. Governance bodies ensure that stakeholder's interests are safe and communicated with them in a transparent manner. Governance bodies set the tone at the top to develop an ethical and accountability culture. Governance roles delegate the responsibilities among its members in the shape of committees such as the audit and risk committee and the nomination and remuneration committee. Often, responsibilities are also assigned to the executive/ senior management. This assignment usually occurs when the organization has a strong ethical and accountability culture. The governance function also establishes the organizational risk appetite and maintains oversight over governance, risk, regulation, and controls. The governance body also establishes IA.
Management – first line	Responsible for the implementation of policies and procedures and leads the risk management functionalities, reports directly to the governance body on the planned and actual performance, and ensures that compliance is made with all the regulations and maintains appropriate structure for its operations.
Management – second line	Involve in the reporting, analyzing, and informing of risks, operations, and internal controls. This function is involved in the control self-assessment and informs the executive management about potential improvements.
Internal audit	Reports directly to the governance body and ensures that satisfaction is provided to the shareholders and the organization's vision and mission are achieved.
External assurance provider	These are providing reporting to satisfy regulatory and legislative requirements. This is necessary as many other external stakeholders such as bankers and financial institutions require the opinion of this authority.

Society of Corporate Compliance and Ethics & Health Care Compliance Association (SCCE & HCCA), IA can help organizations identify and mitigate risks, ensure compliance with regulations and standards, establish effective governance structures, measure performance, and manage the change (SCCE & HCCA, 2020) this can also be associated with Industry 5.0 initiatives. By leveraging IA's expertise and guidance in these areas, organizations can successfully transition to Industry 5.0, optimize their operations, and achieve their goals while meeting their ethical and social responsibilities. It is worth noting that Industry 5.0 can play a major role in achieving sustainability.

IA can play a crucial role in supporting Industry 5.0 initiatives (Verma, et al., 2022). IA can help organizations identify and manage risks associated with the adoption of new technologies and processes. By providing guidance on risk management, compliance, governance, performance measurement, and change management best practices, IA can help ensure that Industry 5.0

initiatives are aligned with organizational goals and objectives (Assiri & Humayun, 2023). IA can also help organizations improve sustainability by identifying and assessing risks, optimizing energy consumption, monitoring and improving the sustainability of the supply chain, reducing greenhouse gas emissions, and improving sustainability reporting (Herz, et al., 2023). IA can play a crucial role in helping organizations achieve sustainability goals (Rehman, 2022a). Some of the ways in which IA can be used to support sustainability initiatives are as follows.

- *Risk assessment*: IA and ML can be used to identify and assess risks associated with sustainability. By analyzing data from various sources, such as environmental impact reports and supply chain data, IA can help organizations identify areas where sustainability risks are most significant. ML can also be used to predict future risks and help organizations take proactive measures to mitigate them.
- *Energy management*: IA and ML can be used to optimize energy consumption in buildings and manufacturing facilities. By analyzing data from sensors and other sources, ML algorithms can identify energy usage patterns and help organizations optimize their operations to reduce energy consumption. IA can also help ensure that energy management systems are functioning effectively and efficiently.
- *Sustainable supply chain*: IA and ML can be used to monitor and improve the sustainability of the supply chain. By analyzing data from suppliers and other sources, IA can help identify risks and opportunities for improvement in areas such as carbon emissions, waste reduction, and ethical sourcing. ML can also be used to identify potential issues before they arise, such as supply chain disruptions or breaches of ethical sourcing standards.
- *Emissions management*: IA and ML can be used to monitor and reduce greenhouse gas emissions. By analyzing data from emissions monitoring systems, energy usage data, and other sources, ML algorithms can identify areas where emissions can be reduced and help organizations implement solutions such as renewable energy sources and energy efficiency measures.
- *Industry*: IA and ML can be used to improve sustainability reporting. By analyzing data from various sources, IA can help ensure that sustainability reports are accurate and complete. ML can also be used to automate the reporting process, reducing the time and effort required to prepare reports.

7.2.3.3 Machine learning and Industry 5.0

Industry 5.0 contains a human-focal attitude, which emphasizes the integration of advanced technologies with human skills and capabilities (Longo & Padovano, 2020). Unlike previous industrial revolutions, which were primarily focused on automating and optimizing manufacturing processes, Industry 5.0 aims to create a more collaborative and flexible environment (Leng et al., 2022), where humans and machines work together to achieve common goals – in this context, efficient and effective decision-making (Jarrahi, 2018). Machine learning algorithms can analyze vast amounts of data from sensors, equipment, and other sources to identify patterns and optimize manufacturing processes. However, humans are still needed to interpret and act on this data, make critical decisions, and perform tasks that require creativity, empathy, and other uniquely human skills.

ML is a subfield of artificial intelligence that involves training computer algorithms to identify patterns and make predictions based on data (Myers et al., 2020). In the context of Industry 5.0, ML can be used to analyze vast amounts of manufacturing data, identify patterns, and optimize manufacturing processes (Maddikunta et al., 2022). Some specific ways that ML can help Industry 5.0 are defined in Table 7.7.

As mentioned in Section 7.1.2, the United Nations' SDGs are a set of 17 global goals that were adopted by all United Nations Member States in 2015 as a universal call to action to end poverty, protect the planet, and ensure that all people enjoy peace and prosperity. While the SDGs do not explicitly mention "Industry 5.0", they do address many of the key issues and challenges that are driving the development of this new approach to manufacturing (Kasinathan et al., 2022). SDG 9 – Industry, Innovation, and Infrastructure – recognizes the critical role that technology and innovation play in driving sustainable economic growth and development. This goal specifically calls for the promotion of sustainable industrialization and the adoption of new and innovative technologies that can help address global challenges such as climate change, resource depletion, and economic inequality. Similarly, SDG 8 – Decent

Table 7.7 Machine learning assistance toward Industry 5.0

Machine learning assistance	Details
Predictive maintenance	By analyzing data from sensors and other sources, machine learning algorithms can predict when equipment is likely to fail, allowing for preventative maintenance to be performed before a breakdown occurs.
Quality control	Machine learning can be used to detect defects and anomalies in products, helping to ensure that only high-quality products make it to market.
Supply chain optimization	By analyzing data from the entire supply chain, including suppliers, manufacturers, and distributors, machine learning algorithms can optimize inventory levels, reduce waste, and improve logistics.
Process optimization	Machine learning can be used to optimize manufacturing processes by identifying inefficiencies and bottlenecks and recommending changes to improve efficiency.

Source: Maddikunta et al., (2022)

Work and Economic Growth –highlights the importance of creating sustainable and inclusive economic growth that benefits all members of society. This goal specifically calls for the promotion of full and productive employment and decent work for all, which is closely linked to the development of new technologies and manufacturing processes that can create new job opportunities and improve working conditions.

Both SDG 9 (Industry, Innovation, and Infrastructure) and SDG 8 (Decent Work and Economic Growth) can benefit greatly from the application of ML. ML is a subset of AI that involves using algorithms to learn from data and make predictions or decisions without being explicitly programmed. In the context of SDG 9, ML can be applied to various areas to improve efficiency and sustainability. Smart manufacturing is one area where ML can make a significant impact. By analyzing data from sensors and other sources, ML algorithms can identify patterns and insights that help optimize production processes, reduce waste, and improve product quality (Maddikunta, et al., 2022). This can lead to cost savings and increased productivity, while also minimizing the environmental impact of manufacturing. Predictive maintenance is another area where ML can be highly beneficial (Kasinathan, et al., 2022).

By analyzing data from sensors and other sources, ML algorithms can predict when equipment is likely to fail, allowing for preventative maintenance to be performed before a breakdown occurs. This can save costs, reduce downtime, and increase safety by preventing accidents caused by equipment failures. In the supply chain, ML can help identify inefficiencies and optimize logistics to reduce waste and improve delivery times. By analyzing data on demand, inventory levels, and shipping times, ML algorithms can help optimize routes and delivery schedules, leading to cost savings and reduced emissions from transportation. ML can be applied to sustainable energy management by predicting energy consumption patterns and optimizing energy storage and distribution. This can help reduce greenhouse gas emissions and improve the efficiency of energy systems, contributing to the transition to a more sustainable energy future.

SDG 8 aims to promote sustained, inclusive, and sustainable economic growth, full and productive employment, and decent work for all. ML technology can play a significant role in achieving this goal by improving productivity, optimizing processes, and promoting innovation. ML can be applied to human resources management. ML algorithms can help identify the best candidates for a given job by analyzing resumes, skills, and past performance, leading to better matches between employees and job requirements (Zhao, et al., 2022).

ML can also help in identifying training needs and provide personalized learning opportunities to help employees develop the skills they need for their jobs. ML can also be used to optimize business processes, leading to increased productivity and efficiency. By analyzing data on production processes, supply chain management, and customer interactions, ML algorithms can identify inefficiencies and opportunities for improvement (Zhao, et al., 2022). This can lead to cost savings and increased profitability, while also improving customer satisfaction. In the context of entrepreneurship and innovation, ML can help identify promising startups and innovative ideas by analyzing data on market trends and consumer preferences. This can help investors and entrepreneurs make better decisions about where to invest their time and resources.

The fact cannot be denied that Industry 5.0 is human-centric; however, ML is capable of replacing many human-related jobs. ML and other advanced technologies in Industry 5.0 may lead to some displacement of human workers (Alves et al., 2023). However, it is important to note that ML is not designed to replace human workers entirely, but rather to augment their capabilities and improve the efficiency of manufacturing processes. In many cases, ML algorithms are used to automate repetitive or dangerous tasks that are not well-suited for human workers (Adel, 2022). For example, robots can be used to perform tasks such as welding, painting, or assembly line work that may be hazardous or physically taxing for humans. At the same time, ML can also create new job opportunities for workers with skills in data analysis, programming, and other related areas. As manufacturing processes become more automated and data-driven, there will be an increased demand for workers who can develop and maintain these systems (Uzialko, 2023).

Ethics is a critical challenge that needs to be addressed as these technologies, including ML, are becoming more widespread (Vollmer, et al., 2020). One of the key issues in this regard is how to ensure that ML models are transparent, accountable, and fair so that they can be trusted to make decisions that are consistent with ethical and legal norms. Making ML responsible requires a multi-faceted approach that involves ethical design, rigorous data governance, human oversight, monitoring and accountability, and continuous improvement (Burr & Leslie, 2022).

Implementation of the above-mentioned measures can be regularly monitored by IA. By implementing these measures, IA can ensure that ML is used in a way that is ethical, transparent, and accountable and that it promotes fairness and justice. As ML technologies become more widespread and influential in our society, it is crucial that organizations address the ethical challenges they pose and take proactive steps to ensure that they are used in a responsible and beneficial way. By doing so, organizations – with the assistance of IA – can harness the power of ML to improve lives and address some of the most pressing issues facing the world today.

7.2.3.4 Internal audit, machine learning, and Industry 5.0

Industry 5.0 was formally established in 2021 by the European Commission (Xu et al., 2021). This establishment was made after conducting critical analysis among scientists and industry experts in different branches of research, academia, and technology (Breque et al., 2021). Therefore, digital transformation is slowly taking shape in the Industry 5.0. The emergence of Industry 5.0 is spurring research and innovation focused on human-centric, sustainable industrial practices. The core values of Industry 5.0 can be summarized into three main elements: human-centric, sustainable, and resilient. These three core values have been established to promote the diversity and talents of people, to consider the planet's boundaries and environmental space, and to adopt new technologies with flexibilities (Longo et al., 2020).

It is clear that Industry 5.0 is value-driven not technology-driven (Longo et al., 2020), which leads to a number of changes and responses that could affect the IA readiness in organizations in the future. As one of the main focuses of Industry 5.0 is human-centric, this means that IA audits have to be adjusted to comply with the new changes to create effective and inclusive IA policies, considering this dimension in a way that allows for IA to conduct comprehensive analysis and acquire credible evidence when needed. There is an excessive need for organizations to digitize their operations in a bid to become viable in the long run (Mihardjo et al., 2019). To help curb the issues emerging from the legal and regulatory compliance of digitization, there is a need to conduct audits using IA (Rehman, 2022b). The diversity in computing platforms and the organizational rate of digitization have been proven to be challenging for collection and investigation using internal audits. Overcoming the challenges in the IA process is required so that organizations can be "internally audit ready". This involves the collective capability to process, review, and store digital information.

IA is a rapidly evolving area, and ML is playing an increasingly important role in enabling auditors (Slapni et al., 2022) to gain insights and make more informed decisions. IA and ML include the use of predictive analytics for fraud detection, risk assessment, and financial forecasting. Text analytics is another area where machine learning is being applied to analyze unstructured data and gain insights into employee behavior and customer feedback (Pejić Bach et al., 2019). RPA is being used to automate routine tasks, allowing auditors to focus on more strategic activities (Huang & Vasarhelyi, 2019). Cognitive computing is also being used to gain insights into complex data sets and make more informed decisions. Finally, machine learning is being used in the area of cybersecurity to identify potential threats and vulnerabilities in the organization's IT infrastructure (Pejić Bach et al., 2019).

IA is an important function in organizations, providing independent assurance and consulting services that help organizations achieve their objectives (Okodo et al., 2019). With the advent of Industry 5.0, IA has taken on a new level of importance, as organizations seek to leverage the power of technology to improve their operations and enhance the customer experience.

ML is a type of artificial intelligence that involves the use of algorithms to learn from data and make predictions or decisions (Diligent, 2021). IA and ML can be used for a variety of purposes, including fraud detection, risk assessment, and performance measurement. There are several benefits and challenges of using ML in IA and has highlighted the importance of ensuring that the algorithms are transparent, explainable, and ethical.

IA plays a major role in supporting organizations as they transition to Industry 5.0. This involves not only providing assurance and consulting services but also helping to identify and manage the risks associated with new technologies and processes. IA can contribute to risk management, compliance, governance, performance measurement, and change management in the context of Industry 5.0.

IA plays an important role in data analytics and visualization. With the increasing amount of data being generated by organizations, IA needs to be able to analyze and interpret this data in order to identify patterns (Almasria, 2022) and trends that can help improve operations and enhance the customer experience. Data analytics and visualization tools can be used in IA and they have highlighted the importance of developing the skills and capabilities necessary to use these tools effectively (Idil & Ozoner, 2018).

The environment for Industry 5.0 can be divided into micro- and macro-environments as it creates crucial aspects of sustainable economies, involving the processes of supply and demand and modern industry approaches. Monitoring the industry environment is strategic and necessary. The concepts of IA and internal control are applicable to this monitoring. The application of audit concepts is for audit purposes, and control concepts for control purposes (Kupec et al., 2021).

7.3 ASSOCIATE THEORIES

This section will discuss Stakeholder Theory, RBV of the industry, and EDM Theory as these are related to Industry 5.0, internal audit, and machine learning.

7.3.1 Stakeholder theory

Stakeholder Theory can guide organizations in their implementation of machine learning initiatives to promote sustainable development and build stronger relationships with their stakeholders (Freeman & Dmytriyev, 2017). Internal audits can play a crucial role in ensuring that these initiatives align with Stakeholder Theory and address potential risks and challenges. By doing so, organizations can enhance their reputation as responsible and sustainable businesses and contribute to the betterment of society and the environment.

Stakeholder Theory proposes that organizations and industries have a responsibility to take into account the interests of all stakeholders, including customers, employees, shareholders, regulators, the wider community, and the environment, in addition to their own financial interests (Bridoux & Stoelhorst, 2022). With the emergence of Industry 5.0, the application of machine learning for sustainable development can be viewed as a means for industries to fulfill this responsibility. By leveraging machine learning technology, organizations can reduce their environmental footprint, promote social responsibility, and provide better products and services to their customers (Freeman et al., 2021). However, the implementation of machine learning initiatives also presents potential risks and challenges, including concerns around data privacy, bias in algorithms, and job displacement.

To ensure that machine learning initiatives align with Stakeholder Theory and address potential risks and challenges, internal audits can play a crucial role in overseeing the process. They can ensure that the organization is considering the interests of all stakeholders, including potential environmental and social impacts, in its machine learning initiatives. Internal auditors can use specific audit procedures to assess the impact of machine learning initiatives on different stakeholder groups, such as conducting risk assessments, analyzing data quality, and monitoring compliance with relevant regulations.

Stakeholder Theory provides a framework for understanding the ethical, effective, and practical management of uncertainties related to stakeholder interests and organizational management (Guo et al., 2023). Industry 5.0, with its various technologies such as machine learning, edge computing, big data, IoT, autonomous robots, artificial intelligence, augmented reality, and innovation system integration, presents an opportunity for industries to develop sustainable performance and build long-term and sustainable relationships with their stakeholders, such as suppliers, buyers, and society. Sustainability can be achieved by providing justice and caring for all stakeholders, which includes addressing potential risks and challenges associated with machine learning initiatives (Freudenreich & Schaltegger, 2020).

7.3.2 Resource-based view theory

The RBV of the firm has its roots in the 1970s when scholars began to focus on the internal resources and capabilities of firms as a source of sustainable competitive advantage. The RBV gained significant attention in the 1980s and 1990s, and today it remains one of the most influential theories in strategic management (Brahma & Chakraborty, 2011).

Today, the RBV remains an important theoretical framework in strategic management and has been applied to a wide range of industries and contexts. The framework continues to evolve, with scholars incorporating new insights from fields such as evolutionary biology, psychology, and neuroscience to refine our understanding of the sources of sustained competitive advantage (Freeman et al., 2021).

RBV is related to Industry 5.0, machine learning, and internal audit, as it provides a framework for organizations to identify and leverage their resources and capabilities to achieve sustainable competitive advantages through the implementation of machine learning initiatives. Internal audits can play a crucial role in assessing the organization's resources and capabilities and ensuring that the machine learning initiatives align with the RBV framework (Chen et al., 2022).

The RBV of the industry is a strategic management framework that emphasizes the importance of an organization's internal resources and capabilities in achieving sustainable competitive advantage. According to RBV, resources are valuable, rare, inimitable, and non-substitutable, and therefore, firms must possess and develop these resources to achieve a competitive edge (Rengkung et al., 2018). In the context of Industry 5.0 and machine learning, RBV can help organizations identify and leverage their resources and capabilities to create and implement machine learning initiatives that align with their stakeholders' interests.

Machine learning is a resource that organizations can leverage to develop sustainable competitive advantages, such as improved customer satisfaction, increased efficiency, and reduced costs. Internal audits can play a role in assessing the organization's machine learning capabilities and identifying areas for improvement to ensure that the machine learning initiatives align with the organization's resources and capabilities (Barnes et al., 2021).

RBV can also be used to evaluate the impact of Industry 5.0 on the organization's resources and capabilities. The technologies associated with Industry 5.0, such as machine learning, big data, and IoT, can provide valuable resources and capabilities that organizations can leverage to achieve sustainable competitive advantages. Internal audits can help the organization identify the resources and capabilities needed to implement these technologies successfully and assess the risks associated with these initiatives (Cillo et al., 2022).

7.3.3 Ethical decision-making theory

EDM Theory is highly relevant to the topics of Industry 5.0, internal audit, and machine learning, as it provides a framework for promoting ethical behavior and decision-making within organizations (Dhirani et al., 2023; Yuhertiana et al., 2019; Rodgers et al., 2023).

As organizations increasingly adopt Industry 5.0 technologies such as machine learning, it becomes essential to consider the ethical implications of these technologies on stakeholders. EDM Theory provides a framework for understanding how individuals and organizations can make ethical decisions in the face of competing interests and values. In the context of Industry 5.0, EDM Theory can guide organizations in developing ethical guidelines and decision-making processes that take into account the interests of all stakeholders.

Internal audits also play a critical role in promoting EDM within organizations. By monitoring the organization's operations and ensuring compliance with ethical standards, internal audits can help prevent unethical behavior and promote a culture of EDM (Rehman, 2022b).

Machine learning also presents unique ethical challenges, such as algorithmic bias, privacy concerns, and the potential for misuse. EDM Theory can guide organizations in developing ethical guidelines and policies for the use of machine learning technologies. Internal audits can also help ensure compliance with these guidelines and identify potential ethical issues (Rhim, 2021).

7.4 CONCLUSION AND FUTURE SCOPE

Industries play a crucial role in the economy of any nation. They provide employment opportunities, generate income, contribute to economic growth and development, and promote innovation and technological advancement. The different levels of industries, from primary to quinary, provide insight into the diverse activities that contribute to the economy and how they work together to create value. Industry 5.0, the future of manufacturing and production, envisions a more collaborative and integrated approach to manufacturing that combines advanced technology with human creativity and ingenuity while promoting sustainability and social responsibility. However, this new concept poses many risks, including fraud, that have yet to be fully identified. Understanding the importance of industries and their potential risks is crucial for the sustainable development of any nation.

This chapter utilized Stakeholder Theory, RBV Theory, and EDM Theory. Stakeholder Theory suggests that organizations and industries have a responsibility to consider the interests of all stakeholders, including the environment and society, in addition to their shareholders. RBV Theory suggests that the strategic resources and capabilities of an industry can be a source of sustainable competitive advantage. EDM theory refers to the process of making decisions that are morally and ethically sound. These theories are relevant to the discussion of internal audit and machine learning technology in the text because they provide a framework for understanding the role of internal audit in promoting sustainability, addressing risks, and ensuring EDM in the context of Industry 5.0.

The implementation of machine learning technology in organizations faces several challenges, including a lack of clear strategy and a shortage of skilled personnel. The internal audit function can play a vital role in mitigating the risks associated with the adoption of these technologies. The internal audit can ensure the alignment of machine learning initiatives with the organization's sustainability objectives, identify and mitigate risks, and promote transparency and accountability. Additionally, the internal audit can help identify potential data privacy and security risks and ensure that the organization has effective controls in place to protect data. The digitization of control functions is essential to keep pace with Industry 5.0, and the internal audit can benefit from integrating machine learning into its processes to automate audit procedures, improve skill sets, and address potential issues before they become major problems. Furthermore, considering stakeholder interests and RBV Theory can help industries achieve their sustainability objectives and gain a competitive advantage.

It is argued that the success of Industry 5.0 adoption depends on adopting a holistic approach that considers the interests of all stakeholders, invests in critical resources such as data infrastructure and skilled personnel, and upholds ethical principles. By implementing these practices, organizations can effectively leverage Industry 5.0 technologies to promote sustainable development while minimizing risks and ensuring compliance with ethical standards. This chapter could potentially inspire regulators and policymakers to revise their rules and policies to promote the use of Industry 5.0 for sustainable development while making internal audits and machine learning mandatory resources.

The future scope of industries lies in the adoption of Industry 5.0, a collaborative and integrated approach to manufacturing that combines advanced technology with human creativity and ingenuity while promoting sustainability and social responsibility. However, the successful implementation of Industry 5.0 depends on a holistic approach that considers the interests of all stakeholders, invests in critical resources such as data infrastructure and skilled personnel, and upholds ethical principles. Internal audits can play a vital role in mitigating the risks associated with the adoption of machine learning technology, ensuring alignment with sustainability objectives, identifying and mitigating risks, and promoting transparency and accountability. The integration of machine learning into audit processes can automate procedures, improve skill sets, and address potential issues before they become major problems. The adoption of these practices can inspire regulators and policymakers to promote the use of Industry 5.0 for sustainable development, while making internal audit and machine learning mandatory resources.

REFERENCES

Abel, D., MacGlashan, J., & Littman, M. L. (2016). Reinforcement Learning as a Framework for Ethical Decision Making. AAAI workshop: AI, ethics, and society (Vol. 16, p. 02), 16, (02).

ACFE. (2022). *Occupational Fraud 2022: A Report to the nations*. LA: Association of Certified Fraud Examiners.

Adams, A. C. (2017). The Sustainable Development Goals, Integrated Thinking and the Integrated Report. International Integrated Reporting Council (IIRC), London.

Adel, A. (2022). Future of Industry 5.0 in society: Human-centric solutions, challenges and prospective research areas. *J Cloud Comput (Heidelb)*, 11(1): 40. https://doi.org/10.1186/s13677-022-00314-5

Ahmad, M. A., Eckert, C. and Teredesai, A., 2018, August. Interpretable machine learning in healthcare. In *Proceedings of the 2018 ACM international conference on bioinformatics, computational biology, and health informatics* (pp. 559–560).

Aithal, P. S. (2019). Information communication & computation technology (ICCT) as a strategic tool for industry sectors. *International Journal of Applied Engineering and Management Letters (IJAEML)*, 3(2), 65–80.

Ali, A. (2021). Natural resources depletion, renewable energy consumption and environmental degradation: A comparative analysis of developed and developing world. *International Journal of Energy Economics and Policy*. https://doi.org/10.32479/ijeep.11008

Almasria, N. A. (2022). Corporate Governance and the Quality of Audit Process: An Exploratory Analysis Considering Internal Audit, Audit Committee and Board of Directors. *European Journal of Business and Management Research*, 7(1), 78–99.

Alves, J., Lima, T. M., & Gaspar, P. D. (2023). Is Industry 5.0 a Human-Centred Approach? *A Systematic Review. Process*, 11(1), 193. https://doi.org/10.3390/pr11010193

Amoako, G. K., Bawuah, J., Asafo-Adjei, E., & Ayimbire, C. (2023). Internal audit functions and sustainability audits: Insights from manufacturing firms. *Cogent Business & Management*, https://doi.org/10.1080/23311975.2023.2192313

Arnhof, M., 2016, July. Design of a human settlement on Mars using in-situ resources. *46th International Conference on Environmental Systems*.

Asadikia, A., Rajabifard, A., & Kalantari, M., 2021. Systematic prioritisation of SDGs: Machine learning approach. *World Development*, 140, p. 105269.

Assembly, G. (2015). Sustainable Development Goals. SDGs. Transforming our world: The, 2030. Global Change.

Assiri, M., & Humayun, M. (2023). A blockchain-enabled framework for improving the software audit process. *Applied Sciences*, 13(6), 3437; https://doi.org/10.3390/app13063437

Audretsch, D., Heger, D., & Veith, T. (2015). Infrastructure and entrepreneurship. *Small Business Economics*, 44, 219–230. https://doi.org/10.1007/s11187-014-9600-6

Ayodele, T. O. (2010). Types of machine learning algorithms. *New Advances in Machine Learning*, 3, pp.19–48.

Barnes, S., Rutter, R. N., La Paz, A. I., & Scornavacca, E. (2021). Empirical identification of skills gaps between chief information officer supply and demand: A resource-based view using machine learning. *Industrial Management & Data Systems*, 121(8), 1749–1766.

Barth, F., Eckert, C., Gatzert, N., & Scholz, H. (2022). Spillover effects from the Volkswagen emissions scandal: An analysis of stock and corporate bond markets. *Schmalenbach Journal of Business Research*, 74, 37–76. https://doi.org/10.1007/s41471-021-00121-9

Bednar, P., & Welch, C. (2020). Socio-technical perspectives on smart working: Creating meaningful and sustainable systems. *Information Systems Frontiers*, 22, 281–298, https://doi.org/10.1007/s10796-019-09921-1

Bloem, J., Van Doorn, M., Duivestein, S., Excoffier, D., Maas, R., & Van Ommeren, E. (2014). The fourth industrial revolution. *Things Tighten*, 8(1), pp. 11–15.

Bloomberg. (2022, August 8). PwC hit with $2.1 million fine over BT audit. Retrieved from https://english.alarabiya.net/: https://english.alarabiya.net/business/2022/08/08/PwC-hit-with-2-1-million-fine-over-BT-audit

Brahma, S. S., & Chakraborty, H. (2011). From industry to firm resources: Resource-based view of competitive advantage. *The IUP Journal of Business Strategy*, 8(2), 7–21.

Breque, M., De Nul, L., & Petridis, A. (2021). *Industry 5.0: Towards a Sustainable, Human-Centric and Resilient European Industry*. Luxembourg: European Commission, Directorate-General for Research and Innovation.

Bridoux, F., & Stoelhorst, J. (2022). Stakeholder theory, strategy, and organization: Past, present, and future. *Strategic Organization*, 20(4), 797–809. https://doi.org/10.1177/14761270221127628

Brundtland Commission, (1987). Our Common Future. Oxford University Press: Oxford, London, United Kingdom. See: https://sswm.info/sites/default/files/reference_attachments/UN%20WCED%201987%20Brundtland%20Report.pdf (accessed 18/04/2023).

Bruno, C., & Pedrosa Isabel, M. A. (2020). State of the Art of Artificial Intelligence in Internal Audit Context. *15th Iberian Conference on Information Systems and Technologies (CISTI)*. Sevilla: AISTI. https://doi.org/10.23919/CISTI49556.2020.9140863

Bubilek, O. (2017). *Importance of Internal Audit and Internal Control in an organization – Case Study*. NY: Arcada.

Burkart, N., & Huber, M. F. (2021). A survey on the explainability of supervised machine learning. *Journal of Artificial Intelligence Research*, 70, pp.245–317.

Burr, C., & Leslie, D. (2022). Ethical assurance: A practical approach to the responsible design, development, and deployment of data-driven technologies. *AI and Ethics*, 1–26.

Campbell, E. M., Sittig, D. F., Guappone, K. P., Dykstra, R. H. and Ash, J. S., (2007). Overdependence on technology: An unintended adverse consequence of computerized provider order entry. In AMIA Annual Symposium Proceedings (Vol. 2007, p. 94). American Medical Informatics Association.

Chen, D., Esperança, J. P., & Wang, S. (2022). The Impact of Artificial Intelligence on Firm Performance: An Application of the Resource-Based View to e-Commerce Firms. *Frontiers in Psychology*, 13, https://doi.org/10.3389/fpsyg.2022.884830

Chiang, C. W. and Yin, M., 2021, June. You'd better stop! Understanding human reliance on machine learning models under covariate shift. In *13th ACM web science conference 2021* (pp. 120–129).

Cillo, V., Gregori, G. L., Daniele, L. M., Caputo, F., & Bitbol-Saba, N. (2022). Rethinking companies' culture through knowledge management lens during Industry 5.0 transition. *Journal of Knowledge Management*, 26(10), 2485–2498.

Coelho, P., Bessa, C., Landeck, J., & Silva, C. (2023). Industry 5.0: The Arising of a Concept. *Procedia Computer Science*, 217, pp.1137–1144.

Cohen, B. (2020, May 18). Three Risks in Building Machine Learning Systems. Retrieved from https://insights.sei.cmu.edu/: https://insights.sei.cmu.edu/blog/three-risks-in-building-machine-learning-systems/

Cohen, C. B. (2018). Industry 4.0: Are You Ready for the Fourth Industrial Revolution? See: https://medium.com/@carmitberdugocohen/industry-4-0-are-you-ready-for-the-fourth-industrial-revolution-464de2dea3b1 (accessed 24/09/2022).

Coman, D. M., Ionescu, C. A., Duică, A., Coman, M. D., Uzlau, M. C., Stanescu, S. G., & State, V. (2022). Digitization of accounting: The premise of the paradigm shift of role of the professional accountant. *Applied Sciences*, 12(7), 3359. https://doi.org/10.3390/app12073359

Committee, E. E. (2021). Industry 5.0. Retrieved from Research and innovation: https://ec.europa.eu/info/research-and-innovation/research-area/industrial-research-and-innovation/industry-50_en (Accessed 28 September 2021)

Dat, P. M., Mau, N. D., Loan, B. T., & Huy, D. T. (2020). Comparative china corporate governance standards after financial crisis, corporate scandals and manipulation. *Journal of Security & Sustainability Issues*, 9(3).

Deane, P. M. (1979). *The first industrial revolution*. Cambridge University Press.

Dhirani, L. L., Mukhtiar, N., Chowdhry, B. S., & Newe, T. (2023). Ethical dilemmas and privacy issues. *Emerging Technologies: A Review. Sensors*, 23(3), 1151.

Diligent. (2021, April 8). Understanding the Risks of Machine Learning. Retrieved from https://www.wegalvanize.com/: https://www.wegalvanize.com/risk/understanding-the-risks-of-machine-learning/

Dinu, D.-G., & Nosca, G. (2020). The influence of digital technologies in global business future: The quaternary and quinary economic sectors. *Conference of EuroMed Academy of Business* (pp. 465–467). Bihar: Somaiya Vidyavihar University.

Dinu, D. G., Stoian-Karadeli, A., Mihoreanu, L., Ilić, M., & Păduraru, M. (2022). Competitive intelligence and neuro-technologies: The new strategic tools to boost the digital economy. *International Journal of Business and Management Invention*, 11(7): 39–45, https://doi.org/10.35629/8028-1107013945

Dorminey, J. F., Kranacher, M. J., & Riley Jr, R. A. (2012). The evolution of fraud theory. *Issues in Accounting Education*, 27(2), 555–579.

Dwivedi, A., Agrawal, D., Jha, A., & Mathiyazhagan, K. (2023). Studying the interactions among Industry 5.0 and circular supply chain: Towards attaining sustainable development. *Computers & Industrial Engineering*, 176, 108927, https://doi.org/10.1016/j.cie.2022.108927

Dwivedi, Y. K., Hughes, L., Ismagilova, E., Aarts, G., Coombs, C., Crick, … Janssen. (2019). Artificial Intelligence (AI): Multidisciplinary perspectives on emerging challenges, opportunities, and agenda for research, practice and policy. *International Journal of Information Management*. https://doi.org/10.1016/j.ijinfomgt.2019.08.002

Esmail, Z. M., & Haque, S. I. (2022). The influence of the business risk-based auditing application on the audit process: An empirical investigation in the Yemeni Context. *Journal of Business Strategy Finance and Management*, 4(2), 214.

Espina-Romero, L., Guerrero-Alcedo, J., Avila, N. G., Sánchez, J. G., Hurtado, H. G., & Li, A. Q. (2023). Industry 5.0: Tracking scientific activity on the most influential industries, associated topics, and future research agenda. *Sustainability*, 15(6), 5554. https://doi.org/10.3390/su15065554

Eulerich, A., & Eulerich, M. (2020). What is the value of internal auditing? – A literature review on qualitative and quantitative perspectives. *Maandblad voor Accountancy en Bedrijfseconomie*, 94(3/4): 83–92. https://doi.org/10.5117/mab.94.50375

Fei, W., Opoku, A., Agyekum, K., Oppon, J. A., Ahmed, V., Chen, C., & Lok, K. L. (2021). The critical role of the construction industry in achieving the sustainable development goals (SDGs): Delivering projects for the common good. *Sustainability*, 13(16), 9112. https://doi.org/10.3390/su13169112

Freeman, R. E., & Dmytriyev, S. (2017). Corporate social responsibility and stakeholder theory: Learning from each other. *Symphonya Emerging Issues in Management.* https://doi.org/10.4468/2017.1.02freeman.dmytriyev

Freeman, R. E., Dmytriyev, S. D., & Phillips, R. A. (2021). Stakeholder theory and the resource-based view of the firm. *Journal of Management,* 47(7), 1757–1770. https://doi.org/10.1177/0149206321993576

Freudenreich, B. L.-F., & Schaltegger, S. (2020). A stakeholder theory perspective on business models: Value creation for sustainability. *Journal of Business Ethics,* 166, 3–18.

Froomkin, A. M., Kerr, I., & Pineau, J. (2019). When AIs outperform doctors: Confronting the challenges of a tort-induced over-reliance on machine learning. *Arizona State Law Journal,* 61, 33.

García-Martín, E., Lavesson, N., Grahn, H., Casalicchio, E., & Boeva, V. (2019). How to measure energy consumption in machine learning algorithms. In ECML PKDD 2018 Workshops: Nemesis 2018, UrbReas 2018, SoGood 2018, IWAISe 2018, and Green Data Mining 2018, Dublin, Ireland, September 10–14, 2018, Proceedings 18 (pp. 243–255). Springer International Publishing.

Ge, H. L., Tang, D., & Boamah, V. (2022). Research on digital inclusive finance promoting the integration of rural three-industry. *International Journal of Environmental Research and Public Health,* 19(6), 3363.

Gera, R., Chadha, P., Khera, G. S., & Yadav, R. (2023). A Comprehensive and Narrative Review of Industry 5.0 Technologies: 2018–2022. In A. A. Khosla, *Renewable Energy Optimization, Planning and Control. Studies in Infrastructure and Control* (pp. 237–259). Singapore: Springer.

Ghoddusi, H., Creamer, G. G. and Rafizadeh, N. (2019). Machine learning in energy economics and finance: A review. *Energy Economics,* 81, 709–727.

Ghosal, M. and Chakraborty, A., 2021, August. The growing use of nanotechnology in the built environment: A review. In *IOP Conference Series: Materials Science and Engineering* (Vol. 1170, No. 1, p. 012007). IOP Publishing.

Greengard, S. (2021). *The internet of things.* MIT Press.

Griggs, D., Stafford-Smith, M., Gaffney, O., Rockström, J., Öhman, M. C., Shyamsundar, P., Steffen, W., Glaser, G., Kanie, N. and Noble, I. (2013). Sustainable development goals for people and planet. *Nature,* 495(7441), 305–307.

Guo, H. (2010). Understanding global natural disasters and the role of earth observation. *International Journal of Digital Earth,* 3(3), 221–230.

Guo, L., Sun, D., Warraich, M. A., & Waheed, A. (2023). Does Industry 5.0 model optimize sustainable performance of Agri-enterprises? Real-time investigation from the realm of stakeholder theory and domain. *Sustainable Development.* https://doi.org/10.1002/sd.2527

Hakimah, Y., Pratama, I., Fitri, H., Ganatri, M., & Sulbahrie, R. A. (2019). Impact of Intrinsic Corporate Governance on Financial Performance of Indonesian SMEs. *International Journal of Innovation, Creativity and Change,* 7(1), 32–51.

Herz, R. H., Jr, Hileman, D., Littan, S. H., Monterio, B. J., & Thomson, J. C. (2023). *Achieving effective internal control over sustainability reporting (icsr): Building trust and confidence through the coso internal control—integrated framework.* NY: COSO.

Hopwood, B., Mellor, M., & O'Brien, G. (2005). Sustainable development: Mapping different approaches. *Sustainable Development,* 13(1), 38–52.

Huang, F., & Vasarhelyi, M. A. (2019). Applying robotic process automation (RPA) in auditing: A framework. *International Journal of Accounting Information Systems*, 35, 100433.

İdil, K. A., & Ozoner, K. (2018). Big data analytics in internal audit. *Press Academia Procedia*, 7(1), 260–262.

IGBP (International Geosphere-Biosphere Programme), 1999. "Global Analysis, Integration and Modelling, "(IGBP, International Council of Scientific Unions, (ICSU), Durham, NH, 2000) http://gaim.unh.edu/; H. J. Schellnhuber, Nature 402, C19 (1999); National Research Council, Committee on Global Change Research, Global Environmental Change: Research Pathways for the Next Decade (National Academy Press, Washington, DC, 1999), p. 531, http://www.nap.edu/catalog/5992.html. (accessed 21/04/2023).

IIA. (2020, February 15). The IIA's Three Lines Model: An update of the Three Lines of Defense. Retrieved from https://www.theiia.org/en: https://www.theiia.org/en/content/position-papers/2020/the-iias-three-lines-model-an-update-of-the-three-lines-of-defense/

Iliev, P. (2022). Accounting, Digitalization, Control. *Knowledge: International Journal*, 53(1), 125–129.

Jacobson, M. Z. (2009). Review of solutions to global warming, air pollution, and energy security. *Energy & Environmental Science*, 2(2), 148–173.

Janicke, M., & Jacob, K. (2013). A third industrial revolution. *Long-term Governance for Social-ecological Change*, 47–71.

Jarrahi, M. H. (2018). Artificial intelligence and the future of work: Human-AI symbiosis in organizational decision making. *Business Horizons*, 61(4), 577–586.

Jones, H., & Hobson, P. (2020, April 17). EY whistleblower awarded $11 million after suppression of gold audit. Retrieved from https://www.reuters.com/: https://www.reuters.com/article/gold-whistleblower-ey-idAFL8N2C54EL

Jørgensen, R. (1989). Criteria for dating prehistoric graves: Stone Age, Bronze Age or Iron Age? *Acta Borealia*, 6(2), 28–41.

Kaelbling, L. P., Littman, M. L. and Moore, A. W., 1996. Reinforcement learning: A survey. *Journal of Artificial Intelligence Research*, 4, 237–285.

Kamaruddin, M. I., & Ramli, N. M. (2018). The impacts of internal control practices on financial accountability in Islamic non-profit organizations in Malaysia. *International Journal of Economics, Management and Accounting*, 26(2), 365–391.

Kasinathan, P., Pugazhendhi, R., Elavarasan, R. M., Ramachandaramurthy, V. K., Ramanathan, V., Subramanian, S., ... Rangasamy, S. (2022). Realization of sustainable development goals with disruptive technologies by Integrating Industry 5.0, Society 5.0, smart cities and villages. *Sustainability*, 14(22), 15258. https://doi.org/10.3390/su142215258

Kimani, D., Adams, K., Attah-Boakye, R., Ullah, S., Frecknall-Hughes, J., & Kim, J. (2020). Blockchain, business and the fourth industrial revolution: Whence, whither, wherefore and how? *Technological Forecasting and Social Change*, 161, 120254.

Klein, P. G., Mahoney, J. T., McGahan, A. M., & Pitelis, C. N. (2019). Organizational governance adaptation: Who is in, who is out, and who gets what. *Academy of Management Review*, 44(1), 6–27.

Koolen, D., Sadat-Razavi, N., & Ketter, W. (2017). Machine learning for identifying demand patterns of home energy management systems with dynamic electricity pricing. *Applied Sciences*, 7(11), 1160.

Koptelov, A. (2022, July 15). How to Overcome Machine Learning Risks. Retrieved from https://www.spiceworks.com/: https://www.spiceworks.com/tech/artificial-intelligence/guest-article/how-to-overcome-machine-learning-risks/

Kristensen, I. (2022, February 18). Building the internal-audit function of the future. Retrieved from https://www.mckinsey.com/: https://www.mckinsey.com/capabilities/risk-and-resilience/our-insights/building-the-internal-audit-function-of-the-future

Kupec, V., Písař, P., Lukáč, M., & Bartáková, G. P. (2021). Conceptual comparison of internal audit and internal control in the marketing environment. *Sustainability*, 13(12) 6691. https://doi.org/10.3390/su13126691

Kvamsdal, S. F., Belik, I., Hopland, A. O., & Li, Y. (2021). A machine learning analysis of the recent environmental and resource economics literature. *Environmental and Resource Economics*, 79, 93–115.

Kynčlová, P., Upadhyaya, S., & Nice, T. (2020). Composite index as a measure on achieving Sustainable Development Goal 9 (SDG-9) industry-related targets: The SDG-9 index. *Applied Energy*, 265, 114755. https://doi.org/10.1016/j.apenergy.2020.114755

Larchenko, L. V., Kolyshkin, A. V., & Yakovleva, T. V. (2020). The Reproduction of the Mineral Resource Base as the Basis for the Sustainable Development of the Resource-Producing Regions of the North and urce-Producing Regions of the North and. *IOP Conference Series: Materials Science and Engineering*, Boston: International Science and Technology Conference. https://doi.org/10.1088/1757-899X/753/6/062016

Leng, J., Sha, W., Wang, B., Zheng, P., Zhuang, C., Liu, Q., ... Wang, L. (2022). Industry 5.0: Prospect and retrospect. *Journal of Manufacturing Systems*, 65, 279–295. https://doi.org/10.1016/j.jmsy.2022.09.017

Liu, Z., Adams, M., Cote, R. P., Geng, Y., & Li, Y. (2018). Comparative study on the pathways of industrial parks towards sustainable development between China and Canada. *Resources, Conservation and Recycling*, 128, 417–425, https://doi.org/10.1016/j.resconrec.2016.06.012

Longo, F., Padovano, A., & Umbrello, S. (2020). Value-oriented and ethical technology engineering in Industry 5.0: A human-centric perspective for the design of the factory of the future. *Applied Science*, 10, 4182. https://doi.org/10.3390/app10124182

Longo, F., & Padovano, A. (2020). Voice-enabled Assistants of the Operator 4.0 in the Social Smart Factory: Prospective role and challenges for an advanced human–machine interaction. *Manufacturing Letters*, 26, 12–16, https://doi.org/10.1016/j.mfglet.2020.09.001

Louch, W., O'Dwyer, M., & Kerr, S. (2023, April 15). Dubai court orders KPMG to pay $231mn for Abraaj fund audit failure. Retrieved from https://www.ft.com/: https://www.ft.com/content/0843cd97-ca9b-4f8e-b54e-0a7c779d0e8d

Maddikunta, P. K., Pham, Q.-V., Deepa, N., Dev, K., Gadekallu, T. R., ... Liyanage, M. (2022). Industry 5.0: A Survey on Enabling Technologies and Potential Applications. *Journal of Industrial Information Integration*, 26, 100257, https://doi.org/10.1016/j.jii.2021.100257

Mahesh, B. (2020). Machine learning algorithms-a review. *International Journal of Science and Research (IJSR)*. [Internet], 9, 381–386.

Manzoor, F., Wei, L., Asif, M., Haq, M. Z., & Rehman, H. U. (2019). The contribution of sustainable tourism to economic growth and employment in Pakistan.

International Journal of Environmental Research and Public Health, 16(19), 3785. https://doi.org/10.3390/ijerph16193785

MDGs (Millennium Development Goals), 2000. The eight Millennium Development Goals (MDGs). United Nations, New York, United States. See: https://www.un.org/millenniumgoals/bkgd.shtml (accessed 20/04/2023).

Mehrabi, N., Morstatter, F., Saxena, N., Lerman, K. and Galstyan, A. (2021). A survey on bias and fairness in machine learning. *ACM Computing Surveys (CSUR)*, 54(6), 1–35.

Meier, L. J., Hein, A., Diepold, K., & Buyx, A. (2022). Algorithms for ethical decision-making in the clinic: A proof of concept. *The American Journal of Bioethics*, 22(7), 4–20. https://doi.org/10.1080/15265161.2022.2040647

Mensah, J. (2019). Sustainable development: Meaning, history, principles, pillars, and implications for human action: Literature review. *Cogent Social Sciences*, 5(1), 1653531. https://doi.org/10.1080/23311886.2019.1653531

Mergel, I., Edelmann, N., & Haug, N. (2019). Defining digital transformation: Results from expert interviews. *Government Information Quarterly*, 36(4), 101385. https://doi.org/10.1016/j.giq.2019.06.002

Mihardjo, L., Sasmoko, S., Alamsjah, F., & Elidjen, E. (2019). Digital leadership role in developing business model innovation and customer experience orientation in Industry 4.0. *Management Science Letters*, 9, 1749–1762.

Milberg, W., & Winkler, D. (2013). *Outsourcing Economies; global value chains in capitalist development.* NY: Cambrige University Press.

Mitchell, T. M., 2007. *Machine learning* (Vol. 1). New York: McGraw-Hill.

Mohamad Nasri, N., Husnin, H., Mahmud, S. N., & Halim, L. (2020). Mitigating the COVID-19 pandemic: A snapshot from Malaysia into the coping strategies for pre-service teachers' education. *Journal of Education for Teaching*, 46(4), 546–553.

Mokyr, J. and Strotz, R. H. (1998). The second industrial revolution, 1870-1914. *Storia dell'economia Mondiale*, 21945(1), 22–48.

Morrar, R., Arman, H., & Mousa, S. (2017). The Fourth Industrial Revolution (Industry 4.0): A social innovation perspective. Technology Innovation *Management Review*, 7(11), 12–24.

Mourtzis, D. (2021). Towards the 5th industrial revolution: A literature review and a framework for process optimization based on big data analytics and semantics. *Journal of Machine Engineering*, 21(3), 5–39. https://doi.org/10.36897/jme/141834

Müller, J. M., Buliga, O., & Voigt, K. I. (2021). The role of absorptive capacity and innovation strategy in the design of Industry 4.0 business Models-A comparison between SMEs and large enterprises. *European Management Journal*, 39(3), 333–343.

Myers, T. G., Ramkumar, P. N., Ricciardi, B. F., Urish, K. L., Kipper, J., & Ketonis, C. (2020). Artificial intelligence and orthopaedics: An introduction for clinicians. *The Journal of bone and joint surgery. American*, 102(9), 830.

Myradov, B. (2022). Challenges of Machine Learning for Covid-19 Diagnosis based on Blood Tests. *International Conference on Innovation and Intelligence for Informatics, Computing, and Technologies (3ICT)* (pp. 721–727, Sakheer, Bahrain: IEEE. https://doi.org/10.1109/3ICT56508.2022.9990818

NASA (National Aeronautics and Space Administration) (2022). Global Temperature, NASA, Washington, DC. United States. See: https://climate.nasa.gov/vital-signs/global-temperature/ (accessed 19/04/2023).

Nayyar, A., & Kumar, A. (Eds.) (2020). *A Roadmap to Industry 4.0: Smart Production, Sharp Business and Sustainable Development* (pp. 1–21). Berlin: Springer.

Nilsson, M., Griggs, D., & Visbeck, M. (2016). Policy: Map the interactions between sustainable development goals. *Nature*, 534(7607), 320–322.

Noble, S. M., Mende, M., Grewal, D., & Parasuraman, A., (2022). The Fifth Industrial Revolution: How harmonious human–machine collaboration is triggering a retail and service [r] evolution. *Journal of Retailing*, 98(2), pp. 199–208.

Ntim, C. G. (2018). Defining corporate governance: Shareholder versus stakeholder models. In A. Farazmand, *Global Encyclopedia of Public Administration, Public Policy, and Governance*. Singapore: Springer International Publishing.

Nurafia, A. (2023). Fraud Prevention In The Manufacturing Company Sector (Study On Manufacturing Companies Registered In The Association Of Bonded Zone Companies In Central Java - DIY). *Conference on Economics and Business Innovation* (pp. 3(1), 223–231). PN: CEBI.

NYT. (2023, January 26). *What's Going On in This Graph?|Global Temperature Change*. Retrieved from https://www.nytimes.com/: https://www.nytimes.com/2023/01/26/learning/whats-going-on-in-this-graph-feb-1-2023.html

Ofli, F., Meier, P., Imran, M., Castillo, C., Tuia, D., Rey, N., Briant, J., Millet, P., Reinhard, F., Parkan, M. and Joost, S., 2016. Combining human computing and machine learning to make sense of big (aerial) data for disaster response. *Big Data*, 4(1), pp. 47–59.

Okodo, D., Momoh, M. A., & Yahaya, A. O. (2019). Assessing the Reliability of Internal Audit Functions: The Issues. *Journal of Contemporary Research in Business, Economics and Finance*, 1(1), 46–55.

Opoku, A. (2022). Construction industry and the Sustainable Development Goals (SDGs). In *Research Companion to Construction Economics*. Edward Elgar Publishing.

Paris Agreement, 2015. Paris Agreement. United Nations, New York, United States. See: http://unfccc.int/files/essential_background/convention/application/pdf/english_paris_agreement.pdf (accessed 21/04/2023).

Paschek, D., Mocan, A. and Draghici, A., 2019, May. Industry 5.0—The expected impact of next industrial revolution. In *Thriving on future education, industry, business, and Society, Proceedings of the MakeLearn and TIIM International Conference*, Piran, Slovenia (pp. 15–17).

Pathak, P., Pal, P. R., Shrivastava, M. and Ora, P. (2019). Fifth revolution: Applied AI & human intelligence with cyber physical systems. *International Journal of Engineering and Advanced Technology*, 8(3), 23–27.

Pejić Bach, M., Krstić, Ž., Seljan, S., & Turulja, L. (2019). Text mining for big data analysis in financial sector: A literature review. *Sustainability*, 11(5), 1277.

Philbeck, T. and Davis, N., 2018. The fourth industrial revolution. *Journal of International Affairs*, 72(1), 17–22.

Popović, A. (2020). Implications of the Fourth Industrial Revolution on sustainable development. *Economics of Sustainable Development*, 4(1), 45–60.

Prisecaru, P. (2016). Challenges of the fourth industrial revolution. *Knowledge Horizons. Economics*, 8(1), 57.

Rahman, M. S., Khomh, F., Rivera, E., Guéhéneuc, Y. -G., & Lehnert, B. (2022). Challenges in Machine Learning Application Development: An Industrial Experience Report. *IEEE/ACM 1st International Workshop on Software Engineering for Responsible Artificial Intelligence (SE4RAI)* (pp. 21–28, PA: IEEE. https://doi.org/10.1145/3526073.3527593

Rasborg, K. (2021). What is risk. In Ulrich Beck: theorising world risk society and cos, opolitansim. Singapore: Palgrave Macmillan. https://doi.org/10.1007/978-3-030-89201-2_2

Ray, S. (2019, February). A quick review of machine learning algorithms. In *2019 International conference on machine learning, big data, cloud and parallel computing (COMITCon)* (pp. 35–39). IEEE.

Rehman, A. (2022b). Organizational Corruption Prevention, Internal Audit, and Sustainable Corporate Governance: Evidence from Omani Public Listed Companies. *International Journal of Social Sciences and Economic Review*, 4 (12), 10–20, https://doi.org/10.36923/ijsser.v4i.2.159

Rehman, A., Khan, S., & Hashim, F. (2023). Covid 19 impact on educational institutions: A detailed analysis. In J. R. Szyrocka, J. Żywiołek, A. Nayyar, & M. Naved. (eds), *Advances in Distance Learning in Times of Pandemic* SG: CRC Press- Taylor & Francis. https://doi.org/10.1201/9781003322252-1

Rehman, A. A. (2022a). With Application of Agency Theory, Can Artificial Intelligence Eliminate Fraud Risk? A Conceptual Overview. In B. H. Alareeni, *Artificial Intelligence and COVID Effect on Accounting. Accounting, Finance, Sustainability, Governance & Fraud: Theory and Application* (pp. 115–127,). Singapore: Springer.

Rengkung, L. R., Pangemanan, L. R., & Sondak, L. W. (2018). Competitiveness of Small and Medium Firms (SMEs) in facing ASEAN economic community. *International Research Journal of Business Studies*, 10(2), 123–133.

Rhim, J. L. (2021). A deeper look at autonomous vehicle ethics: An integrative ethical decision-making framework to explain moral pluralism. *Frontiers in Robotics and AI*, 8, 632394.

Rodgers, W. M., Stefanidis, A., Degbey, W. Y., & Tarba, S. Y. (2023). An artificial intelligence algorithmic approach to ethical decision-making in human resource management processes. *Human Resource Management Review*, 33(1), 100925.

Rodrik, D. (2018). Populism and the economics of globalization. *Journal of International Business Policy*, 1, 12–33.

Roy, A., & Pramanick, K. (2019). Analysing progress of sustainable development goal 6 in India: Past, present, and future. *Journal of Environmental Management*, 232, pp. 1049–1065.

Sachs, J., Lafortune, G., Kroll, C., Fuller, G., & Woelm, F. (2022). From Crisis to Sustainable Development: The SDGs as Roadmap to 2030 and Beyond. Sustainable Development Report 2022. Cambridge: Cambridge University Press. See: https://dashboards.sdgindex.org/ (accessed 20/04/2023).

Sachs, J., Schmidt-Traub, G., Mazzucato, M., Messner, D., Nakicenovic, N., & Rockström, J. (2019). Six transformations to achieve the Sustainable Development Goals. *Nature Sustainability*, 2(9), 805–814. https://doi.org/10.1038/s41893-019-0352

Sachs, J. D. (2015). *The age of sustainable development*. New York, United States: Columbia University Press.

Sarkodie, S. A. (2022). Winners and losers of energy sustainability—Global assessment of the Sustainable Development Goals. *Science of the Total Environment*, 831, 154945. https://doi.org/10.1016/j.scitotenv.2022.154945

SCCE, & HCCA. (2020). *Compliance risk management: Applying the COSO ERM framework*. NY: COSO.

SDGR (Sustainable Development Goals Report) (2022). Sustainable Development Goals Report, 2022, United Nations, New York, United States. See: https://unstats. un.org/sdgs/report/2022/ (accessed 21/04/2023).

Shaw, J., Rudzicz, F., Jamieson, T., & Goldfarb, A. (2019). Artificial intelligence and the implementation challenge. *Journal of Medical Internet Research*, 21(7), e13659. https://doi.org/10.2196/13659

Shinde, B. (2021). *Artificial Intelligence Adoption in Internal Audit Processes*. NY: ISACA.

Singh, S., Tabassum, N., Darwish, T. K., & Batsakis, G. (2018). Corporate governance and Tobin's Q as a measure of organizational performance. *British Journal of Management*, 29(1), 171–190.

Sinha, A., Sengupta, T., & Alvarado, R. (2020). Interplay between technological innovation and environmental quality: Formulating the SDG policies for next 11 economies. *Journal of Cleaner Production*, 242, 118549.

Sithic, H., & T. Balasubramanian. (2013). Survey of insurance fraud detection using data mining techniques. *International Journal of Innovative Technology and Exploring Engineering*, 2(3), 62–65.

Slapni, S., Vuko, T., Cular, M., & Drascek, M. (2022). Effectiveness of cybersecurity audit. *International Journal of Accounting Information Systems*, 44, 100548, https://doi.org/10.1016/j.accinf.2021.100548

Solangi, Z. A., Solangi, Y. A., Chandio, S., bin Hamzah, M. S., & Shah, A. (2018, May). The future of data privacy and security concerns in Internet of Things. In *2018 IEEE International Conference on Innovative Research and Development (ICIRD)* (pp. 1–4). IEEE.

Soori, M., Arezoo, B., & Dastres, R. (2023). Machine learning and artificial intelligence in CNC machine tools, A review. *Sustainable Manufacturing and Service Economics*. https://doi.org/10.1016/j.smse.2023.100009

Stearns, P. N. (2020). *The industrial revolution in world history*. Routledge.

Syam, N., & Sharma, A. (2018). Waiting for a sales renaissance in the fourth industrial revolution: Machine learning and artificial intelligence in sales research and practice. *Industrial Marketing Management*, 69, 135–146.

Taub, B. (2023, February 27). How the Biggest Fraud in German History Unravelled. Retrieved from https://www.newyorker.com: https://www.newyorker.com/magazine/2023/03/06/how-the-biggest-fraud-in-german-history-unravelled#:~:text=Then%2C%20on%20June%2018%2C%202020,or%20it%20had%20never%20existed

Tizghadam, A., Khazaei, H., Moghaddam, M. H., & Hassan, Y. (2019). Machine learning in transportation. *Journal of Advanced Transportation*, 2019.

Umar, T., & Umeokafor, N. (2022). Exploring the GCC progress towards United Nations sustainable development goals. *International Journal of Social Ecology and Sustainable Development (IJSESD)*, 13(1), 1–32. https://doi.org/10.4018/IJSESD.2022010105

UN. (2022). *The Sustainable Development Goals Report*. United Nations.

UN SDGs (United Nations Sustainable Development Goals), 2015. About the Sustainable Development Goals. United Nations, New York, United States. See: https://www.un.org/sustainabledevelopment/sustainable-development-goals/ (accessed 20/04/2023).

UNSC (United Nations Statistical Commission), 2019. Interlinkages of the 2030 Agenda for Sustainable Development, prepared by the Interlinkages Working Group of the Inter-Agency and Expert Group on Sustainable Development Goal Indicators (IAEG-SDGs) United Nations Statistical Commission (UNSC). 2019. See: https://unstats.un.org/unsd/statcom/50th-ession/documents/BG-Item3a-Interlinkages-2030-Agenda-for-Sustainable-Development-E.pdf (accessed 20/03/2023).

Usama, M., Qadir, J., Raza, A., Arif, H., Yau, K. L. A., Elkhatib, Y., Hussain, A., & Al-Fuqaha, A., 2019. Unsupervised machine learning for networking: Techniques, applications and research challenges. *IEEE Access*, 7, 65579–65615.

Uzialko, A. (2023, February 21). How Artificial Intelligence Will Transform Businesses. Retrieved from https://www.businessnewsdaily.com/: https://www.businessnewsdaily.com/9402-artificial-intelligence-business-trends.html

Vadasi, C., Bekiaris, M., & Andrikopoulos, A. (2020). Corporate governance and internal audit: An institutional theory perspective. *Corporate Governance: The International Journal of Business in Society*, 20(1), 175–190.

Van der Sommen, F., de Groof, J., Struyvenberg, M., van der Putten, J., Boers, T., Fockens, K., Schoon, E. J., Curvers, W., Mori, Y., Byrne, M., & Bergman, J. J., 2020. Machine learning in GI endoscopy: Practical guidance in how to interpret a novel field. *Gut*, 69(11), 2035–2045.

Verma, A., Bhattacharya, P., Madhan, N., Trivedi, C., Bhushan, B., Tanwar, … Sharma, R. (2022). *Blockchain for Industry 5.0: Vision, Opportunities, Key Enablers, and Future Directions*. IEEE, 69160–69199, https://doi.org/10.1109/ACCESS.2022.3186892

Vollmer, S., Mateen, B. A., Bohner, G., Király, F. J., Ghani, R., Jonsson, P., & … Hemingway, H. (2020). Machine learning and artificial intelligence research for patient benefit: 20 critical questions on transparency, replicability, ethics, and effectiveness. *Research Methods and Reporting*. https://doi.org/10.1136/bmj.l6927

Wang, F., Elbadawi, M., Tsilova, S. L., Gaisford, S., Basit, A. W., & Parhizkar, M. (2022a). Machine learning to empower electrohydrodynamic processing. *Materials Science & Engineering C*. https://doi.org/10.1016/j.msec.2021.112553

Wang, Y., Tian, J., Ones, D. S., & Landers, R. N. (2022b). Using natural language processing and machine learning to replace human content coders. *Psychological Methods*, 7(4), 1–17.

Waqas, A., Halim, H., & Ahmad, N. (2022). Design leadership and SMEs Sustainability; Role of Frugal Innovation and Technology Turbulence. *International Journal of Systematic Innovation*, 7(4), 1–17. https://doi.org/10.6977/IJoSI.202212_7(4).0001

Weart, S. R. (2003). *The discovery of global warming*. Harvard University Press.

WEF (World Economic Forum), 2018. This is how to sustainably feed 10 billion people by 2050. World Economic Forum, Colony, Switzerland. See: https://www.weforum.org/agenda/2018/12/how-to-sustainably-feed-10-billion-people-by-2050-in-21-charts/ (accessed 21/04/2023).

Williams, M. (2022). Elizabeth Holmes and Theranos: A play on more than just ethical failures. *Business Information Review*, 39(1), 23–31, https://doi.org/10.1177/02663821221 0888

Xu, M., David, J. M., & Kim, S. H., 2018. The fourth industrial revolution: Opportunities and challenges. *International Journal of Financial Research*, 9(2), 90–95.x

Xu, X., Lu, Y., Vogel-Heuser, B., & Wang, L. (2021). Industry 4.0 and Industry 5.0— Inception, conception and perception. *Journal of Manufacturing Systems*, 61, 530–535. https://doi.org/10.1016/j.jmsy.2021.10.006

Yuhertiana, I., Patrioty, C. N., & Mohamed, N. (2019). The moderating effect of organizational changes on the influence of ethical decision making on public sector internal auditor performance. *Contemporary Economics*, 13(4), 480–494.

Zhao, C. S., Chen, X., Liu, J., Zheng, C., Qu, S., Zou, J.-P., & Xu, M. (2022). Quantifying the impacts of COVID-19 on sustainable development goals using machine learning models. *Fundamental Research*, https://doi.org/10.1016/j.fmre. 2022.06.016

Zhou, Z. H. (2021). *Machine learning*. Springer Nature.

Zhu, T., Ye, D., Wang, W., Zhou, W., & Philip, S. Y. (2020). More than privacy: Applying differential privacy in key areas of artificial intelligence. *IEEE Transactions on Knowledge and Data Engineering*, 34(6), 2824–2843.

Chapter 8

Analysis of Industry 5.0 contributions to sustainable development

From Welfare to Wellbeing

Renuka Sharma

Amity University, Gurugram, India

8.1 INTRODUCTION

Since the First Industrial Revolution, humanity has understood the possibility of employing technology as a tool for growth (Industry 1.0). Around the year 1780, mechanical power was developed using necessary resources viz. water, steam, and fossil fuels. This was the start of the First Industrial Revolution. Industrialists who employed mass manufacturing and assembly lines during the second such revolution in the 1870s welcomed electrical energy (Industry 2.0). Automation was incorporated into industrial enterprises utilising electronics and information technology during the Third Industrial Revolution (Industry 3.0) in the 1970s (IT). With the help of the Internet of Things (IoT), cloud computing, and artificial intelligence (AI), the Fourth Industrial Revolution (Industry 4.0) is making Smart Cyber-Physical Systems (CPS), which act as a real-time interface between the virtual and physical worlds, conceivable (Mourtzis, 2016, 2022 ElMaraghy et al., 2021).

The previous decade has seen a fast transition in technology, industries, and social patterns and processes, conceptualised by the term "Industry 4.0" (I4.0). I4.0 is designed to boost production efficiency and higher service and product quality. Pillar technologies, including Big Data Analytics (BDA), AI, and digital twins, are emerging and developing in this regard (Rüßmann et al., 2015). Because engineers have mostly concentrated on the development of technology, manufacturing and systems of production and networks, there are certain limitations even with the breakthroughs and opportunities provided by the I4.0 framework (digitalisation), favouring industrial sustainability and worker wellbeing over industrial flexibility and efficiency (Xu et al., 2021).

In light of this, a new age of industrial revolution is about to begin. Engineers can completely exploit the existing technology to progress mankind and democratise manufacturing in this new century. Many nations, including the European Union, Japan, and the USA, are now working to construct the human-centric period, also known as Industry 5.0, which also refers to Society 5.0. Industry 4.0 is continuing technical development, and

Society 5.0 (including Industry 5.0) is currently being prepared. It is important to emphasise this to dispel misconceptions about Industry 5.0's status as a standalone revolution in industry.

According to the European Commission's 2021 report, the Fifth Industrial Revolution, also known as Industry 5.0 (I5.0), would comprise autonomous manufacturing that uses AI and human intelligence as its core technologies, both in and out of the production process. As a result, many business owners and technical trailblazers are already anticipating Industry 4.0, even though it has not yet received widespread adoption. Furthermore, it is anticipated that the 4.59 billion social media users in 2022 will have significantly increased to 5.85 billion users by 2027 because of the Internet's and its related technologies' rapid expansion.

Small- and medium-sized businesses have yet to benefit from Industry 4.0 technology fully, but business executives have already begun to imagine what Industry 5.0 may look like. To provide individualised products, their aggregate perspective underscores the necessity to reintroduce the human touch to digital technologies. Industry 5.0 aims to create a relationship between the digital world and human potential for critical and creative thought, while Industry 4.0 is about digitalisation, being adaptable to dynamic market scenarios, and emphasising sustainability (Venaik et al., 2023). The history of previous industrial revolutions has also shown that disruptive technologies take decades or even centuries to emerge and that industries must gradually adapt before moving from one "revolution" to another.

It is appropriately called a "revolution" since the degree of technology of the invention can completely transform all industrial processes rather than just bringing about little changes. The next great thing, however, would be a human collaboration with digital technology with an added importance to sustaining the economy, society, and the environment. Hence, "Industry 4.0" or "Sustainable Industry 4.0" would be preferable to "Industry 5.0" when referring to the approaching revolution. However, given that Industry 4.0 technologies have just recently begun to penetrate small- and medium-sized businesses and developing countries, rather than the giant multinational enterprises of industrialised countries, it is still too soon to predict the forthcoming wave.

Industry 5.0 recognises that by mandating that production consider the limitations of our planet and making employee wellbeing a priority, the manufacturing sector may serve societal purposes beyond employment and growth and become a sustainable source of development. Industry 5.0 adds to the technological update needed by businesses to be a reliable system for people looking for a fulfilling and healthy profession. It prioritises the welfare of its workers and employs cutting-edge technology to produce income that goes beyond employment and development while respecting the limits of the world. It empowers employees and satisfies their evolving training and skill needs. It recruits elite talent and intensifies industrial competitiveness. Therefore, Industry 5.0 is based on ideas like human-centricity, ecological consciousness, and benefits for society rather than being heavily focused on technology. This shift in viewpoint is supported by the notion that technology may be created to promote moral ideals and that ethical objectives may guide technical innovation rather than the other way around.

8.1.1 Industry 4.0

The adoption of cutting-edge technical solutions is the foundation of Industry 4.0. In this case, it is especially important that the production processes have been improved and that advanced hardware and software have been incorporated. Implementing the Industry 4.0 idea aims to bring about changes in working practices, the sector's employer role, and not only technology. Regarding industrial management, Industry 4.0 is characterised as an intelligent, real-time, digital network focused primarily on people (Dombrowski et al., 2017). Through real-time data interchange, it promotes greater digitisation of the whole value chain and the exchange of information between systems, things, and people (Hecklau et al., 2016).

The following are the three used to characterise Industry 4.0 (Mrugalska & Wyrwicka, 2017):

- Intelligent devices enable this since they have a memory to record operational data and standards independently, coordinating manufacturing processes and soliciting the necessary resources.
- Decentralised self-organisation takes the role of the conventional production hierarchy in Intelligent Machines, enabling a flexible and adaptable production line.
- An enhanced operator with knowledge automation encourages a flexible and adaptable component in the production system.

The CPS, Internet of Services (IoS), and IoT, which will be crucial instruments in its deployment, are the main components of Industry 4.0 (Wan et al., 2016). The Industry 4.0 framework is built on the CPS system, which integrates hardware and software into an electrical or mechanical system with a specific function.

The idea of an industry that emphasises preserving the environment and society has lately gained traction under the name Industry 5.0. The proponents of Industry 5.0 think that the Industry 4.0 framework needs to be revised to attain sustainable growth. According to the literature, Industry 4.0 paid emphasis on productivity driven by technology (Ghobakhloo et al., 2021b). Industry 4.0's internal productivity mechanism, which unintentionally enhances a few indicators of microenvironmental sustainability, such as production effectiveness or emission reduction (Ng et al., 2022) is unable to get around the fact that modern production and consumption economic models are profit-driven (Sindhwani et al., 2022). Critics contend that

Industry 4.0 is consistent with established neoliberal capitalism models that prioritise shareholder interests and profitability, amplifying some of the socio-environmental issues already present, such as regional disparities, environmental degradation, robot collaboration for an intelligent business ecosystem that promotes resource efficiency and the bioeconomy, broadening the scope of Industry 5.0 to include economic sectors other than manufacturing (Breque et al., 2021).

Sustainable manufacturing must be made possible with the introduction of Industry 5.0. All these goals may be fulfilled when certain firms embrace Industry 4.0 ideas associated with digitalising production processes. Global rivalry and the necessity to swiftly respond to constantly shifting market demands drive industrial production today (Wolniak et al., 2019, Olkiewicz et al., 2019, Zhong et al., 2017). The First Industrial Revolution simplified the activities in manufacturing facilities, laying the groundwork for modern production; the second revolution. In the Second Revolution, electricity was introduced to industry; in the third, industrial employees' uniform jobs were automated (Pilloni, 2018, Zunino et al., 2020). In contrast to its predecessors, the Fourth Industrial Revolution—also known as Industry 4.0—encompasses many aspects of society (Herceg et al., 2020), among them is the risky trend of ageing populations and the resulting decline in the labour force (Nanterme & Daugherty, 2017, Zhou et al., 2015, Saniuk et al., 2020, Baden-Filler & Haefliger, 2013). This revolution is also related to the fact that, despite the use of outsourcing and the "lean" manufacturing approach, there are no longer any prospects to increase the profit of industrial production (Lee et al., 2014, Longo et al., 2017, Beier et al., 2020, Fareri et al., 2020). Industry 4.0 refers to product developments, such as the switch from mass manufacturing to customised production, which increases the production processes' flexibility and makes it possible to meet the specific requirements of varied clients more effectively (Wolniak et al., 2019).

8.1.2 Industry 5.0

With Industry 5.0, there are several alternative perspectives. Some futurists contend that whereas Industry 4.0 primarily focuses on connecting gadgets, Industry 5.0 emphasises human and machine cooperation on production floors (Johansson, 2017). Gotfredsen (2016) listed the advantages of teamwork between machines and humans. Instead of typical robotic manufacturing, there will be a creative human touch. There will be entirely novel job vacancies. On the manufacturing floor, the roles that humans play will be better. Rendall (2017) asserted that where Germany is leading the Fourth Industrial Revolution, North America is best positioned to lead Industry 5.0.

Rendall (2017) shared man–machine cooperation for Industry 5.0 is a notion. On the Internet, there are several conversations and blogs about

Industry 5.0. (Johansson, 2017) contends the proximity of the two industrial revolutions so close in time may be seen as a single phenomenon. As a result, connectivity and machine–human cooperation may be part of Industry 4.0. Robotics and AI are one of the foundations of Industry 4.0. The interaction of humans and machines is a logical extension of robotics and AI. The mere advent of human–machine collaboration on manufacturing floors cannot justify a new industrial revolution.

Collaborative industries, often referred to as "Industry 5.0," reflect manufacturers' rising belief that they must meet consumers' growing desire for greater personalisation. Eighty-five per cent of firms believe that by 2020, coexisting robots and humans will be commonplace in production. Also, most firms say it is now a crucial component of their company strategy.

8.1.3 Advantages of a collaborative workforce with machines

The following points demonstrate the diverse advantages of a Collaborative Workforce with machines.

- The ability to personalise, bringing in the imaginative human aspect
  Robots are adept at producing conventional goods using standardised methods in high production volumes but may need assistance when customising or personalising each product. Keeping a human touch in the production processes is, therefore, essential. Only when human inventiveness is impacting the industrial processes can automation be used to its fullest potential. Conventional industrial robots used in automated manufacturing will frequently only obey commands after considerable and time-consuming programming effort. However, collaborative robots, or "cobots," operate alongside people. Because while the robot produces the product or prepares it for human touch, people may handle modifications, man and machine can coexist in this situation. This gives the worker control over the cobot, allowing them to utilise it as a multifunctional tool, much like a screwdriver, packing tool, or palletiser.
- Job creation
  According to a recent Deloitte analysis on skills gaps, there will be 3.4 million open positions over the next ten years but only 1.4 million competent candidates to fill them. Several of these roles are ideal for robots; cobots may collaborate with human workers. Cobots are helpful because they may replace risky, boring, and menial professions while allowing human workers to move into higher-paying ones. A company's inability to remain competitive is the most significant danger to job security. Automation gives businesses a competitive edge by increasing productivity, lowering costs, and delivering more consistent product quality. Robots are predicted to provide 70,000–90,000 new employment globally in the food business between 2017 and 2020.

Unlike other industrial robots, cobots can improve productivity and foster the development of new talents, leading to more job possibilities inside firms. These machines assist humans in manufacturing and processing jobs and do not take their place.

- Better roles for human workers
  The cobot is intended to take over laborious or even hazardous activities rather than replace the human worker. Consequently, human workers can use their imagination to take on more challenging tasks. For instance, when robots take over simple assembly activities, staff may switch to more complex duties that call for creativity from humans. For instance, at the Australian plastic injection moulding business Prysm Industries, employees used to name items every 6–7 seconds while standing still for hours. The firm may automate this monotonous task using a robotic arm from Universal Robots, freeing workers to handle other responsibilities. Employees may now concentrate on operating numerous more pieces of equipment and can do more diversified activities around the production floor because the labelling chores are handled. The flexibility to work on other tasks has enhanced employee happiness, and they feel proud of themselves for learning how to operate the robot. Also, Prysm saves $550 on labelling costs each day the robot is in use.

There are many chances to increase industrial productivity and Innovation with the help of a connected and collaborative workforce. Additionally, it provides the potential to increase workplace safety and satisfaction while giving human employees more fascinating job duties and fostering employment growth. Finally, less adaptable rivals that are too sluggish to change will stay caught up as industrial processes progress to become more innovative and networked. Manufacturers need to understand that collaborative factories have the potential to lower labour costs in more competitive markets, in addition to improving operational efficiency and the other advantages already outlined.

The three pillars of human centricity, environmental sustainability, and economic resilience are among Industry 5.0's basic ideals. On these ideas, there is broad agreement (Akundi et al., 2022). Nevertheless, academic and industry professionals cover the definition, nature, and constraints of this phenomenon from numerous perspectives. Industry 5.0 is centred on human-centric production, and technology advancements in social smart factories should promote human wellness. (Longo et al., 2020, Lu et al., 2022). When seen from the standpoint of environmental sustainability, Industry 5.0 is a metaphor for the idea of human–robot collaboration for an intelligent corporate environment that fosters resource efficiency and economics (Sindhwani et al., 2022).

Industry 5.0 includes additional economic areas outside manufacturing, such as agriculture and healthcare. As an alternative, the European Commission views Industry 5.0 as a socioeconomic structural transformation towards a sustainable future industry. According to the European Commission, there has never been a historical shift from Industry 4.0 to Industry 5.0. Industry 5.0 develops and, in many ways, improves the Industry 4.0 paradigm rather than displaces it by including socio-environmental components (Breque et al., 2021).

Industry 5.0's definition, aim, and restrictions are not apparent, and the literature also falls short of clearly describing how this phenomenon may succeed in promoting sustainable development. In contrast, a lot of studies have been done on Industry 4.0. Increasing industrial productivity (Hughes et al., 2022), supply chain innovation (Hahn, 2020), sustainable manufacturing (Ng et al., 2022), and sustainable energy are some of the objectives of Industry 4.0 (Ghobakhloo & Fathi, 2021) that may be accomplished using the techniques outlined in the literature. Although academia has recently made great strides in understanding Industry 5.0's guiding principles, these first breakthroughs have concentrated on the microeffects of this phenomenon.

Fatima et al. (2022) showed how Industry 5.0 may lead to production and warehouse automation, while (Sharma & Arya 2022) addressed how it might improve cutting-edge city monitoring systems. According to the academic community, Industry 5.0 will help with a variety of sustainable development initiatives (Grabowska et al., 2022; Renda et al., 2022). Understanding how a paradigm-shifting event like Industry 5.0 and its underlying sustainability functions relate to micro- and macroeconomic, environmental, and social sustainability goals is crucial.

8.1.4 Evolution from Industry 4.0 to Industry 5.0

Industry 5.0 seeks to redefine and expand the purpose of digital and intelligent technologies besides creating goods and services to make money. Instead, it places a focus on attaining actual prosperity, which calls for improving the social and physical environment. "Industry 5.0," a concept that recognises "the power of industry to achieve social goals beyond jobs and growth to become a resilient provider of prosperity," places an emphasis on the wellbeing of industrial employees while mandating that output respect the constraints of our earth. The Operator 4.0 concept already complies with Industry 5.0's requirements for social sustainability and human-centricity (Romero et al., 2016, 2020, Longo et al., 2020).

"Industry 5.0" is said to have been coined by Rada. One of the cornerstones of Industry 5.0 is the use of collaborative robotics to lower risk. Since the activities are being carried out by a human operator, these robots can recognise and understand the objectives and expectations of the jobs. The idea is that these robots will watch people perform tasks, learn how they do them, and then help human operators finish the job. Additionally, Industry 5.0 recommends using AI in daily activities to maximise human potential. Industry 5.0 is aggressively deploying robotics, artificial intelligence, the IoT, and augmented reality for the convenience and profit of human labour (Skobelev & Borovik, 2017).

Industry 5.0 recognises that by mandating that manufacturing takes our planet's constraints into consideration and emphasising employee wellbeing, the sector may serve societal purposes beyond providing jobs and economic progress, and evolve into a sustainable source of development. Industry 5.0 adds to the technology upgrade businesses need to be a reliable system for people looking for a fulfilling and healthy profession. To generate income above and beyond employment and growth while respecting the limits of the world, it prioritises employee wellbeing and uses new technologies. It increases worker control and satisfies their evolving training and skill needs.

It enhances industry rivalry and draws top talent. The three goals of the Commission in putting Industry 5.0 into practice are "an economy that works for people," "European Green Deal," and "Europe fit for the digital era." As a result, the emphasis should be more on the values of human-centricity, environmental responsibility, and social benefit. The idea that technology may be used to advance moral principles and that ethical goals may lead to technological innovation rather than the other way around supports this change in perspective (Müller, 2020).

Rada offers a different vision for Industry 5.0 (Rada, 2015, 2017). Rada argued that the objective of Industry 5.0 is "to effectively utilise a workforce made up of both machines and people while working with the environment. It switches back and forth between a real and a virtual world. He added an explanation of Industry 5.0 (Rada, 2017). This vision's central idea is industrial upcycling. Waste reduction is the primary goal of this vision. Rada also emphasises how important it is to incorporate people into the manufacturing process. He condemns the attempt to incorporate 1s and 0s into any living thing, which is the present digitisation movement (Rada, 2015).

Rada claimed that the six R methodology and L.E.D. principles are part of Industry 5.0. They are as follows.

Recognise: We must first recognise the benefits that industrial upcycling provides. The first essential step is being aware.

Reconsider: Our operating and production practices need to be reviewed and updated. Redesigning processes is a crucial initial step in the industrial upcycling process.

Realise: After considering the potential and business processes, we must realise business process innovation or improvement.

Reduce: The methodology's core goal is to employ fewer resources to produce effective results.

Reuse: The system also places reusing resources at its core that were thought to be usable before process optimisation.

Recycle: Recycling as much as possible is a crucial objective of the Industrial Upcycling effort. The best scenario is, of course, zero waste.

The technology of Industry 5.0 that enables an industry that is competitive, inventive, resilient, focused on society, and respectful of the planet's boundaries is referred to as "Industry 5.0." This industry also strives to have as little negative environmental impact as possible. In the domains of technology, socioeconomics, regulations, and governance, it offers several new issues. Considering this, two virtual workshops were organised on July 2 and July 9, 2020. The idea of Industry 5.0 was investigated by financing for participants, organisations, and Research and Technology Organisations (RTOs) across Europe. The goal was to talk about the supporting technology, and potential challenges and to get input on the overall concept. Participants also agreed that rather than depending exclusively on individual technologies, a comprehensive approach was required to handle the challenges' complexity (European Commission, 2021). The following are some of the Key Enabling Technologies (KET) that are facilitating the shift from Industry 4.0 to the idea of Industry 5.0:

- Human-centered technologies and solutions that link and combine the skills of humans and robots.
- Intelligent materials and bio-inspired technologies that allow recyclable materials with greater features and incorporated sensors.
- Modelling entire systems using real-time digital twins and simulation.
- Methods for managing the compatibility of systems and information that are cybersecure in their data analysis, transfer, and storage.
- Artificial intelligence can identify causal connections in complex, dynamic systems and generate insightful data.
- The KET will consume a lot of energy, necessitating energy-efficient and dependable autonomy technologies.

When combined with cognitive abilities and innovation, a number of enabling technological trends, such as Edge Computing (EC), Digital Twin (DT), Internet of Everything (IoE), BDA, Cobots, 5G, and blockchain, can help industries increase production and deliver customised products more quickly. With the use of these enabling technologies, the 5.0 Industry production model accentuates the relationship between people and machines. The primary technologies that will enable Industry 5.0 are shown in Figure 8.1.

8.1.5 The cornerstone of Industry 5.0

The following points illustrate the cornerstone of Industry 5.0.

- **Enterprise agility:** Businesses have used agile extensively as a methodology, structure, set of procedures, and technology for their gain. What is truly meant is to promote agile as a culture throughout the organisation rather than only focusing on one function or division. Enterprise

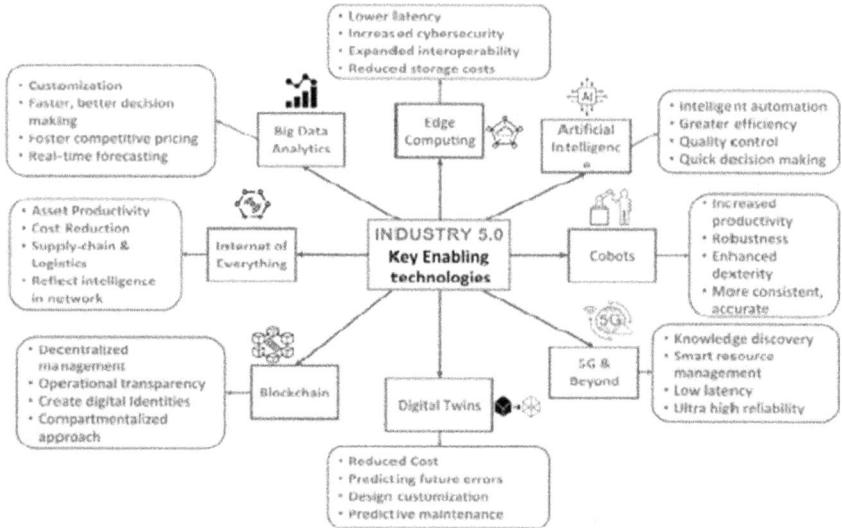

Figure 8.1 Key enabling technologies in the transition from Industry 4.0 to Industry 5.0.

agility aids in handling such urgent situations that any firm cannot afford to delay while dealing with today's continuously evolving client expectations.

- **Mass customisation:** Develop technology platforms that can extend to any size and scale to give every consumer the most customised options for products and services while respecting their needs in terms of price and comfort.

- **Cultural collaboration:** Acknowledging that not every struggle can be won at home. So, working together across many geographies and finding common ground by thinking beyond commercial and political benefits is essential. More cooperation would result in better ideas and more expertly crafted top goods/services.

- **Cyber-physical cognitive systems:** Integrating cognitive (human) and cyber-physical (artificial) intelligence skills would make it easier to get the most out of machines regarding performance and function.

- **Transparency:** Cheap company strategies for generating margins have no place in the world we live in today, which is quickly interconnected. Increasing openness would assist the company in attracting dependable staff members and clients.

- **Customer obsession:** The company's top priority should be easing consumer pain rather than concentrating on business limits. Every day should be started and finished with the customer in mind. Designing and redesigning organisational structures and procedures with a focus on customer goals would be the key to success. To guarantee that goods and services live up to expectations, customers should be included at every step of creation.

- **People-centric:** Any company's first supporters and prospective consumers outside the corporate walls are its employees. Giving employees a fair environment for advancement in all areas will aid employee retention and overall business success.
- **Green manufacturing:** Implementing environmentally friendly, low-polluting manufacturing techniques will help preserve the environment and ensure that future generations have a healthy environment.

It is past due for companies to begin thinking about incorporating cognitive sense into their CPS, which by fusing the capabilities of artificial and human intelligence, would live the purpose of industrial digital transformation and deliver some promising solutions to the masses all over the world. Because Industry 5.0 is constructed on the basis of human–machine collaboration, the fear of job loss due to rapid automation would be somewhat reduced. It would aid in encouraging upskilling of the labour force to fill specific new roles that have never been. Also, since they would be focused on providing real value to their consumers, it would help firms recoup their return on investments more quickly by reducing overhead expenditures for sales, marketing, and other unnecessary operations.

8.1.6 Industry 5.0 and sustainable development goals

The Sustainable Development Goals (SDGs) are made up of 17 articles, 169 objectives, and 232 charts that cover economic and environmental topics, including climate change, sources of water, marine and terrestrial life, as well as societal aspects, including gender equality, equality for all people, and human rights (Biggeri et al., 2019). These objectives are expected to be gradually attained until 2030.

The listed 17 objectives are as follows

1. **No Poverty (SDG1):** The goal is to eradicate severe poverty, make essential resources and services accessible to people, and shield them from unforeseen economic and environmental calamities (Campagnolo & Davide, 2019).
2. **Zero Hunger (SDG2):** The goal is to end global poverty and ensure that everyone has access to healthy, sufficient food throughout the year (Byerlee & Fanzo, 2019).
3. **Good Health and Wellbeing (SDG3):** It strives to provide individuals with the basic healthcare they need, including therapy, medical care, and affordable drugs (Wakunuma et al., 2020).
4. **Quality Education (SDG4):** The goal is to promote universal, lifelong, and open access to education for people of all ages (McKay, 2020).
5. **Gender Equality (SDG5):** The mission of the United Nations is to advance gender equality in local communities and raise the position of women and girls in society (Yount et al., 2018).

6. **Clean Water and Sanitation (SDG6):** Water is one of the essential elements on the planet for sustaining life and all living things. Hence, it is intended to implement new policies to properly utilise technology and provide people with fresh drinking water (Fehri et al., 2019).

7. **Affordable and Clean Energy (SDG7):** Its objective is to promote the production of energy from non-CO2-emitting sources, such as geothermal, hydro, solar, wind, and sea waves (Nam-Chol & Kim, 2019).

8. **Decent Work and Economic Growth (SDG8):** The primary goals are to encourage economic expansion and provide people with adequate employment; the ideal work and development model is dependable, promotes gender equality, entails little to no risk, and properly rewards employees for their efforts (Rai et al., 2019).

9. **Industry, Innovation, and Infrastructure (SDG9):** It strives to improve human wellbeing by producing innovations, leveraging those breakthroughs in the industrial sector to make eligible items that meet human requirements, and building infrastructure that is considerably safer, of higher quality and more robust (Cervelló-Royo et al., 2020).

10. **Reduced Inequalities (SDG10):** It aims to lessen disparities between and within nations and shield individuals from prejudice based on age, gender, ethnicity, religion, political views, and disability (Horn & Grugel, 2018).

11. **Sustainable Cities and Communities (SDG11):** Most people on Earth reside in urban areas. It aims to improve transportation options, set up sewage systems and other facilities that affect human health, and implement policies and practices that lead to the creation of decent, sustainable cities (Rozhenkova et al., 2019).

12. **Responsible Consumption and Production (SDG12):** To respect the future and the environmental implications of the things we use now, it is essential to utilise food, water, household goods, technology, power, and all fossil fuels responsibly (Sala & Castellani, 2019).

13. **Climate Action (SDG13):** To lessen climate change, it strives to minimise energy consumption and the usage of fossil fuels reliant on them, build carbon-free cities, prepare for natural disasters brought on by climate change, and come up with countermeasures (Ma et al., 2021).

14. **Life under Water (SDG14):** Its objectives are to guarantee the efficient utilisation of oceans and seas as well as those resources in order to make up for losses, build a solid foundation, and utilise resources sustainably (Okafor-Yarwood, 2019).

15. **Life on Land (SDG15):** It emphasises the preservation of Earth's ecosystems and species and their sustainable exploitation (Opoku, 2019).

16. **Peace, Justice, and Strong Institutions (SDG16):** Many people lose their lives yearly due to international or splinter group gunfights. Conflicts like this prohibit offering assistance to the populace and creating a more sensible and calmer atmosphere (Calvo et al., 2019).

17. **Partnerships for the Goals (SDG17):** By promoting efficient international collaboration and communication, it aims to aid global development (Lamichhane et al., 2021).

8.1.6.1 Objectives of the chapter

The objectives of the chapter are as follows:

- To articulate the relationship between Industry 4.0 and Industry 5.0;
- To demonstrate Industry 4.0 to Industry 5.0 transition;
- And, to emphasise the significance of sustainability in Industry 5.0.

8.1.6.2 Organisation of the chapter

The rest of the chapter is structured as follows: Section 8.2 enlightens on the Literature Review. Section 8.3 emphasis on Industry 4.0 and Industry 5.0 relationship. Section 8.4 stresses organisational agility. Section 8.5 enlightens the core values of Industry 5.0. Section 8.6 stresses Industry 5.0 potential technologies and applications. Section 8.7 shows the linkage between Sustainability and Industry 5.0. The Roadmap to Industry 5.0-driven Sustainability is discussed in Section 8.8. Section 8.9 highlights the challenges of Industry 5.0. Sections 8.10 and 8.11 overlay discussion and Implications. And, finally, Section 8.12 concludes the chapter with future scope and limitations.

8.2 LITERATURE REVIEW

8.2.1 Industry 4.0

Beier et al. (2020), Saniuk et al. (2020), Longo et al. (2017), Biel and Glock (2016), Zhou et al. (2015), Lee et al. (2014), Baden-Filler and Haefliger (2013) witnessed Industry 4.0, the linking of machines and gadgets through the Internet and the digitisation across all production techniques is critical. Enabling factories to be sufficiently intelligent to forecast, recognise, and address issues is crucial to implementing innovative industries.

Müller and Voigt (2018), Lu (2017), Posada et al. (2015) discussed technology developments are simply one aspect of Industry 4.0. It is now also linked to the challenge of obtaining traceability and trackability when combining various devices. There are ways to separate the idea of Industry 4.0 in the

272 Modern Technologies and Tools Supporting the Development of Industry 5.0

following ways. The data may be easily accessed from any location and transferred across devices. Many components of the supply chain exchange information in real-time.

- Common digitalisation is the process of making all individuals and all gadgets digitalised and constantly connected to one another and to one another's equipment.
- Autonomous manufacturing systems: Building smart manufacturing facilities that can autonomously manage production processes and adapt quickly to changes in the demands of industrial processes. There is practically no need for human intervention in the manufacturing processes of intelligent factories.
- Customisation of the product: Supplying the consumer with a product that is specifically tailored to their requests.
- Robotisation: Putting in place flexible production systems based on industrial robots and using robots designed to work in conjunction with people.
- Implementation of a CPS-based architecture.
- Proliferating the use of disruptive innovations may significantly boost a business's socioeconomic and operational system efficacy.

Schwab (2020), Kagermann (2015) and Lasi et al. (2014) elucidated that the concept of Industry 4.0 was created in the second decade of the 21st century in Germany. Industry 4.0 aims to transform existing factories into autonomous, self-adaptive, social, and technical systems (smart factories), which will enable the development of intelligent value chains. This has already received several scientific investigations and implementations in many countries, creating a worldwide megatrend. These difficulties influence the current business and production model. Oesterreich and Teuteberg (2016) highlighted the development of a digital value chain has made it feasible for this new industrial paradigm to be conceptualised technically as an increase in the production environment's digitisation and automation as well as an increase in communication. Kagermann et al. (2013) stated the three characteristics of integration describe the essential elements of the Industry 4.0 concept: (1) Horizontal integration via value networks, (2) vertical integration and networked production systems, and (3) end-to-end digital integration of engineering across the whole value chain.

Hermann et al. (2016) said that "Industry 4.0" is a phrase that describes a group of technologies and ideas that cover the whole value chain of enterprises. According to the author, the first four components of Industry 4.0 are CPS, IoT, IoS, and Smart Factory. These components highlight the idea of intelligent manufacturing and the integration of its parts along the value chain using essential technological enablers. The modular structure of smart factories under the Industry 4.0 framework allows for decentralised decision-making while controlling and monitoring operations. While

Internet-based internal and inter-organisational services are provided by IoS technology, IoT technology enables real-time interaction between each CPS and operators in the smart factory.

Weyer et al. (2015) conceptualised that this new industrial paradigm supports the creation of intelligent settings that can unite the actual and virtual worlds through using CPS, integrating devices, and other technologies. Machines, production units, and products activate operations and regulate one another independently. Author highlighted that there are three paradigms that encompasses the core elements of Industry 4.0: smart machine, smart product, and enhanced operator. Two crucial elements of Industry 4.0 are the human–machine interface and the emergence of new employment kinds, in addition to the relevance of smart products and smart machines/factories.

Posada et al. (2015) took it to a step further by listing and synthesising the main issues that Industry 4.0 addresses: (1) wide-spread adoption of mass customisation enabled by IT; (2) automatic and flexible production system adaptation to changing requirements; (3) tracking and self-awareness of parts and products, and their capacity to communicate within their environment; (4) increased human–machine interface, coexistence with robots, and the emergence of new modes of interaction and operation. Production optimisation facilitated by the IoT; and (5) the communication within smart factories, as well as the introduction of new services and business models that have an impact on the entire value chain, round out the list of five. In summary, Industry 4.0 has enormous potential and can affect every aspect of the value chain, including the production process, product quality, stakeholder relationships, and the introduction of new business models and operational methods.

8.2.2 Industry 5.0

Grybauskas et al. (2022), Maddikunta et al. (2022), and Xu et al. (2021) described in detail how Industry 4.0 would harm sustainability. While many academics concur that Industry 5.0 depends on Industry 4.0's technology components, it also involves dramatic advances in energy transition technologies, smart materials, and cognitive artificial intelligence (CAI), to mention a few.

Kumar et al., (2021), Nahavandi (2019), and Özdemir and Hekim (2018) described Industry 5.0 as an incremental development of Industry 4.0 that is symmetrical and evolutionary in character and could offer a method to overcome the constraints of the Industry 4.0 innovation ecosystem. They counter that a new generation of disruptive technologies that encourage synergistic human–machine integration while improving working conditions, employment, and production are to blame. Industry 5.0 heralds the rise of industrial operations that are centred on people. The focus on productivity in Industry 4.0 has drawn criticism from academics. In contrast to past industrial revolutions, the socio-technical phenomenon known as "Industry 5.0" is driven by stakeholders and increasingly replaces conventional profit- and consumption-driven economic models with circular, regenerative, sustainable, and robust value-creating economic models.

Longo et al. (2020), Javaid and Haleem (2020), Aslam et al. (2020), Di Nardo et al. (2020), Demir et al. (2019), Welfare et al. (2019), and Nelles et al. (2016) elaborated that humans and machines cohabit in the contemporary world, and intelligent technologies connect humans to smart industries. Technology, mass customisation, and sophisticated manufacturing are all undergoing rapid development. Due to advancements in artificial intelligence and the ability to connect robots to the human mind, robots are becoming ever more crucial. Using artificial intelligence (AI) algorithms, human–machine interactions will probably get more advanced as we approach the Fifth Industrial Revolution. This will lead to greater automation, better automation, and the strength of human minds combined. That suggests that, in contrast to predictions made during the Industry 4.0 period, robots won't soon take over manufacturing operations. Productivity will undoubtedly increase as a result of Industry 5.0's merger of the best aspects of the human and machine worlds. A brand-new paradigm called "Industry 5.0" emphasises collaboration and human–machine interaction.

Onday (2019), Wang et al., (2018a), and Zhang et al. (2015) introduced a new paradigm for how businesses should operate. The IoT facilitates the transition from inter-personal to inter-societal cooperation through the interconnection of growing, interoperable information and communication technologies (ICT). This interconnectedness and interoperability have helped to alleviate any restrictions brought on by the nature of the Internet and digital technologies, accelerating the growth of the "smart society." As the knowledge society is being recognised and incorporated into the online society enabled by technology, Industry 5.0 also has a beneficial influence by resolving social difficulties and fostering abundance for society.

Bansal (2018) and Chandola (2015) stated that Customers' experiences had changed in Industry 5.0 based on social growth, from mass customisation to personalisation of customisation. This has happened due to the development of technological platforms that allow for scale expansion and maximum product customisation while still providing customers with flexibility. Using technology to gather accurate information and provide feedback to the business is essential for decision-making. As a result, the creation of commercial and social value is significantly impacted by the use of Big Data to support the evolution of the customer journey in Society 5.0 (Chandola, 2015). The point out of experiences might supply different business capacities using technology.

Johansson (2017), Rendall (2017) and Gotfredsen (2016) emphasised Industry 5.0 cooperation on production floors between people and machines.

The advantages of a collaborative workforce between humans and machines are listed by (Gotfredsen, 2016). Instead of typical robotic manufacturing, there will be a creative human touch. There will be new employment openings. On the manufacturing floor, the roles that humans play will be better. The return of human interaction to factory floors is Industry 5.0. The Fifth Industrial Revolution, Industry 5.0, will be best led by North America, while Germany is leading the fourth. The concept of man–machine collaboration for Industry 5.0 is shared by many other authors. The mere advent of human–machine collaboration on manufacturing floors cannot justify a new industrial revolution.

The following are some ways in which Industry 5.0 varies from Industry 4.0:

- Sustainable development and productivity-driven competitiveness are both valued in Industry 5.0.
- By promoting human-centric approaches to technical progress, Industry 5.0 empowers the human workforce.
- Industry 5.0 promotes technical Innovation in environmental sustainability (such as smart renewable systems).
- Industry 5.0 encourages the prioritisation of stakeholders in the management of sustainable performance, innovation growth, and technology governance.
- Industry 5.0 expands the reach of corporate responsibility to include the entire value chain by utilising specific technology and operational concepts.

8.2.3 Transition from Industry 4.0 to Industry 5.0

The integration of smart devices and systems as well as the modification of industrial processes to boost productivity are referred to as "Industry 4.0" concepts. Resilient technology, innovative working methods, and the importance of people in the workplace are all included in Industry 4.0 (Luthra & Mangla, 2018). The improvements brought about by Industry 4.0 cover a broad spectrum of developments at the level of plants and factories belonging to various industries and services, as well as the running of entire communities. Industry 4.0 includes technologies like the IoT, cloud computing, cognitive computing, and artificial intelligence, a project to automate and exchange data in industrial technologies and processes (Lee et al., 2015; Solanki and Nayyar, 2019; Sahoo et al., 2021).

Industry 4.0 advocates that the generation of industrial value is concentrated on sustainability to evolve towards more sustainably producing industrial value. Sustainability concerns in Industry 4.0 are highly valued by consumers, business models, and the economy (Romerao et al., 2020). By providing information about each manufacturing location, it may be possible to enhance processes, resource utilisation, and energy consumption throughout the value network (Davies et al., 2017). It is projected that the cutting-edge technological techniques of Industry 4.0 will enhance the environmental performance of products over the course of their lifetimes. Additionally, this suggests an increase in demand for intelligent products and industrial methods (Stock & Seliger, 2016).

The increasing use of Industry 4.0 technology and the commercial focus on the dehumanisation of manufacturing processes are causing many concerns for workers, society, and even governments. Industry 4.0 places an emphasis on integrated, complex, and robust manufacturing technologies that are equipped with sensors to track machine activity and communication networks to make it easier to submit data and run sophisticated simulations. Numerous scientific studies (Romero et al., 2016) emphasise the need to take into account how people will play a key role in future industrial progress. Resilience should be present across the whole value chain, company processes, and even business conceptions. Thus, a conversation on the idea of Industry 5.0 started in 2019. This idea suggests bringing back human interaction in the industrial sector, i.e., encouraging collaboration between workers and intelligent production systems and fusing automation's speed and precision with workers' cognitive abilities and critical thinking (Longo et al., 2020).

The focus of Industry 5.0 is the relationship between humans and robots. In the current world, people live alongside machines and are connected to intelligent factories through smart devices (Demir et al., 2019). Rapid change is taking place in the fields of technology, mass customisation, and sophisticated manufacturing. Due to advancements in artificial intelligence and the ability to connect robots to the human mind, robots are becoming ever more crucial (Welfare et al., 2019). In the modern world, robots collaborate with humans rather than as competitors (Nahavandi, 2019).

The Fifth Industrial Revolution will probably promote using artificial intelligence (AI) algorithms to create more sophisticated human–machine interactions. This will result from more integration, quicker and better automation, and the strength of human minds (Aslam et al., 2020). That also indicates that contrary to what was anticipated in the Industry 4.0 era, robots will not soon take over production facilities (Javaid & Haleem, 2020). I5.0's integration of the best elements of the human and machine worlds should increase productivity (Di Nardo et al., 2020; Nelles et al., 2016). To coordinate business and supply chain activities, digitalisation in Industry 5.0 should be complete. From the megatrend known as Industry 4.0, Industry 5.0 was born. Artificial intelligence infiltration into everyday life is the growing vision of the Industry 5.0 paradigm. Researchers suggested that "Society 5.0" (Super Smart Society) would likely be a more accurate word than "Industry 5.0" (Elim & Zhai, 2020). Unlike Industry 4.0, which only applies to the industrial industry, Society 5.0 addresses societal issues by integrating the real and virtual worlds. In a society known as "Society 5.0," cutting-edge technologies are actively employed in daily life, business, the medical field, and other areas, not just for advancement but also for everyone's benefit and convenience (John et al., 2020).

The primary focus of Industry 5.0 is human–robot interaction. Individuals and machines coexist in the modern world, and smart devices connect individuals to intelligent industries (Romero et al., 2016, Felsberger & Reiner, 2020). EC has used the word "resilient" in its work. The term "resilient" is being used increasingly often. Particularly considering the COVID-19 pandemic's effects on communities, economies, and businesses, the "Industry 5.0" idea is thought to be helpful. In a globalised and technology society that is unstable (unpredictable), the relevance of sustainability for firms concentrating just on profit is increasingly harder to sustain. For an industry to fully develop, social, environmental, and societal concerns must be taken into account. Industry 5.0's core is composed of the interplay of the three sectors: technical, ecological and social. Technology was at the vanguard of Industry 4.0, but the next iteration will centre on factory employees who perceive rising automation as a danger to their employment. The focus of Industry 5.0 is on the significance of technology for industrial (commercial) growth while fusing business objectives with social objectives at work (emphasising the use of cutting-edge technology or human–machine interactions to increase worker safety both inside and outside, for instance).

The central nervous system of an industrial process is a CPS. Social and ecological frameworks are intertwined, and the upkeep of these frameworks (respect, appreciation, and compliance) by Industry 4.0 technology promotes economic growth. Cooperation is centred on the human individual. Industry 4.0 combines automation and technology, exchanges data, develops CPS, and introduces new manufacturing techniques. Products also address the possibility of personalisation and changes in value chains. The megatrend known as "digital transformation" includes all of these. Industry 4.0 development has been recognised to have its origins in digitalisation (Kagermann, 2015).

ICT and operational technologies (OT) were to work closely as part of Industry 4.0. The interdependence between IC and OT in Industry 4.0 was supposed to accomplish these technologies. The need for a digital society, economies, and industries is acknowledged by Industry 5.0 (Kent & Kopacek, 2020). To coordinate operations in Industry 5.0, digitalisation must be through enterprise and supply chains.

8.3 INDUSTRY 4.0 AND 5.0 RELATIONSHIP

Industry 4.0 has brought several benefits to corporate life, including cost reduction, zero error using equipment relying on artificial intelligence, and providing timely customised items to consumers (Mi'skiewicz & Wolniak, 2020). One central question is how and by whom those items made by smart factories will be purchased to make money (Kaygin et al., 2019). Japan has adopted the United Nations SDGs and the Society 5.0 initiative, which will be the top priority in solving this issue. According to Society 5.0, artificial intelligence will resolve several issues, including air pollution, unemployment, and poverty (De Pascale et al., 2021).

It is widely recognised that the social structure of society has undergone major change as a result of the industrial revolution, which started in 1784 (Chang et al., 2020). The word "Society 5.0" first appears in the Fifth Science and Technology Basic Plan, dated December 18, 2015, which is where we first come across it. Research, technology, and innovation were to be employed to promote Society 5.0. It becomes clear that technologies such as cloud computing, robots, augmented reality, Big Data, CPS, and artificial intelligence are utilised by both Society 5.0 and Industry 4.0 (Pereira et al., 2020). Modelling and planning tasks are completed in the management of a similar system, under the guidance and supervision of the government, businesses, and academic organisations. Unlike Society 5.0, which takes the entire society, including the industry, as its practising area, Industry 4.0 simply chooses a specific sector of the industry to focus on (Kansha & Ishizuka, 2019). In contrast, Industry 4.0 is primarily concerned with deploying more cost-effective production techniques (Deguchi et al., 2020).

Industry 4.0 and SDGs 7, 8, 9, 11, 12, and 13 are directly related as a result of factors like Economic expansion, the invention of new working paradigms, highly effective manufacturing, safe and shrewd waste management, and judicious and sustainable usage of infrastructural systems are all examples of sustainable energy policy (Dantas et al., 2021). The SDGs are nonetheless completely realised in Society 5.0.

It is paradoxical that Japan, a developed nation with cutting-edge technology and a high standard of living, felt the need to design a growth strategy like Society 5.0.

Several countries, including the USA, France, China, South Korea, and Italy, formed their application plans like Industry 4.0 to purchase inexpensively and market using smart technologies (Wang et al., 2020). The planned plans are variables that lessen Japanese firms' ability to compete in global markets and make it challenging for them to do so (Cabinet Office Government of Japan, 2020).

8.4 ORGANISATIONAL AGILITY

Since Industry 5.0 emphasises human–machine collaboration, the latter is the source of all creative inspiration. As a result, the centre of agility in version 5.0 was the human. The capacity of the company to modify its strategic direction to better align with its core business and produce value is referred to as organisational agility (Doz & Kosonen, 2010).

Agility may also be characterised as a component of decision-making flexibility regarding simplicity and speed (Hugos, 2009). According to

Vokurka and Fliedner (1998), agility is the capacity to adjust to changes in the market while providing high-quality services at a cheap cost and in a shorter amount of time with fluctuating product volumes. The ability of the company to respond to changes by offering an effective and efficient capacity to produce value is what we refer to in this study as organisational agility.

Moreover, organisational agility concentrates on resource deployment to be more effective and efficient in dealing with Society 5.0. Building the organisation around people, culture, and process could be the best course of action (Carvalho et al., 2017). People are the foundation of a company's capacity to seize opportunities and take risks to grow the business. As part of the organisation's culture and lean process, procedures may be streamlined and made more efficient, which might boost the company's profitability (Ozkeser, 2018).

Making better judgements and assisting academics in understanding Society 5.0 involves a thorough understanding of organisational agility focused on humans at the heart of the enterprise.

8.5 CORE VALUES OF INDUSTRY 5.0

Industry 5.0's three core principles are Human Centricity, Sustainability, and Resilience and are considered interdependent with one another.

The human-centric strategy allows for the shift from technology-driven development to a totally human-centric and society-centric approach that centres manufacturing on the basic needs and interests of people.

Industrial staff will take on new tasks as a result of the shift in value from seeing employees as "cost" to "investment." The technology employed in production is adaptable to the demands and diversity of industrial employees because technology serves people and societies. Employees' should prioritize physical, mental, and emotional wellbeing and safeguard their fundamental rights. The career possibilities and work–life balance of industrial workers must be improved by constant retraining and upskilling (Breque et al., 2021).

Industry must be sustainable in order to adhere to environmental restrictions. To create a circular economy that is more effective and efficient with resources, it is required to create circular processes that reuse, repurpose, and recycle natural resources while minimising waste and impact on the environment (Breque et al., 2021).

The notion of resilience holds that industrial production must become more resilient in order to be better equipped to resist interruptions and preserve crucial infrastructure in times of disaster. The sector's future depends on its ability to adapt quickly to (geo) political developments and natural disasters (Breque et al., 2021).

8.6 INDUSTRY 5.0'S POTENTIAL TECHNOLOGIES AND APPLICATIONS

The previous sections make it abundantly clear that the main goal of Industry 4.0 has been to increase profitability via the use of various digital technologies, with an emphasis on the quality of goods and process efficiency (Kumar & Nayyar, 2020; Kumar et al., 2020a; Kumar et al., 2020b). Unfortunately, it has disregarded chiefly and failed to recognise the necessity of the human intellect. The effects of digital technology on society and the environment. It is anticipated that Industry 5.0 will integrate the two crucial components currently lacking: human inclusion and sustainable development. Moreover, it is anticipated to offer the adaptability and agility different businesses need to swiftly adapt to shifting consumer preferences and market situations. Even though most business leaders and academics have identified exploiting human creativity as Industry 5.0's main differentiator, few other scholars contend this is the case. Figure 8.2 illustrates the main distinctions between the various industrial revolutions.

1. **Collaborative robots (Cobots):** Robots of the fifth industrial generation are anticipated to work together with him and with other humans. Due to the development of digital technologies like artificial intelligence, machine learning, and traditional robotics, next-generation collaborative robots, or "cobots," are now a reality. These cobots can perceive their environment, adjust to it, and learn on the go. Their adaptability to various task needs might lessen reliance on traditional industrial robots, which substantially impact ownership costs in the manufacturing industry (Lu et al., 2021; Elena, 2020). Cobots can now make efficient judgements in real-time because of recent developments in technologies like

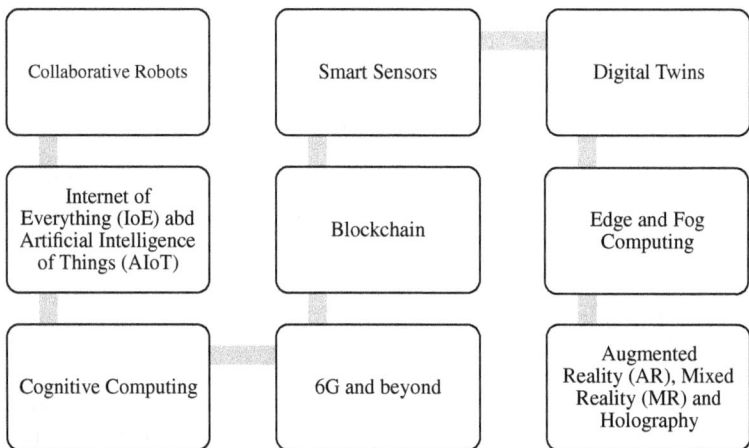

Collaborative Robots	Smart Sensors	Digital Twins
Internet of Everything (IoE) abd Artificial Intelligence of Things (AIoT)	Blockchain	Edge and Fog Computing
Cognitive Computing	6G and beyond	Augmented Reality (AR), Mixed Reality (MR) and Holography

Figure 8.2 The prospective technologies of Industry 5.0.

distributed artificial intelligence, edge computing, parallel processing, and linked data.

2. **Smart sensors:** It is a piece of technology that monitors any changes to physical properties and transmits those changes via required adjustments to electrical output. In response to a change in temperature, a thermocouple sensor, for example, generates a proper output voltage. Future smart factories will need smart sensor systems that can autonomously execute data gathering, data conversion, data processing, and data connection with an external system like a cloud server. A traditional sensor, in comparison, just includes the fundamental components for sensing. Such improved capabilities are made possible by a base sensing device, CPU, communication, and memory modules integrated into a single system.

3. **Digital twins:** Digital twins are exact digital copies of physical systems that function as their virtual counterparts. The fundamental idea behind twinning is to emulate and comprehend real-time behaviour by taking data collected from the original model and reflecting it on its duplicate model (Padmakumar & Shunmugesh, 2022) Mirroring, shadowing, and threading are the three key features of digital twins (Jiang Yuchen et al., 2021). Digital twins can be used in conjunction with other digital technologies like IoT, Big Data, Artificial Intelligence, and Machine Learning to collect data in real-time, monitor system health, predict remaining useful life, and carry out preventive maintenance because they are an extremely accurate replica of a physical system.

4. **Internet of Everything and Artificial Intelligence of Things:** Artificial Intelligence of Things (AIoT), a new technical subject in industrial automation, results from the deadly amalgamation of artificial intelligence with the IoT. A typical IoT system is made up of sensors that can produce a considerable volume of unstructured data. A linked physical device with AIoT can solve issues and make previously unattainable decisions for IoT-capable instruments.

5. **Blockchain:** The integration of blockchain technology with cutting-edge information technologies like the IoT, cloud computing, and Big Data may change how data is handled in almost all industrial areas. The centralised nature of the present IoT architecture makes it vulnerable to security risks. Using blockchain architecture will thereby provide safe, private, and encrypted data transport to ensure data security (Li & Zhang, 2017; Raja Santhi and Muthuswamy, 2022).

6. **Edge and fog computing:** When it comes to exploiting the computing resources accessible inside a local network, edge computing and fog computing are fundamentally equivalent. The primary contrast between cloud, edge, and fog computing is the location of data processing In contrast, edge computing places data processing near to the IoT sensors on the hardware that houses the sensors or at a gateway

device. On the other side, fog computing uses processors that are remote from the LAN equipment's sensors to do operations.

7. **Cognitive computing:** Intelligent computing is the name given to a group of technological platforms that handle complicated issues by simulating human brain processes using digital models. The purpose of cognitive computing is to support humans in decision-making by digesting data, not to replace them. A system with cognitive computing skills can communicate with people, understand context-sensitive language, evaluate data using prior knowledge, and draw conclusions based on communication (Gupta et al., 2018).

8. **6G and beyond:** There needs to be more than innovations in digital and computational technologies to meet the demands of a fully linked digital world; a significant improvement in communication technologies is also essential. The telecom companies have already begun their research operations towards the next technical advance needed to reach the 6G targets, which are scheduled for a commercial debut by 2030, even if the next generation of 5G technologies is only around the horizon. The market for intelligent devices and services is always expanding, and 6G and beyond can help with those needs.

The use of augmented reality (AR) technology to overlay digital data on top of real-world applications gives consumers a composite perspective. It is a 3D interactive environment that merges the physical and digital worlds (Carmigniani & Furht, 2011). Although they are not considered to be Industry 4.0 or Industry 5.0 technology enablers, AR, MR, and holography—collectively referred to as "Extended reality technologies" (Gerencer, 2021)—are one of the technologies that will likely gain from the adoption of digital and have a big impact on the development of future smart factories.

8.7 LINKAGE BETWEEN SUSTAINABILITY AND INDUSTRY 5.0

8.7.1 Industry commitment to sustainable development

Resources will be used by Industry 5.0, which promises to adapt them to the industrial sector's current demands. Human–machine collaboration leads to adaptable business models (Ciccarelli et al., 2022). Overproduction also has to be under control if waste is to be eliminated. Local production and fresh ideas (Javaid & Haleem, 2020; Saraswa et al., 2021) made the economy viable. With Industry 5.0, corporate technologies are reversing the trend. It results in the adoption of sustainable measures, such as reduced waste creation and management, which can improve corporate performance (Muthuswamy, 2022). Industry 5.0 places a conscious emphasis on cutting-edge research and places knowledge at the fore of progress in order to be practical. It may be identified by a commitment that goes beyond merely creating goods for profit. The three pillars of Industry 5.0 are resilience, human-centricity, and sustainability.

8.7.1.1 Human productivity

As a result of new technology, humans are once again at the centre of industry. Collaborative robots do risky and repetitive jobs when people concentrate on innovation and workable business solutions (Yu et al., 2018). Developing skills boosts workplace productivity when workers are motivated to put in the effort and see benefits to individuals. The technique places a higher priority on meeting human needs than on using production procedures. Producers need to be aware of what technology is capable of for individuals and how it may be changed to match worker demands rather than the other way around.

For sustainable growth, intelligent business is transforming from Industry 4.0 to Industry 5.0. As a creative example of Smart Industry 4.0, a novel, ground-breaking model of cooperation between people, machines, robots, and the environment, as well as synergies from such cooperation, additions to or significant expansions of Industry 4.0 by the environmental and social dimension, and foundations for human rights could be used to describe the upcoming Fifth Industrial Revolution (Nayyar & Kumar, 2020; Nayyar et al., 2020; Müller et al., 2018; Müller, 2019). The following advantages are anticipated from implementing the technical and economic growth plan created in this way: radical advances based on digitisation, artificial intelligence, and robotics-inclusive workplaces, and the welfare of industrial employees. Production and consumption habits that are more flexible and sophisticated in their supply networks.

With a focus on sustainability, Industry 5.0 must be all about using it wisely. It also entails finding ways to do so quickly and with the least amount of information possible. Furthermore, these technologies' effects must focus on safeguarding society and the environment in addition to production and industrial facilities. Hence, the three pillars of sustainability—economic, social, and environmental—can be used to classify sustainability. Yet, a few authors have included other pillars like institutional, cultural, and technical in addition to the three already mentioned (Purvis et al., 2019).

This is in line with data from the World Economic Forum, which estimates that small to medium-sized IoT efforts make up 75% of the market and frequently neglect sustainability in favour of more profitable goals like energy efficiency, productivity, competitiveness, and cost reduction by substituting low- and medium-skilled human employees on the production floor with industrial robots and automation, which have an impact on social sustainability, technologies have the potential to generate social upheaval. The current difficulties that businesses must overcome in order to concurrently attain the three sustainability pillars of economic, social, and environmental sustainability are referred to as the "sustainability trilemma." For example, businesses that emphasise increasing profits (economic sustainability) through excessive automation risk compromising social sustainability by eliminating jobs.

Social sustainability is yet another crucial but much-disregarded element. Research has also demonstrated that the workplace substantially affects employees' welfare, affecting social wellbeing. It can increase productivity, product quality, and efficiency by emphasising human factors such as employee wellbeing and work–life balance, creating a healthy work environment, and offering appropriate employee help. So, balancing economic development, environmental deterioration, and social wellbeing is necessary. However, by utilising next-generation. This may lead to huge reductions in the amount of processing power and energy needed (Arias et al., 2018).

Eco-friendly products that are an alternative to plastics can be developed with the aid of AI and related technologies. However, as these technologies advance and develop over time, meeting all three significant sustainability pillars at once may be feasible. A few authors have also made the case that adopting lean manufacturing techniques like Just-in-Time (JIT), Total Quality Management (TQM), Total Preventative Maintenance (TPM), and Human Resource Management (HRM) might assist businesses in addressing a variety of sustainability-related problems (Resta et al., 2016).

Even though a few businesses are already utilising some of these next-generation technologies, the benefits can only be realised when all the technologies are used simultaneously. Big Data, for instance, is a digital waste without AI or ML. Hence, companies aiming to increase economic sustainability must plan to deal with the loss of jobs to low- and medium-skilled workers. Prior to using this new technology, industries must consider personnel training. The development of transdisciplinary competences, reskilling and upskilling programmes, university collaborations to support new learning, and creative training methodologies (such as online and flexible learning modules) are a few examples. Because green energy is far more expensive than non-renewable resources, firms attempting to fulfil environmental sustainability goals using renewable energy sources may need help. Hence, industries must adopt a culture of sustainable awareness that emphasises all three sustainability components, and future research must be directed in that direction. In other words, businesses must place more emphasis on "how the business is done" than "how much is made" in order to preserve sustainable practices and a circular economy.

8.8 THE ROADMAP TO INDUSTRY 5.0-DRIVEN SUSTAINABILITY

The main objective of the research was to provide a roadmap that outlined how Industry 5.0 would promote sustainable growth, particularly in terms of resilience, human centricity, and environmental sustainability.

By performing a content-focused literature analysis, we could pinpoint the sustainability features of Industry 5.0. The functionality of Industry 5.0 for sustainability is shown by these findings, as well as the necessary functional development steps. However, it cannot be considered a sustainability blueprint since it lacks the knowledge of the contextual links between each pair of activities or the significance of each function in advancing Industry 5.0's SDGs.

Participants in Industry 5.0 who are interested in changing industries towards SDGs may utilise the path depicted in the following image as a model. This road map places a focus on using Industry 5.0 characteristics to advance sustainable goals (Majerník et al., 2022). As a result, this pathway aids in bridging the gap between Industry 5.0's stated sustainability goals and the underlying interactions and behaviours that make them possible. This roadmap outlines Industry 5.0's future ambitions for the industry, including its projects, partnerships, and goals. It provides leaders, Industry 5.0 participants, and decision-makers with the whole picture. Illustrations of the sequence in which the various Industry 5.0 functions ought to be developed or financed in order to maximise the system's sustainability advantages. With regard to societal and environmental boundaries, the current research intends to execute the Industry 5.0 industrial transformation framework to achieve sustainable development. The 16 sustainability responsibilities of Industry 5.0, which were created after conducting a literature analysis, serve as the active component of the strategy roadmap. In order to achieve the targeted goal, activities (or functions) should be developed progressively, as shown by the roadmap.

- Value network integration, intelligent automation, sustainable thinking, renewable integration, circular intelligent goods, and operational and resource efficiency are the main factors advancing the environmental sustainability goals;
- System interoperability and integration, data openness and sharing, service orientation, open, sustainable innovation, skill upskilling and reskilling, employee technical help, and customisation are the main forces behind Industry 5.0's human-centricity ambitions.

8.9 CHALLENGES OF INDUSTRY 5.0

It is easier to ignore any potential issues with Industry 5.0. The problems that need to be fixed in order for Industry 5.0 improvements to prosper for the company are currently being discovered.

1. In order to cooperate with makers of advanced robots and intelligent machines, humans must have competent abilities (Narvaez Rojas et al., 2021). Human employees frequently struggle to gain technical

capabilities in addition to the soft skills required (Liu et al., 2020). Among the challenging duties in the new professions that demand a high level of technological competence are managing translation and programming industrial robots.

2. Making use of advanced technology requires more time and effort from human workers. Collaborative robots, artificial intelligence, software-connected factories, real-time data, and the IoT are all necessary components of Industry 5.0. (Sun et al., 2020, Tripathy & Pattanaik, 2020, Wang et al., 2018b, Abdelmageed & Zayed, 2020).

3. It's essential to make investments in cutting-edge technology. The UR Cobot is pricey. Training people for new occupations has become more expensive. Companies are finding it challenging to upgrade their manufacturing lines for Industry 5.0 (Yin et al., 2020). Adopting Industry 5.0 is expensive since it necessitates sophisticated machinery and highly trained personnel in order to increase output and efficiency.

4. While the development of Industry 5.0 depends on ecosystem trust, security poses a challenge. Authentication is employed on a large scale in the business to communicate with various devices and defend against potential quantum computing applications while deploying IoT nodes (Ghobakhloo et al., 2021a; Kent & Kopacek, 2020; Rashidi et al., 2020). The use of automation and artificial intelligence in Industry 5.0 presents difficulties for businesses and calls for trustworthy security (Li et al., 2019, Tanveer, 2022, Tönnissen & Teuteberg, 2020, Tijan et al., 2019). Since ICT systems are at the heart of Industry 5.0 applications, strict security requirements are required to avoid security risks.

5. Humans are anticipated to contribute high-value activities to manufacturing regulations throughout the upcoming industrial revolution. Legalisation and standardisation will prevent significant problems between enterprises, society, and technology (Mourtzis, 2016).

8.10 DISCUSSION

The development strategy known as "Industry 4.0" confronts us as a way to attract a talented workforce to a country like Germany that feels inept and to bring back the production that went to far-off Eastern nations. When Industry 4.0 was first implemented, the goal was to reduce the amount of human labour and replace it with alternative technologies like sensors, robots, and artificial intelligence. Nevertheless, it was planned that through Society 5.0, the technology utilised in Industry 4.0 could be used by all of society by the SDGs, and objectives would be established in that manner. Even though they employ and build technology similarly, Industry 4.0 is focused on industrial output, whereas Society 5.0 is focused on the entire society. Industry 4.0 argues for using new technology to produce goods more cheaply, accurately, and quickly (Birkel & Müller, 2021). In order to keep their competitive edge and adapt to changing market conditions, organisations must implement Industry 4.0 by changing their organisational structures (Cirillo et al., 2021). Organisational structure modifications benefit enterprises and nations from an economic, ecological, and social standpoint (Rahman et al., 2020).

Reduced carbon emissions, improved recycling procedures, the production of more environmentally friendly goods, and more opportunities for workers to participate in management roles as opposed to physically demanding and contaminating jobs are just a few benefits provided by Industry 4.0 (Birkel & Müller, 2021; Rahman et al., 2020; Müller & Voigt, 2018; Chen et al., 2021). These changes are intended to benefit everyone, but Japan and Germany in particular. This will decrease the cost of goods made in Japan and Germany, increasing competitiveness and consumer pleasure. On the other hand, the impoverished nations would be most affected negatively by Industry 4.0 and Society 5.0.

8.11 IMPLICATIONS

The research tried to illustrate how Industry 5.0 may achieve its desired ideals for sustainable development. The study initially examined the ambiguities around this idea and created a model to use as a guide to comprehend this occurrence comprehensively. The reference model demonstrated how Industry 5.0 is a socio-technical phenomenon that depends on stakeholder cooperation and technology innovation to move beyond economic development and address contemporary environmental and social challenges. Disruptive and enabling technologies, both of which have evolved since the Third Industrial Revolution and have seen significant commercialisation, are primarily responsible for the creation of Industry 5.0's new technologies, which include the revolutionary technological advancements like CAI or adaptable robots. The research made an effort to demonstrate how Industry 5.0 may fulfil its principles for sustainable development. The research first looked at the ambiguity around this notion and developed the technological component of Industry 5.0, which also requires developing the fundamental principles of techno-functional design, such as decentralisation or real-time capabilities. These regulations enable the appropriate operation and usefulness of Industry 5.0's technology components. The effects of Industry 5.0 go well beyond the industrial sector and the deterioration of production facilities, in contrast to Industry 4.0, which usually emerges in smart factories. When deciding on the elements and scope of Industry 5.0, the sociocultural aspect of this phenomenon serves as an example of the change towards the stakeholder viewpoint. The hyperconnected value ecosystem of intelligent clients, stakeholders (including governments or labour unions), suppliers, adaptable industrial facilities, and technology providers must all be digitalised and integrated. Depending on the commercial and industrial environment, the particularity of Industry 5.0 components may change. For instance, intelligent patients, pharmaceutical companies, healthcare providers, hospitals, and insurance companies are some of the elements of Health 5.0, also known as the Healthcare Industry 5.0.

This chapter aimed to construct a thorough literature assessment and offer an incomplete picture of the more valuable components and technologies for the transition from Industry 4.0 to Industry 5.0 by developing a conceptual model in an area that has not yet been fully examined in the literature. In order to create circular intelligent goods that can self-map their whole lifespan and integrate into the circular economy, firms need to use Industry 5.0 technologies, concepts, and functionalities. In the long run, these features enable organisations to improve resource and operational efficiency, boosting economic development and lowering waste, energy use, and emissions. Finally, by doing these tasks, industrial value chains can acquire the required adaptability, resources, competencies, expertise, and innovative capacities to respond to market structural adjustments brought on by disruptions and build the resilience they need.

8.12 CONCLUSION, LIMITATIONS, AND FUTURE RESEARCH DIRECTIONS

With an emphasis on the future of Industry 5.0 and Society 5.0, a thorough review of the literature was conducted as part of this study project. This chapter sought to build a comprehensive literature assessment and provide a partial picture of the more valuable elements and technologies for the transition from Industry 4.0 to Industry 5.0 by developing a conceptual model in a field that has not yet been fully explored in the literature.

The current move to a digital industrial foundation, according to experts, has resulted in social and environmental problems. Industry 5.0 has evolved to support Industry 4.0 and address these issues. Sustainable development has been promoted by Industry 5.0. However, it must be made clear how Industry 5.0 will include concepts from sustainable development. The ideals of sustainable development, notably human centricity, environmental sustainability, and resilience, have the potential to be advanced by Industry 5.0. It's important to remember that several of the traits, like the ability to communicate in real-time and the ability to keep track of corporate activities, have been connected to technical developments since the Third Industrial Revolution and are not unique to Industry 5.0. However, the societal and technological advances enabled by Industry 5.0 have redefined the breadth (quality) and depth (scope) of these commitments.

Since Industry 4.0 is about digitisation, Industry 5.0 will concentrate on creating the partnership between the digital world, critical and creative thinking skills in people, being adaptable to changing market scenarios, and emphasising sustainability. In this chapter, Industry 5.0 has first been explored from an academic and industrial perspective, and then the elements of sustainability in Industry 5.0 have been discussed. It has also been discussed additional crucial Industry 5.0 enabling technologies. Additional research should be done on the notion of Industry 5.0, the human role in industrial digitalisation processes, and the development of economy 4.0 and society 5.0 for sustainable development, according to the study. Some people can object to the lack of study on the social, ecological, and economic impacts of the Industry 4.0 idea. Regarding Society 5.0's potential future behaviours, there are none.

Despite our best efforts, much still needs to be determined about how this phenomenon will develop. In order to further our work in the following ways, further research is encouraged. Owing to methodological limitations, we could only establish the contextual links between the functions by using the opinions of European experts. Although Industry 5.0 is primarily a European effort, it will have the same worldwide impact as Industry 4.0. Reconciling the perspectives of a broader range of specialists from diverse socioeconomic settings and resynthesising the connections between Industry 5.0's sustainable development functions might be a significant direction for future study. Despite the fact that Industry 5.0 is seen to have the ability to enable sustainable growth, this is not an autonomous occurrence. Our model described a hypothetical ideal situation in which all necessary components for creating Industry 5.0 were present.

To develop impartial and fair technical progress of value networks, this situation demands inclusive technology diffusion activities across a variety of disciplines and sectors. Industry 5.0 is only now beginning. Therefore, there is no plan in place for the creation of assistive technology. Future study is encouraged to examine the best ways to create a roadmap for Industry 5.0 technology implementation to support projects including collaborative technology diffusion across regions, open innovation, and shared research. Industry 5.0, a mostly technological phenomenon, further relies on value-adding network-wide connection, data-driven processes, and integration. Cybersecurity is a significant facilitator of this phenomenon because of Industry 5.0's rapid digital transition. Given the magnitude and complexity of cyber risks under Industry 5.0, further research on strong passwords is required.

According to the reference model, encouraging policies should be implemented in a variety of fields, such as taxation, government operations, energy, society, culture, innovation, and education are necessary to make the transition to this phenomenon. Individual parties (such as governments or labour unions) are unable to develop these policies collectively and synergistically; therefore, Industry 5.0 stakeholders must frequently coordinate on a bilateral basis.

It must be made clear how stakeholders may collaborate to create relevant laws and regulations that promote Industry 5.0. Future research must thus focus on how public–private stakeholder involvement may be used to develop Industry 5.0 supporting policies, initiatives, and regulatory frameworks. This encourages companies to align their operations with Industry 5.0's SDGs, which might be done by applying a revolutionary corporate governance paradigm. Even though the need for Corporate Governance 5.0 is widely acknowledged, additional knowledge about the development of such sustainable corporate governance frameworks is still required. By doing in-depth evaluations of how companies may incorporate Industry 5.0's principles for sustainable growth into their corporate strategies and using quantitative indicators to track their progress towards those objectives, future research might fill this knowledge vacuum.

The technologies covered in the section on Industry 5.0 enablers can appear to be modest upgrades or newer versions of Industry 4.0 technologies. Although somewhat accurate, a thorough examination of each technology would show that they place a major emphasis on considering people, the environment, and social responsibility. The two additional buzzwords of the decade that industries are attempting to adopt are ESG (Environmental, Social, and Governance) and sustainability (Padmakumar & Shunmugesh, 2022). Although they are distinct, according to the dictionary, ESG is a progression of sustainability. Both may be referred to as "doing good to the environment and society" in basic words, and there is no better way to do good to society than by incorporating humans. A survey of several industrial sectors (Brozzi et al., 2020) has shown that social and environmental sustainability are frequently neglected while economic sustainability is given greater attention, which stresses the necessity for enterprises to implement Industry 5.0 technology.

The development of sustainability is ESG. Simply put, both may be referred to as "doing good to the environment and society," and when it comes to society, nothing is more critical than the involvement of humans. Also, research reveals that no studies have evaluated how these technologies would affect social sustainability. Thus, that area has to be the focus of further study.

Instead of using Industry 5.0, use Sustainable Industry 4.0. It is premature to refer to the upcoming transformation as Industry 5.0, however, without fully understanding what is in store for us over the next few decades, given that Industry 4.0 technologies are just now beginning to penetrate small- and medium-sized businesses and developing countries in developed countries. It highlights once more that "Industry 4.0S" is the appropriate designation for emerging technologies.

BIBLIOGRAPHY

Abdelmageed, S., & Zayed, T. (2020). A study of literature in modular integrated construction-critical review and future directions. *Journal of Cleaner Production*, 277, 124044.

Akundi, A., Euresti, D., Luna, S., Ankobiah, W., Lopes, A., & Edinbarough, I. (2022). State of industry 5.0-analysis and identification of current research trends. *Applied System Innovation*, 5(1), 27.

Arias, R., Lueth, K. L. Rastogi, A. (2018). The effect of the Internet of Things on sustainability. *World Economic Forum*.

Aslam, F., Aimin, W., Li, M., & Rehman, K. (2020). Innovation in the era of IoT and industry 5.0: Absolute Innovation Management (AIM) Framework. *Information*, 11(2), 124.

Baden-Filler, C., & Haefliger, S. (2013). Business models and technological innovation. *Long Range Planning*, 46(6), 419–426.

Bansal, S. (2018). Industry 5.0 – Next Generation Customer Experience Redefined? Retrieved from https://www.thedigitaltransformationpeople.com/channels/customer engagement/industry-5-0-next-generation-customer-experience-redefined/

Beier, G., Ullrich, A., Niehoff, S, Reißig, M., & Habich, M. (2020). Industry 4.0: How it is defined from a sociotechnical perspective and how much sustainability it includes–A literature review. *Journal of Cleaner Production*, 259, 120856.

Biel, K., & Glock, C. H. (2016). Systematic literature review of decision support models for energy-efficient production planning. *Computers and Industrial Engineering*, 101(C), 243–259.

Biggeri, M., Clark, D. A., Ferrannini, A., & Mauro, V. (2019). Tracking the SDGs in an 'integrated' manner: A proposal for a new index to capture synergies and trade-offs between and within goals. *World Development*, 122, 628–647.

Birkel, H., & Müller, J. M. (2021). Potentials of industry 4.0 for supply chain management within the triple bottom line of sustainability—A systematic literature review. *Journal of Cleaner Production*, 289, 125612.

Breque, M., De Nul, L., & Petridis, A. (2021). *Industry 5.0: towards a sustainable, human-centric and resilient European industry*. Luxembourg, LU: European Commission, Directorate-General for Research and Innovation.

Brozzi, R., Forti, D., Rauch, E., & Matt, D. T. (2020). The advantages of industry 4.0 applications for sustainability: Results from a sample of Manufacturing Companies. *Sustainability*, 12(9), 3647.

Byerlee, D., & Fanzo, J. (2019). The SDG of zero hunger 75 years on: Turning full circle on agriculture and nutrition. *Global Food Security*, 21, 52–59.

Cabinet Office Government of Japan. Report on the 5th Science and Technology Basic Plan. Available online: https://www8.cao.go.jp/cstp/kihonkeikaku/5basicplan_en.pdf (accessed on 22 December 2020).

Calvo, T., Razafindrakoto, M., & Roubaud, F. (2019). Fear of the state in governance surveys? Empirical evidence from African countries. *World Development*, 123, 104609.

Campagnolo, L., & Davide, M. (2019). Can the Paris deal boost SDGs achievement? An assessment of climate mitigation co-benefits or side-effects on poverty and inequality. *World Development*, 122, 96–109.

Carmigniani, J., & Furht, B. (2011). Augmented reality: An overview. In: Furht, B. (ed.) *Handbook of Augmented Reality*. Springer, New York, NY.

Carvalho, A. M., Sampaio, P., Rebentisch, E., Carvalho, J. Á., & Saraiva, P. (2017). Operational excellence, organisational culture and agility: the missing link? *Total Quality Management and Business Excellence*, 30, 1495–1514.

Cervelló-Royo, R., Moya-Clemente, I., Perelló-Marín, M., & Ribes-Giner, G. (2020). Sustainable development, economic and financial factors that influence the opportunity-driven entrepreneurship. An fsQCA approach. *Journal of Business Research*, 115, 393–402.

Chandola, V. (2015). *Digital Transformation and Sustainability Study and Analysis.* Cambridge, Massachusetts.

Chang, V., Xu, Y. K., Zhang, J., & Xu, Q. (2020). Research on intelligent manufacturing development approach for China's local valve industry. *Smart and Sustainable Built Environment*, 10(2), 293–321.

Chen, M., Sinha, A., Hu, K., & Shah, M. I. (2021). Impact of technological innovation on energy efficiency in industry 4.0 era: Moderation of shadow economy in sustainable development. *Technological Forecasting and Social Change*, 164, 120521.

Ciccarelli, M., Papetti, A., Germani, M., Leone, A., Rescio, G. (2022).Human work sustainability tool. *Journal of Manufacturing System*, 62, 76–86.

Cirillo, V., Rinaldini, M., Staccioli, J., & Virgillito, M. E. (2021). Technology vs. workers: The case of Italy's Industry 4.0 factories. *Structural Change and Economic Dynamics*, 56, 166–183.

Dantas, T., De-Souza, E., Destro, I., Hammes, G., Rodriguez, C., & Soares, S. (2021). How the combination of Circular Economy and Industry 4.0 can contribute towards achieving the Sustainable Development Goals. *Sustainable Production and Consumption*, 26(4), 213–227.

Davies, R., Coole, T., & Smith, A. (2017). Review of socio-technical considerations to ensure successful implementation of Industry 4.0. *Procedia Manufacturing*, 11, 1288–1295.

De Pascale, A. Arbolino, R., Szopik-Depczyńska, K., Limosani, M., & Ioppolo, G. (2021). A systematic review for measuring circular economy: The 61 indicators. *Journal of Cleaner Production*, 281, 124942.

Deguchi, A., Hirai, C., Matsuoka, H., Nakano, T., Oshima, K., Tai, M., & Tani, S. (2020). What Is Society 5.0? In *Society 5.0: A People-Centric Super-Smart Society*; Springer, 1–23. ISBN 9789811529894.

Demir, K. A., Döven, G., & Sezen, B. (2019). Industry 5.0 and human-robot co-working. *Procedia Computer Science*, 158, 688–695.

Di Nardo, M., Forino, D., & Murino, T. (2020). The evolution of man-machine interaction: the role of human in industry 4.0 paradigm. *Production and Manufacturing Research*, 8(1), 20–34.

Dombrowski, U., Richter, T., & Krenkel, P. (2017). Interdependencies of Industrie 4.0 & Lean Production Systems – A use cases analysis. *Procedia Manufacturing*, 11, 1061–1068.

Doz, Y. L., & Kosonen, M. (2010). Embedding strategic agility A leadership agenda for accelerating business model renewal. *Long Range Planning*, 43(2 and 3), 370–382.

Elena, F. (2020). What are cobots and how will they impact the future of manufacturing? Ericsson blog, https://www.ericsson.com/en/blog/2020/5/what-are-cobots-and-the-future-of-manufacturing

Elim, H. I., & Zhai, G. (2020). Control system of multitasking interactions between society 5.0 and industry 5.0: A conceptual introduction & its applications. *Journal of Physics: Conference Series*, 1463(1), 012035.

ElMaraghy, H., Monostori, L., Schuh, G., & El Maraghy, W. (2021). Evolution and future of manufacturing systems. *CIRP Annals*, 70(2), 635–658.

European Commission. Industry 5.0. (2021). Towards a Sustainable, Human Centric and Resilient European Industry. https://ec.europa.eu/info/publications/industry-50_en

Fareri, S., Fantoni, G., Chiarello, F., Coli, E., & Binda, A. (2020). Estimating Industry 4.0 impact on job profiles and skills using text mining. *Computers in Industry*, 118, 103222.

Fatima, Z., Tanveer, M. H., Waseemullah, Z. S., Naz, L. F., Khadim, H., & Tahir, M. (2022). Production plant and warehouse automation with IoT and industry 5.0. *Applied Sciences*, 12(4), 2053.

Fehri, R., Khlifi, S., & Vanclooster, M. (2019). Disaggregating SDG-6 water stress indicator at different spatial and temporal scales in Tunisia. *Science of the Total Environment*, 694, 133766.

Felsberger, A., & Reiner, G. (2020). Sustainable Industry 40 in production and operations management: A systematic literature review. *Sustainability*, 12, 7982.

Gerencer, T. (2021). What Is Extended Reality (XR) and How Is it Changing the Future? 3 April 2021, https://www.hp.com/us-en/shop/tech-takes/what-is-xr-changing-world

Ghobakhloo, M., & Fathi, M. (2021). Industry 4.0 and opportunities for energy sustainability. *Journal of Cleaner Production*, 295, 126427.

Ghobakhloo, M., Fathi, M., Iranmanesh, M., Maroufkhani, P., & Morales, M. E. (2021a). Industry 4.0 ten years on: A bibliometric and systematic review of concepts, sustainability value drivers, and success determinants. *Journal of Cleaner Production*, 302, 127052.

Ghobakhloo, M., Iranmanesh, M., Grybauskas, A., Vilkas, M., & Petraitė, M. (2021b). Industry 4.0, innovation, and sustainable development: A systematic review and a roadmap to sustainable innovation. *Business Strategy and the Environment*, 30(8), 4237–4257.

Gotfredsen, S. (2016). Bringing back the human touch: Industry 5.0 concept creating factories of the future. Retrieved from http://www.manmonthly.com.au/features/bringing-back-the-human-touchindustry-5-0-concept-creating-factories-of-the-future/

Grabowska, S., Saniuk, S., & Gajdzik, B. (2022). Industry 5.0: improving humanization and sustainability of industry 4.0. *Scientometrics*, 127, 3117–3144.

Grybauskas, A., Stefanini, A., & Ghobakhloo, M. (2022). Social sustainability in the age of digitalization: a systematic literature review on the social implications of industry 4.0. *Technology in Society*, 70(8), 101997.

Gupta, S., Kar, A. K., Baabdullah, A., & Al Khowaiter, W. A. (2018). Big data with cognitive computing: A review for the future. *International Journal of Information Management*, 42, 78–89.

Hahn, G. J. (2020). Industry 4.0: A supply chain innovation perspective. *International Journal of Production Research*, 58(5), 1425–1441.

Hecklau, F., Galeitzke, M., Flachs, S., & Kohl, H. (2016). Holistic approach for human resource management in Industry 4.0. *Procedia CIRP*, 54, 1–6.

Herceg, I. V., Ku, V., Mijuškovi, V. M., & Herceg, T. (2020). Challenges and driving forces for Industry 4.0 implementation. *Sustainability*, 12(10), 4208.

Hermann, M., Pentek, T., & Otto, B. (2016). "Design Principles for Industrie 4.0 Scenarios: A Literature Review" *2016 49th Hawaii International Conference on System Sciences (HICSS)*.

Horn, P., & Grugel, J. (2018). The SDGs in middle-income countries: Setting or serving domestic development agendas? Evidence from Ecuador. *World Development*, 109, 73–84.

Hughes, L., Dwivedi, Y. K., Rana, N. P., Williams, M. D., & Raghavan, V. (2022). Perspectives on the future of manufacturing within the industry 4.0 era. *Production Planning Control*, 33(2–3), 138–158.

Hugos, M. H. (2009). Business Agility: sustainable prosperity in relentlessly competitive world (1st ed.). New Jersey: John Wiley & Sons, *IncIn IFIP international conference on advances in production management systems* (pp. 677–686). Cham: Springer.

Javaid, M., & Haleem, A. (2020). Critical components of Industry 5.0 towards a successful adoption in the field of manufacturing. *Journal of Industrial Integration and Management*, 5(3), 327–348.

Jiang Yuchen, Y., Shen, L., Kuan, L. H., & Okyay, K. (2021). Industrial applications of digital twins, *Philosophical Transactions of the Royal Society A Mathematical Physical and engineering Sciences*, 379(2207), 20200360.

Johansson, H. (2017) Profinet Industrial Internet of Things Gateway for the Smart Factory. Master's Thesis in Embedded Electronic System Design, Department of Computer Science and Engineering, Chalmers University Of Technology, University Of Gothenburg Gothenburg, Sweden.

John, K. K., Adarsh, S. N., & Pattali, V. (2020). Workers to super workers: A brief discussion on important technologies for industry 5.0 manufacturing systems. *In AIP conference proceedings*, 2311(1), 070025.

Kagermann, H. (2015). Change through Digitization—Value creation in the age of Industry 4.0. *Management of Permanent Change*, 2015, 23–45.

Kagermann, H., Wahlster, W., & Helbig, J. (2013). Recommendations for Implementing the Strategic Initiative INDUSTRIE 4.0, München.

Kansha, Y., & Ishizuka, M. (2019). Design of energy harvesting wireless sensors using magnetic phase transition. *Energy*, 180, 1001–1007.

Kaygin, E., Zengin, Y., & Topçuoglu, E. (2019). Endüstri 4.0′ akademik bakıₛs. Atatürk Univ. Iktis. ᐟIdari Bilim. Derg., 33, 1065–1081.

Kent, M. D., & Kopacek, P. (2020). Do we need synchronization of the human and robotics to make industry 5.0 a success story? In: *The International Symposium for Production Research*. Springer, 302–311.

Kumar, A., & Nayyar, A. (2020). Si 3-Industry: A sustainable, intelligent, innovative, internet-of-things industry. *A roadmap to Industry 4.0: Smart production, sharp business and sustainable development*, 1–21. https://doi.org/10.1007/978-3-030-14544-6

Kumar, A., Srikanth, P., Nayyar, A., Sharma, G., Krishnamurthi, R., & Alazab, M. (2020a). A novel simulated-annealing based electric bus system design, simulation, and analysis for Dehradun Smart City. *IEEE Access*, 8, 89395–89424.

Kumar, A., Rajalakshmi, K., Jain, S., Nayyar, A., & Abouhawwash, M. (2020b). A novel heuristic simulation-optimization method for critical infrastructure in smart transportation systems. *International Journal of Communication Systems*, 33(11), e4397.

Kumar, R., Gupta, P., Singh, S., & Jain, D. (2021). Human Empowerment by Industry 5.0 in Digital Era: Analysis of Enablers. *2nd International Conference on Future Learning Aspects of Mechanical Engineering, FLAME 2020*. Springer Science and Business Media Deutschland GmbH, 401–410.

Lamichhane, S., Egilmez, G., Gedik, R., Bhutta, M. K. S., & Erenay, B. (2021). Benchmarking OECD countries' sustainable development performance: A goal-specific principal component analysis approach. *Journal of Cleaner Production*, 287, 125040.

Lasi, H., Fettke, P., Kemper, H.-G.; Feld, T., & Hoffmann, M. (2014). Industrie 4.0. *Business and Information Systems Engineering*, 56(4), 239–242.

Lee, J., Bagheri, B., & Kao, H. (2015). Research letters: A cyber-physical systems architecture for Industry 4.0-based manufacturing systems. *Manufacturing Letters*, 3, 18–23.

Lee, J., Kao, H.-A., & Yang, S. (2014). Service innovation and smart analytics for industry 4.0 and big data environment. *Procedia Cirp*, 16, 3–8.

Li, B., Boiarkina, II, Yu, W., Huang H. M., Munir, T., Wang, G. Q., & Young, B. R. (2019) Phosphorous recovery through struvite crystallization: Challenges for future design. *Science of the Total Environment*, 648, 1244–1256.

Li, C., & Zhang, L. J. (2017). A blockchain based new secure multilayer network model for Internet of Things, *Proceedings of the IEEE International Congress on Internet of Things (ICIOT)*, IEEE, 33–41.

Liu, Y., Yuan, X., Xiong, Z., Kang, J., Wang, X., & Niyato, D. (2020). Federated learning for 6G communications: Challenges, methods, and future directions. *China Communications*, 17(9), 105–118.

Longo, F., Nicoletti, L., & Padovano, A. (2017). Smart operators in industry 4.0: A human-centered approach to enhance operators' capabilities and competencies within the new smart factory context. *Computer and Industrial Engineering*, 113, 144–159.

Longo, F., Padovano, A., & Umbrello S. (2020). Value-oriented and ethical technology engineering in Industry 5.0: A human-centric perspective for the design of the factory of the future. *Applied Sciences*, 10(12), 4182.

Lu, Y. (2017). Industry 4.0: A survey on technologies, applications and open research issues. *Journal of Industrial Information Integration*, 6, 232–248.

Lu, Y., Adrados, J. S., Chand, S. S., & Wang, L. (2021). Humans are not machines—anthropocentric human–machine symbiosis for ultra-flexible smart manufacturing. *Engineering*, 7(6), 734–737.

Lu, Y., Zheng, H., Chand, S., Xia, W., Liu, Z., Xu, X., & Bao, J. (2022). Outlook on human-centric manufacturing towards industry 5.0. *Journal of Manufacturing Systems*, 62, 612–627.

Luthra, S., & Mangla, S. K. (2018). Evaluating challenges to Industry 4.0 initiatives for supply chain sustainability in emerging economies. *Process Safety and Environment Protection*, 117, 168–179.

Ma, W., De Jong, M., De Bruijne, M., & Mu, R. (2021). Mix and match: Configuring different types of policy instruments to develop successful low carbon cities in China. *Journal of Cleaner Production*, 282, 125399.

Maddikunta, P. K. R., Pham, Q. V., Deepa, N., Dev, K., Gadekallu, T. R., & Liyanage, M. (2022). Industry 5.0: A survey on enabling technologies and potential applications. *Journal of Industrial Information Integration*, 26, 100257.

Majerník, M., Daneshjo, N., Malega, P., Drábik, P., & Barilová, B. (2022). Sustainable Development of the Intelligent Industry from Industry 4.0 to Industry 5.0, *Advances in Science and Technology Research Journal*, 16(2), 12–18.

McKay, V. I. (2020). Learning for development: Learners' perceptions of the impact of the Kha Ri Gude Literacy Campaign. *World Development*, 125, 104684.

Mi'skiewicz, R., & Wolniak, R. (2020). Practical application of the Industry 4.0 concept in a steel company. *Sustainability*, 12(14), 5776.

Mourtzis, D. (2016). Challenges and future perspectives for the life cycle of manufacturing networks in the mass customisation era. *Logistics Research*, 9, 2.

Mourtzis, D. (2022). *Design and Operation of Production Networks for Mass Personalization in the Era of Cloud Technology*, 1st ed.; Elsevier: Amsterdam, The Netherlands.

Mrugalska, B., & Wyrwicka, M. K. (2017). Towards lean production in Industry 4.0, *Procedia Engineering*, 182, 466–473.

Müller, J. (2020). Enabling Technologies for Industry 5.0: Results of a Workshop with Europe's Technology Leaders; Publications Office: Luxembourg, European Commission; Directorate-General for Research and Innovation, 8–10.

Müller, J. M. (2019). Assessing the barriers to Industry 4.0 implementation from a workers' perspective. *IFAC PapersOnline*, 52, 2189–2194.

Müller, J. M., & Voigt, K.-I. (2018). Sustainable industrial value creation in SMEs: A comparison between Industry 4.0 and Made in China 2025. *International Journal of Precision Engineering and Manufacturing-Green Technology*, 5, 659–670.

Müller, J. M., Buliga, O., & Voigt, K. I. (2018). Fortune favors the prepared: How SMEs approach business model innovations in Industry 4.0. *Technological Forecasting and Social Change*, 132, 2–17.

Muthuswamy, P. (2022). Investigation on sustainable machining characteristics of tools with serrated cutting edges in face milling of AISI 304 Stainless Steel. *Procedia CIRP*, 105, 865–871.

Nahavandi, S. (2019). Industry 5.0-a human-centric solution. *Sustainability*, 11(16), 4371.

Nayyar, A., Rameshwar, R., & Solanki, A. (2020). Internet of Things (IoT) and the digital business environment: a standpoint inclusive cyber space, cybercrimes, and cybersecurity. *The Evolution of Business in the Cyber Age*, 10, 9780429276484-6.

Nam-Chol, O., & Kim, H. (2019). Towards the 2 °C goal: Achieving Sustainable Development Goal (SDG) 7 in DPR Korea. *Resources, Conservation and Recycling*, 150, 104412.

Nanterme, P., & Daugherty, P. (2017). *Technology for People: The Era of the Intelligent Enterprise*. Accenture: Dublin, Ireland.

Narvaez Rojas, C., Alomia Peñafel, G. A., Loaiza Buitrago, D. F., & Tavera Romero, C. A. (2021). Society 5.0: A Japanese concept for a superintelligent society. *Sustainability*, 13(12), 6567.

Nayyar, A., & Kumar, A. (Eds.). (2020). *A roadmap to industry 4.0: smart production, sharp business and sustainable development*, 1–21, Berlin: Springer.

Nelles, J., Kuz, S., & Mertens, A. (2016). Human-centered design of assistance systems for production planning and control. In *Conference: IEEE International Conference on Industrial Technology (ICET) Location*: Taipei, Taiwan, 21–2099.

Ng, T. C., Ghobakhloo, M., Iranmanesh, M., Maroufkhani, P., & Asadi, S. (2022). Industry 4.0 applications for sustainable manufacturing: A systematic literature review and a roadmap to sustainable development. *Journal of Cleaner Production*, 334, 130–133.

Oesterreich, T. D., & Teuteberg, F., (2016). Understanding the implications of digitisation and automation in the context of Industry 4.0: A triangulation approach and elements of a research agenda for the construction industry, *Computers in Industry*, 83, 121–139.

Okafor-Yarwood, I. (2019). Illegal, unreported and unregulated fishing, and the complexities of the sustainable development goals (SDGs) for countries in the Gulf of Guinea. *Marine Policy*, 99, 414–422.

Olkiewicz, M., Wolniak, R., Grebski, E. M., & Olkiewicz, A. (2019). Comparative analysis of the impact of the business incubator center on the economic sustainable development of regions in USA and Poland. *Sustainability*, 11(1), 173.

Onday, O. (2019). Japan's Society 5.0: Going Beyond Industry 4.0. *Business and Economics Journal*, 10(2), 1–6.

Opoku, A. (2019). Biodiversity and the built environment: Implications for the Sustainable Development Goals (SDGs). *Resources, Conservation and Recycling*, 141, 1–7.

Özdemir, V., & Hekim, N. (2018). Birth of industry 5.0: Making sense of big data with artifcial intelligence, "the internet of things" and next-generation technology policy. *Omics: A Journal of Integrative Biology*, 22(1), 65–76.

Ozkeser, B. (2018). Lean Innovation Approach in Industry 5.0. *In the Eurasia Proceedings of Science, Engineering & Mathematics*, 2, 422–428.

Padmakumar, M., & Shunmugesh, K. (2022). Artificial intelligence based tool condition monitoring digital twins and industry 4.0 applications. *International Journal on Interactive Design and Manufacturing*.

Pereira, A. G., Lima, T. M., & Charrua-Santos, F. (2020). Industry 4.0 and Society 5.0: Opportunities and threats. *International Journal of Recent Technology and Engineering*, 8(5), 3305–3308.

Pilloni, V. (2018). How data will transform industrial processes: Crowdsensing, crowdsourcing and big data as pillars of Industry 4.0. *Future Internet*, 10(3), 24.

Posada, J., Toro, C., Barandiaran, I., Oyarzun, D., Stricker, D., de Amicis, R., Pinto, E. B., Eisert, P., Döllner, J., & Vallarino, I. Jr. (2015). Visual computing as a key enabling technology for Industrie 4.0 and Industrial Internet. *IEEE Computer Graphics and Applications*, 35(2), 26–40. https://doi.org/10.1109/MCG.2015.45

Purvis, B., Mao, Y., & Robinson, D. (2019). Three pillars of sustainability: In search of conceptual origins. *Sustainability Science*, 14, 681–695.

Rada, M. (2017). Industry 5.0 definition, 3 February 2017. Retrieved from https://www.linkedin.com/pulse/industrial-upcycling-definition-michael-rada

Rada, M. (2015). Industry 5.0 - from virtual to physical, 1 December 2015, https://www.linkedin.com/pulse/industry-50-from-virtual-physical-michael-rada, Accessed on 1 June 2017.

Rahman, S. M., Perry, N., Müller, J. M., Kim, J., & Laratte, B. (2020). End-of-Life in industry 4.0: Ignored as before? *Resources Conservation and Recycling*, 154, 104539.

Rai, S. M., Brown, B. D., & Ruwanpura, K. N. (2019). SDG 8: Decent work and economic growth—A gendered analysis. *World Development*, 113, 368–380.

Raja Santhi, A., & Muthuswamy, P. (2022). Influence of Blockchain technology in manufacturing supply chain and logistics. *Logistics*, 6(1), 15.

Rashidi, M., Mohammadi, M., Sadeghlou Kivi, S., Abdolvan, M. M., Truong Hong, L., & Samali, B. (2020). A decade of modern bridge monitoring using terrestrial laser scanning: Review and future directions. *Remote Sensing*, 12(22), 3796.

Renda, A., Schwaag Serger, S., Tataj, D., Morlet, A., Isaksson, D., Martins, F., & Giovannini, E. (2022). *Industry 5.0, a transformative vision for Europe: governing systemic transformations towards a sustainable industry*. European Commission, Directorate-General for Research and Innovation.

Rendall, M. (2017). The New Terminology: CRO and Industry 5.0. Retrieved from ttps://www.automation.com/automation-news/article/the-new-terminology-cro-and-industry-50

Resta, B., Dotti, S., Gaiardelli, P., & Boffelli, A. (2016). Lean Manufacturing and sustainability: an Integrated View. In: et al. In: *Advances in Production Management Systems. Initiatives for a Sustainable World. APMS 2016. IFIP Advances in Information and Communication Technology*, vol. 488. Springer, Cham.

Romero, D., Bernus, P., Noran, O., Stahre, J., & Fast-Berglund, Å. (2016). The operator 4.0: Human cyberphysical systems & adaptive automation towards human-automation symbiosis work systems. In *IFIP international conference on advances in production management systems* (pp. 677–686). Cham: Springer.

Romero, D., Stahre, J., & Taisch, M. (2020). The Operator 4.0: Towards socially sustainable factories of the future. *Computers and Industrial Engineering*, 139, 2–5.

Rozhenkova, V., Allmang, S., Ly, S., Franken, D., & Heymann, J. (2019). The role of comparative city policy data in assessing progress toward the urban SDG targets. *Cities*, 95, 102357.

Rüßmann, M., Lorenz, M., Gerbert, P., Waldner, M., Justus, J., Engel, P., & Harnisch, M. (2015). Industry 4.0: The Future of Productivity and Growth in Manufacturing Industries. *Boston Consultant Group*, 9, 54–89.

Sahoo, K. S., Tiwary, M., Luhach, A. K., Nayyar, A., Choo, K. K. R., & Bilal, M. (2021). Demand–supply-based economic model for resource provisioning in Industrial IoT Traffic. *IEEE Internet of Things Journal*, 9(13), 10529–10538.

Sala, S., & Castellani, V. (2019). The consumer footprint: Monitoring sustainable development goal 12 with process-based life cycle assessment. *Journal of Cleaner Production*, 240, 118050.

Saniuk, S., Grabowska, S., & Gajdzik, B. (2020). Social expectations and market changes in the context of developing the Industry 4.0 concept. *Sustainability*, 12(4), 1362.

Saraswa, V., Jacobberger, R. M., & Arnold, M. S. (2021). Materials science challenges to graphene nanoribbon electronics. *ACS Nano*, 15(3), 3674–3708.

Schwab, K. The Fourth Industrial Revolution: What It Means, How to Respond. Available online: https://www.weforum.org/agenda/2016/01/the-fourth-industrial-revolution-what-it-means-andhow-to-respond/ (accessed on 10 January 2020).

Sharma, R., & Arya, R. (2022). UAV based long range environment monitoring system with industry 5.0 perspectives for smart city infrastructure. *Computers and Industrial Engineering*, 168, 108066.

Sindhwani, R., Afridi, S., Kumar, A., Banaitis, A., Luthra, S., & Singh, P. L. (2022). Can industry 5.0 revolutionize the wave of resilience and social value creation? A multi-criteria framework to analyze enablers. *Technology in Society*, 68 (C), 101887.

Skobelev, P. O., & Borovik, S. Y. (2017). On the way from Industry 4.0 to Industry 5.0: From digital manufacturing to digital society. *Industry 4.0*, 2, 307–311.

Stock, T., & Seliger, G. (2016). Opportunities of sustainable manufacturing in Industry 4.0. *Procedia CIRP*, 40, 536–541.

Solanki, A., & Nayyar, A. (2019). Green internet of things (G-IoT): ICT technologies, principles, applications, projects, and challenges. In *Handbook of research on big data and the IoT* (pp. 379–405). IGI Global.

Sun, W., Wang, Q., Zhou, Y., & Wu, J. (2020). Material and energy flows of the iron and steel industry: Status quo, challenges and perspectives, *Applied Energy*, 268, 114946.

Tanveer, M. (2022). Supply chain and logistics operations management under the era of advanced technology. In: *Integrating Blockchain technology into the circular economy*. IGI Global, 126–136.

Tijan, E., Aksentijević, S., Ivanić, K., & Jardas, M. (2019). Blockchain technology implementation in logistics. *Sustainability*, 11(4), 1185.

Tönnissen, S., & Teuteberg F. (2020). Analysing the impact of blockchain technology for operations and supply chain management: An explanatory model drawn from multiple case studies. *International Journal of Information Management*, 52(C), 101953.

Tripathy, H. P., & Pattanaik, P. (2020). Birth of industry 5.0: "the internet of things" and next-generation technology policy. *International Journal of Advanced Research Engineering Technology*, 11(11), 1904–1910.

Venaik, A., Jain, S., & Nayyar, A. (2023). Industry 4.0—Its Advancement and Effects on Security of Whistle-Blowers on Dark Web. In *Industry 4.0 and the Digital Transformation of International Business*, 103–121, Singapore: Springer Nature Singapore.

Vokurka, R. J., & Fliedner, G. (1998). The journey toward agility. *Industrial Management and Data Systems*, 98(4), 165–171.

Wakunuma, K., Jiya, T., Aliyu, S. (2020). Socio-ethical implications of using AI in accelerating SDG3 in Least Developed Countries. *Journal of Responsible Technology*, 4, 100006.

Wan, J., Tang, S., Shu, Z., Li, D., Wang, S., Imran, M., & Vasilakos, A. V. (2016). Software-defined industrial Internet of Things in the context of Industry 4.0, *IEEE Sensors Journal*, 16, 7373–7380.

Wang, F. Y., Yuan, Y., Wang, X., & Qin, R. (2018a). Societies 5.0: A New Paradigm for Computational Social Systems Research. *IEEE Transactions on Computational Social Systems*, 5(1), 2–8.

Wang, Q., Liu, X., Liu, Z., & Xiang, Q. (2020). Option-based supply contracts with dynamic information sharing mechanism under the background of smart factory. *International Journal of Production Economics*, 220, 107458.

Wang, Y., Han, J. H., & Beynon-Davies P. (2018b). Understanding blockchain technology for future supply chains: a systematic literature review and research agenda. *Supply Chain Management*, 24(1), 62–84.

Welfare, K. S., Hallowell, M. R., Shah, J. A., & Riek, L. D. (2019). Consider the human work experience when integrating robotics in the workplace. In *2019 14th ACM/IEEE international conference on human-robot interaction (HRI)*, 75–84, IEEE.

Weyer, S. Schmitt, M. O., & Gorecky, D. (2015). Towards Industry 4.0 - Standardization as the crucial challenge for highly modular, multi-vendor production systems. *IFAC-PapersOnLine*, 48(3), 576–584.

Wolniak, R., Grebski, M. E., & Skotnicka-Zasadzien, B. (2019). Comparative analysis of the level of satisfaction with the services received at the business incubators (Hazleton, PA, USA and Gliwice, Poland). *Sustainability*, 11(10), 2889.

Xu, X., Lu, Y., Vogel-Heuser, B., Wang, L. (2021). Industry 4.0 and Industry 5.0—Inception, conception and perception. *Journal of Manufacturing Systems*, 61, 530–535.

Yin, Z., Zhu, L., Li, S., Hu, T., Chu, R., Mo, F., Hu, D., Liu, C., & Li, B. (2020). A comprehensive review on cultivation and harvesting of microalgae for biodiesel production: Environmental pollution control and future directions. *Bioresource Technology*, 301, 122804.

Yount, K. M., Crandall, A., & Cheong, Y. F. (2018). Women's age at first marriage and long-term economic empowerment in Egypt. *World Development*, 102, 124–134.

Yu, M., Lou, S., & Gonzalez-Bobes, F. (2018). Ring-closing metathesis in pharmaceutical development: Fundamentals, applications, and future directions. *Organic Process Research and Development*, 22(8), 918–946.

Zhang, Y., Wen, J., Qiuli, Q. I. N., Hao, Y. U., Leminen, S., Rajahonka, M., & Chan, H. C. Y. (2015). How smart, connected products are transforming companies. *Blog.Prossess.Com*, 4(4), 5–14.

Zhong, R. Y., Xu, X., Klotz, E., & Newman, S. T. (2017). Intelligent manufacturing in the context of Industry 4.0: A review. *Engineering*, 3(5), 613–630.

Zhou, K., Liu, T., & Zhou, L. (2015). Industry 4.0: Towards future industrial opportunities and challenges. In *Proceedings of the IEEE 2015 12th International Conference on Fuzzy Systems and Knowledge Discovery (FSKD)*, Zhangjiajie, China, 15–17, 2147–2152.

Zunino, C., Valenzano, A., Obermaisser, R., & Petersen, S. (2020). Factory communications at the dawn of the fourth industrial revolution. *Computer Standards and Interfaces*, 71, 103433.

Chapter 9

Reverse logistics and green supply chain in terms of sustainability

An application in an enterprise

Ayşenur Erdil

Istanbul Medeniyet University, Türkiye

9.1 INTRODUCTION

Over several periods, businesses have been exposed to Supply Chain Management (SCM). It has gotten more emphasis, and new ideas have emerged to adapt to an evolving world. Supply chain studies have been researched from a variety of viewpoints, including operation studies, optimization, modeling, and management. As a result of technological developments, related supply chain concepts like Material Requirements Planning (MRP), Manufacturing Resource Planning (MRP II), Just-in-Time, Enterprise Resource Planning (ERP), Lean Manufacturing, Customization, Six Sigma, and Radio Frequency Identification (RFID) have undergone continuous transformations. The term "green supply chain management" (GSCM) refers to one of the most current trends in recent years. The attention to protecting the environment is growing in importance over time, owing mostly to climate change and humans' ongoing damage to the environment. Currently, as the effects of global warming begin to influence everyday life, citizens, and consequently businesses, are attempting to do their share to ensure the rescue of the environment. Designers are discussing the incentives and motivations that cause entities to distinguish themselves, regardless of whether human beings are dealing with an individual or a firm. Although people might well be driven by emotional rewards to begin assisting in the slowing of climate change, corporations require more complicated and interest-based motives, such as money benefits, reputation, brand recognition, or compliance with applicable laws. Such ecological activities, nevertheless, are not confined to groups. Application at the organizational level is inadequate to accomplish environmental objectives. Cooperation among a network of companies or between governing agencies and groups is prevalent. Cooperation is an essential element of SCM that may occur both inside and between enterprises. Since SCM encompasses all product-related business tasks, implementing environmental initiatives at the supply chain layer may be more useful than at the organizational level. The supply chain takes into account the complete product chain from manufacturers, suppliers, distributors, transporters, retailers, and shops to consumers. As a

DOI: 10.1201/9781003489269-9

result, incorporating environmental ideas into SCM is more effective and gives greater advantages to the entire supply chain while also helping individuals (Azevedo et al., 2011; Srivastava, 2007; Sarkis, 2003).

The concept of reverse logistics is general; it is defined as processes of efficient and effective planning, implementation, and control of secondary material warehouses, material flow, and related information in order to recover or destroy the material in the opposite direction of the traditional supply chain. However, this concept is also known as environmentally friendly logistics in terms of recycling and re-producing unwanted materials (waste materials, boxes, bottles, paper, etc.) (Koban and Keser, 2013; Fleischmann et al., 1997).

Green logistics is concerned with environmental protection. Preserving natural sources and discovering renewable sources involves decreasing carbon emissions and waste, as well as lowering a corporation's total carbon footprint. The reasons companies choose to *go green* are that there is now a market in this area and green strategies give the company a competitive advantage. Customers today want businesses to be green. In his study, Srivastava (2007) concluded that customers demand greener products and packaging for the environment. Author stated that some consumers arte willing to pay higher prices for environmentally friendly products and want to know more about content, usage waste, and recyclability.

Reverse Logistics (RLs) is one of the important processes in the supply chain in the current period. Reverse logistics studies are important for businesses due to ecological and economic reasons, corporate and social responsibilities, laws, sustainable development for industry, protection of natural resources, and less material and resource consumption. In this case, systematic handling of reverse logistics has become an ecological, saving, and legal obligation. The process of collecting returned, expired, or unused products due to the purchase of a new product has a strategic importance as it can create a serious cost for the manufacturer. It is important to focus on how the products will be collected, then grouping, sorting, and maintenance of the collected products, and through which channels the grouped and separated products, raw material sources, or wastes will be transported to the target point. Reverse logistics is involved in many industries such as automotive, electronics, computer, chemical, pharmaceutical, and medical. RL is the control activities by planning and applying an effective method to ensure that the raw material, semi-processed, final product, and its information are disposed of properly, from the point of consumption to the point of production. The most well-known idea about reverse activities is that the used product is returned from the end consumer to the manufacturer. RLs are also known as environmentally friendly logistics by recycling and reusing (waste material, cans, bottles, paper, etc.) that is not intended to be used, and its evaluation (Kaçtıoğlu and Şengü, 2010; Kaymak, 2010; Dinç et al., 2008).

If the concept of reverse logistics is evaluated as a process, it is a process in which a flow, in which each stage is planned, is applied and controlled in order to reconvert and reproduce any product, information, or material from the consumption area where it reaches the end consumer to the starting area. For a more comprehensive explanation, it is necessary to exemplify the term in question and if we need to specify the purposes of reverse logistics, we can say that the primary purpose of the companies is to provide the management of all goods and services in all their lives and all life cycles in their lives. It is obvious that the concept of forward logistics, that is, familiar logistics, is actually quite different from the concept of reverse logistics. The important difference that should be mentioned at this point is recycling. The mentioned recovery may be the company that produces the product itself or the recycling facility located in the supplier. RLs have been defined as a concept that organizes the processes that include recycling, remanufacturing, resale, reuse, and destruction processes to provide gains to businesses. Carrying out the necessary purchasing activities in order to organize the waste materials, and outdated products and make them reusable before and after production; the management of the processes related to the return or return of the products is organized with the reverse logistics system (Çetik and Batuk, 2013; Kocabasoglu et al., 2007; James et al., 2000; see Figure 9.1).

Universal Collection Center Operations: These include collecting used products from product owners, assessing the situation of the returns through investigation and/or separation, remanufacturing the rates of return to acquire their residual balance, disposing of returns that are found to be impossible to recover due to financial and/or technological purposes, and redistributing the recoverable products. Product Recovering Options are listed below (Pokharel and Mutha, 2009; Aras et al., 2008).

   (i) Repairing: When a product is maintained and ready for sale refurbishing: when an item is sanitized and fixed to bring it to a "like new" condition with more labor required in the restoration
  (ii) Remanufacturing: Remanufacturing is comparable to refurbishment but involves more substantial effort; it sometimes necessitates totally dismantling the product.
 (iii) Cannibalization: Cannibalization occurs when pieces or components from one product are removed and utilized to restore or construct another piece of the same product.
  (iv) Recycling: Recycling is the process of reducing a product to its fundamental ingredients, which are then reused.

There are similarities and distinctions between Green Supply Chain (GSC), Green Transportation, and Reverse Logistics. Reverse logistics focus on product re-evaluation and reuse. Green logistics and GSC are concerned with issues such as selecting a more ecologically friendly mode of transportation and decreasing packaging. With green logistics, the company considers more than just profit. More crucial for the corporation is the company's image (Rogers and Tibben-Lembke, 2001).

0	0	0	0	0	0	0	0	0	0	0	0
1	1	1	1	1	1	1	1	1	1	1	1
2	2	2	2	2	2	2	2	2	2	2	2
3	3	3	3	3	3	3	3	3	3	3	3
4	4	4	4	4	4	4	4	4	4	4	4
5	5	5	5	5	5	5	5	5	5	5	5
6	6	6	6	6	6	6	6	6	6	6	6
7	7	7	7	7	7	7	7	7	7	7	7
8	8	8	8	8	8	8	8	8	8	8	8
9	9	9	9	9	9	9	9	9	9	9	9

Figure 9.1 Scope of reverse logistics in terms of waste management.

Source: Figure is modified from literature of Kaçtıoğlu and Şengül (2010) and Pokharel and Mutha (2009).

9.1.1 Objectives of the chapter

The current study's goal is to demonstrate GSCM as a significant research field for construction management. Primarily and foremost, the differences between SCM and GSCM are presented and the literature research on GSCM and its subsystems are integrated, which includes waste management, reverse supply, and remanufacturing. Furthermore, determinants of GSCM have been described, which are used to evaluate the success of GSCM in any industry or marketplace. In this study, a mathematical model approach to decision-making for the issue is provided and the obstacles to GSCM are discussed. The present study helps in identifying the initial phase drivers of its adoption, especially for attaining objectives and sustainability of GSCM and RLs. The study discusses the following research issues: (i) To identify the product recovering options and factors of RLs and GCSC toward attaining sustainability and sustainable development of companies,

(ii) to find the inter-relationship among the identified potential drivers, to evaluate the success of GSCM in any industry or marketplace, and (iii) to analyze the obstacles to RLs and GSCM for a mathematical model approach to decision-making for the sustainability of a business.

9.1.2 Organization of chapter

The chapter is organized as follows: Section 9.2 presents information about research methodology, the importance of GSCM for the development of the industry, and the company as a significant scientific scope for future research in development organization and process for prioritization of the identified product recovering option, drivers, and factors of GSCM and RLs. In addition, the section elaborates and comprises the subjects to make a distinction between SCM, RLs and its subsystems, and GSCM, the research on the management of waste, reverse supplying, green logistics, and remanufacturing in GSCM and its subsystems. Section 9.3 presents GSCM criteria, which could potentially be included to evaluate the efficacy of GSCM in almost any industry and environment. This section also covers and highlights the GSCM's difficulties and obstacles, determining the right amount, cost reduction–optimization strategies, and ideal locations for central return centers are crucial considerations in RLs and GSCs. Section 9.4 suggests and discusses a mathematical model approach to decision-making (multi-criteria decision-making) for the problem of GSCM to increase the productivity of the business and the sector. Finally, Section 9.5 concludes the findings of this study along with limitations and future scope.

9.2 LITERATURE REVIEW

Due to rising environmental management concerns, several writers have alluded to the GSC throughout the last decade. The introduction of corporation environmental management, ecologically minded industrial strategy, and SCM research in the early 1990s fueled the emergence of GSC research (Zhu and Sarkis, 2006).

9.2.1 Supply Chain Management

SCM is the control of product and service transport facilities, and it encompasses all procedures that convert raw resources into finished goods. It entails purposefully simplifying a company's supply-side activities in order to increase customer involvement and obtain a competitive edge in the market place.

The inside of Sustainable Supply Chain Management (SSCM) comes in second, with the downstream SSCM making the most contribution to the company's performance. The upstream SSCM makes the smallest contribution. Additionally, the link between SSCM firm performance and publication reputation may evolve. They first integrate SSCM with the innovation component in addition to expanding the foundation for SSCM and company performance. These results also assist managers in better understanding how to compromise while using SSCM techniques (Wang et al., 2023).

The mediated channel via which the circular economy capabilities of technological advances are used can support the beneficial impacts of supply chain risk management and cooperation while reducing the negative consequences of combining supply chains. These mediating pathways distinguish enterprises with uneven development and are more prominent in populations with slow growth. It offers a chance to employ digital technology to lessen the detrimental effects of the integration of supply chains on circular economy capabilities while enhancing the good effects of supply chain risk management and cooperation (Yuan and Pan, 2023).

Researchers and decision-makers have focused a lot of emphasis on GSCM in the commercial and manufacturing sectors. It comprises both conceptual and practical techniques, ranging from lean manufacturing, integrated quality management, and relationship with suppliers' leadership at the level of the micro to institutional pressure at the organizational scale. As a consequence of social pressure or to meet their ethical and ecological commitments, an increasing number of firms are embracing sustainable growth strategies. The purpose of this research is to examine existing research and provide a theoretical structure that incorporates an efficient developing theories methodology between the three essential features of advantages in competitiveness, business performance, and worldwide SCM methodologies. The investigation used bibliometric analyses to conduct a full review of GSCM works and examine its scientific structure, published trends, and topics (Saini et al., 2023).

More expansive sustainability concerns, including climate change, have been overlooked and even worsened by concentrating on decreasing interruptions in supply chains, enhancing the resilience of supply chains, and increasing supply chain sustainability outcomes. In order to improve sustainability leadership by recognizing the interests of supply chains and associated issues related to sustainability, the present chapter first discusses the complementary relationships among the theory of stakeholders and the practice of oversight of SSCM from the point of view of systems. Next, it cultivates a novel supply chain concept. By connecting stakeholder theory with SSCM, it is feasible to gain a comprehensive awareness of sustainability concerns and spot possibilities to enhance one's choices, deeds, and present consumption and production behaviors. The supply chain perspective centers on the procedure of identifying stakeholders around the good or service being offered rather than maintaining a firm- or client-centered viewpoint. It clearly states participants in the Supply Chain (SC), whether they are located forward, downward, inside, or outside the focus company. All businesses and individuals are urged to use systems approach techniques in order to develop a sustainable mentality and increase their capacity to address ethical and environmental problems. The suggested SC perspective aids in efficiently recognizing stakeholders and comprehending sustainability concerns associated with manufacturing and eating for managers, politicians, trainers, consultants, shoppers, and individuals. Specifically, for the ethical decision-making and behavioral sciences, this expands the present understanding of sustainable management through a supply chain viewpoint and offers up new study topics (Fritz, 2022).

A multi-level organizational framework is created using an incorporated, two-stage technique that combines interpretative structural modeling and modeling of structural equations in order to determine the connection between Industry 4.0 technology, GSC procedures, and GSC effectiveness. The research showed that Industry 4.0 technology had an indirect impact on GSC performance using GSC practices and that this relationship was stronger than the direct impact of Industry 4.0 and GSC procedures in the automobile supply networks. Regarding the efficient adoption of GSC procedures, future supply chains ought to concentrate on advancing and connecting technologies like the Internet of Things (IoT), Cyber-Physical Systems (CPS), and Blockchain. Technological disruptions have a significant impact on GSC procedures, particularly reverse logistics and green buying, which are essential for driving improvements in GSC performance. Organizations can profit from making decisions for greater environmental efficiency by recognizing and connecting important Industry 4.0 technologies with GSC standards (Ghadge et al., 2022).

Hugos (2010) defined Supply Chain Management as "the coordination of manufacturing, storage, placement, and transportation among supply chain actors to achieve the greatest mix of speed and flexibility for the marketplace being serviced." The SCM idea arose in the 1980s. Previously, the method was classed as part of the foundation's operating and logistics division. Porter (1985) stands out amongst the foundational philosophers who created the principles integrated into what is recognized as SCM, having laid the basis through which all others have developed. With the development of the value chain concept for industry, Porter launched a three-decade-long movement that transformed the approach of exploiting the supply chain to improve profits (Hugos, 2010; Porter, 1985; see Figure 9.2). Green SCM identifies the disproportionate influence on the environment of a foundation's supply chain procedures as Figure 9.2.

SCM may be divided into five important regions which the administrator should comprehend and steer. These are presented below (Hugos, 2010; Porter, 1985).

> *Production*: Production is divided into two categories in SCM – production planning, which needs a grasp of the product technique and desired usage, and production scheduling, which includes the SCM group to possess a good comprehension of supply and demand and demand satisfaction (Hugos, 2010).

Question 1 :		A	B	C	D
Question 2 :		A	B	C	D
Question 3 :		A	B	C	D
Question 4 :		A	B	C	D
Question 5 :		A	B	C	D
Question 6 :		A	B	C	D
Question 7 :		A	B	C	D
Question 8 :		A	B	C	D
Question 9 :		A	B	C	D

Figure 9.2 Green supply chain management.

Source: This figure is adapted from literature of Hugos (2010).

Inventory: Inventory is used in SCM to provide a pillow of material expected to perform each phase of the manufacturing operation. Inventory's significance could contribute to a prolonged commitment to raw resources, necessitating a tracking mechanism that maintains appropriate stock volumes. Current SCM advancements have concentrated on developing lean thinking in the profession (Hugos, 2010; Vollmann et al., 2005).

Location: Executives should concentrate on where production and supplies must be put, utilizing accurate methods that result in a high-performing SCM. The technology sector pioneered the "just-in-time" or "build-to-order" supply chain approach. Supplies are purchased as required from firms with warehousing near the production site in a particular model (Hugos, 2010; Christensen et al., 2005).

Transportation: According to Hugos (2010), transportation is the procedure whereby the product is transferred from one manufacturing warehouse to another or is supplied to a customer, and it plays a significant part in SCM. To be competitive in this field, it is necessary to give the most convenient methods of accessing while maintaining cost management and productivity improvement.

Information: A leader may successfully organize product choices focused on supply demands with excellent, accurate information (Vollmann et al., 2005). Each of these major roles has contributed to the advancement of SCM. Considering each phase of the supply chain enables executives to make production design decisions. Executives who implement Porter's attitudes reform may ensure that the corporation's profitability expands. These stages also assist executives in properly directing the traditional SCM methodology toward green supply-chain planning.

Today's business community and other sectors of society have made sustainability a very popular term. It is hard, for instance, to pass by a supermarket without spotting at least one magazine front depicting renewable energy resources, global warming concerns, or the recognizable polar bear resting on a thin layer of ice. A range of variables, including the supply and demand dynamics affecting energy consumption, a better understanding of the science behind climate change, and improved organization-wide transparency with respect to their environmental and social policy initiatives, are fueling the growing importance of sustainability. The majority of the research on supply chains that many might well take into account to be a component of Corporate Social Responsibility (CSR) and sustainability has taken place in isolation, with little or no acknowledgment of the interconnections between subjects such as the surroundings, diversity, human rights, charitable work, and safety, and the reality that these are truly elements of the greater, more holistic notions of CSR and sustainability just like they pertain to supply chain (Carter and Jennings, 2002).

The rapidly expanding research on technologies associated with Industry 4.0 and their consequences for supply networks contain both excellent insights and significant complexity. While previous Systematic Literature Reviews (SLRs) started consolidating the research, an SLR that concurrently (a) encompasses several core technologies associated with Industry 4.0, (b) composes their beneficial and detrimental effects on the performance of supply chains in a broad notion, and (c) profiles for the important variables that promote or inhibit these potential ramifications has still been lacking. By performing such an SLR, researchers participate in the establishment of an accumulating body of understanding (Rada et al., 2022). Predictive analysis and operational intelligence are used in Industry 5.0 to construct frameworks that strive to make more accurate and less volatile judgments. The bulk of the manufacturing procedure would be computerized in Industry 5.0 since real-time data from devices would be gathered in partnership using highly trained people. In comparison to Industry 4.0, Industry 5.0 is viewed as the next industrial revolution. Its goal is to employ the inventiveness of human specialists in conjunction with efficiency, intelligent, and precise devices in order to get wealth and user-preferred producing processes (Maddikuntaa et al., 2022).

The research of Xu et al. (2021) seeks to establish the terrain co-occupied by Industry 4.0 and Industry 5.0, as well as to offer some light on potential responses to these problems and maybe others. To keep things simple, this platform offers no endeavor to describe or explain what an Industrial Revolution, or Industry 4.0 or Industry 5.0, is. Industry 5.0 is the product of foresight research, a method of defining how industry and rising society trends and requirements would cooperate. As a result, Industry 5.0 supplements and expands on the key characteristics of Industry 4.0. This could assist in distinguishing Industry 5.0 from the others, recognizing that the earlier industrial breakthroughs are the historical progression of their counterparts. Along with Industry 5.0, a more immature word, the key linked ideas are regarded as widely recognized. The discussion over Revolution versus Evolution remains currently underway and falls outside the topic of this chapter (Xu et al., 2021).

The study of Tayal et al. (2021) combined information systems, Blockchain (BC), and SCM. The suggested approach addresses the limitations of the Food Supply Chain (FSC) and would inspire academics and businesses to embrace and accomplish efficient and successful FSC objectives. BC connections have spread across a wide range of industries. Authors received a lot of interest because of their ability to create secured individual and commercial networks in a distributed fashion. The traditional FSC does not fulfill consumer demands for quality, safety, dependability, transparency, and traceability. This study describes a novel three-stage method for constructing an interconnecting BC using the FSC that incorporates principal component analysis (PCA), Total Interpretive Structural Model (m-TISM), and Matrice d'Impacts Croisés Multiplication Appliquée à un Classement analysis (MICMAC). The competitive profile matrix is utilized to generate the MICMAC analysis, which can be utilized to determine the driving and reliance power on the inter-related drivers (Dwivedi et al., 2023; Tayal et al., 2021).

In order to deliver the initial conversation on Industry 5.0, researchers want to present a survey-based instructional on possible applications and infrastructure networks of Industry 5.0 in this study. Researchers outline various research problems and outstanding concerns which need to be addressed further in order to achieve Industry 5.0. SCM is the regulation of the transportation of products and services, and it includes all processes that transform raw materials into completed commodities. It comprises deliberately streamlining a corporation's supply-side procedures in order to improve customer engagement and gain a competitive advantage in the market system (Nahavandi, 2019). Individuals would cooperate and work alongside robotic systems without concern of employment instability, culminating in value-added services. Nahavandi (2019) explained the principles, characteristics, and current technology of Industry 5.0.

Within the study of Sahoo et al. (2021), researchers offered a demand-supply-based economic model for capturing interactions in Enterprise Resource Planning – Internet of Things (ERP-IoT) systems that operate like cloud micro services. The goal was to boost the usefulness of service providers. This costing model was built on demand and supply fundamentals, with prices given to ERP-IoT flows based on actual bandwidth requirements and resource availability. From the perspective of link utilization, flow queuing utilization, service providers, and customer satisfaction, as well as service providers' efficiency, the suggested technique surpasses the relevant components. The suggested queue scheduler optimizes the assignment of queues to ERP-IoT traffic patterns by taking service-level agreement (SLA) criteria into account (Sahoo et al., 2021).

Technologies such as 5G, Machine Learning (ML), the IoT, and Edge Computing (EC) that are linked with human intelligence and development could assist companies in fulfilling demand and offering individualized and customized goods at a quicker rate (Li, 2020). This helps SCM incorporate mass customization, a key concept in Industry 5.0, into their production operations. Finally, Industry 5.0 is a concept that aims to continually align the work environment and efficiency of humans and robots. Industry 5.0, enabled by a range of developing applications and supporting technology, is predicted to boost industrial output and consumer satisfaction. Researchers actually discussed a variety of obstacles and unresolved topics, such as security, confidentiality, and human–robot collaboration in manufacturing. Scalability and skilled labor should be addressed in order to better implement the notion of Industry 5.0 in coming decades (Maddikuntaa et al., 2022).

To secure business resilience in the post-COVID-19 future, decision-makers should rearrange their existing SC infrastructures, with the help of cutting-edge technology development and deployment. Lately, Industry 5.0 (I5.0) has acquired prominence as a framework that provides distinguishing elements for the development of robust and inclusive processes while maintaining long-term SC sustainability. This investigation seeks to investigate the problems of applying I5.0 when controlling the impact of SC disruptions caused by the COVID-19 pandemic in a developing economy. This study used both qualitative and quantitative methods. This research is intended to assist managers and decision-makers in effectively overcoming the problems of adopting I5.0 in SCs, hence improving SC sustainability (Karmaker et al., 2023).

Blockchain, the IoT, and Artificial Intelligence (AI) Technology for SCM highlights the challenges that supply chain operators face, as well as possible alternatives to such issues and challenges. This would then be the first manuscript that attempts to identify the impact of COVID-19 on supply chain structures comprising diverse decision-makers such as manufacturers, distributors, and producers, and it would then lay the groundwork for future studies' chances that would permit unregulated corporate investment and wealth. This would provide the basis for researchers, scientists, and educators concerned with applying current SCM innovations such as the IoT, AI, and Blockchain. Apart from researchers, undergraduates and professional college pupils may utilize this chapter as a reference guide. This chapter also offers the public audience a broad viewpoint on themes such as computer science, the food industry, hotel administration, design, healthcare, inventory control, and agricultural areas (Chawla et al., 2023).

9.2.2 Green supply chain management

GSCM is the concept of combining sustainable ecological activities-processes within the traditional supply chain. These processes include product design, material sourcing and selection, production and manufacture, operation, and end-of-life management. In order to boost revenues in an already competitive market, enterprises have increasingly focused on measures to manage expenses. One approach to achieve this aim is to transition from a traditional supply chain to a green one.

The research of Das and Nayyar (2020) examined the conduct of founding fathers contributing to the advancement of green steadfastness in the Indian retail market, utilizing similar research of these measurement methods in four general merchandise layouts: nourishment, do-it-yourself (DIY), electronic and family unit apparatuses, and shape and running shoes. The results show that in this expanding business industry, social predecessors differ throughout the analyzed retail arrangements in generating green fidelity, which refers to a challenge for shopkeepers in their quest to recruit, satisfy, and attach customers to their retail arrangements and establishments.

Previous research on GSCM was collated and categorized by Srivastava (2007). GSCM is defined as incorporating an environmental cognition procedure into supply chain planning, which contains product design, material procurement and classification, manufacturing processes, final product transporting to customers, and product end-of-life strategic planning after its forecasted lifetime. According to this description, GSCM refers to a broad spectrum of manufacturing processes, from product design through recycling or destruction (Srivastava, 2007).

Being green thereby influences ecologically conscious customers' perceptions of the firm (H'Mida, 2009), which benefits the businesses and, as a result, their supplier chains. Ecologically concerned customers can choose green businesses, boosting the business's (and its supply chain's) competitiveness and business success (Kumar and Malegeant, 2006; Rao and Holt, 2005). Consequently, in order for the GSC to be relevant and lucrative, the initiatives must be properly identified and compensated by customers and investors.

Pati et al. (2008), while trying to minimize the amount of waste paper of unsuitable quality as well as the total system cost, maximize the amount of other scanned recycled waste paper. Authors established a multi-objective programming model to do it.

Louwers et al. (1999) and Matthew et al. (2004) worked on carpet recycling; for example, in their study, authors developed a facility location and assignment problem for the collection, reprocessing, and redistribution of carpet materials and applied the model in Europe and the USA.

Extreme climate change that has affected all of mankind in recent years is mostly linked to human activities. The ongoing climate warming, which offers huge challenges and threats to the whole human race in the form of global warming, earthquakes, hurricanes, storm surges, and floods, is mostly caused by harmful pollutants. The purpose of this study is to propose a foundational framework for GSCM/SSCM which adds to the Systems Theory (ST) and Knowledge-Based Theory (KBT), as well as a thorough outline of future research opportunities. The researchers have put up a fundamental paradigm for a sustainable supply chain network and have provided future research suggestions in the conclusion. The present study is an endeavor to construct a conceptual framework centered on KBT. The research contributes to the expansion of previous efforts that require a theory-focused perspective (Dubey et al., 2017).

9.2.2.1 Importance of green supply-chain management

As GSCM expanded in popularity and drew more emphasis from scholars, the initial research in the field concentrated on the approach's major influence on the well-being of the surroundings in which organizations operate. Most of the original research concentrated primarily on the green paradigm as an economic preservation strategy. Porter and van der Linde (1995a, 1995b) addressed the fundamentals of the green revolution, which included (a) boosting supply economies, (b) lowering waste, and (c) enhancing productivity. According to Srivastava (2007), three methods for GSCM have appeared: reactive, proactive, and importance. The reactive method takes the least amount of supply development for the industry and often entails revising product marking and investigating strategies to reduce the environmental effect of production. The comprehensive approach is a mid-level activity in which businesses spend minimal money in an attempt to be environmentally conscious and focus research and design on producing green and sustainable goods while building a recycling program. The last tactic is value-seeking, in which firms emphasize the use of ISO regulations and a green purchasing strategy.

The 4R1D (reduce, reuse, recycle, reclaim, and degradable) idea must be followed when incorporating sustainable ecological operations into conventional supply chains from manufacturing to operations to end-of-life management. Reducing the environmental impact of factors including pollutants, deforestation, ozone layer loss, and global warming is the goal of supply chain sustainability. Effective packaging solutions can include utilizing proper-sized boxes for shipments, avoiding using big containers for smaller shipments, and using reusable sheets for plastic wrap (What Is Green Supply Chain Management?, 2023).

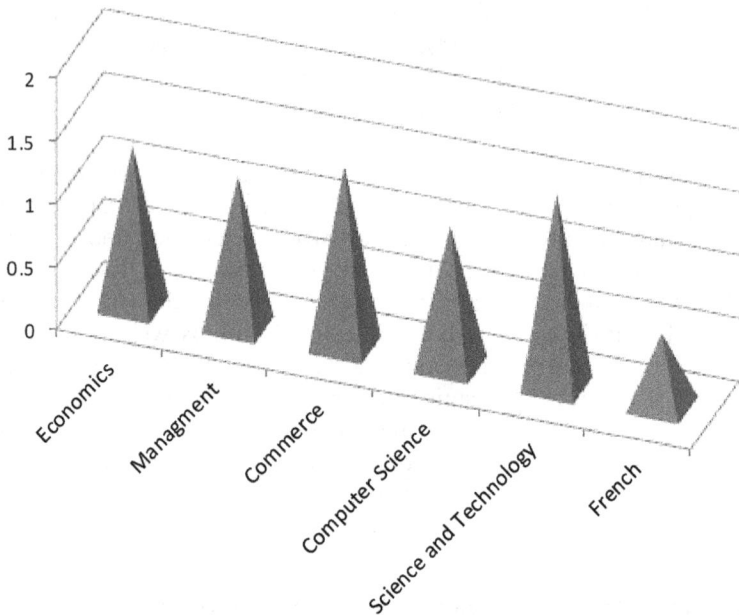

Figure 9.3 A conceptual approach for understanding the impact of green initiatives on supply chain operation.

Source: Modified by author from Azevedo et al. (2011).

According to these three methods and paradigms, a conceptual approach for understanding the impact of green initiatives on supply chain operation is presented in Figure 9.3.

Stevels (2002) highlighted the advantages of GSCM to many supply chain responsibilities, encompassing the environment and society, in perspective of three subgroups: tangible, immaterial, and emotional. Additionally, a GSCM benefits matrix was created for each classification and position (Table 9.1).

9.2.2.2 Green supply approaches to manufacturing and remanufacturing

Businesses must place an emphasis on a green supply strategy, particularly when considering the current manufacturing methods, which include finding strategies to reduce the amount of energy and materials consumed during manufacturing. With the aim of integrating recyclable substances into the production process, that procedure would enable businesses to monitor the entire usage of fresh raw materials connected with the manufacturing procedure. The procedure opens up additional avenues for analysis and breakdown utilizing three methods: economic assessment, energy, and life cycle assessment (Tan and Khoo, 2005; Singhvi et al., 2004; Türkay et al., 2004).

Table 9.1 Benefits of participating in a green supply chain

Within classification	Advantage				
	Manufacturer	Societies	Provider	Environmental	Consumer
Component	Decreased pricing	Minimal resource usage	Decreased pricing expense	Lessening ecological impact	Reducing operation costs
Completely irrelevant	More convenient to produce	Improved conformity	Fewer rejections	Trying to overcome skepticism and intolerance	Delightful cuisine
Feelings	Clearer view	Business on the appropriate transport	Clearer view	Stakeholders' incentive	Mood and standards of living

Source: Modified by author from Stevels (2002).

As rules are implemented, GSCM is required to safeguard the environment; recycling material is a crucial element of the strategy. According to Srivastava (2007), recycling is primarily motivated by economic and legislative concerns and is accomplished in order to reclaim the material composition of old and nonfunctional products. The overwhelming costs connected with recycling are related to the acquisition of recyclables as well as transportation and processing (Sheu et al., 2005). Consumer devices and autos are two businesses that have conquered the cost obstacles connected with recycling.

Remanufacturing was described by Chung et al., (2008) as the method of using recycled components in the manufacturing technique. The recycling of technology and automobiles enables businesses to restore goods to like-new condition for continued consumer usage. Even though the quality of the refurbished product is equivalent to that of the new product, since these items are not new, the method enables a person to purchase an item that might otherwise stand out of his or her price range. Sundin and Bras (2005) examined the procedures required to remanufacture a product and came to the conclusion that the product must be cleaned and updated.

9.2.2.3 Green design – eco-design

Starting with the product design, it follows the idea of the product's lifetime. Srivastava (2007) asserts that life cycle evaluation and analysis are both heavily emphasized in the research on green design. The design group could alter the production process' basic components to become less hazardous and more ecologically beneficial while creating a product. Design for the ecology, often known as Eco-design, is another term for green design. The hybrid automobile is one of these eco-friendly goods. Automotive producers have been developing new engines that use less or no gasoline as a result of rising demand and dwindling petroleum availability. The hybrid automobile has developed over time. McAuley (2003) explored the usage of lighter and fewer components in automotive design. The manufacturing firm requires close collaboration with its suppliers while creating a product. Stevels (2002) conducted an investigation on the supplier–manufacturer collaboration in eco-designs and additionally demonstrated two effective green supply agendas between manufacturers and suppliers.

9.2.2.4 Green drivers

In research, a variety of motivators have indeed been proposed to explain what makes businesses want to engage in green operations, including economic considerations, legal requirements, social duty, ethical considerations, and shareholder engagement (Walker et al., 2008).

Depending on scientific research and the recommendations of a decision-making committee that comprised industry leaders, the many factors crucial to the adoption of GSCM methods have been determined. The key factors that this research uncovered (Diabat and Govindan, 2011; Walker et al., 2008) are as follows: suppliers' ecological cooperation, manufacturers and suppliers collaborating to decrease and remove products, incorporating environmental quality management into the planning and operating processes, accrediting sustainable environmental strategies of suppliers, lowering energy usage, resource and package reuse and recycling, and ecological partnership with consumers.

9.2.2.5 Challenges and barriers to green supply chain application

The GSC has been increasingly implemented by businesses in recent years as a result of heightened awareness of environmental problems. Some businesses began it earlier. For both small and large businesses, GSCM implementation faces challenges and barriers. Numerous investigations have discovered these challenges. These are obstacles to GSCM are listed below (Mathiyazhagan et al., 2013; Diabat and Govindan, 2011; Walker et al., 2008).

The manufacturer's design is complex enough – it cannot be reused or recycled, expensive expenditures with a low return on investing, a lack of understanding concerning reverse logistics development, a scarcity of innovative technologies, materials, and processes, the inability to get borrowed funds to promote eco-friendly items and methods, absence of senior management engagement in implementing GSCM, financial limitations, consumer apathy and pressure about GSCM, absence of CSR, absence of environmental awareness, difficulty in measuring and overseeing manufacturers' environmental policies, inadequate government backing for environmentally beneficial measures, inadequate technological knowledge, deteriorating manufacturers' ecological consciousness, reduced participation in sustainability initiatives and gatherings, absence of belief in the ecological advantages, impression of being out of accountability, specialists with little experience to green systems, corporate practices that limit product/process sustainability, expensive disposal of toxic materials, a shortage of training courses/consultancy/institutions to teach, monitor, and coach industry-specific development, disappointment anxiety, absence of collaboration between organizations, lack of supplier engagement and unwillingness to share information, and inadequate protection of the environment procedures.

The GSCM method is by which the product is transferred from one manufacturing storage to another or transferred to a purchaser. To be competitive in this field, it is necessary to give the most effective methods of admission while maintaining cost management and productivity improvement (Mathiyazhagan et al., 2013; Walker et al., 2008).

Description of GSCM barriers are highlighted as follows (Mathiyazhagan et al., 2013; Walker et al., 2008; Zhu et al., 2008; Srivastava, 2007; Stevels, 2002).

Manufacturers struggle to keep up with environmentally sensitive suppliers, and distributors worry about upholding environmental principles in their own fields. The main cause of inefficiencies and disruption in supply chain relationships is believed to be measurement mismatch. The tools that are absolutely important for the appropriate governance of commercial operations, including the environment in which they function, are regulation and legislation. Environmental rules and regulations are a critical underpinning that businesses must perform. It encompasses the sense of rejection in implementing a GSC, which might result in lost earnings for the business or the concern of defective products, resulting in the loss of commercial advantage. Environmental assessment is required for the implementation and maintenance of the green paradigm in enterprises. The foundation's work force is understaffed, and/or the staff are of poor quality. Fundamentally, a major hindrance to the Small- and Medium-sized Enterprise (SME) industry's improvement in ecological efficiency is a shortage of human resources, shortage of environmental knowledge inside the company (human resources), failure to develop a pollution-free product alternatively to meet environmental standards, lack of funding for improvements in technology and other things (especially long term), and industry-wide technical assistance not maintained (Mathiyazhagan et al., 2013; Walker et al., 2008; Srivastava, 2007; Stevels, 2002).

A foundation's belief is that taking action toward environmental awareness is not their obligation. Lack of adequate organizational systems and broad ignorance of supply chain concepts have also been noted as impediments to GSCM adoption. Lack of awareness regarding environmental consequences or underestimation of environmental repercussions is typical among SMEs; one reason for this is that the legal gateway is frequently larger. SMEs are widely recognized for having insufficient human resources, both in terms of quantity and technical expertise, to undertake environmental management. Green idea implementation requires significant expenditure, yet the Return on Investment (ROI) is modest. Industries are having difficulty obtaining bank funds for environmental efforts. The absence of money for environmental initiatives or the realization that the ROI term following the implementation of GSC is quite long. Carrying over GSCM activities necessitates greater efforts as well as a higher cost, with fewer evident economic returns from these programs (Mathiyazhagan et al., 2013; Walker et al., 2008; Srivastava, 2007; Stevels, 2002).

9.2.3 Reverse logistics

Reverse logistics is the movement of resources from the site of consumption to the place of production in order to reuse diverse items that have reached the end of their useful lives, reduce environmental harm, and safeguard natural resources. Regarding the connection and differences between forward logistics and reverse logistics, first of all, the difference in direction as product provider and user is mentioned. Then, it is clear that the demand forecasting of the reverse logistics concept, which is different from the familiar concept of logistics, is quite difficult. Of course, it is unnecessary to look for the quality at which the product was first produced. In the normal process, the last stage of the product, that is, the consumers, is not clear, but on the contrary, the person who returns the product is clear. Finally, the reverse process includes a guarantee process. The main purpose here is to divide the product into parts and to find the part that can create the remaining value. It is also possible that many alternatives at this point are declared as follows (Çetik and Batuk, 2013; Pokharel and Mutha, 2009; Rogers and Tibben-Lembke, 2001; Rogers et al., 1998).

- In case the product does not work at all, necessary repairs are made and delivered to the user.
- The product can be corrected, cleaned, and delivered to the consumer under warranty.
- The product can be disassembled and rebuilt, and in this way, it is sent to transportation networks.
- As a last alternative, the product is completely disassembled and the parts that can be useful are determined, cleaned, and sent to the necessary places to be used again in the production phase.
- It is necessary to mention two different products that can be explained in this concept; the first are idle parts that are now deemed to be completely useless and must be properly disposed of. The second is the parts to be used in the production phase of the new products to be produced during the remanufacturing phase.

9.2.3.1 Reverse logistics activities

Products that come back to manufacturers for reuse go through various processes. These vary depending on what the returned product will be used for. There are different methods of repairing the product, destroying it, regaining some or all of it, or selling it without processing. These activities are briefly described below (Çetik and Batuk, 2013; Coşkun, 2011; Rogers and Tibben-Lembke, 2001; see Figure 9.4).

> *Product renewal*: This is to bring the used product to a certain quality and to prolong the product usage period, for example, due to the low performance of the car's engine, the replacement of the part and, as a result, the prolongation of the service life with the increase in the engine quality of the car.
>
> *Reproduction*: It is the process of checking the condition of the collected used products or components and replacing the old, broken, or non-functioning parts with new ones (Paksoy, 2012). Salvage objects are overused or destroyed and cannot be offered as new. Salvage objects decrease value in proportion to their usage or damage. The most challenging aspect of salvage management is assessing its worth.
>
> *Repair*: It is the process to make the product work or use again.
>
> *Defectives*: These have been determined to be actually faulty by the store or the client. In many circumstances, a company may notify the manufacturer of a fault, and the manufacturer would refund the merchant with a replacement product or payback in the type of a check or credit (Çetik and Batuk, 2013; Coşkun, 2011; Rogers and Tibben-Lembke, 2001).

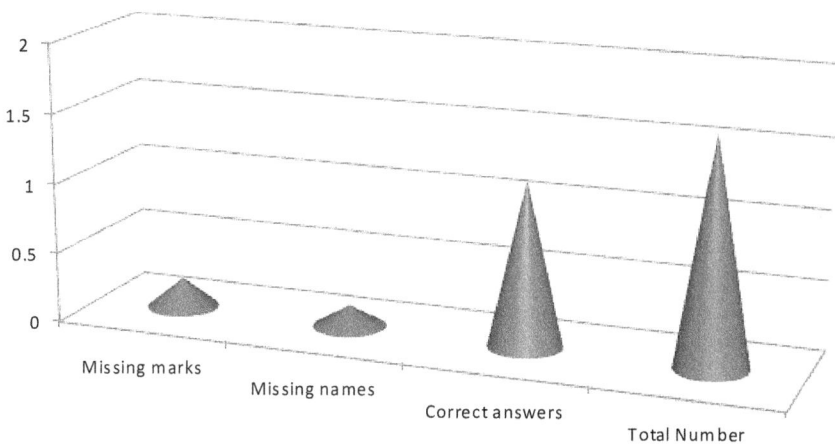

Figure 9.4 Flow diagram of reverse logistics operations.

Source: Krumwiede and Sheu (2002).

Surplus Surplus refers to high-quality products that the firm has a surplus of. The reason for this is that the company may have exaggerated demand or order volume. It's also conceivable that this is the outcome of a hyperactive producer (Çetik and Batuk, 2013; Coşkun, 2011; Rogers and Tibben-Lembke, 2001).

Partial use of the product: It is the use of only a certain part of the product in the improvement process. The aim here is to ensure that the part of the product that can be reused is reused. A product with this feature is carefully examined and disassembled. Its parts, which can be reused later, are used in the repair, refurbishment, or remanufacturing of other materials (Coşkun, 2011).

Buy out: This arises when one manufacturer purchases an existing inventory of a counterpart's product from a store; this purchase relieves storage space so that the producer may place its product where the vendor's product was formerly, and thus decreases the retailer's vulnerability producer (Çetik and Batuk, 2013; Coşkun, 2011; Rogers and Tibben-Lembke, 2001).

Recycle: One of the important issues in reverse logistics activities is transformation. In other reverse logistics activities, the aim is to preserve the properties of the used products as much as possible. In recycling, the returned products are first disassembled and disassembled. The separated parts are classified according to their properties, and in the next step, the base material is created and used in the production of new units' producer (Çetik and Batuk, 2013; Coşkun, 2011; Rogers and Tibben-Lembke, 2001).

Returns: Returns are items that the client has purchased and utilized. Consumer exchanges are often processed in the same way as salvage or excess merchandise. Even if a returned item is not damaged, it could not indeed typically be distributed as the most reliable supplier and producer (Çetik and Batuk, 2013; Coşkun, 2011; Rogers and Tibben-Lembke, 2001).

Incineration and burial: This process is not preferred much, but it is among the alternatives in reverse logistics activities. These are the activities carried out to destroy the product itself or the hazardous wastes generated during its production with appropriate technology without harming the environment. In this section, it is checked whether the product contains harmful substances. (Paksoy, 2012). This process is generally preferred for products that will not bring economic benefits (Coşkun, 2011).

In order to examine the viability of adopting reverse logistics in third-party vendors, including shipping firms, a framework for making decisions for reverse logistics is created. The fundamental process diagram for the operations of reverse logistics is shown in Figure 9.4. From the bottom-left to the top-right in the representation, the level of complexity of activities and the amount of information retrieved rise. The typical sequence of return in terms of quantity, quality, and expected arrival is of utmost significance (Krumwiede and Sheu, 2002).

Unsustainable economic operations cause environmental deterioration, which could jeopardize a developing country's long-term development and economic competitiveness. Sustainable business practices could assist businesses to succeed and increase the condition of life in emerging markets (Schmidheiny, 1992).

The conceptual foundation for the strategic positioning precursors and reverse logistics results of sustainable supply chain activities was offered by the researchers. The study hypotheses take into account information from a variety of sources. In the study of Hsu et al. (2016), the investigation technique and the findings of the data analysis were explained. The present study is an endeavor to construct a conceptual paradigm centered on knowledge-based thought. The research contributes to the expansion of previous efforts that require a theory-focused perspective (Hsu et al., 2016).

9.2.3.2 Closed-loop supply chains and reverse logistics

Closed loops, as defined by Wells and Seitz (2005), typically consist of two supply chains: a forward chain and a reverse chain, in which a recovered product re-enters the typical front chain (see Figure 9.5). In a closed-loop supply chain (CLSC), reverse logistics is utilized to feed things that have served their purpose back into the forward logistics process. A producer develops a product, distributes it to clients, and then enables them to return it when they no longer require it. Via repairing, reselling, or constituent reusing, these old items eventually constitute half of the product's inventory. These processes are a critical component of the circular economy,

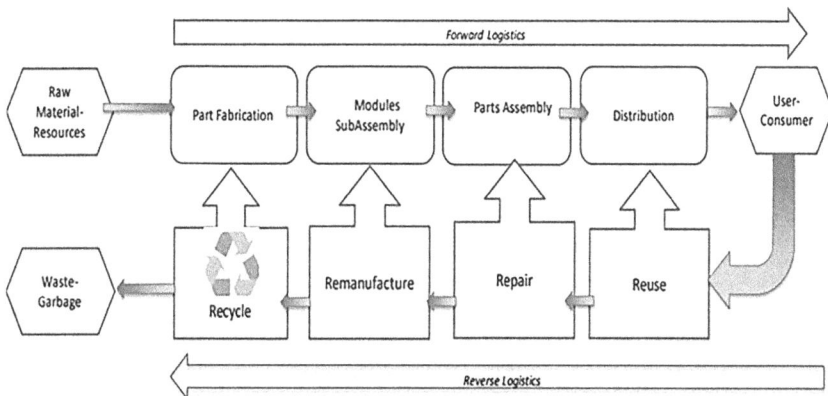

Figure 9.5 Closed-loop supply chains. This figure is modified from literature of Defee et al. (2009), Kocabasoglu et al. (2007), and Wells and Seitz (2005).

which aims to remove waste. CLSCs, which employ garbage as producing supplies, are the starting point for reducing waste. CLSCs concentrate on gathering customer goods returns, recovering remaining value, and reselling them. According to the existing research on core recruitment management, academics have not studied the behavioral aspects of customers (What Is a Closed-Loop Supply Chain and How Can it Help?, 2022; Defee et al., 2009). The CLSC is the same as Circular Supply Chain Management (CSCM).

The term "circular supply chain management" describes cooperative forward and reverse supply chains that are accomplished through planned business ecosystem integration for value creation from goods and services, contaminants, and usable end-of-life tends to flow over extended life loops that improve associations' economic, social, and environmental sustainability (Batista et al., 2018). The Circular Supply Chain (CSC) idea advises a change in business operations from linear to circular systems in order to guarantee the sustainability of the world. CSC comprises recovery techniques that boost value and lower supply chain carbon footprints. A network of connected devices and procedures throughout the supply chain is required for the I5.0 transformation to enable intelligent manufacturing depending on the unique requirements of each customer. It has the potential to be a useful instrument for attaining sustainability through a circular economy (Dwivedi et al., 2023). Based on a survey of the research and contact with industry specialists and operators, the current study outlines 16 possible drivers in the context of establishing I5.0 and CSC convergence. To evaluate the criticality intensity of components, a mixed model is constructed using the improved m-TISM and the Cross-Impact Matrix Multiplication Applied to Classification (MICMAC). Whereas MICMAC organizes precise evaluations of the drivers' dependence and driving strength, m-TISM organizes the binary exchanges between the drivers. A framework for addressing the factors of I5.0 and CSC in order to achieve sustainable growth is produced by the research (Dwivedi et al., 2023).

Defee et al. (2009) provided justification for this procedure by stating that strategic and operational choices throughout a corporation's forward and backward supply chains must be integrated. This explanation makes clear that the goal of a CLSC is to encourage recirculating so that waste may be reduced.

In terms of the environment and environmentally responsible business operations, the CLSC is the desired configuration, since it fulfills the purpose of conserving the environment by reducing waste created. As a result, if a CLSC is desired, reverse logistics, a kind of GSCM, is the best strategic approach for accomplishing this objective, because in order to reduce waste, a reverse circulation must be presented into the supply chain and intended to produce the most beneficial usage materials/products obtained, whether with the maintenance and reuse, remodeling, remanufacturing, cannibalization, or recycling (Kumar and Malegeant, 2006; Beamon, 1999). Due to this new intimate link, this research examines the research on CLSCs and reverse logistics concurrently.

Closed-loop supply networks complicate the existing supply chain since they generally necessitate investment, and the newly added supply chain functions are fraught with uncertainty (Kocabasoglu et al., 2007).

9.2.4 Waste management

Waste management entails the control of items that have been recycled in order to reduce the quantity of materials that are really destroyed and disposed of. Due to the general rising awareness of businesses in the European Union, including in the EU nations and in the nations that produce to the EU, waste management is a prominent theme in literature (Zhu et al., 2007).

The waste hierarchy prioritizes waste avoidance. To achieve the government's goal of a zero-waste economy, waste prevention should be supported and enhanced. Waste prevention, according to the Waste Framework Directive of 2008 (WFD), encompasses waste avoidance, reducing waste at the source or in the operation, and product recycling at the end of life (Waste Framework Directive of 2008 [WFD], 2023).

According to Andiç et al. (2012), some research findings in the previous research emphasized on comprehending the essence of the waste (Kahhat et al., 2008; Mazzanti and Zoboli, 2008; Pongrácz and Pohjola, 2004; Pitt and Smith, 2003; Burnley, 2001; Cardinali, 2001), whereas others attempt to recommend alternatives to the waste issue (Chakrabarti et al., 2009; Bovea and Powell, 2006; Joseph, 2006; Mohan et al., 2006; Wright et al., 2005; Boyle, 2000; Haastrup et al., 1998).

9.3 APPLICATION AREAS IN THE INDUSTRY – EXAMPLES

Corporations develop their strategy and long-term reverse logistics plan after realizing the advantages of reverse logistics. Reverse logistics is used in a variety of sectors and marketplaces, besides electronics, computers, pesticides, medicines, internet commerce, construction machines, and medical vehicles. There are several significant businesses, such as BMW, Delphi, DuPont, General Motors, HP, Dell, Caterpillar, Xerox, Canon, IBM, Ford, and Phillips, that use reverse logistics (Nakıboğlu, 2007; McAuley, 2003).

> *Dell*: If any order is canceled or returned, almost 90% of the returns, which makes eight hundred units, are quickly made usable by refurbishing and reselling if necessary. All repurposed products are put up for sale through the Global Dell Outlet and Dell guarantee all of them just like new. The remaining 10%, which cannot be renewed or resold, is recycled responsibly (Application Areas in the Industry, www.dell.

com, 2022a). The success of Dell, one of the leading computer manufacturers, in recycling is another example that can be given to the concept of reverse logistics.

Considering that a significant amount of waste will occur at the end of the life of automobiles, it causes environmental pollution. The selection of parts in the production of the car is a serious step in minimizing the harmful effect on the environment at the end of the car's life. The fact that these materials are composed of recyclable materials is of great importance in terms of the environment. If it is not possible, different material types should be sought instead of plastic material. Tires, which are part of automobiles, are a serious cause of environmental pollution when they are not recycled. According to the information of the Tire Industrialists' Association (LASDER), around 200,000 tons of worn tires are produced in Türkiye every year. New areas of use can be created for tires that have completed their time. Truck and bus tires can be coated to be reused. Tires with high calorific value are used in cement plants, etc. When they are burned in furnaces at high temperatures, energy is produced and waste generation is prevented. Rubber is also used in the construction of asphalt, in making sound-proofing materials/walls, and in making the tartan floor in sports and parking areas (Dinç and Erol Genevois, 2023).

Caterpillar: Remanufacturing is a form of production where the entire life cycle of any product is ended, the product is somehow returned to the place of purchase or to any dealer, and the product is returned to the production facility with reverse logistics systems. In the recycling system, which is very sustainable and useful, the mentioned metals go to a foundry, melt, and can be recycled by converting the returned part into another component. If all these stages are carried out, the product does not have to use that energy in the first production stage. This is because it already uses metals or materials in their original form (Application Areas in the Industry, www.caterpillar.com, 2022b). If we need to think of Caterpillar from the machinery industry branch and make an analysis in this area, we can think of the metals used by the company in the first stage of the product or the materials that make up its skeleton as a mold.

Migros: It has become one of the major corporations in the nation that has been successfully conducting GSCM operations for many years. The corporation strives to lessen its environmental impact on both the manufacturing and the supply chain sectors. In terms of energy savings and minimizing greenhouse gas emissions, the organization also aims to improve logistics and transportation performance. It attempts to establish the optimal pathways for distributing trucks by evaluating the distances among distribution facilities and retailers and, as a result, defines the placement of prospective distribution facilities (Sustainability, 2023).

A milk producer strives to decrease the range traveled as well as greenhouse gas emissions by developing optimal pathways for transporting milk from suppliers situated in various places. Furthermore, it attempts to cut carbon energy consumption and emissions by guaranteeing that vehicles transferring items from manufacturers to distributing hubs equally bring goods back to manufacturers. The corporation focuses on measures to minimize greenhouse gas emissions through its manufacturing processes, create sustainable energy using refuse, and improve resource performance (Sustainability Reports, 2021).

FedEx: The leader in international package shipment, FedEx, has proposed a number of efforts to lessen its effect on the environment (Green Supply Chain - Examples of Companies, 2023).

Ford Motor Company: The utilization of recycled components in the manufacture of vehicles has expanded because of Ford's relationship to its value chain. The corporation also provides recycling standards to global providers and designers, enabling the dismantling of the business's vehicles. According to research, 4 billion pounds of recycled content are used in the manufacture of vehicles (Green Supply Chain - Examples of Companies, 2023).

Kodak: Kodak used Design for Environment to enable the photographic equipment and packing to be reused, recycled, and disassembled to make it easier to use the reversed supply chain for extracting resources. It collaborated with significant groups to establish recycling initiatives, combine logistical resources, and split the monetary benefit in order to keep improving the reverse supply chain (Green Supply Chain - Examples of Companies, 2023).

Walmart: Walmart uses the Packaging Sustainability scorecard for providers, which determines the "green quotient" for each item of packaging according to a variety of factors such as emissions of greenhouse gases, production-related material utilized, product-to-packaging proportion, cube expenditure (i.e., efficient use of space within the storage area), use of recycled chemicals creative thinking, the quantity of renewable energy utilized in producing the wrapping, greenhouse gas emission levels associated with mode of transport and packaging supplies, and so on (Green Supply Chain - Examples of Companies, 2023).

Many companies strategically apply the concept of reverse logistics in which a flow where each stage is planned is implemented and controlled for the purpose of recycling and repetitive production. To explain the general approaches of these companies, which come to mind when reverse logistics is mentioned, and the measures and policies they have taken, with examples; by making environmentally compatible plans, designs that will minimize air and noise pollution of the product, and that will provide the highest gain in recycling when the product's useful life expires should be applied. Now, corporate companies are very meticulous about the use of environmentally harmful products by creating recycling programs (Azevedo et al., 2011; Nakıboğlu, 2007).

9.4 MATERIALS AND METHOD

The data utilized for the present research came from the computerized databases of important corporate and public organizations, as well as their reports on operations. Relevant data that was not obtainable in records was acquired through interviews of individuals and customers. While being used, the information gathered was standardized. Data normalization is a method of scaling that seeks to improve the precision of the values in the data collection when doing numerical computations. This constitutes a key part of the research. The structure consists of amenities, gathering locations, and reversal sites. The model's goal is to lower total costs associated with renting, inventory utilizing, handling materials, choosing orders, and shipment. In accordance with the model, three distinct decisions are considered.

The first and most fundamental decision is to find which collection stations and centralized returning hubs ought to be established. In the context of this preliminary judgment, the other two possibilities are considered. The second choice is that the places of collection which have been established are open; the total amount of periods remains to be determined; and the third decision is connected to the allocation of consumers to the collection centers and collection facilities to central return facilities.

A firm, a white goods producer based in the Marmara area, distributes and collects the products it manufactures using the prevailing supply chain infrastructure. The firm is seeking responses; a commitment has been undertaken to develop a framework for calculating the entire cost of transportation if the containers of white products are returned to the plant manufacturer to which they correspond. What ought to be the quantity of goods in the storage facilities, as well as the supply and demand, and what ought to be the total cost of transportation in order to decrease the expense if the products generated are gathered in a storage facility and are to be bought to an appropriate location and provided to the manufacturing facilities in full trucks?

9.4.1 Methodology

The utilization of mathematical models, especially optimizing configurations, to aid in decision-making is known as mathematical programming (MP). There are different kinds of mathematical programming: Ng (2008) created a weighted linear programming model for the supplier selection issue, with the goal of increasing the provider score. It is similar to AHP in that decision-makers are involved in setting the relative important weightings of criteria. For the supplier selection problem, Hong et al. (2005) introduced a mixed-integer linear programming model. The model was designed to identify the ideal number of suppliers and the appropriate order amount in order to maximize revenue. The evolution of providers' supply capabilities and customer requirements through time was studied.

To assess and choose suppliers, Karpak et al. (2001) developed a goal programming (GP) approach. The model took into account three objectives: cost, quality, and delivery dependability. The model was designed to find the ideal number of items to order while taking into account customer demand and supplier capacity restrictions.

9.4.2 Case study

In the study by Min et al. (2006), a model prepared was used. The main layout of the reverse logistics network concept is depicted in this model. This network is made up of facilities, collecting points, and turnaround points. The model's purpose is to reduce overall expenses, which involve renting, inventory handling, handling of materials, order picking, and shipping. Three different decisions are attempted, based on the model. The first and most basic choice is to decide which collecting stations and central return centers will be opened. The other two options are investigated in light of this initial decision. The second decision is that the collection centers are open, the number of days is determined, and the final decision is related to the assignment of customers to collection centers and collection centers to the central return center. If there is no capacity limitation, each client is distributed to the nearest collection location, and each collecting point is allocated to the nearest centralized return location.

The company, which is a white goods manufacturer operating in the Marmara region, distributes and collects the products it produces in the existing transportation supply system. The business is looking for answers.

An attempt has been made to establish a model for what the total transportation cost will be in case the pallets of white goods are sent back to the factory manufacturer they belong to (it is important to balance supply and demand).

If the products produced are collected in a warehouse to be installed in a suitable place and sent to the factories in full trucks, what should be the product quantity in the warehouses and the supply and demand quantities, and what will be the total transportation cost in a way that will minimize the cost?

Depending on the given constraints, a mathematical model was established and solved in the GAMS software program, and using this developed model, the situations containing the above-mentioned constraints were analyzed and the transportation model that minimized the total transportation costs and its results were obtained as shown in Tables 9.2–9.4.

Constrains:

i)	Production Capacity Factory 1	200
ii)	Production Capacity Factory 2	300
iii)	Production Capacity Factory 3	100
iv)	Shipping Capacity	200
v)	Demand Customer 6	400
vi)	Demand Customer 7	180
vii)	No waste (Production = Demand)	
viii)	Disabling Transitions	
ix)	No Inventory in Transposition 4	
x)	No Inventory in Transposition 5	
xi)	No negative variables	

Table 9.2 Shipping cost per unit

From (i)	C_{ij} shipping cost from i to j							
	i \ j	1	2	3	4	5	6	7
	Plant 1	0,0	5,0	3,0	5,0	5,0	20,0	20,0
	Plant 2	9,0	0,0	9,0	1,0	1,0	8,0	15,0
	Plant 3	0,4	8,0	0,0	1,0	0,5	10,0	12,0
	Warehouse 1	0,0	0,0	0,0	0,0	1,2	2,0	12,0
	Warehouse 2	0,0	0,0	0,0	0,8	0,0	2,0	12,0
	Demander 1	0,0	0,0	0,0	0,0	0,0	0,0	1,0
	Demander 2	0,0	0,0	0,0	0,0	0,0	7,0	0,0

Table 9.3 Capacity constraints

From (i)	X_{ij} shipping quantity from i to j								Total
	i \ j	1		2	3	4	5	6	Total
	Plant 1	0,0	0,0	180,0	0,0	0,0	0,0	0,0	180,0
	Plant 2	0,0	0,0	0,0	120,0	0,0	180,0	0,0	300,0
	Plant 3	0,0	0,0	0,0	80,0	200,0	0,0	0,0	**280,0**
	Warehouse 1	0,0	0,0	0,0	0,0	0,0	200,0	0,0	200,0
	Warehouse 2	0,0	0,0	0,0	0,0	0,0	200,0	0,0	200,0
	Demander 1	0,0	0,0	0,0	0,0	0,0	0,0	180,0	180,0
	Demander 2	0,0	0,0	0,0	0,0	0,0	0,0	0,0	0,0
	Total	0,0	0,0	180,0	200,0	200,0	580,0	180,0	
	Min Z	= 3260			[$]				

Table 9.4 Solution values table

		To (J)							
		1	2	3	4	5	6	7	
		Plant 1	Plant 2	Plant 3	Warehouse 1	Warehouse 2	Demander 1	Demander 2	Total output
	1 Plant 1	0	0	180	0	0	0	0	180
	2 Plant 2	0	0	0	120	0	180	0	300
	3 Plant 3	0	0	0	80	200	0	0	100
From	4 Warehouse 1	0	0	0	0	0	200	0	200
(i)	5 Warehouse 2	0	0	0	0	0	200	0	200
	6 Demander 1	0	0	0	0	0	0	180	
	7 Demander 2	0	0	0	0	0	0	0	0
	Total input	0	0	180	200	200	580	180	

Variables:

    i, j: The network's generic (suppliers, transit points, and demanders) node; Generic (Suppliers, Transshipment points, Demanders) node of the network; i: From node; j: To node

    x_{ij}: Transportation amount from node i to node j (ton); Shipping quantity from node i to node j [ton]

    c_{ij}: Transportation expense/cost from node i to node j ($/ton) – Unit shipment Cost from node i to node j

Objective Function:

$$\min(z) = \sum_{j=1}^{7}\sum_{i=1}^{7}\left(x_{ij} * c_{ij}\right)$$

Constrains:

$$\sum_{j=1}^{7}x_{1j} - \sum_{i=1}^{7}x_{i1} \leq 200 \quad \text{Observed Production Capacity of Factory 1:} \quad (9.1)$$

$$\sum_{j=1}^{7}x_{2j} - \sum_{i=1}^{7}x_{i2} \leq 300 \quad \text{Observed Production Capacity of Factory 2:} \quad (9.2)$$

$$\sum_{j=1}^{7}x_{3j} - \sum_{i=1}^{7}x_{i3} \leq 100 \quad \text{Observed Production Capacity of Factory 3:} \quad (9.3)$$

$$x_{ij} \leq 200 \;\; \forall i, j \;\; \text{Shipping Capacity:} \quad\quad\quad\quad\quad\quad\quad\quad\quad (9.4)$$

$$\sum_{i=1}^{7}x_{i6} - \sum_{j=1}^{7}x_{6j} = 400 \quad \text{Observed Demand of Customer 6:} \quad\quad (9.5)$$

$$\sum_{i=1}^{7}x_{i7} - \sum_{j=1}^{7}x_{7j} = 180 \quad \text{Observed Demand of Customer 7:} \quad\quad (9.6)$$

$$\sum_{i=1}^{7}\sum_{j=1}^{7}x_{ij} - x_{12} - x_{13} - x_{21} - x_{23} - x_{31} - x_{32} = \left(\sum_{i=1}^{7}x_{i6} - \sum_{j=1}^{7}x_{6j}\right)$$

$$+ \left(\sum_{i=1}^{7}x_{i7} - \sum_{j=1}^{7}x_{7j}\right) \text{No waste}\left(\text{Production} = \text{Demand}\right): \quad (9.7)$$

if $\left(c_{ij} = 0\right)$ ARC from i to j is Disabled $\forall i, j$ Disabling Transitions: (9.8)

Not just transition but also persistence in node 6
Not just transition but also persistence in node 7

$$\sum_{i=1}^{7} x_{i4} - \sum_{j=1}^{7} x_{4j} = 0 \ \text{No inventory in Transposition 4}: \qquad (9.9)$$

$$\sum_{i=1}^{7} x_{i5} - \sum_{j=1}^{7} x_{5j} = 0 \ \text{No inventory in Transposition 5}: \qquad (9.10)$$

$$x_{ij} \geq 0; c_{ij} \geq 0 \forall i, j \ \text{No Negative Variables}: \qquad (9.11)$$

According to this linear modeling, Constraint (1) minimizes transportation costs. Constraint (2) states that every vehicle must start its route from the factory. Constraint (3) states that each vehicle must complete its route at the factory. Constraint (4) states that it is absolutely necessary to come to each dealer from another place, and Constraint (5) states that one must go to another place from each dealer. Constraint (6) ensures that the same vehicle is linked between dealers i and j. In this case, reverse logistics – SCM may choose to close some collection centers and merge some collection centers in order to reduce costs and increase efficiency.

However, customer dissatisfaction that may arise as a result of such an improvement should be carefully analyzed. The modeled values are as follows: the best capacity production of Plant 1 is 180 tons; the best capacity production of Plant 2 is 300 tons; the best capacity production of Plant 3 is 100 tons; disabling transitions is 0; permanence in node 6, not only transition is 400 tons; permanence in node 7, not only transition is 180 tons; there is no waste (Production = Demand); therefore, the capacity of total demand equals the capacity of total production which is 580 tons. According to the solution values, the cost has the optimum value. This minimizes the cost. Minimum Z (cost) is 3250 $. Thus, the model of this problem reaches the optimum solution values.

9.5 CONCLUSION, LIMITATIONS, AND FUTURE RESEARCH

Responsibility for the environment is one of today's most pressing issues. Environmental consciousness has grown during the previous few generations. Green concepts have spread to a variety of industries, particularly the supply chain. GSCM has grown in popularity in recent years. This concept encompasses all stages of production, from the beginning to the end of the life cycle, or from product design through recycling. GSCM is used in various economic areas outside manufacturing, such as governance and services. Descriptions of Green SCM have been provided in this research. The research was conducted in order to comprehend the aspects, drivers, and challenges of GSCM.

RLs and green logistics applications have started to increase in Türkiye in recent years. RLs and green logistics are implemented by logistics companies operating in Türkiye in order to survive in a competitive environment and contribute to sustainable development. Based on the reverse logistics and green logistics indicators, the research reveals that green logistics activities are observed in companies in Türkiye and the measures and actions implemented by some companies in line with sustainable development are presented with examples in the research. In terms of future studies, this method can be applied in different sectors depending on different variables determined. Thus, the importance of the concept of sustainability is emphasized for different sectors. Apart from this, this issue can be handled with the network structure of businesses operating in different sectors under deterministic assumptions as a solution approach. The solution approach can be evaluated by taking the data and the detected variables as input to the modeling and making sensitivity analyses over different scenarios. In addition, the model could be made more dynamic using decision support systems in future studies. Apart from this, the application part of the study can be expanded with different decision-making techniques as a method of decision-making at the strategic level during the decision-making phase.

The current research is confined to product recovery options for RLs and GCSM for long-term development. Additional recovering options, variables, and drivers may be gathered for future research, and an empirical statistical study might be carried out to obtain insight into the relationships between the discovered drivers and recovering options. Experts from many fields may be recruited in future research to provide their recommendations. A disadvantage is the omission of organizational-specific elements such as firm size, organizational structure, region, and so on, which alter the inclinations and connection of the drivers. For future study, it is advised that the work be extended to the industrial generic domain. The chapter also identifies critical future research directions, stressing possible facilitators of Industry 5.0 improvement in the framework of GSCM as well as incorporating industrial advancements such as Industry 5.0, as industry initiatives play a significant role in simplifying efforts toward sustainability. Future research demands a platform for national and international collaboration, industry associations, and civic groups collaborating on global initiatives to solve the drivers for RL, GSCM, which is incorporated into Industry 5.0, and sustainability objectives. This research may be expanded to discover the finest strategies and practices for implementing numerous sectors and industries.

REFERENCES

Andiç, E., Yurt, Ö., & Baltacıoğlu, T. (2012). Green supply chains: Efforts and potential applications for the Turkish market. *Resources, Conservation and Recycling*, 58, 50–68.

Application Areas in the Industry (2022a), www.dell.com, Access date: 12. 10. 2022.

Application Areas in the Industry (2022b), www.caterpillar.com, Access date: 12. 10. 2022.

Aras, N., Aksen, D., & Gönül Tanuğur, A. (2008). Locating collection centers for incentive-dependent returns under a pick-up policy with capacitated vehicles. *European Journal of Operational Research*, 191(3), 1223–1240.

Azevedo, S. G., Carvalho, H., & Cruz Machado, V. (2011). The influence of green practices on supply chain performance: A case study approach. *Transportation Research Part E: Logistics and Transportation Review*, 47(6), 850–871.

Batista, L., Bourlakis, M., Smart, P., & Maull, R. (2018). In search of a circular supply chain archetype – a content-analysis-based literature review. *Production Planning & Control*, 29(6), 438–451.

Beamon, B.M. (1999), Designing the Green Supply Chain. *Logistics Information Management*, 12(4), 332–342.

Bovea, M.D., & Powell, J.C. (2006). Alternative scenarios to meet the demands of sustainable waste management. *Journal of Environmental Management*, 79(2), 115–132.

Boyle, C. (2000). Solid waste management in New Zealand. *Waste Management*, 20(7), 517–526.

Burnley, S. (2001). The impact of the European landfill directive on waste management in the United Kingdom. *Resources, Conservation and Recycling*, 32(3–4), 349–358.

Cardinali, R. (2001). Waste management: A missing element in strategic planning. *Work Study*, 50(5), 197–201.

Carter, C. R., & Jennings, M. M. (2002). Logistics Social Responsibility: An Integrative Framework. *Journal of Business Logistics*, 23(1), 145–180.

Çetik, M.O., & Batuk, S.(2013). Tersine Lojistikte Teknoloji Kullanımı ve Yaşanan Sorunlar, *Sosyal ve Beşeri Bilimler Dergisi*, 5(1), 364–375.

Chakrabarti, S., Majumder, A., & Chakrabarti, S. (2009). Public–community participation in household waste management in India: An operational approach. *Habitat International*, 33(1), 125–130.

Chawla, P., Kumar, A., Nayyar, A., & Naved, M. (Eds.). (2023). *Blockchain, IoT, and AI Technologies for Supply Chain Management*. CRC Press.

Christensen, W.J., Germain, R., & Birou, L. (2005). Build-to-order and just-in-time as predictors of applied supply chain knowledge and market performance. *Journal of Operations*, 23, 470–481.

Chung, S. L., Wee, H. M., & Yang, P. C. (2008). Optimal policy for a closed-loop supply chain inventory system with remanufacturing. *Mathematical and Computer Modeling*, 48, 867–881.

Coşkun, A. (2011). The factors that affect the reverse logistic activities of producers: A practice in house appliance industry. 26, 29, 30, 38, 48–52. Nevşehir: Nevşehir University Institute of Social Sciences, Department of Business Administration, Master's Thesis.

Das, S., & Nayyar, A. (2020). Effect of consumer green behavior perspective on green unwavering across various retail configurations. In *Green marketing as a positive driver toward business sustainability* (pp. 96–124). IGI Global.

Defee, C.C., Esper, T., & Mollenkopf, D. (2009). Leveraging closed-loop orientation and leadership for environmental sustainability. *Supply Chain Management: An International Journal*, 14(2), 87–98.

Diabat, A., & Govindan, K. (2011). An analysis of the drivers affecting the implementation of green supply chain management. *Resources, Conservation and Recycling*, 55(6), 659–667.

Dinç, K., Erol, S., & Yücer, Ü. (2008). Tersine Dağıtım Sisteminde Yeni Bir Model. Mühendislik ve Teknoloji Sempozyumu. 328. Ankara: Çankaya Üniversitesi.

Dinç, M. & Erol Genevois, M. (2023). Design of Reverse Logistics Network for Waste Tire Incineration in Cement Factories. Journal of Polytechnic-Politeknik Dergisi, 1–1 (Early View).

Dubey, R., Gunasekaran, A., & Papadopoulos, T. (2017). Green supply chain management: Theoretical framework and further research directions. *Benchmarking: An International Journal*, 24(1), 184–218.

Dwivedi, A., Agrawal, D., Jha, A., & Mathiyazhagan, K. (2023). Studying the interactions among Industry 5.0 and circular supply chain: Towards attaining sustainable development, *Computers & Industrial Engineering*, 176, 108927, 1–17.

Fleischmann, M., Bloemhof-Ruwaard, J. M., Dekker, R., van der Laan, E., van Nunen, J. A. E. E., & Van Wassenhove, L. N. (1997). Quantitative models for reverse logistics: A review. *European Journal of Operational Research*, 103(1), 1–17.

Fritz, M. M.C. (2022). A supply chain view of sustainability management, *Cleaner Production Letters* 3, 1–13.

Ghadge, A., Mogale, D.G., Bourlakis, M., Maiyar, L.M., & Moradlou, H. (2022). Link between Industry 4.0 and green supply chain management: Evidence from the automotive industry, *Computers & Industrial Engineering*, 169, 1–14.

Green Supply Chain - Examples of Companies, https://www.mbaskool.com/business-articles/operations/16360-green-supply-chain-examples-of-companies.html, Access date: 12.04.2023.

Haastrup, P., Maniezzo, V., Mattarelli, M., Mazzeo Rinaldi, F., Mendes, I., & Paruccini, M. (1998). A decision support system for urban waste management. *European Journal of Operational Research*, 109(2), 330–341.

Hong, P., Liu, X.S., Zhou Q., Lu, X., Liu, J.S., & Wong, W.H. (2005). A boosting approach for motif modeling using ChIP-chip data. *Bioinformatics*, 21(11), 2636–2643.

Hsu, C.-C., Tan, K.-C., & Mohamad Zailani, S. H. (2016). Strategic orientations, sustainable supply chain initiatives, and reverse logistics. *International Journal of Operations & Production Management*, 36(1), 86–110.

Hugos, M. (2010). *Essentials of supply chain management*, 2nd edition, (304 p). New York, NY: John Wiley.

H'Mida, S. (2009). Factors contributing in the formation of consumers' environmental consciousness and shaping green purchasing decisions (pp. 957–962). In: *Proceedings of the international conference on computers & industrial engineering*.

James, R. Stock, & Douglas M. Lambert (2000). *Strategic Logistics Management*, 4th Edition, (pp.1–896). Irwin: McGraw Hill.

Joseph, K. (2006). Stakeholder participation for sustainable waste management. *Habitat International*, 30(4), 863–871.

Kaçtıoğlu, S, & Şengül, Ü. (2010). Erzurum Kenti Ambalaj Atıklarının Geri Dönüşümü İçin Tersine Lojistik Ağı Tasarımı ve Bir Karma Tamsayılı Programlama Modeli. *Atatürk Üniversitesi Iktisadi ve Idari Bilimler Dergisi*, 24(1), 89–112.

Kahhat, R., Kim, J., Xu, M., Allenby, B., Williams, E., & Zhang, P. (2008). Exploring e-waste management systems in the United States. *Resources, Conservation and Recycling*, 52(7), 955–964.

Karmaker, L.C., Baria, M. A.B.M., Anama, Z.M., Ahmed, T., Alia, M.S., Pacheco, J.A., & Moktadir, A.M. (2023). Industry 5.0 challenges for post-pandemic supply chain sustainability in an emerging economy, 250, 108806, 1–12.

Karpak, B., Kumcu, E., & Kasuganti, R. R. (2001). Purchasing materials in the supply chain: Managing a multi-objective task. *European Journal of Purchasing & Supply Management*, 7(3), 209–216.

Kaymak, G. (2010). Tıbbi Atık sterilizasyon Sisteminde Tersine Lojistik Uygulaması (pp. 23–26). İstanbul: Marmara University Institute of Social Sciences, Department of Business Administration, Master's Thesis.

Koban, E., & Keser, H.Y. (2013). Dış Ticarette Lojistik (352 p.). Ekin yayınevi yayınları, Bursa.

Kocabasoglu, C., Prahinski, C., & Klassen, R. (2007). Linking forward and reverse supply chain investments: The role of business uncertainty. *Journal of Operations Management*, 25(6), 1141–1160.

Krumwiede, D. W., & Sheu, C. (2002). A model for reverse logistics entry by third-party providers. *Omega*, 30(5), 325–333.

Kumar, S., & Malegeant, P. (2006). Strategic alliance in a closed-loop supply chain, a case of manufacturer and eco-non-profit organization. *Technovation*, 26(10), 1127–1135.

Li, L. (2020). Education supply chain in the era of Industry 4.0, *Systems Research and Behavioral Science* 37(2) 37 (4), 579–592.

Louwers, D., Kip, B. J., Peters, E., Souren, F., & Flapper, S. D. P. (1999). A facility location allocation model for reusing carpet materials. *Computers & Industrial Engineering*, 36(4), 855–869.

Maddikuntaa, R.K.P., Pham, Q.V., Prabadevi, B., Deepa, N., Dev, K., Gadekallu, R.T., Rubya, R., & Liyanage, M. (2022). Industry 5.0: A survey on enabling technologies and potential applications, Journal of Industrial Information *Integration*, 25, 100257, 1–19.

Mathiyazhagan, K., Govindan, K., NoorulHaq, A., & Geng, Y. (2013). An ISM approach for the barrier analysis in implementing green supply chain management, *Journal of Cleaner Production*, 47, 283–297.

Matthew, J. Realff, Jane C. Ammons, & David J. Newton (2004) Robust reverse production system design for carpet recycling, *IIE Transactions*, 36(8), 767–776.

Mazzanti, M., & Zoboli, R. (2008). Waste generation, waste disposal and policy effectiveness. *Resources, Conservation and Recycling*, 52(10), 1221–1234.

McAuley, J. W. (2003). Global sustainability and key needs in future automotive design. *Environmental Science and Technology*, 37(23), 5414–5416.

Min, H., Jeung Ko, H., & Seong Ko, C. (2006). A genetic algorithm approach to developing the multi-echelon reverse logistics network for product returns. *Omega* 34(1), 56–59.

Mohan, R., Spiby, J., Leonardi, G. S., Robins, A., & Jefferis, S. (2006). Sustainable waste management in the UK: The public health role. *Public Health*, 120(10), 908–914.

Nahavandi, S. (2019). Industry 5.0 - A human centric solution, *Sustainability*, 11(16), 4371, 1–13.

Nakıboğlu, G. (2007). Tersine Lojistik: Önemi ve Dünyadaki Uygulamaları. *Gazi Üniversitesi İktisadi ve İdari Bilimler Fakültesi Dergisi*, 9(2), 181–196.

Ng, W. L. (2008). An efficient and simple model for multiple criteria supplier selection problem. *European Journal of Operational Research*, 186(3), 1059–1067.

Paksoy, T. (2012). Lojistik ve Tedarik Zinciri Yönetimi, www.turanpaksoy.com, Access Date: 11.05.2015.

Pati, R., Vrat, P., & Kumar, P. (2008). A goal programming model for paper recycling system, *Omega*, 36(3), 405–417.

Pitt, M., & Smith, A. (2003). Waste management efficiency at UK airports. *Journal of Air Transport Management*, 9(2), 103–111.

Pokharel, S., & Mutha, A. (2009). Perspectives in reverse logistics: A review. *Resources, Conservation and Recycling*, 53(4), 175–182.

Pongrácz, E., & Pohjola, V. J. (2004). Re-defining waste, the concept of ownership and the role of waste management. *Resources, Conservation and Recycling*, 40(2), 141–153.

Porter, M. E. (1985). *The Competitive Advantage: Creating and Sustaining Superior Performance* (557 p.). New York, NY: Free Press.

Porter, M. E., & van der Linde, C. (1995a). Green and Competitive: Ending the Stalemate. *Harvard Business Review*, 73(5), 120–134.

Porter, M. E., & van der Linde, C. (1995b). Toward a New Conception of the Environment-Competitiveness Relationship. *Journal of Economic Perspectives*, 9(4), 97–118.

Rada, F.F., Oghazi, P., Palmie, M., Chirumallad, K., Pashkevich, N., Patele, C.P., & Sattarif, S. (2022). Industry 4.0 and supply chain performance: A systematic literature review of the benefits, challenges, and critical success factors of 11 core Technologies, 105, 268–293.

Rao, P., & Holt, D. (2005). Do green supply chains lead to competitiveness and economic performance? *International Journal of Operations & Production Management*, 25(9), 898–916.

Rogers, D. S., & Tibben-Lembke, R. (2001). An Examination of Reverse Logistics Practices. *Journal of Business Logistics*, 22(2), 129–148.

Rogers, D. S., Ronald S., & Tibben-Lembke, R. (1998). *Going Backwards: Reverse Logistics Trends and Practices* (pp. 1–283). University of Nevada, Reno Center for Logistics Management.

Sahoo, K. S., Tiwary, M., Luhach, A. K., Nayyar, A., Choo, K. K. R., & Bilal, M. (2021). Demand–Supply-Based Economic Model for Resource Provisioning in Industrial IoT Traffic. *IEEE Internet of Things Journal*, 9(13), 10529–10538.

Saini, N., Malik, K, & Sharma, S. (2023). Transformation of Supply Chain Management to Green Supply Chain Management: Certain investigations for research and applications, *Cleaner Materials*, 7, 100172, 1–14.

Sarkis, J. (2003). A Strategic Framework for Green Supply Chain Management, *Journal of Cleaner Production*, 11, 397–409.

Schmidheiny, S. (1992). *Changing course: A global business perspective on development and the environment*, (448 p). The MIT Press; English Language edition.

Sheu, J.-B., Chou, Y.-H., & Hu, C.-C. (2005). An integrated logistics operational model for green-supply chain management. *Transportation Research Part E: Logistics and Transportation Review*, 41(4), 287–313.

Singhvi, A., Madhavan K. P., & Shenoy, U. V. (2004). Pinch analysis for aggregate production planning in supply chains. *Computers & Chemical Engineering*, 28, 993–999.

Srivastava, S. K. (2007). Green supply-chain management: A state-of-the-art literature review. *International Journal of Management Reviews*, 9(1), 53–80.

Stevels, A. (2002). Green supply chain management much more than questionnaires and ISO 14.001. *IEEE International Symposium on Electronics and the Environment*, San Francisco, CA, 96–100.

Sundin, E., & Bras, B. (2005). Making functional sales environmentally and economically beneficial through product remanufacturing. *Journal of Cleaner Production*, 13(9), 913–925.

Sustainability (2023). https://www.migroskurumsal.com/en/sustainability/our-reports#surdurulebilirlik Access date: 12.01.2023

Sustainability Reports (2021). https://www.sutas.com/assets/uploads/hakkimizda/Sutas_Sustainability_Report_2021.pdf, Access date: 12.01.2023.

Tan, R. B. H., & Khoo, H. H. (2005). An LCA study of a primary aluminum supply chain. *Journal of Cleaner Production*, 13(6), 607–618.

Tayal, A., Solanki, A., Kondal, R., Nayyar, A., Tanwar, S., & Kumar, N. (2021). Blockchain-based efficient communication for food supply chain industry: Transparency and traceability analysis for sustainable business. *International Journal of Communication Systems*, 34(4), e4696.

Türkay, M., Oruç, C., Fujita, K., & Asakura, T. (2004). Multi-company collaborative supply chain management with economical and environmental considerations. *Computers & Chemical Engineering*, 28(6–7), 985–992.

Vollmann, T. E., Berry, W. L., & Whybark, D. C. (2005). *Manufacturing planning and control systems*. New York: Irwin/McGraw-Hill.

Walker, H., Di Sisto, L., & McBain, D. (2008). Drivers and barriers to environmental supply chain management practices: Lessons from the public and private sectors. *Journal of Purchasing and Supply Management*, 14(1), 69–85.

Wang, J., Zhu, L., Feng, L. and Feng, J. (2023). A meta-analysis of sustainable supply chain management and firm performance: Some new findings on sustainable supply chain management, *Sustainable Production and Consumption*, 38, 312–330.

Waste Framework Directive of 2008 (WFD) (2023), https://www.eea.europa.eu/policy-documents/waste-framework-directive-2008–98-ec, Access Date: 12.01.2023.

What Is Green Supply Chain Management? (2023), https://www.gep.com/knowledge-bank/glossary/what-is-supply-chain-management#:~:text=Green%20supply%20chain%20management%20(GSCM,recycle%2C%20reclaim%20and%20degradable, Access Date: 12.01.2023.

Wells, P., & Seitz, M. (2005). Business models and closed-loop supply chains: A typology. *Supply Chain Management: An International Journal*, 10(4), 249–251.

What Is a Closed-Loop Supply Chain and How Can it Help? https://revolutionized.com/closed-loop-supply-chain/, Access Date: 12.05.2022.

Wright, E., Azapagic, A., Stevens, G., Mellor, W., & Clift, R. (2005). Improving recyclability by design: A case study of fibre optic cable. *Resources, Conservation and Recycling*, 44(1), 37–50.

Xu, X., Lu, Y., Vogel-Heusers, B., & Wang, L. (2021). Industry 4.0 and Industry 5.0-Inception, conception and perception, *Journal of Manufacturing Systems*, 61, 530–535.

Yuan, S., & Pan, X (2023). The effects of digital technology application and supply chain management on corporate circular economy: A dynamic capability view, *Journal of Environmental Management*, 341, 1–13.

Zhu, Q., Sarkis, J., & Lai, K. (2007). Initiatives and outcomes of green supply chain management implementation by Chinese manufacturers. *Journal of Environmental Management*, 85(1), 179–189.

Zhu, Q., Sarkis, J., & Lai, K. (2008). Green supply chain management implications for "closing the loop". *Transportation Research Part E: Logistics and Transportation Review*, 44(1), 1–18.

Zhu, Q., & Sarkis, J. (2006). An inter-sectoral comparison of green supply chain management in China: Drivers and practices. *Journal of Cleaner Production*, 14(5), 472–486.

Chapter 10

Influence of Internal Audit on Implementation of Environmental, Social, and Governance Factor
Case of Public Listed Companies in Oman

Ali Rehman
A'Sharqiyah University, Ibra, Oman

Yuvaraj Ganesan
Universiti Sains Malaysia, Penang, Malaysia

Nahad Al-Maskari
A'Sharqiyah University, Ibra, Oman

10.1 INTRODUCTION

Organizations worldwide are facing increasing pressure to prioritize ESG (Environmental, Social, and Governance) aspects alongside profits. However, many organizations continue to prioritize profits over ESG factors, including some of the largest and most influential companies such as Exxon Mobil, Chevron, Toyota, Southern Company, Sempra Energy, Conoco Philips, Glencore, British Petroleum, and OMV Petroleum (Armstrong, 2021). Despite being from various sectors and countries that promote and fund ESG initiatives, these organizations have negative impacts on environmental policies. While regulatory requirements for ESG are swelling every day, the disclosure requirements for ESG compliance are not sufficient. Thus, internal auditing activity (IA) is crucial in ensuring ESG compliance.

IA is an objective assurance and independent authority available within an organization and designed to add value (Rehman, 2021). IA is the agent to the principal (shareholders) that provides satisfaction and assurance for the risk management, control, governance, and achievement of organizational mission and vision. However, IA is not part of the formation of the mission, vision, and goals of an organization. In this case, if organizations opt not to include the ESG as part of their mission, then it would be difficult for IA to provide any sort of assurance and IA would be concentrated only on the assurance of internal controls and governance issues (Stojanović & Andrić, 2016).

Industry 5.0, also referred to as the "human-centered industry," signifies the latest stage in the progression of industry and manufacturing. It emphasizes

DOI: 10.1201/9781003489269-10

the human experience, with technology serving a supporting role. Within IA, organizations can utilize IA's expertise to facilitate their transition to Industry 5.0. IA's crucial role involves helping organizations identify and mitigate risks, ensuring compliance, establishing effective governance structures, measuring performance, and managing change, as highlighted in a recent report by the Society of Corporate Compliance and Ethics & Health Care Compliance Association (SCCE & HCCA) (Kuldova, 2022). By leveraging IA's guidance in areas such as risk assessment, energy management, sustainable supply chain, emissions management, and reporting, organizations can successfully navigate the transition, optimize operations, meet ethical and social responsibilities, and achieve sustainability objectives.

Industry 5.0, the next stage of industrial evolution, shifts its focus from purely technological advancements to a more human-centric approach (De Giovanni, 2023). It integrates ESG initiatives with technological innovation. Organizations prioritizing ESG initiatives aim to meet sustainability standards, including addressing climate change and waste management. ESG investments help industries reduce energy consumption, manage waste, comply with regulations, and enhance overall performance (Madsen & Slåtten, 2023). Industry 5.0 emphasizes environmental sustainability, social inclusivity, fair labor practices, and responsible governance, emphasizing the well-being of workers and the planet (Kasinathan et al., 2022). By embracing Industry 5.0 and prioritizing ESG initiatives, companies can create a positive image, attract customers and investors, and contribute to a more sustainable and socially conscious future (Fontaine, 2013).

Industry 5.0 represents a departure from Industry 4.0, focusing on integrating ESG initiatives into technological advancements (Ghobakhloo et al., 2022). ESG encompasses environmental sustainability, waste management, and adherence to regulations. By implementing ESG initiatives, industries can reduce operational costs, improve employee motivation, and gain support from governments and stakeholders (Kuo et al., 2016). Industry 5.0 places a strong emphasis on societal well-being, promoting inclusivity, diversity, and fair labor practices. Effective governance is crucial for responsible practices in Industry 5.0, with considerations such as data privacy, cybersecurity, and corporate transparency. This transformation in industry governance centers on human-centric strategies that prioritize talent, diversity, empowerment, and creating unique value for employees. Embracing Industry 5.0 and ESG initiatives enables organizations to navigate the challenges of sustainability, societal impact, and ethical practices while fostering a culture of innovation and social responsibility (Mourtzis et al., 2022).

Much focus is provided by the United Nations (UN) and European Union (EU) on the issues of ESG and now the environment is one of the main factors of governance. Regulatory requirements for ESG are swelling with every passing day and organizations are pushed to provide disclosures in their financial statements which are related to ESG. In August 2010, the

International Integrated Reporting Council (IIRC) was established with the aim of providing a universally accepted framework and disclosures related to ESG information (Soh & Martinov-Bennie, 2015). However, after a decade of IIRC provided framework, organizations are still in the early stages of the implementation of ESG and they are not able to assess the effectiveness of the ESG disclosures on organizational performance and its behavior toward society (IIA, 2021). There are more than 600 ESG assessment methods available but none of them can be considered a standard system applicable to all organizations. In accordance with recent research, it was identified that due to several evaluation standards, investors assess ESG differently as compared to asset managers and ratings materially differ from one assessment method to the other (Wong & Petroy, 2020).

The environmental part of ESG is focused on the climate only. In the USA, the Securities and Exchange Commission (SEC) proposed changes to the disclosure requirements which will demonstrate the climate-related risk that can adversity impact businesses. Under the new requirements organizations which follow the USA SEC's requirements are necessitated to provide information related to the climate risk, the material impact of this risk on organizational operations and financial performance, and how such risk can impact organizations' strategies. SEC believes that such disclosures will help investors to make appropriate decisions for their future investments (SEC, 2022). With the focus on climate only, control environment is completely ignored (Rehman, 2023). Control environment is defined by the Committee of Sponsoring Organizations of the Treadway (COSO) as

> the set of standards, processes, and structures that provide the basis for carrying out internal control across the organization. The board of directors and senior management establish the tone at the top regarding the importance of internal control including expected standards of conduct.
>
> (Schandl & Foster, 2019)

IA is required to monitor, provide recommendations, and give assurance on the control environment. A strong and auditable control environment ensures that organizations are operating in accordance with the approved policies, regulatory requirements, and industry standards (Barišić & Tušek, 2016; Diamond, 2013; Fourie & Ackermann, 2013). The control environment assists governance management in carrying out their oversight responsibilities. It is highly recommended that the environmental factor of ESG should not only focus on the climate factor but also emphasize on control environment. If a control environment is properly implemented, it can cover the issues related to climate and its related impact on organizational performance (Rehman, 2023).

The social part of ESG focuses on the promotion of equality among employees and society, relationship with laborers, and investment in human-related capital (Commission, 2021). Social issues are often confused with or perceived as corporate social responsibility (CSR). The social aspect of ESG considers society as a whole and tries to create a positive impact this includes data protection, respecting privacy, fair dealing of business affairs, and helping the community to grow with the growth of the organization; furthermore, the social aspect of ESG is controlled and regulated. CSR deals with the organizational commitment of what they choose to commit; fulfillment of this commitment is entirely dependent upon the executive/senior management (Polley, 2022; Sila & Cek, 2017).

The social aspect of ESG has existed in our society for generations; however, few considerations are provided by the governments or the corporate world. Many policies were drafted, and provisional laws were created but to date, there is no visible impact available. In accordance with the study, it was identified that there is no relationship between the social responsibility of an organization with its financial performance (Han et al., 2016). The social aspect is included in the organizational budget or fewer contributions are made for the sake of attracting the investment and to please shareholders (Ellemers & Chopova, 2021). The purpose of society was completely ignored and financial gains were Fraud cases like Formosa Plastics and Atlantic Richfield are recent evidence where organizations are harming society for the sake of their profits and the legislative authorities are busy imposing petty fines and not mitigating the root cause (Patton & Reisch, 2021; Jobin, 2021; Justice, 2020).

In the current business environment, it is necessary for organizations to identify which social aspects are necessary for their organization that can have a positive impact on its employees, customers, and society. Clear messages and communication should be made with the stakeholders and not only with the shareholders with the embedded accountability. The role of IA is very critical in the manner of auditing and providing opinions for the social aspect of ESG. In order to ensure routine and rigorous investigation and reporting of incidents, IA should clearly define risks and materiality. It is essential that the IA can identify the tools it needs to accurately collect and measure performance data, with adequate segregation between information providers and approvers. In general, companies should align their policies and assurance activities such as IA with their business risks and strategic goals. It is particularly important for IA to take necessary accountability steps as a growing number of companies release annual reports and incorporate ESG-related key performance metrics into executive compensation contracts and shareholders reports (IIA, 2021).

The governance aspect of ESG deals with the governance of the organization which can have an impact on the company and society. Governance of ESG is defined as decision-making and segregation of duties among the board of directors (BOD) and executive management with the focus of making the organization more responsible toward environmental and social aspects. Governance in this scenario also impacts organizational culture and responsible investment (Global S&P, 2020; Miler, 2021; Bank, 2020). It is worth noting that to date, there is no specific definition of the governance aspect of ESG, and it cannot be utilized interchangeably with the corporate governance available and imposed by many governments as codes of corporate governance.

Corporate governance has standard definitions across several codes developed by many countries; however, all the definitions explain the policies, processes, and controls toward individuals and businesses in the development of the organization and its related economic system. Corporate governance also obliges organizations to demonstrate relevant information that is comparable and enables stakeholders to make appropriate decisions (Rehman & Hashim, 2018; Broni & Velentzas, 2012). Codes of corporate governance in Oman were first issued in the year 2002 and revised codes were issued in the year 2016 (Rehman, 2021). Fewer changes to these codes are made as and when required. Public listed companies in Oman are necessitated to follow the codes and the Commercial Companies Law for better control and disclosure purposes. Codes in Oman require that all boards of directors must have expertise in finance and accounts, understand general business trends, be aware of industry best practices, have proper experience in the business of the company, and have the ability to contribute as an effective steward of the organization (CMA, 2016). Under the code, the BOD is not required to have knowledge of ESG, and they are not answerable for the impact on the environment or social aspects of their businesses.

The role of IA in corporate governance is obvious and regulators in Oman have given more authority to IA to perform their work with complete independence and objectivity (Rehman, 2021). IA is required to perform audits and provide meaningful recommendations which enhance the organizational value. IA's key focus area is governance, risk, and compliance within organizations which enables it to provide opinions and recommendations. IA reports to the Audit and Risk Committee (ARC) and provides overall opinion during the financial yearend of the company. Capital Market Authority (CMA) is the regulatory authority in Oman that governs and manages the codes of corporate governance. IA in public listed companies is obliged to follow the guidelines provided by the CMA and the standards provided by the Institute of Internal Auditors (IIA) (CMA, 2021; Thottoli, 2022). There are several other mandatory guidelines issued by many other authorities but in Oman, IA is required to follow only the guidelines provided by CMA and IIA. To date, both CMA and IIA are silent about the guidance or standards of ESG and the role of IA in it. IIA is considering adding or amending their framework to reflect ESG in IA's charter; whereas there is no visible effort available from CMA that can demonstrate that they are planning to amend codes of corporate governance or guidelines for IA to include ESG parameters or ESG-related clauses.

With the application of stewardship theory and agency theory, this chapter intends to define the impact of IA on ESG within Omani public listed companies. IA is the agent for its principle under agency theory (Adams, 1994; Colbert & Jahera, 1988), and investors or shareholders are responsible stewards of the organization (Chevrollier et al., 2020). Stewardship theory provides a supportive basis for change and has the potential to initiate significant changes in the behavior of the investors (Klettner, 2021). This study identified the potential improvements within the codes of corporate governance and the framework of IA. Recommendations were also be provided for the inclusion or incorporation of the ESG clauses within organizational policies and the inclusion of a control environment along with the climate factor. This study can be beneficial to regulators, standard-setting bodies, and organizations.

This study aims to fill the research gap by examining the impact of IA on ESG implementation and offering recommendations for the integration of ESG clauses into organizational policies, amendments to corporate governance codes, and the advancement of knowledge regarding ESG factors specific to the Omani context. By addressing these objectives, this research aims to contribute to the existing body of knowledge, inform future research and practice, and promote alignment with international ESG standards and regulations. Furthermore, by examining the experiences of companies in Oman, this research can provide valuable insights that can be used for research and practice in other regions with similar social, economic, and regulatory contexts such as Gulf Corporation Council (GCC) countries.

Till date, there is no study conducted for the Omani public listed companies which identify the impact of IA on ESG. This study is an effort to enhance the existing body of knowledge and, if adopted by CMA, will enable the Omani listed companies to be aligned with the regulations of many other countries of the world such as the European Union and the USA. This alignment will also assist in capturing the ESG funds committed by several authorities.

10.1.1 Objectives of the chapter

The objectives of the chapter are as follows:

- To identify the impact of IA on ESG implementation within Omani public listed companies;
- To offer recommendations for the integration of ESG clauses into organizational policies and amendments to corporate governance codes;
- To advance knowledge regarding ESG factors specific to the Omani context;
- To contribute to the existing body of knowledge, inform future research and practice, and promote alignment with international ESG standards and regulations;
- And, to provide valuable insights for research and practice in other regions with similar social, economic, and regulatory contexts, such as GCC countries.

10.1.2 Organization of the chapter

The rest of the chapter is organized as: Section 10.2 defines the literature review, Section 10.3 defines associated theories, Section 10.4 describes methodologies, Section 10.5 defines results and discussion, and, finally, Section 10.6 concludes the chapter with future scope.

10.2 LITERATURE REVIEW

This section will discuss the relevant literature and associated theories. This section will also identify the gaps within literature and ESG-related codes and shed light on the correlation of ESG and IA with Industry 5.0.

10.2.1 Environmental, social, and governance (ESG) factors

Organizational sustainability is now linked with the ESG factors of governance. ESG highlights the facts that are relevant to the organizational survival and its future viability to support society. The The environmental factor focuses on the climate aspect which can be impacted by the organization and how organizations are reducing environmental emissions. Social factors identify how organizations are socially active in generating ethical values, respecting and promoting their employees, and assuring the proper human rights. The governance factor of ESG is related to the organizational efforts to maintain good investor relations through proper management systems and efficient processes (Dicuonzo et al., 2022). ESG-related matters will be going to impact the organizations positively and negatively (OCEG, 2021). ESG will impact positively when organizations make their ESG efforts more visible and not only limited to financial statement disclosures. The negative impact will be obvious when organizations fail to demonstrate their efforts toward ESG. In accordance with recent research, it was identified that organizations are more focused on the disclosure of the governance factor, and very little transparency is provided for the environmental factor, whereas social factor information varies materially from organization to organization and from sector to sector (Tamimi & Sebastianelli, 2017). These factors are discussed in detail in the below sub-sections.

10.2.2 Environmental factor

The environmental factor in ESG highlights the importance of protecting the environment and minimizing environment-related damages. Environment caters for the controlling of pollution, protection of natural resources, and having a positive impact on climate change (Lee & Suh, 2022).

Organizations are required to be knowledgeable about energy conservation or consumption, water usage, greenhouse gas emissions, and reduction of carbon footprints. With this knowledge, organizations might end up earning more profits and at the same time create a positive impact on the climate.

Organizations like Google set the example of achieving environmental sustainability by investing in energy-efficient data centers, having green transportation, and utilizing recyclable materials to build their offices (Team, 2022). There are other examples of a similar sort where garment industries utilize recyclable materials to produce their products; however, there is no established link between environmentally friendly organizations and their increased earnings (Lee & Suh, 2022). In the year 2016, Volkswagen admitted to conducting environmental fraud (Jong & Linde, 2022) and was penalized for USD 2.8 billion; however, its sales grew from USD 216 billion to USD 250 billion from the year 2016 to 2021 (Carlier, 2022). The environmental factor is a sentiment that is difficult to measure and differs from person to person.

The environmental factor only relates to the climate and does not consider the organizational environment which impacts the social and governance factors. It is important to note that the satisfaction of all stakeholders is necessary including customers and employees and without emphasizing the organizational environment ESG cannot succeed. There are no specific environmental measures available for the organizations and there is no clear link between the environmental factor of ESG with social and governance factors. Available measures are only for specific industries and restricted to a few countries (Worker, 2015; Lee & Suh, 2022). Furthermore, there is no application or software available which can measure the environmental impact of the organizations, and hence reliable reporting cannot be made properly (Parmelee, 2021).

Despite the non-availability of a standard definition, basis for measurement, or disclosure, it is understood by the organizational executives that climate change is not a distant threat anymore, and mitigating actions are required to reduce or diminish any adverse effects created by their organization. Many organizations are facing the threats of scarcity of resources and natural disasters; however, a gap exists between organizational feelings and actual actions (Parmelee, 2021; Mazzotta, 2021). It is required that countries translate their environmental policies into the codes of corporate governance and develop related standards. Compliance with the codes and standards can be assured via regulators and IA. This could be the first step toward environmental policy implementation. Organizations are facing pressure from their customers to be environmentally friendly; however, in the absence of any measures or specific requirements, it would be difficult to identify whether customers' needs were met or not. Furthermore, it is also required to develop standardized mechanisms to measure the environmental effect regardless of the nature of the organization and the country of its operations (Lee & Suh, 2022).

10.2.3 Social factor

Social factors deal with the organizational relationship with their employees, their behavior toward the society in which they operate, and their contribution to the political environment. The social factor can also be termed as social investment toward sustainability. The social factor of ESG has a financial impact and could also impact the organizational reputation. The social aspect can also be termed as corporate social responsibility; however, the social of ESG is different from CSR (Park et al., 2022; Pelosi & Adamson, 2016). When organizations are socially active, they think beyond numbers invest in their human capital, and also try to build the society (Escrig-Olmedo et al., 2019).

CSR is required to be focused on the social well-being of the organization which increases the organizational value (Feroz & Kumar, 2012). Organizations are obliged by the law and codes of corporate governance to be socially responsible and for this reason, many directives have been issued and implemented in many countries, whereas the social aspect of ESG tries to link it with the environmental and governance aspects of the organization (Gerard, 2019). Social factors play a major role in the organizations' demonstration of their non-financial performance and therefore it is the focus of many potential investors; however, there is no clarity available in the shape of standards on what should be considered a socially active organization (Waas, 2021) and the social aspect is not integrated into many organizational business models (Pelosi & Adamson, 2016).

There is a lack of board oversight on the social aspects and less concentration is provided by the executive management on this factor; furthermore, organizations do not have any data to identify and mitigate social risk (Pelosi & Adamson, 2016). This highlights the fact that there is a lack of accountability and organizations are not eager to build positive associations with the communities in which they are operating.

Similar to the environmental aspect, there is no definitive measurement available for the social aspect (Wong & Petroy, 2020; Berg et al., 2019). The social factor is utilized by the organizations just to fulfill the disclosure needs as organizations do not see any potential benefits toward investment in the social aspect (Becchetti et al., 2022). A similar context was provided by Zeilina and Toth (2021) where it was stated that the environment is the more important factor for organizations as compared to the social factor; organizations do not have any data to act on the social contribution and organizations fail to measure the qualitative data as social aspect cannot be measured in quantitative terms. If the social factor is properly implemented, then it can assist organizations in reducing economic uncertainties, eliminating bias in bonus distributions and salary increments, mitigating the risk of human trafficking, and helping society progress toward sustainability. Social factors help in creating societies with socio-economic equality, diversity, and inclusion and provide rights to all individuals.

10.2.4 Governance factor

Governance factors refer to the organizational governance available within the organization. Governance refers to the policies and procedures that are adopted by the organizations enabling them to achieve their mission and vision. Governance management refers to those who are responsible for the implementation of policies and the achievement of good/mature governance (Rehman & Hashim, 2018).

For ESG, the governance factor is working toward the achievement of social and environmental factors. Poor/immature governance can lead to many environmental or social frauds (Global S&P, 2020; Jong & Linde, 2022). Governance in an organization gives authority and creates account-abilities for the BOD and executive management (EM). The policies are approved by the BOD and implemented by EM. For the assurance part of governance, the ARC and IA play a vital role. Assurance is provided to the shareholders to ensure that the established policies are achieving the desired results and EM is implementing the policies properly (Rehman, 2021).

For the development of sustainable business practices, good/mature corporate governance is required. Mature corporate governance is essential for effective capital allocation, preserving long-term growth, employment sustainability, and providing better services for society and its related development (Khan, 2019). The governance factor assists in creating openness and trust, shareholders' rights, and control executives' payments, and developing stakeholder engagement plans (Li et al., 2021).

Governance reporting provides information related to the economic, environmental, and social aspects of the organizations (Pritchard & Çalıyurt, 2021). These aspects can also be considered as dimensions for sustainable corporate governance. These dimensions were further extended by adding two more aspects namely political and territorial dimensions. In the year 2019, two additional dimensions were added namely corporate governance and the nature of the product (Dvořáková & Zborkova, 2014; Lombardi et al., 2019).

Corporate sustainability can be achieved by integrating all seven dimensions of sustainable development. However, several scholars believe that the economic dimension provides the strongest financial foundation and prevents organizations from collapsing due to financial issues. Through organizations' governance practices and economic development, companies can achieve corporate governance sustainability (Rehman & Hashim, 2021) and it can be established that by achieving sustainable corporate governance, organizations can achieve environmental and social goals.

In the current business environment, sustainability reports do not address the governance aspect of ESG, and their prime focus is on the environmental aspect only. This highlights the issue that complete metrics of ESG cannot be obtained and still rating agencies are gauging these factors in silos (OCEG, 2021).

10.2.5 ESG and Industry 5.0

The concept of Industry 5.0 has sparked numerous discussions, leaving people curious about what lies ahead. While Industry 4.0 primarily centered on technological advancements (Kumar & Nayyar, 2020) such as AI, blockchain, and the Internet of Things in various industries, Industry 5.0 brings a humanistic approach to the machine-dependent nature of Industry 4.0 (Pramanik et al., 2022). Industry 5.0 represents a leap forward, merging the realms of technology, humanity, society, environment, and sustainability. In Industry 5.0, the focus shifts toward integrating ESG initiatives into technologically innovative.

Organizations that prioritize the development of ESG initiatives aim to adhere to a set of standards that uphold their socially conscious image in the eyes of investors and customers. Kotsantonis and Serafeim (2019) explained that ESG initiatives encompass environmental criteria that allow organizations to operate sustainably on a global and local scale, including corporate policies geared toward addressing climate change. Another influential factor within ESG initiatives is waste management, considering the increasing pollution levels worldwide and their impact on human lifestyles in different countries (Li et al., 2021).

Since early industrialization, industries have been expanding and developing; however, this has also led to an increase in solid waste generated from these industries. This necessitates the implementation of environmental measures throughout the United Nations Sustainable Development Goals 2030 (UN SDG). Waste management involves the proper utilization of waste materials through recycling or reusing, thereby reducing landfill volumes and waste (Pedersen et al., 2021). ESG initiatives encompass investments in waste management, which not only mitigate environmental impact but also reduce raw material costs through recycling. Managing the entire supply chain, from manufacturing to packaging and delivery, is crucial for minimizing carbon footprint and waste generation. Consequently, ESG investments help assess and improve overall industry performance, ultimately enhancing the sustainability of the entire industry area.

Implementing ESG propositions facilitates the reduction of energy consumption and water usage, enabling industries to lower their overall operational costs. Organizations that prioritize sustainability in their operations can gain greater flexibility through deregulation, earning governmental support and subsidies (Phillips et al., 2019). Moreover, by embracing ESG initiatives industries can also boost employee motivation levels (Sachin and Rajesh, 2022). Industries are increasingly leveraging benefits from suppliers, compliance, and enhanced visibility to promote their operations in an eco-friendly manner (Stranieri et al., 2019). Therefore, prioritizing ESG initiatives in industrial operations not only showcases progress in environmental sustainability but also creates a positive image that attracts customers and investors.

10.2.5.1 Environmental transformation in Industry 5.0

Industry 5.0 is not solely focused on efficiency and productivity like its predecessor, Industry 4.0 (Madsen & Slåtten, 2023). Industry 5.0 embraces sustainability as a core principle, placing the well-being of workers and the planet at the center of its vision (Xu et al., 2021). It examines the integration of technologies and practices that minimize ecological footprints and promote renewable energy, efficient waste management, and eco-friendly manufacturing processes. Additionally, it highlights the importance of adhering to environmental regulations and standards, addressing climate change, and preserving natural resources within the context of Industry 5.0 (Maddikunta et al., 2022).

Embracing Industry 4.0 has been a significant focus for many companies, driven by technological advancements and automation. However, the next revolution, Industry 5.0, is already emerging, bringing with it a paradigm shift toward environmental sustainability (Broo et al., 2022). Industry 5.0, according to the European Union, envisions an industry that goes beyond efficiency and productivity to emphasize its role in society (Saniuk et al., 2022). It prioritizes the well-being of workers and utilizes new technologies to create prosperity while respecting the planet's ecological limits. This represents a radical departure from the traditional profit-oriented approach, highlighting the urgent need to repurpose the core objectives of the industry toward a more environmentally conscious and sustainable future.

10.2.5.2 Social transformation in Industry 5.0

Social factors are an integral part of Industry 5.0, as it emphasizes the human aspect of technological innovation. Industry 5.0 fosters inclusivity, diversity, and fair labor practices, ensuring that technological advancements benefit all members of society. Furthermore, the sub-section explores the impact of Industry 5.0 on employment, skill development, and the redistribution of wealth. It also investigates the potential social challenges and ethical considerations that arise from the implementation of Industry 5.0 (Carayannis & Morawska-Jancelewicz, 2022; Carayannis & Morawska, 2023; Kasinathan et al., 2022).

Industry 5.0 goes beyond economic value and places a strong emphasis on societal value and well-being. This shift from a welfare-oriented approach to one focused on well-being reflects a growing awareness of the importance of considering the social impact of businesses (Cenci & Cawthorne, 2020). While concepts such as CSR, ESG, and the Triple Bottom Line have existed for some time, Industry 5.0 takes this idea further by centering the definition of industry around people and the planet rather than solely on profits and growth (De Giovanni, 2023). This aligns with a broader trend in recent years on business and economics that integrates social value and the well-being of humans and society.

10.2.5.3 Governance transformation in Industry 5.0

Effective governance is crucial in harnessing the benefits of Industry 5.0 and ensuring responsible and ethical practices (Ghobakhloo et al., 2023). The role of governments, regulatory bodies, and policymakers in shaping the development and deployment of Industry 5.0 technologies is critical (Saniuk et al., 2022) and this can include legal and ethical considerations related to data privacy and cybersecurity, and intellectual property rights are given additional value within Industry 5.0. Furthermore, transparency, accountability, and corporate governance are the major pillars in facilitating the successful implementation of Industry 5.0 initiatives.

Industry 5.0's impact extends to the governance of organizations (Olaizola et al., 2022). A human-centric strategy is at the core of Industry 5.0, promoting talent, diversity, and empowerment (Huang et al., 2022). This shift in perspective implies a departure from viewing people as mere resources to seeing them as the ultimate beneficiaries of organizational efforts. In a job market where attracting and retaining talent has become increasingly challenging, organizations need to adapt their strategies to prioritize the well-being and unique value creation of their employees (Earle, 2003). By placing human-centricity at the forefront, Industry 5.0 challenges the traditional notion of a strategy focused solely on gaining a competitive advantage for customers. Instead, it calls for strategies that prioritize a competitive advantage through creating unique value for employees, fostering a culture of empowerment, and supporting individual growth and fulfillment.

10.2.6 Internal audit

Internal audit has evolved significantly over the past century, from its origins as a compliance-focused function to a strategic and value-added service. In the early 20th century, internal audit was primarily focused on ensuring compliance with accounting standards and preventing fraud. However, with the growth of large corporations and the increased complexity of business operations, the role of internal audit expanded to include broader risk management and control functions.

The IIA was established in 1941, as a professional organization dedicated to promoting the importance and value of internal audit. The organization initially focused on providing education and training to internal auditors, as well as establishing professional standards for the industry. The IIA also developed a code of ethics, which set the standard for the ethical conduct of internal auditors. Over the years, the IIA has grown to become a global organization with more than 200,000 members in over 170 countries. The organization has also expanded its scope beyond traditional internal audit functions, to include areas such as risk management, governance, and control.

The International Professional Practices Framework (IPPF) for the IIA is one of the most significant developments in the history of internal auditing;

it is a comprehensive framework that provides guidance and standards for the practice of internal auditing. The framework was established in 1999 and has been revised multiple times since then to account for changes in the profession and business environment. The IIA is currently in the process of introducing a new IPPF. The Topical Requirements and Guidance is a newly proposed IPPF element that is undergoing preliminary consideration and development for future IPPF (Ruud et al., 2011).

The IIA officially defined the internal audit function as an independent, objective assurance and consulting activity designed to add value and improve an organization's operations. The function is aimed at aiding an organization to accomplish its objectives by bringing a systematic, disciplined approach to evaluate and improve the effectiveness of risk management, control, and governance processes. The internal audit function is responsible for providing assurance by thoroughly reporting and assessing the organization's control process, risk management as well as the effectiveness of corporate governance to achieve organizational objectives. It is essential for the internal audit function to provide assurances that are on par with the organizational strategies and follow the Institute of Internal Audit standards. For an internal audit function to perform at its best implementation, it should be free from any influences (Christopher et al., 2009). Having an independent internal audit function, an internal audit is able to carry out its assessment objectively, assisting management in internal control, organization risk management, and governance processes (Chang et al., 2019).

In recent years, the role of internal audit has continued to evolve, with a greater focus on strategic risk management and the use of data analytics and technology (Foronda et al., 2023). Internal auditors are increasingly expected to provide insights and recommendations to management and the BOD, based on their analysis of organizational risks and opportunities. In line with the development, the IIA revised the Internal Audit Competency Framework in 2022. This is a comprehensive tool that helps internal auditors assess their skills and identify areas for development, while also providing organizations with guidance on how to develop and manage internal audit talent. The new IA competency framework consists of four dimensions, namely Professionalism, Performance, Leadership & Communication, and Environment. The framework provides an explicit and concise professional development plan for internal auditors at all career levels. The framework identifies four knowledge areas centered on diverse standards, situation-specific functions, and key proficiencies, with three distinct competency levels that progress from general awareness to applied knowledge to expert practitioner. The comprehensive and concurrent strategy defines and imparts the knowledge and skills required to navigate a successful career in internal auditing, with an emphasis on best practices and practical applications (Auditors, 2023).

The history of IA and the IIA's role has been instrumental in shaping the profession and promoting its value within organizations. The Standards have been contributing to providing a framework for IAs to follow, and the IIA's ongoing efforts to promote good governance and responsible business practices have helped to ensure that IA continues to play a vital role in organizations today.

10.2.6.1 Internal audit and Industry 5.0

Industry 5.0, also known as the "human-centered industry," represents the latest stage in the evolution of industry and manufacturing. In this new era, the focus shifts toward the human experience, with technology playing a supportive role. Within the context of IA, organizations can leverage IA's expertise to support their transition to Industry 5.0. According to a recent report by the Society of Corporate Compliance and Ethics & Health Care Compliance Association (SCCE & HCCA), IA can play a vital role in helping organizations identify and mitigate risks, ensure compliance with regulations and standards, establish effective governance structures, measure performance, and manage change.

By leveraging IA's guidance in these areas, organizations can successfully navigate the transition to Industry 5.0, optimize their operations, and achieve their goals while meeting ethical and social responsibilities. It is important to note that Industry 5.0 can also contribute significantly to achieving sustainability objectives. IA can play a crucial role in supporting Industry 5.0 initiatives, particularly in the following areas.

- **Risk assessment:** IA can help organizations identify and manage risks associated with the adoption of new technologies and processes. By analyzing data from various sources, such as environmental impact reports and supply chain data, IA can assist in identifying areas where sustainability risks are most significant. This enables organizations to take proactive measures to mitigate these risks.
- **Energy management:** IA can contribute to optimizing energy consumption in buildings and manufacturing facilities. By analyzing data from sensors and other sources, IA can help identify energy usage patterns and guide organizations in optimizing their operations to reduce energy consumption. IA can also ensure the effectiveness and efficiency of energy management systems.
- **Sustainable supply chain:** IA can monitor and improve the sustainability of the supply chain by analyzing data from suppliers and other sources. This helps identify risks and opportunities for improvement in areas such as carbon emissions, waste reduction, and ethical sourcing. Proactive measures can be taken to address potential issues and promote a more sustainable supply chain.

- **Emissions management:** IA can help organizations monitor and reduce greenhouse gas emissions. By analyzing data from emissions monitoring systems and energy usage data, IA can identify areas where emissions can be reduced and assist in implementing solutions such as renewable energy sources and energy efficiency measures.
- **Reporting:** IA can contribute to improving sustainability reporting by ensuring the accuracy and completeness of sustainability reports. By analyzing data from various sources, IA helps verify the integrity of the information included in the reports. Furthermore, IA can leverage machine learning algorithms to automate the reporting process, reducing the time and effort required to prepare reports.

IA plays a vital role in supporting organizations' transition to Industry 5.0 and their sustainability initiatives. Through effective risk assessment, energy management, supply chain sustainability, emissions management, and reporting, IA helps organizations navigate the challenges and seize the opportunities presented by Industry 5.0 while promoting a more sustainable future.

10.2.7 Integration of internal audit and ESG

To effectively address the complexities associated with ESG practices, IA should be positioned as enablers of learning and change, while also considering the professional demands for independence and objectivity. To maximize their value, IA should engage in discussions with senior management and the board to determine how best to leverage IA in successfully managing ESG-related risks. As ESG practices continue to be a hot topic for forthcoming generations, IA must seize opportunities to enhance their contributions to these efforts. Although external consultants may be considered, their usage should be selective and wise due to the ongoing, complex nature of ESG-related challenges. In essence, the ESG "Garden" is never finished, and there is no panacea or expert who can single-handedly address these challenges other than IA (Lenz & Hoos, 2023).

Due to the most significant financial frauds of the last century, IA emerged as the most advanced and excellent form of control. There are several studies available that assure IA is an effective and central part of corporate governance (Soh & Martinov-Bennie, 2015; Harasheh & Provasi, 2022). It is possible to enhance the reliability of financial statements by implementing effective internal controls and embedding IA as a control-assuring activity/function (Rehman, 2021). Several countries have introduced new directives requiring organizations to disclose nonfinancial disclosures (NFD). NFD among many aspects also includes ESG reporting. IA serves as a primary assurance for the quality of NFD and the integration of ESG by providing assurance on the functionality, reliability, and credibility of the internal controls implemented for NFD and providing assurance for their related disclosures in financial statements (Harasheh & Provasi, 2022).

IA can play a crucial role in ensuring that the company's ESG practices are implemented effectively. This is done by monitoring, assessing, and reporting on the company's ESG risks and performance. The role of IA in ESG practices is important because it helps the company identify areas where it may need to improve its ESG practices, such as reducing its environmental footprint or enhancing its social and governance practices. Additionally, an internal audit can help ensure that the company is complying with applicable laws and regulations related to ESG practices. Having an effective IA function that is focused on ESG practices has several benefits, such as improved risk management, enhanced stakeholder trust and confidence, and increased transparency and accountability.

IA is an essential function in any organization. With the ever-growing concerns for the integrity of organizational operations, IA plays a vital role in providing their independent opinion which develops the shareholders' and stakeholders' satisfaction. ESG factors and their utilization in organizations require maturity in reporting as many of the organizations focus mainly on the climate factor of their organization and very little importance is given to the social and governance aspects (OCEG, 2021). Moreover, there is no clear link available to how achieving the environmental goals will help in achieving the social and governance goals (Winteroll, 2020). ESG should be linked with each other so that three factors enhance each other's efficiency and the decline of one factor will also decline the other factors. This scenario itself requires some standard ESG framework; however, assurance of IA is required to ensure that the provided indices are correct and that they are measuring what they are designed for.

IA not only provides reporting requirements but also acts as an agent of change. IA also provides training on many issues to the BOD, ARC, and EM. There are no requirements for the BOD to have knowledge of ESG-related matters (Pelosi & Adamson, 2016); furthermore, there is no standard framework available that can measure ESG performance. Although there are many regulatory requirements and agencies available such as environmental authorities, labor ministries, and ISO standards; however, the closest framework for the measurement could be considered as COSO framework (Annandale et al., 2022). In this scenario, IA is the best-suited function that can educate the governance management and assist organizations in achieving their ESG-related goals by implementing the COSO framework and embed in its audit plans.

It is globally accepted and acknowledged that internal controls are required for the effective implementation of sustainable corporate governance and its efficient utilization (Rehman, 2021). These internal controls enhance the reliability of reports and ensure compliance with laws and regulations (Harasheh & Provasi, 2022). Once the controls are developed for ESG then assurance on these controls is required. IA can provide this assurance by ensuring that compliance is made and the developed controls are achieving desired results. IA also provides meaningful recommendations

(Dzikrullah et al., 2020) which assists in updating the controls in a timely manner to avoid any potential risk.

IA must work with agility and shift from their traditional ways of auditing. Although there are no standards made by the IIA that can cater to ESG, with a professional attitude and effective leadership qualities, IA can deliver the requirements. By adopting new technologies and methods, IA's function/department is capable of demonstrating readiness for identifying risks and opportunities. IA is continuously required to identify skills and should define a plan to fill these gaps by continually assessing its staff competencies. This will allow the IA to add value and assist in achieving ESG objectives (Foundation, 2021).

ESG has become a critical aspect of business operations. However, organizations must not focus solely on compliance, as it may not achieve the desired results. Instead, they should concentrate on creating a control environment that enhances the controls around their ESG actions. This is where IA can play a crucial role by providing assurance and advice on governance, risk, and compliance. The control principles that are necessary to mitigate the risk of fraud or ethical malpractices include a commitment to ethics and integrity, board independence, organizational governance structure, risk assessment, and control activities. IA can ensure that these principles are in place and effective in enhancing the organization's ESG practices. Therefore, IA is the only activity available in the organization that can provide assurance for the control environment and support the ESG initiatives (Rehman, 2023).

10.2.8 ESG in Oman

Similar to many countries, Oman implemented codes of corporate governance in the year 2002. These codes were updated in year 2016 CMA, and they are mandatory on all public listed companies. CMA introduced several changes after the second update and included many codes related to IA and safeguarding of minority shareholders (CMA, 2016; Sanyal & Hisam, 2017). Codes developed by CMA cover E and S factors in their 13 principles under CSR and define Corporate Social Responsibility as "linked to the company purposes and activities. The company shall seek to exercise its role as good citizen and to mitigate any adverse impact of its activities on the national economy, community or environment at large" (CMA, 2016).

In the year 2015, a charter for the Oman Center for Governance and Sustainability was established where rules were suggested for the organizations to be socially active and responsible (Sustainability, 2015). CMA via its codes made it obligatory for public listed companies to develop policies for CSR, its related budget support of the board, community engagements/developments targeted by organizations, and NFD related to CSR (CMA, 2016).

Recently Oman also established the Environment Authority (EA) via Royal Decree No. 106/2020. EA is responsible for the development of policies, procedures, and plans directed toward the protection of environment, preserving natural resources, combating pollution, and achievement of UN SDG goals related to the environment. EA is tasked to develop an environment management system and to control all environmental issues. However, there is no link created between the CSR defined by CMA which is also targeting the environment and EA which is responsible for the development of policies. Furthermore, it is obligatory for all organizations to provide for their 20% of the social responsibility budget for the institutions under Oman Charitable Association (Team O. O., 2021).

As of October 2022, there are 114 companies listed in the Muscat Stock Exchange distributed into the sectors defined in Table 10.1.

From the above 114 companies, researchers performed an analysis of disclosures made by companies listed under the industrial sector to identify the level of disclosures for ESG for the last three years. The reason for selecting the industrial sector is that they are more prone to environmental damage. Although this is out of the scope of this research, however, it would be beneficial to provide such a study for this research.

After analyzing all 39 companies, it was identified that none of the companies are providing disclosure for CSR which is inclusive of E and S factors. All companies provided detailed disclosures related to governance. This can be considered as non-compliance with the codes of corporate governance and there are no visible efforts available from EA to issue any notice to them. CMA and EA both are working in silos and there is no link or relationship between them. This creates the necessity for another empirical research to be conducted in this area in which non-compliances can be identified and recommendations can be developed.

In terms of environmental performance, Oman is ranked 110 out of 180 countries (Barbuscia, 2021); however, Oman is aiming to be among the top 20 countries by the year 2040. For the social factor and based on the Government's efforts only, Oman is ranked 70 out of 180 countries (Economics, 2021). In ranking related to governance factors, Oman's score is below moderate (Economics, 2021). Furthermore, other rating agencies

Table 10.1 Number of companies listed in Muscat Stock Exchange

Sectors	Number of companies
Financial	35
Mutual Funds	2
Industry	39
Service	38
Total	114

Source: MSX (2022)

ranked Oman's banks as high risk for ESG reporting and ESG-related policy implementation (Sustainanalytics, 2021).

10.3 ASSOCIATED THEORIES

Internal audit is the control activity available within an organization and works toward the attainment of governance, risk, and compliance features of an organization. ESG is the strategic direction and governance factor which obliges organizations to be socially active and ethically strong to develop satisfied stakeholders. Control and strategic orientation of the organizations can be viewed from the lens of both agency theory and stakeholder theory. This study utilized stewardship theory and agency theory to define the influence of internal audit over ESG. These theories are explained below.

10.3.1 Stewardship theory

Stewardship theory was introduced by Donaldson and Davis in the year 1989. This theory emphasizes that the governance management of an organization will act as a responsible steward toward developing a positive relationship between stakeholders' satisfaction and the success of the organization. Organizational management is responsible for making the correct decisions that enhance the long-term value of the organization and can create an environment of sustainability (Hernandez, 2012).

Organizations implement ESG as they perceive that it is the right thing to do, and it also creates strategic direction. Stewardship theory focuses on identifying and solving societal issues and motivates governance management to report NFD. Stewardship theory obliges the commitment of organizational management to dedication, passion, and care. This theory emphasizes support for innovation and encourages communication between governance management and stakeholders (Chevrollier et al., 2020).

Stewardship theory creates commitment as governance management is obligated to develop and implement policies. Policies are directed toward the achievement of organizational mission, vision, and goals; accordingly, if ESG is embedded in mission, vision, and objectives, then organizational management will develop policies that will be inclusive of ESG and will be directed toward stakeholders (Aguilera et al., 2007; Davis et al., 1997). Employee empowerment is supported by stewardship theory, which prevents employees from becoming unmotivated due to external pressures (Reeve & Deci, 1996). Once employees are empowered, they can also emphasize the implementation of ESG.

It is clear from all of these arguments that when an organization develops strong strategic values, it is highly likely that it will perform well in terms of ESG in the future, as stewardship theory is a long-term approach

(Chakrabarty & Wang, 2012). High ESG performance will result from the implementation of stewardship theory (Chevrollier et al., 2020).

10.3.2 Agency theory

Agency theory defines the relationship between the principal and the agent. A principal is a person or a group of persons who appoints agents to perform the task on their behalf. In the corporate world, a principal is the shareholders who appoint the BOD to act on their behalf. The BOD hires executive/senior management to run the organizational affairs which should be aligned with the interests of the BOD. In this scenario, the BOD becomes the principal as representative of shareholders and executive management is the agent that performs the tasks assigned by the BOD.

It is a well-known fact that IA is one of the pillars of good corporate governance (Rehman, 2021). As per the codes of corporate governance developed by many countries including Oman, IA is hired by the BOD (CMA, 2016) to ensure that governance is assured, and tasks performed by executive management are in accordance with their instructions or not (Adams, 1994). In this scenario, IA becomes the agent for the BOD who is hired to perform the task on behalf of the BOD (Jahera & Colbert, 1988).

IA performs the assignments and engagements that are approved by the board or its related committee and provides their independent opinion accordingly. IA tasks as an agent support the organizational cause in the achievement of organizational mission, vision, and related objectives.

10.4 METHODOLOGY

In order to determine the appropriate sample size for this study, G*Power3 was employed. The analysis indicated that a sample size of 70 organizations would be required to adequately represent the population of 114 companies. However, for the present study, questionnaires were distributed to all 114 companies listed on the Muscat Stock Exchange, and responses were received from 70 companies. The respondents, who were carefully selected based on their ability to represent their respective companies, were requested to complete a questionnaire consisting of 26 questions and provide five demographic details. Data collection was conducted using an internet-based tool. It is important to note that the questionnaire was specifically targeted toward one individual per company, resulting in only one response being received from each company. The research methodology employed in this study was descriptive in nature, and the questionnaire items were adapted from relevant studies in the field. Adopted questions are from the studies conducted by the Open Compliance and Ethics Group (OCEG), Deloitte, and the Internal Auditor Foundation (Parmelee, 2021; OCEG, 2021; Foundation, 2021).

10.5 RESULTS AND DISCUSSION

Table 10.2 shows the demographic information from 70 respondents, which included data on gender, position, highest qualification, years of experience, and sector of work. The results showed that there was significant male dominance in the group, with 92% of the respondents being male and only 8% being female. This indicates that the group is primarily made up of male individuals, with only a small representation of female individuals.

Meanwhile, regarding the positions held by respondents, 72% of the respondents held the position of CAE/Head of Internal Audit, 18% were on the ARC, and 10% were on the BOD. This shows that the majority of the respondents were in leadership positions and had a significant role in decision-making processes. On the other hand, the highest qualification held by the respondents was also analyzed, with 35% holding the highest qualification of CA/CPA/CIMA/ACCA, 20% holding the CIA qualification, 21% holding a Master degree, 13% holding a Bachelor degree, 3% holding a PhD, and 8% holding others. This data demonstrates that the majority of the respondents had professional qualifications and a higher level of education.

In terms of years of experience, 41% of the respondents had more than 20 years of experience, 24% had 16–20 years of experience, 20% had 11–15 years of experience, 7% had 6–10 years of experience, and 8% had 1–5 years

Table 10.2 The profile of respondents

Details	Description	Percent (%)
Gender	Male	92
	Female	8
Position	CAE/Head of Internal Audit	72
	Audit and Risk Committee	18
	Board of Director	10
Highest Qualification	CA/CPA/CIMA/ACCA	35
	CIA	20
	PhD	3
	Masters	21
	Bachelor	13
	Others	8
Year of Experience	More than 20 years	41
	16–20 years	24
	11–15 years	20
	6–10 years	7
	1–5 years	8
Sector	Financial	39
	Service	34
	Industrial	27

of experience. This data highlights that the group has a significant number of individuals with extensive experience in their field. Finally, the data also revealed that the group represents a diverse range of industries, with 39% of the respondents working in the financial sector, 34% in the service sector, and 27% in the industrial sector.

The results of the survey provide a comprehensive picture of the demographic information of the group and demonstrate that the group is male-dominant, led by individuals with professional qualifications and extensive experience, and represents a diverse range of industries.

The results of a survey as shown in Figure 10.1 regarding a company's ESG program and its level of maturity have been analyzed; 56% of the respondents answered "Yes" to the question of whether the company has a formal, documented ESG program in place, while 41% answered "No" and 3% answered "Not aware."

Figure 10.2 depicts the respondents who answered "Yes" to the presence of an ESG program, 6% believed that the program is highly mature and has a strong impact on the company's operations; 29% believed that the program is moderately mature and has a significant impact on the company's

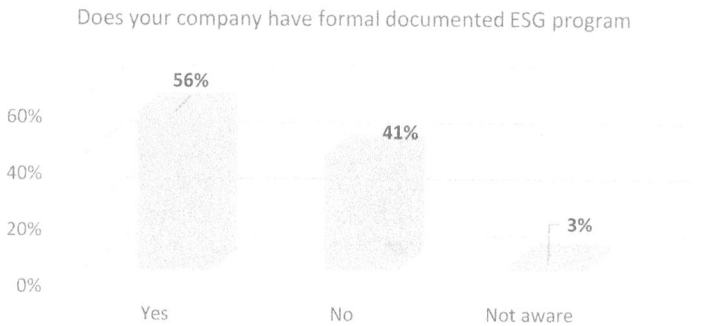

Figure 10.1 Does your company have a formal documented ESG program?

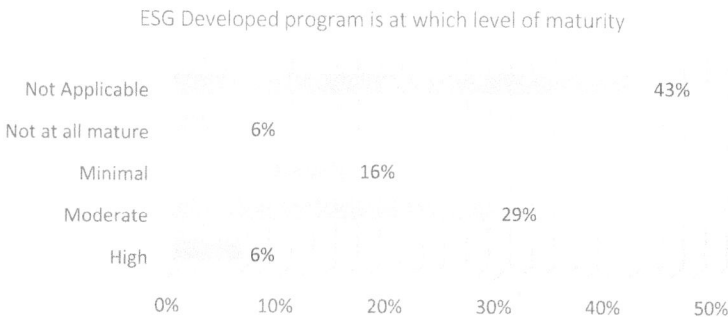

Figure 10.2 The maturity level of the ESG-developed program.

Does Your Company:

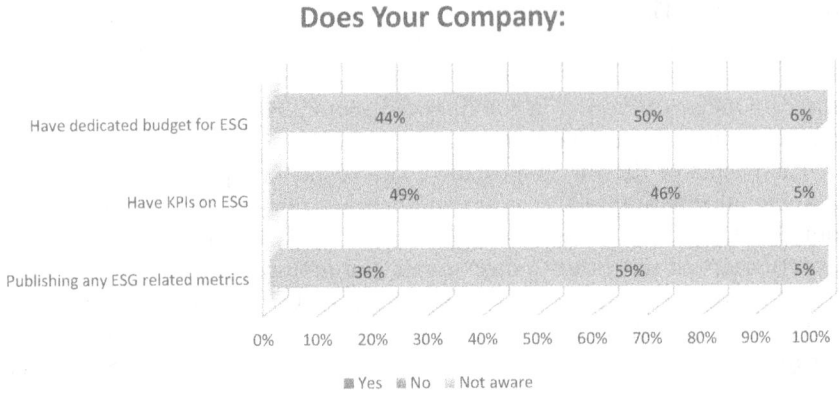

Figure 10.3 Companies that have dedicated budgets and KPIs for ESG and publishing ESG metrics.

operations; 16% believed that the program is in the early stages of development and has a limited impact on the company's operations; 6% believed that the program is not mature and has no significant impact on the company's operations; 43% of the respondents answered "Not Applicable" to the question about the level of maturity of the ESG program, indicating that the question is not applicable to their knowledge of the company's ESG program.

Figure 10.3 shows that a majority of companies (59%) have not published any ESG-related metrics. This could indicate that many companies are not yet fully aware of the importance of measuring and reporting their ESG impact. Nevertheless, the fact that almost half of the companies (49%) have Key Performance Indicators (KPIs) on ESG suggests that some companies are taking steps to track their ESG performance. Furthermore, the fact that 44% of companies have a dedicated budget for ESG is a positive sign, as it indicates that some companies are willing to invest resources in sustainable business practices. However, the fact that 50% of companies do not have a dedicated budget for ESG suggests that there is still a long way to go before ESG practices become mainstream. The results suggest that while some companies are taking ESG seriously, many others are still in the early stages of adopting ESG practices. The results of this survey could be useful for companies looking to benchmark their own ESG performance against their peers and identify areas for improvement.

Figure 10.4 summarizes the results of a survey conducted among 70 respondents regarding the initiators of inquiries related to ESG in a company. From Figure 10.4, the largest group of respondents who initiated inquiries for ESG in the company were those in the "None" category, accounting for 44% of the respondents. The second largest group of initiators were investors, accounting for 34% of respondents. This suggests that

Inquiries in your company for ESG is initiated by

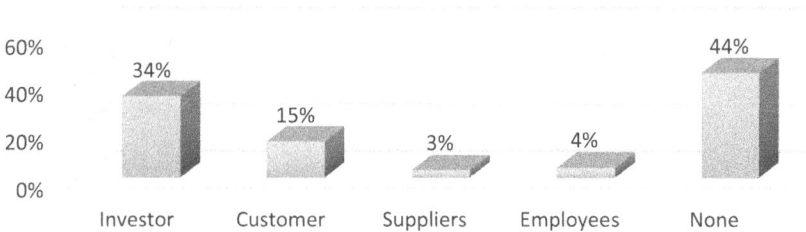

Figure 10.4 The company's adoption of ESG practices is influenced by which stakeholders.

investors are increasingly interested in a company's ESG performance and are more likely to inquire about ESG practices. The Customers group accounted for 15% of respondents, while Employees and Suppliers accounted for 4% and 3%, respectively. This suggests that customers and employees are relatively less likely to initiate inquiries about ESG practices, and suppliers are the least likely group to do so. Figure 10.4 highlights the importance of ESG considerations for companies and the need to engage with a diverse set of stakeholders on ESG-related matters. The four categories of stakeholders amounted to more than 50% therefore suggests that companies need to proactively address ESG issues and engage with a broad range of stakeholders to build trust and demonstrate their commitment to responsible business practices.

Figure 10.5 depicts the impact of ESG on various areas of their business which is measured by four categories of impact, ranging from high to no impact. According to Figure 10.5, the area of the business most impacted by ESG is brand and reputation, with 39 respondents indicating a high impact, followed by customer satisfaction, with 33 respondents indicating a high impact. This indicates that ESG considerations are increasingly important for a company's brand and reputation, as well as its ability to satisfy customers. The area of the business with the least impact from ESG is employee satisfaction, with only 20 respondents indicating a high impact. This may indicate that employees are relatively less affected by a company's ESG practices, although it is worth noting that 27 respondents indicated a minimal impact, indicating that ESG may still be a consideration for some employees. Figure 10.5 also shows that financial outcomes are impacted by ESG, with 30 respondents indicating a high impact, although 19 respondents indicated only a minimal impact, suggesting that the impact of ESG on financial outcomes may vary depending on the company and industry. Overall, they revealed that ESG considerations are increasingly important for various areas of a company's business, particularly brand and reputation and customer satisfaction. Further, the a need for companies to proactively

ESG is impacting which area of your business:

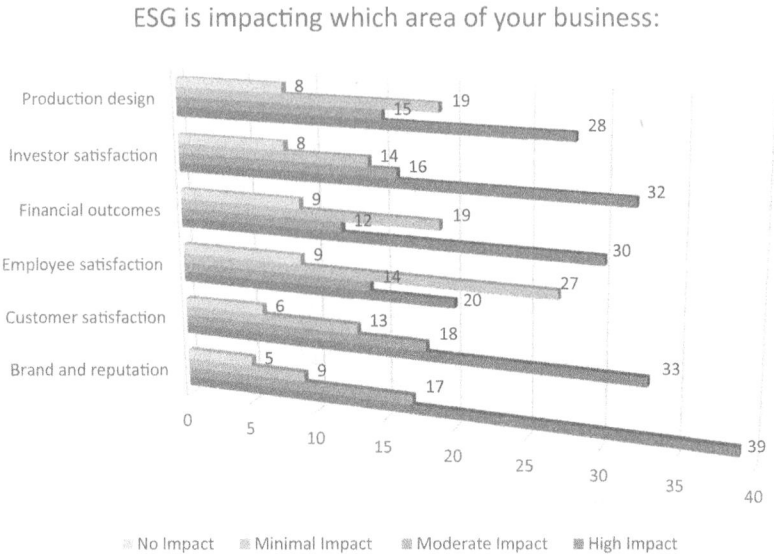

Figure 10.5 Impacting area of business by ESG.

address ESG issues to protect their reputation and maintain customer loyalty.

Figure 10.6 summarizes the results of a survey conducted among 70 respondents regarding ESG issues that are negatively impacting their companies. Three categories of responses: "Yes," indicating that the issue is negatively impacting the company; "No," indicating that the issue is not negatively impacting the company; and "Not aware," indicating that the respondent is not aware of the issue or unsure whether it is negatively impacting the company. Figure 10.6 highlights the ESG issue that is most commonly identified as negatively impacting companies is the cost of resources/raw materials, with 49 respondents indicating a negative impact. This is followed closely by reputation, with 49 respondents also indicating a negative impact. Therefore, companies need to carefully consider the potential negative impact of their ESG practices on their reputation and the cost of resources.

Meanwhile, the operational impacts, such as those related to climate change, floods, and wars, were identified as negatively impacting companies by 47 respondents. This suggests that companies need to be aware of and manage the risks associated with such events. The ESG issue that was least commonly identified as negatively impacting companies was employee health, with only 16 respondents indicating a negative impact. However, it is worth noting that 47 respondents indicated that they were not aware of whether employee health was negatively impacting their company, suggesting that this issue may not be receiving sufficient attention from companies.

Which of the following ESG issues are negatively impacting your company

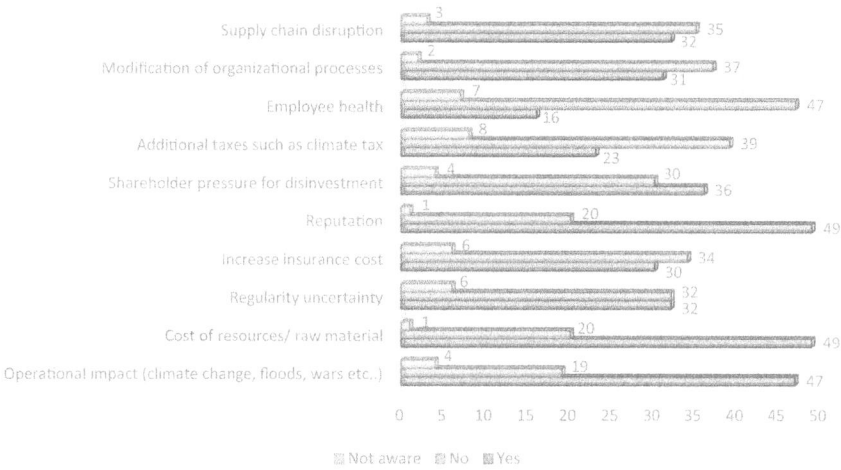

Figure 10.6 The ESG issues are negatively impacting the company.

The result depicts that there is a need for companies to be aware of and manage the potential negative impacts of ESG issues on their business. Further, it also implies that there may be a lack of awareness among companies regarding certain ESG issues, such as employee health, and the need for more attention to be given to these issues.

Three categories of responses revealed the consideration of ESG issues in various aspects of their companies: "Yes," indicating that ESG issues are considered in the respective area; "No," indicating that ESG issues are not considered in the respective area; and "Not aware," indicating that the respondent is unaware of whether ESG issues are considered in the respective area. Figure 10.7 demonstrates that the ESG issue that is most frequently considered is ESG in relation to investors investing in the company, with 50% of respondents indicating that this is taken into account. This suggests that companies may be more concerned with satisfying their investors' ESG criteria than with their own ESG considerations. ESG is also taken into account by a sizeable proportion of respondents when making investments (41%) and selecting vendors (36%). This indicates that ESG factors are becoming increasingly significant in the decision-making processes of companies.

In contrast, only 16% of respondents indicated that ESG is considered when determining executive management salaries and company employee salaries. This could suggest that companies have not yet fully integrated ESG factors into their compensation structures. Figure 10.7 illustrates the growing significance of ESG considerations in various aspects of businesses. In addition, it suggests that companies may need to integrate ESG factors more

Consideration of ESG

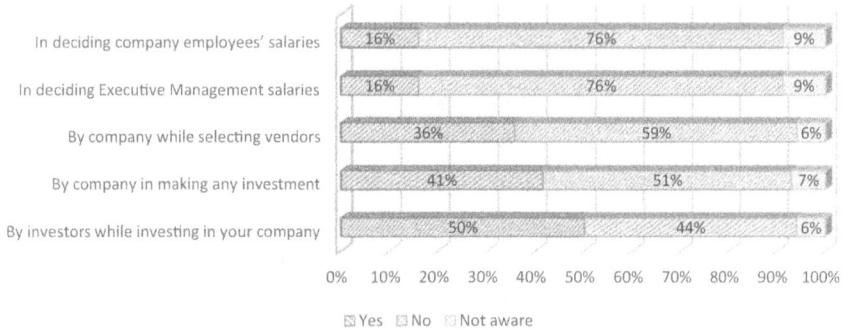

In deciding company employees' salaries: 16% | 76% | 9%
In deciding Executive Management salaries: 16% | 76% | 9%
By company while selecting vendors: 36% | 59% | 6%
By company in making any investment: 41% | 51% | 7%
By investors while investing in your company: 50% | 44% | 6%

0% 10% 20% 30% 40% 50% 60% 70% 80% 90% 100%

☒ Yes ☐ No ☐ Not aware

Figure 10.7 **ESG consideration: Responses and frequency analysis.**

deeply into their decision-making processes, particularly in regard to compensation structures.

According to Figure 10.8, 79% of respondents stated their company uses software for ESG reporting. This indicates that businesses recognize the importance of ESG reporting and are using tools to streamline the reporting process. Despite this, just 20% of respondents said their company uses software to collect ESG data. This demonstrates that companies may be able to improve their ESG data collection methods by adopting software. Figure 10.8 depicts the various levels of software usage in firms for ESG reporting and data collection. It also argues that the mechanisms for collecting ESG data may be improved.

The Company Utilizing any Software for

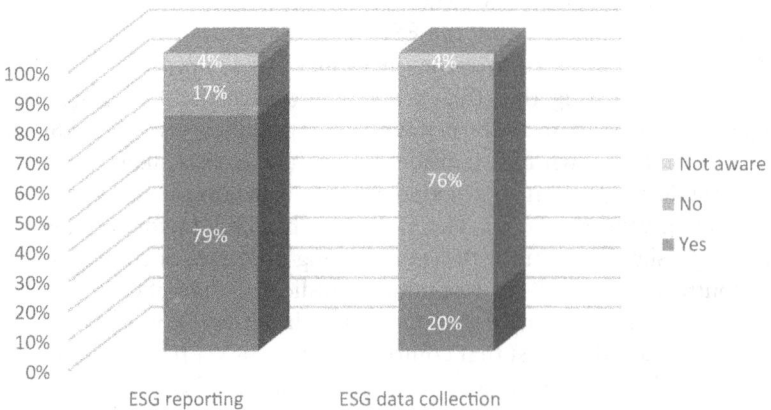

ESG reporting: 4%, 17%, 79%
ESG data collection: 4%, 76%, 20%

Not aware / No / Yes

Figure 10.8 **The company utilizing software in ESG data collection and reporting.**

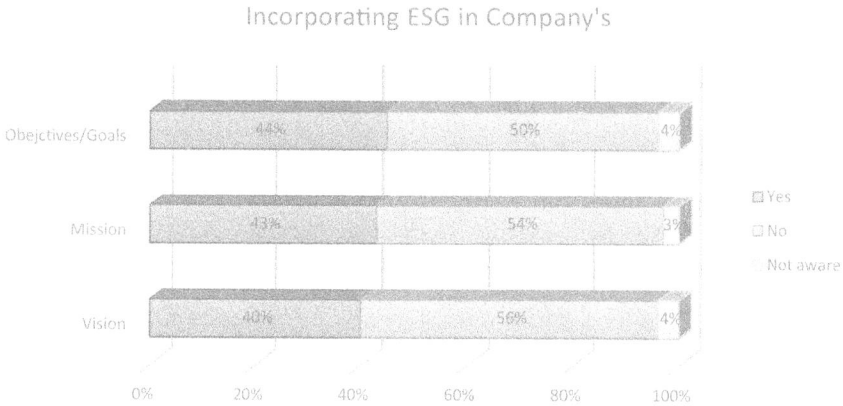

Figure 10.9 Incorporate ESG in the company's vision, mission, and objectives/goals.

Figure 10.9 reveals that the biggest percentage of respondents (44%) indicated that ESG is included in their company's objectives/goals, followed by ESG's incorporation into their mission (43%) and vision (40%) These findings indicate that many companies recognize the significance of ESG and are incorporating it into their fundamental principles and goals. A large number of respondents, however, claimed that their company's vision, mission, and objectives/goals do not include ESG. This suggests that ESG integration into the strategic planning of businesses is not yet complete. Figure 10.9 illustrates the varied levels of ESG incorporation in the vision, purpose, and objectives/goals of companies. It also implies that companies are becoming increasingly aware of the significance of ESG, although there is still space for improvement in terms of properly incorporating ESG into strategic planning.

Table 10.3 shows the IA department competency level in various knowledge areas related to their internal audit department which was rated by 70 respondents. The competency levels are categorized into five levels: Expert, Advance, Applied Awareness, General Knowledge, and No Knowledge. Among the knowledge areas listed, the highest competency levels were in the "Mission of internal audit" (33 respondents rated the IA department are Expert; Advance is 30 respondents) and Organizational independence (26 respondents rated as Expert; Advance level is rated by 37 respondents) categories.

These areas are critical for an effective internal audit department as they set the tone for the department's purpose and independence from management. In contrast, the lowest competency levels were in the areas of Social Responsibility (Expert (14), Advance (28)) and Sustainability (Expert (9), Advance (30)), indicating that many respondents may not have a good understanding of the importance of ESG issues for their internal audit function. Overall, Table 10.3 suggests that respondents have a good

Table 10.3 Internal audit department competency level related to the knowledge area

Area	Expert	Advance	Applied awareness	General knowledge	No knowledge
Ethical behavior	27	34	8	1	0
Individual objectivity	28	35	5	2	0
Mission of internal audit	33	30	6	1	0
Due professional care	24	34	10	2	0
Organizational independence	26	37	5	2	0
Internal control	22	37	7	4	0
Internal audit charter	27	35	4	3	1
Audit engagement	21	44	3	2	0
Professional development	16	32	17	4	1
Reporting	23	32	13	2	0
Engagement planning	27	33	8	2	0
Engagement outcome	21	34	11	4	0
Audit planning	27	33	8	2	0
Soft skills	22	24	22	2	0
Business processes	21	27	21	1	0
Relationship building	21	25	20	4	0
Strategic planning	19	34	9	7	1
Strategic management	19	30	14	7	0
Quality assurance and improvement program	16	23	22	9	0
Fraud audits	20	33	10	6	1
Fraud investigations	22	32	7	8	1
Social responsibility	14	28	20	8	0
Sustainability	9	30	19	12	0
Agile auditing	10	25	22	13	0
Data analytics	9	35	21	5	0
IT controls	7	21	32	10	0

understanding of the core competencies required for an internal audit function, such as ethical behavior, individual objectivity, and internal control. However, there may be room for improvement in some of the more specialized areas, such as agile auditing and data analytics, which are becoming increasingly important in today's business environment. Companies may need to invest in training and development programs to ensure that their internal audit teams have the necessary skills and knowledge to meet these emerging challenges.

The study also has a survey regarding the internal audit function of the company. Figure 10.10, presents the results of a survey conducted on 70 respondents regarding the necessary skills of their internal audit function to effectively audit specific areas. The survey asked respondents to indicate

The following areas that internal audit function have the
necessary skills to effectively audit

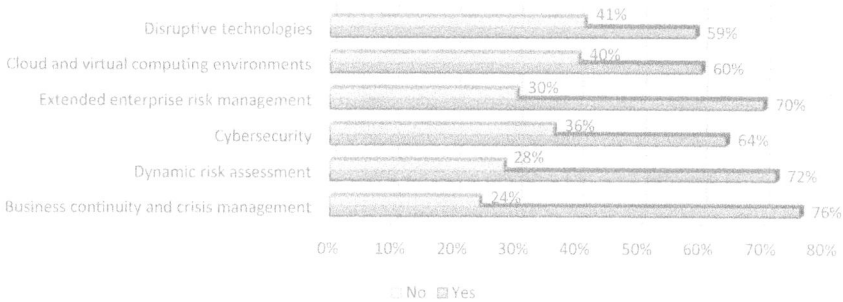

	No	Yes
Disruptive technologies	41%	59%
Cloud and virtual computing environments	40%	60%
Extended enterprise risk management	30%	70%
Cybersecurity	36%	64%
Dynamic risk assessment	28%	72%
Business continuity and crisis management	24%	76%

Figure 10.10 Areas that the internal audit function has the necessary skills to
effectively audit.

whether their internal audit function has the necessary skills to audit in the
areas listed.

Figure 10.10 shows that the majority of respondents indicated that their
internal audit function is able to conduct audits in the areas of business con-
tinuity and crisis management (76%), dynamic risk assessment (71%), and
extended enterprise risk management (70%). Cybersecurity, cloud and virtual
computing environments, and disruptive technologies, on the other hand,
received a lower proportion of positive responses: 64%, 60%, and 59%,
respectively. These results emphasize the need for internal audit functions to
continuously develop and improve their skills in order to keep up with the
ever-changing business environment. As the world becomes increasingly digital and interconnected, it is unsurprising that areas related to technology, such
as cybersecurity, cloud and virtual computing environments, and disruptive
technologies, have received lower proportions of positive responses. This suggests that internal audit functions may need to invest more resources in developing and enhancing their technology-related skills and knowledge in order
to keep up with the rapid pace of technological progress.

Notably, while the majority of respondents indicated that their internal
audit function has the necessary skills to audit in the areas of business con-
tinuity, crisis management, and dynamic risk assessment, a significant number of respondents indicated that their internal audit function lacks the
necessary skills in these areas (24% and 29%, respectively). This indicates
that there is still room for improvement in these areas, and internal audit
functions should consider investing additional resources in the development
of these skills. The results of this survey shed light on the areas in which
internal audit functions may need to concentrate their efforts in order to
advance their skills and knowledge. To effectively support the organization's
objectives and mitigate risks, it is crucial that internal audit functions

How mature your internal audit department towards ESG audits

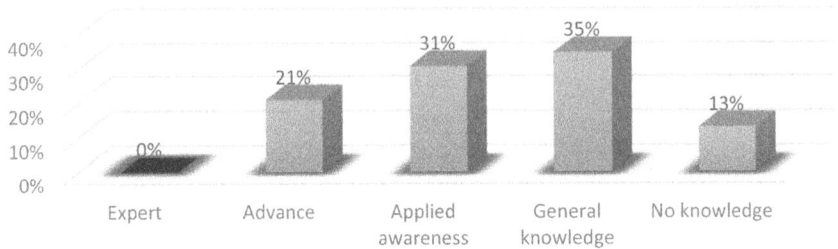

Figure 10.11 The maturity level of the internal audit department toward ESG audits.

continually assess their skills and identify areas where they need to develop and enhance their capabilities.

Figure 10.11 depicts the distribution of respondents based on the level of ESG audit maturity of their internal audit department. Respondents were asked to categorize the ESG audit maturity of their department into five levels: Expert, Advanced, Applied Awareness, General Knowledge, and No Knowledge. No respondent categorized their internal audit department as "Expert" in ESG audits, as indicated by the results. The majority of respondents (35%) rated the maturity level of their internal audit department's knowledge of ESG audits as "General knowledge." Moreover, 31% of respondents indicated that their internal audit department has "Applied Awareness" regarding ESG audits. Around 21% of the respondents categorized their department as having an "Advanced" level of maturity toward ESG audits; 13% of respondents indicated that their internal audit department has "No Knowledge" of ESG audits.

This number suggests that the majority of internal audit departments have a fundamental understanding of ESG audits. Nonetheless, few possess advanced knowledge or expertise in this field. This finding underscores the need for internal audit departments to develop their knowledge and expertise in ESG audits in order to better address the associated risks and opportunities.

10.6 CONCLUSION AND FUTURE SCOPE

Integrating internal audit with ESG practices is essential for enhancing effective corporate governance. This integration improves the accuracy of financial statements and provides assurance regarding the quality of nonfinancial disclosures. In addition, the role of IA in providing independent opinions and training on ESG-related matters is essential for achieving stakeholder satisfaction and ESG-related objectives. To achieve these goals, IA must be adaptable, adopt new technologies and methods, and continuously assess the competencies of its employees. In addition, organizations should contemplate incorporating ESG-related issues into their internal audit processes and incorporating IA as a control assurance activity. Organizations should develop a standard ESG framework, such as the COSO framework, and collaborate with IA to ensure compliance with the framework and attain desired ESG-related goals. IA should identify skill gaps and create plans to fill them, adopt new technologies and methods, and abandon conventional auditing practices.

The descriptive analysis yielded valuable insights into the demographic characteristics of the respondents and their perspectives on ESG practices. The results revealed a notable gender imbalance, with male participants significantly outnumbering their female counterparts. Moreover, a majority of respondents possessed professional qualifications and possessed considerable industry experience. The sample encompassed a diverse range of sectors, with the financial industry exhibiting the highest representation.

Regarding ESG considerations, the analysis indicated that a significant proportion of companies had implemented formal ESG programs, although the maturity levels of these programs varied. The survey explored the integration of ESG considerations into business decisions, as well as their alignment with the company's vision and objectives. Furthermore, the utilization of software for ESG reporting was examined. Additionally, the analysis shed light on the specific ESG issues that had the greatest impact on businesses and highlighted the adverse consequences associated with certain ESG factors. The study also assessed the competency levels of the internal audit department across various knowledge domains.

In conclusion, the survey results indicate that while some companies take ESG seriously, the adoption of ESG practices by most companies is still in its infancy. The fact that the majority of companies have not published any ESG-related metrics suggests that many are still unaware of the significance of measuring and reporting their ESG impact. However, some companies have ESG-related KPIs and an ESG-specific budget is a positive sign, indicating that some companies are willing to invest in sustainable business practices. Companies seeking to benchmark their ESG performance against their peers and identify areas for improvement may find the results of this survey useful. Investors are becoming increasingly interested in a company's ESG performance and are more likely to inquire about ESG practices, as evidenced by the findings of this survey. ESG considerations are becoming increasingly vital to a company's brand, reputation, and ability to satisfy customers. Companies must carefully consider the potential negative effects of their ESG practices on their reputation and resource costs.

On the basis of the findings, it is recommended that companies adopt and implement a formal, documented ESG program and develop KPIs for ESG metrics. This can be accomplished by incorporating ESG into their strategic planning to ensure that ESG considerations are integrated into their core principles and objectives. This may involve establishing ESG KPIs and collaborating with internal audits to ensure compliance with ESG frameworks. Additionally, companies should allocate specific budgets for ESG practices and report their ESG impact. In addition, companies should engage with a variety of stakeholders, including investors, customers, employees, and suppliers, on ESG-related issues. In addition, companies should carefully consider the potential negative impact of their ESG practices on their reputation and the cost of their resources, as well as the operational risks associated with climate change, floods, and wars. In addition, companies should recognize the significance of ESG factors for various aspects of their operations, particularly brand and reputation and customer satisfaction, and take proactive measures to address these issues.

IA should evaluate their current skill sets and identify areas in which they need to develop and improve their capabilities, particularly in technology and ESG audits. In addition, departments of internal audit should invest in training and development programs to enhance their skills and knowledge in these areas. Due to ESG audits, the majority of IA have a fundamental understanding of ESG audits, whereas only a small percentage have advanced knowledge or expertise in this area. Therefore, IA should invest in developing their knowledge and expertise in ESG audits to better address the risks and opportunities associated with these audits. Finally, IA should evaluate their ESG audit maturity level on a regular basis and strive to attain a higher level of expertise in this crucial area. This enables IA to more effectively support their companies' objectives and mitigate risks.

The study refers to the agency theory in the context of defining the influence of IA over ESG practices. The agency theory is a concept that describes the relationship between the principals (e.g., shareholders) and agents (e.g., managers) of an organization. It highlights the potential conflicts of interest that arise between these two parties due to their differing objectives. In this context, IA acts as an agent of the shareholders, providing independent opinions and training on ESG-related matters to ensure that the organization's ESG practices align with its stakeholders' interests. Therefore, the integration of IA with ESG practices is critical to ensure effective corporate governance and mitigate potential agency problems.

The stewardship theory suggests that governance management is inherently motivated to act in the best interest of their stakeholders and will take responsibility for the success or failure of their actions. In the context of the study, governance managements that adopt ESG practices and develop and implement ESG-related initiatives are acting as stewards of their stakeholders. By taking a proactive approach to sustainability and addressing ESG concerns, companies can enhance their reputation, brand, and customer satisfaction. Furthermore, by evaluating their ESG, governance management can act as stewards of their companies, ensuring that they are well-positioned to support ESG-related initiatives and mitigate risks.

The study can be beneficial to regulators, policymakers, governance management, and standard-setting bodies in several ways. First, it highlights the importance of integrating IA with ESG practices for effective corporate governance and to mitigate potential agency problems. This information can be used by regulators and policymakers to develop guidelines and regulations that encourage companies to adopt ESG practices and ensure that IAs are adequately trained and equipped to address ESG-related risks and opportunities. Second, the study emphasizes the need for companies to develop and implement a formal, documented ESG program and to report their ESG impact. This information can be used by standard-setting bodies to develop ESG frameworks and guidelines for reporting ESG-related metrics. Finally, the study provides insights into the current state of ESG practices and the challenges companies face in implementing ESG-related initiatives. This information can be used by governance management to develop strategies and initiatives that address these challenges and enhance their ESG performance. Overall, the study can be a valuable resource for stakeholders interested in promoting sustainable business practices and improving ESG performance.

REFERENCES

Adams, M. (1994). Agency Theory and the Internal Audit. *Managerial Auditing Journal*, 9 (8), 8–12. https://doi.org/10.1108/02686909410071133

Aguilera, R., Rupp, D., Williams, C., & Ganapathi, J. (2007). Putting the S back in corporate social responsibility: a multilevel theory of social change in organizations. *Academy of Management Review*, 32 (3), 836–863. https://doi.org/10.5465/AMR.2007

Annandale, D., Fuente, L.D., Hileman, D., Hjelm, C. L., & Olson, E. (2022). *The ESG Risk Landscape Part 2 – Implementation, reporting, and internal audit's role*. FL: Institute of Internal Auditors.

Armstrong, M. (2021, November 4). *Lobbying for Profits over Planet*. Retrieved from www.statista.com:https://www.statista.com/chart/26125/companies-exerting-most-negative-influence-on-climate-policy/

Auditors, I. O. (2023). *Standards Mapping*. LA: Institute of Internal Auditors (IIA).

Bank, D. (2020, Feb 14). *What is the "G" in ESG?* Retrieved from https://deutsche wealth.com/en: https://deutschewealth.com/en/our-capabilities/esg/what-is-esg-investing-wealth-management/corporate-governance-g-in-esg-governance.html

Barbuscia, D. (2021, September 6). *Oman working on ESG framework to widen funding base - sources*. Retrieved from https://www.reuters.com: https://www.reuters.com/business/sustainable-business/oman-working-esg-framework-widen-funding-base-sources-2021-09-06/#:~:text=In%20its%20Vision%202040%20%2D%20an,110%20out%20of%20180%20countries

Barišić, I., & Tušek, B. (2016). The importance of the supportive control environment for internal audit effectiveness – the case of Croatian companies. *Economic Research-Ekonomska Istraživanja*, 29 (1), 1021–1037. https://doi.org/10.1080/1331677X.2016.1211954

Becchetti, L., Bobbio, E., Prizia, F., & Semplici, L. (2022). Going Deeper into the S of ESG: A Relational Approach to the Definition of Social Responsibility. *Sustainability*, 14, 9668. https://doi.org/10.3390/su14159668

Berg, F., Kölbel, J. F., & Rigobon, R. (2019). Aggregate Confusion: The Divergence of ESG Ratings. *Forthcoming Review of Finance*, https://doi.org/10.2139/ssrn.3438533

Broni, G., & Velentzas, J. (2012). Corporate Governance, Control and Individualism as a Definition of Business Success. The Idea of a "Post - Heroic" Leadership. *Procedia Economics and Finance*, 1, 61–70. https://doi.org/10.1016/S2212-5671(12)00009-3

Broo, D. G., Kaynak, O., & Sait, S. M. (2022). Rethinking engineering education at the age of industry 5.0. *Journal of Industrial Information Integration*, 25, 100311.

Carayannis, E. G., & Morawska, J. (2023). University and Education 5.0 for Emerging Trends, Policies and Practices in the Concept of Industry 5.0 and Society 5.0. In: Machado, C.F., Davim, J.P. (eds) Industry 5.0. Springer, Cham. https://doi.org/10.1007/978-3-031-26232-6_1

Carayannis, E. G., & Morawska-Jancelewicz, J. (2022). The futures of Europe: Society 5.0 and Industry 5.0 as driving forces of future universities. *Journal of the Knowledge Economy*, 1–27.

Carlier, M. (2022, June 10). *Volkswagen AG's sales revenue 2006-2021*. Retrieved from https://www.statista.com/: https://www.statista.com/statistics/264349/sales-revenue-of-volkswagen-ag-since-2006/

Cenci, A., & Cawthorne, D. (2020). Refining value sensitive design: A (capability-based) procedural ethics approach to technological design for well-being. *Science and Engineering Ethics*, 26 (5), 2629–2662.

Chakrabarty, S., & Wang, L. (2012). The long-term sustenance of sustainability practices in MNCs: a dynamic capabilities perspective of the role of R&D and internationalization. *Journal of Business Ethics*, 110 (2), 205–217. https://doi.org/10.1007/s105

Chang, Y.-T., Chen, H., Cheng, R. K., & Chi, W. (2019). The impact of internal audit attributes on the effectiveness of internal control over operations and compliance. *Journal of Contemporary Accounting & Economics*, 15 (1), 1–19. https://doi.org/10.1016/j.jcae.2018.11.002

Chevrollier, N., Zhang, J., Leeuwen, T.V., & Nijhof, A. (2020). The predictive value of strategic orientation for ESG performance over time. *Corporate Governance*, 20 (1), 123–142. https://doi.org/10.1108/CG-03-2019-0105

Christopher, J., Sarens, G., & Leung, P. (2009). A critical analysis of the independence of the internal audit function: evidence from Australia. *Accounting, Auditing & Accountability Journal*, 22 (2), 200–220. https://doi.org/10.1108/09513570910933942

CMA. (2016). *Code of Corporate Governance for Public Listed Companies*. Muscat: Capital Market Authority.

CMA. (2021). *Decision No. 27/2021 Issuing the Regulation for Public Joint Stock Companies*. Muscat: Capital Market Authority.

Colbert, J. L., & Jahera, J. S. (1988). The Role of the Audit and Agency Theory. *Journal of Applied Business Research*, 4 (2), 7–12. https://doi.org/10.19030/jabr.v4i2.6427

Commission, E. (2021, Sep 11). *Overview of sustainable finance*. Retrieved from https://ec.europa.eu: https://ec.europa.eu/info/business-economy-euro/banking-and-finance/sustainable-finance/overview-sustainable-finance_en

Davis, J., Schoorman, F., & Donaldson, L. (1997). Toward a stewardship theory of management. *The Academy of Management Review*, 22 (1), 20–47.

De Giovanni, P. (2023). Sustainability of the Metaverse: A transition to Industry 5.0. *Sustainability*, 15 (7), 6079.

Diamond, J. (2013). Internal Control and Internal Audit. In R. H. Allen, *The International Handbook of Public Financial Managemen* (pp. 374–395). London: Palgrave Macmillan. https://doi.org/10.1057/9781137315304_18

Dicuonzo, G., Donofrio, F., Ranaldo, S., & Dell'Atti, V. (2022). The effect of innovation on environmental, social and governance (ESG) practices. *Meditari Accountancy Research*, 30 (4), 1191–1209. https://doi.org/10.1108/MEDAR-12-2020-1120

Dvořáková, L., & Zborkova, J. (2014). Integration of Sustainable Development at Enterprise Level. *24th DAAAM International Symposium on Intelligent Manufacturing and Automation* (pp. 686–695). Vienna: Procedia Engineering. https://doi.org/10.1016/j.proeng.2014.03.043

Dzikrullah, A. D., Harymawan, I., & Ratri, M. C. (2020). Internal audit functions and audit outcomes: Evidence from Indonesia. *Cogent Business & Management*, 7 (1), 1750331. https://doi.org/10.1080/23311975.2020.1750331

Earle, H. A. (2003). Building a workplace of choice: Using the work environment to attract and retain top talent. *Journal of Facilities Management*, 2 (3), 244–257. https://doi.org/10.1108/14725960410808230

Economics, W. (2021, Dec 16). *Oman's Average Years of Schooling and Other Social Factors*. Retrieved from https://www.worldeconomics.com/: https://www.worldeconomics.com/ESG/Social/Average-Years-Of-Schooling/Oman.aspx

Ellemers, N., & Chopova, T. (2021). The social responsibility of organizations: Perceptions of organizational morality as a key mechanism explaining the relation between CSR activities and stakeholder support. *Research in Organizational Behavior*, 41, 100156. https://doi.org/10.1016/j.riob.2022.100156

Escrig-Olmedo, E., Fernández-Izquierdo, M. Á., Ferrero-Ferrero, I., Rivera-Lirio, J. M., & Muñoz-Torres, M. J. (2019). Rating the Raters: Evaluating how ESG Rating Agencies Integrate Sustainability Principles. *Sustainability*, 11, 915. https://doi.org/10.3390/su11030915

Feroz, S., & Kumar, M. R. (2012). Corporate Social Responsibility Initiatives in Oman Organizations– A review. *The first international conference on corporate social responsibility, business and human rights (Iccsrbhr 2012)* (pp. 1–6). Jakarta: Human Rights Resource Center.

Fontaine, M. (2013). Corporate social responsibility and sustainability: the new bottom line?. *International Journal of Business and Social Science*, 4 (4), 110–119.

Foronda, R. Á., De-Pablos-Heredero, C., & Rodríguez-Sánchez, J.-L. (2023). Implementation model of data analytics as a tool for improving internal audit processes. *Frontiers in Psychology*, 14, 1140972. https://doi.org/10.3389/fpsyg.2023.1140972

Foundation, I. A. (2021). *Assesing Internal Audit Competancy: Minding the Gaps to Maximize Insight*. Florida: Institute of Internal Auditor Foundation.

Fourie, H., & Ackermann, C. (2013). The impact of COSO control components on internal control effectiveness: An internal audit perspective. *Journal of Economic and Financial Sciences*, 6 (2), 495–518. https://doi.org/10.4102/jef.v6i2.272

Gerard, B. (2019). ESG and Socially Responsible Investment: A Critical Review. *Beta*, 33 (1), 61–83. https://doi.org/10.18261/issn.1504-3134-2019-01-05

Ghobakhloo, M., Iranmanesh, M., Morales, M. E., Nilashi, M., & Amran, A. (2022). Actions and approaches for enabling Industry 5.0-driven sustainable industrial transformation: A strategy roadmap. *Corporate Social Responsibility and Environmental Management*, 1473–1494. https://doi.org/10.1002/csr.2431

Ghobakhloo, M., Iranmanesh, M., Tseng, M. L., Grybauskas, A., Stefanini, A., & Amran, A. (2023). Behind the definition of Industry 5.0: a systematic review of technologies, principles, components, and values. *Journal of Industrial and Production Engineering*, 1–16.

Global S&P (2020, Feb 24). *What is the "G" in ESG?* Retrieved from https://www.spglobal.com: https://www.spglobal.com/en/research-insights/articles/what-is-the-g-in-esg

Han, J.-J., Kim, H. J., & Yu, J. (2016). Empirical study on relationship between corporate social responsibility and financial performance in Korea. *Asian Journal of Sustainability and Social Responsibility*, 1, 61–76. https://doi.org/10.1186/s41180-016-0002-3

Harasheh, M., & Provasi, R. (2022). A need for assurance: Do internal control systems integrate environmental, social, and governance factors? *Corporate Social Responsibility and Environmental Management*, 1–18. https://doi.org/10.1002/csr.2361

Hernandez, M. (2012). Toward an understanding of the psychology of stewardship. *Academy of Management Review*, 37 (2), 172–193. https://doi.org/10.5465/amr.2010.0363

Huang, S., Wang, B., Li, X., Zheng, P., Mourtzis, D., & Wang, L. (2022). Industry 5.0 and Society 5.0—Comparison, complementation and co-evolution. *Journal of manufacturing systems*, 64, 424–428.

IIA. (2021). *American Corproate Governance Index; Grappling with Fatigue Factor*. Florida: Institute of Internal Auditors.

Jahera, J. S., & Colbert, J. L. (1988). The Role of the Audit and Agency Theory. *Journal of Applied Business Research*, 4 (2). https://doi.org/10.19030/jabr.v4i2.6427

Jobin, P. (2021). Our 'good neighbor' Formosa Plastics: petrochemical damage(s) and the meanings of money. *Environmental Sociology*, 7 (1). https://doi.org/10.1080/23251042.2020.1803541

Jong, W., & Linde, V. (2022). Clean diesel and dirty scandal: The echo of Volkswagen's dieselgate in an intra-industry setting. *Public Relations Review*, 48 (1). https://doi.org/10.1016/j.pubrev.2022.102146

Justice, D. O. (2020). *In Settlement with United States and Montana, Atlantic Richfield Agrees to Framework for Cleanup of Mining Contamination in Butte, Montana*. Montana: U.S. Attorney's Office District of Montana.

Kasinathan, P., Pugazhendhi, R., Elavarasan, R. M., Ramachandaramurthy, V. K., Ramanathan, V., Subramanian, S., … Alsharif, M. H. (2022). Realization of sustainable development goals with disruptive technologies by integrating Industry 5.0, Society 5.0, smart cities and villages. *Sustainability*, 14 (22), 15258.

Khan, M. (2019). Corporate governance, ESG, and stock returns around the world. *Financial Analysts Journal*, 75 (4), 103–123. https://doi.org/10.1080/0015198X.2019.1654299

Klettner, A. (2021). Stewardship codes and the role of institutional investors in corporate governance: An international comparison and typology. *British Journal of Management*. https://doi.org/10.1111/1467-8551.12466

Kotsantonis, S., & Serafeim, G. (2019). Four things no one will tell you about ESG data. *Journal of Applied Corporate Finance*, 31 (2), 50–58.

Kuldova, T. Ø. (2022). The Anti-policy Syndrome. In: *Compliance-Industrial Complex*. Palgrave Macmillan, Cham. https://doi.org/10.1007/978-3-031-19224-1_2

Kumar, A., & Nayyar, A. (2020). si 3-Industry: A sustainable, intelligent, innovative, internet-of-things industry. *A roadmap to Industry 4.0: Smart production, sharp business and sustainable development*, 1–21.

Kuo, T. C., Kremer, G. E. O., Phuong, N. T., & Hsu, C. W. (2016). Motivations and barriers for corporate social responsibility reporting: Evidence from the airline industry. *Journal of Air Transport Management*, 57, 184–195.

Lee, M. T., & Suh, I. (2022). Understanding the effects of Environment, Social, and Governance conduct on financial performance: Arguments for a process and integrated modelling approach. *Sustainable Technology and Entrepreneurship*, 1 (1), 1–12. https://doi.org/10.1016/j.stae.2022.100004

Lenz, R., & Hoos, F. (2023). The future role of the internal audit function: Assure. build. *The EDP Audit, Control, and Security Newsletter*, 67 (3), 39–52. https://doi.org/10.1080/07366981.2023.2165361

Li, T.-T., Wang, K., Sueyoshi, T., & Wang, D. D. (2021). ESG: research progress and future prospects. *Sustainability*, 13, 11663. https://doi.org/10.3390/su132111663

Lombardi, R., Trequattrini, R., Cuozzo, B., & Cano-Rubio, M. (2019). Corporate corruption prevention, sustainable governance and legislation: First exploratory evidence from the Italian scenario. *Journal of Cleaner Production*, 217, 666–675. https://doi.org/10.1016/j.jclepro.2019.01.214

Maddikunta, P. K. R., Pham, Q. V., Prabadevi, B., Deepa, N., Dev, K., Gadekallu, T. R., … Liyanage, M. (2022). Industry 5.0: A survey on enabling technologies and potential applications. *Journal of Industrial Information Integration*, 26, 100257.

Madsen, D. Ø., & Slåtten, K. (2023). Comparing the evolutionary trajectories of Industry 4.0 and 5.0: A management fashion perspective. *Applied System Innovation*, 6 (2), 48.

Mazzotta, M. (2021, Feb 21). *2021 Climate Check: Business' views on environmental sustainability*. Retrieved from https://www2.deloitte.com: https://www2.deloitte.com/global/en/pages/risk/articles/2021-climate-check-business-views-on-environmental-sustainability.html

Miler, O. (2021, Oct 11). *Governance Investing: What does the 'G' in ESG mean?* Retrieved from https://www.linkedin.com/pulse/: https://www.linkedin.com/pulse/governance-investing-what-does-g-esg-mean-olga-miler-/

Mourtzis, D., Angelopoulos, J., & Panopoulos, N. (2022). A Literature Review of the Challenges and Opportunities of the Transition from Industry 4.0 to Society 5.0. *Energies*, 15 (17), 6276.

MSX. (2022, Oct 15). *Companies*. Retrieved from Muscat Stock Exchange: https://www.msx.om/companies.aspx

OCEG. (2021). *ESG Planning and Performance Survey*. FL: Open Ethics and Compliance Group.

Olaizola, I. G., Quartulli, M., Garcia, A., & Barandiaran, I. (2022). Artificial Intelligence from Industry 5.0 perspective: Is the Technology Ready to Meet the Challenge? *Proceedings* http://ceur-ws.org ISSN, 1613, 0073.

Park, J., Choi, W., & Jung, S.-U. (2022). Exploring Trends in Environmental, Social, and Governance Themes and Their Sentimental Value Over Time. *Front. Psychol*, 13, 890435. https://doi.org/10.3389/fpsyg.2022.890435

Parmelee, M. (2021). *2021 Climate Check: Business' Views on Environmental Sustainability; Disruptive 2020 slows climate action, but executives determined to act.* LA: Deloitte.

Patton, J., & Reisch, N. (2021). *Formosa Plastics Group: A Serial Offender of Environmental and Human Rights (A Case Study).* Delware: Center of International Environmental Law.

Pedersen, L. H., Fitzgibbons, S., & Pomorski, L. (2021). Responsible investing: The ESG-efficient frontier. *Journal of Financial Economics*, 142 (2), 572–597. https://doi.org/10.1016/j.jfineco.2020.11.001

Pelosi, N., & Adamson, R. (2016). Managing the "S" in ESG: The Case of Indigenous Peoples and Extractive Industries. *Journal of Applied Corporate Finance*, 28 (2). https://doi.org/10.1111/jacf.12180

Phillips, S., Thai, V. V., & Halim, Z. (2019). Airline value chain capabilities and CSR performance: The connection between CSR leadership and CSR culture with CSR performance, customer satisfaction and financial performance. *The Asian Journal of Shipping and Logistics*, 35 (1), 30–40. https://doi.org/10.1016/j.ajsl.2019.03.005

Polley, C. (2022, Feb 10). *ESG vs. CSR: what's the difference?* Retrieved from https://thesustainableagency.com: https://thesustainableagency.com/blog/esg-vs-csr/#:~:text=In%20short%2C%20CSR%20is%20a,a%20company's%20overall%20sustainability%20performance

Pramanik, P. K., Mukherjee, B., Pal, S., Kumar, B., Upadhyaya, & Dutta, S. (2022). Ubiquitous manufacturing in the age of Industry 4.0: A state-of-the-art primer. In A. Nayyar, & A. Kumar. (eds), *A Roadmap to Industry 4.0: Smart Production, Sharp Business and Sustainable Development* (pp. 73–111). Singapore: Springer.

Pritchard, G. Y., & Çalıyurt, K. T. (2021). Sustainability Reporting in Cooperatives. *Risk*, 9 (6), 117. https://doi.org/10.3390/risks9060117

Reeve, J., & Deci, E. (1996). Elements of the competitive situation that affect intrinsic motivation. *Personality and Social Psychology Bulletin*, 22 (1), 24–33. https://doi.org/10.1177/0146167296221003

Rehman, A. (2021). Can Sustainable corporate governance enhance internal audit function? Evidence from Omani Public Listed companies. *Journal of Risk and Financial Management*, 14 (11), 537. https://doi.org/10.3390/jrfm14110537

Rehman, A. (2023, April 23). A strong control environment is needed to support environmental, social, and governance initiatives. *Internal Auditor*, 57–61.

Rehman, A., & Hashim, F. (2018). Corporate governance maturity and its related measurement framework. *Proceedings of the 5th International Conference on Accounting Studies (ICAS 2018)* (pp. 16–28). Penang: UUM.

Rehman, A., & Hashim, F. (2021). Can forensic accounting impact sustainable corporate governance? *Corproate Governance*, 21 (1), 212–227. https://doi.org/10.1108/CG-06-2020-0269

Ruud, F., Friebe, P., Schmitz, D., & Isufi, S. (2011). International Professional Practices Framework – Overview of the current guidance of the Institute of Internal Auditors. In *Global Management Challenges for Internal Auditors: ECIIA Yearbook of Internal Audit 2010/11.* (pp. 17–24). Berlin: Erich Schmid.

Sachin, N., & Rajesh, R. (2022). An empirical study of supply chain sustainability with financial performances of Indian firms. *Environment, Development and Sustainability*, 24 (5), 6577–6601. https://doi.org/10.1007/s10668-021-01717-1

Saniuk, S., Grabowska, S., & Straka, M. (2022). Identification of social and economic expectations: Contextual reasons for the transformation process of Industry 4.0 into the Industry 5.0 Concept. *Sustainability*, 14 (3), 1391.

Sanyal, S., & Hisam, M. W. (2017). Corporate governance in emerging economies: A study of the Sultanate Of Oman. *International Journal of Business and Administration Research Review*, 3 (20), 27–31.

Schandl, A., & Foster, P. L. (2019). *COSO Internal Control – Integrated Framework: An Implementation Guide for the Healthcare Provider Industry*. Durham: Crowe.

SEC. (2022, March 21). *SEC Proposes Rules to Enhance and Standardize Climate-Related Disclosures for Investors*. Retrieved from https://www.sec.gov: https://www.sec.gov/news/press-release/2022-46

Sila, I., & Cek, K. (2017). The Impact of environmental, social and governance dimensions of corporate social responsibility on economic performance: Australian Evidence. *Procedia Computer Science*, 120, 797–804. https://doi.org/10.1016/j.procs.2017.11.310

Soh, D. S., & Martinov-Bennie, N. (2015). Internal auditors' perceptions of their role in environmental, social and governance assurance and consulting. *Managerial Auditing Journal*, 30 (1), 80–111. https://doi.org/10.1108/MAJ-08-2014-1075

Stojanović, T., & Andrić, M. (2016). Internal Auditing and Risk Management in Corporations. *Strategic Management*, 21 (3), 31–42.

Stranieri, S., Orsi, L., Banterle, A., & Ricci, E. C. (2019). Sustainable development and supply chain coordination: The impact of corporate social responsibility rules in the European Union food industry. *Corporate Social Responsibility and Environmental Management*, 26 (2), 481–491. https://doi.org/10.1002/csr.1698

Sustainability, O. C. (2015). *Establishing the Oman Centre for Governance and Sustainability and Promulgating its Charter*. Muscat: Capital Market Authority.

Sustainanalytics. (2021, Oct 19). *Company ESG Risk Ratings*. Retrieved from https://www.sustainalytics.com/: https://www.sustainalytics.com/esg-rating/national-bank-of-oman/1018123150

Tamimi, N., & Sebastianelli, R. (2017). Transparency among S&P 500 companies: an analysis of ESG disclosure scores. *Management Decision*, 55 (8), 1660–1680. https://doi.org/10.1108/MD-01-2017-0018

Team, C. (2022, June 6). *The 3 Pillars of ESG*. Retrieved from https://www.countable.com: https://www.countable.com/blog/the-three-pillars-of-esg

Team, O. O. (2021, Dec 26). *Muscat: The Ministry of Commerce, Industry, and Investment Promotion (MOCIIP) has issued a decision to deduct 20 percent of the budget of institutions and companies allocated to social responsibility programs for the benefit of the Oman Charitable Association*. Retrieved from https://www.omanobserver.om: https://www.omanobserver.om/article/1111670/business/economy/20-of-csr-budget-to-be-marked-for-oman-charitable-association

Thottoli, M. M. (2022). A study on listed companies' compliance with value-added tax: the evolving role of compliance officer in Oman. *Public Administration and Policy: An Asia-Pacific Journal*, 25 (1), 89–98. https://doi.org/10.1108/PAP-02-2021-0017

Waas, B. (2021). The "S" in ESG and international labour standards. *International Journal of Disclosure and Governance*, 18, 403–410, https://doi.org/10.1057/s41310-021-00121-5

Winteroll, M. (2020). *Sustainable Finance - Environment, Social and Governance Criteria Feed Back from Deutsche Börse Group*. Frankfurt: European Commission.

Wong, C., & Petroy, E. (2020). *Rate the Raters 2020: Investor Survey and Interview Results*. LA: SustainAbility.

Worker, J. (2015, May 20). *The Best and Worst Countries for Environmental Democracy*. Retrieved from https://www.wri.org: https://www.wri.org/insights/best-and-worst-countries-environmental-democracy

Xu, X., Lu, Y., Vogel-Heuser, B., & Wang, L. (2021). Industry 4.0 and Industry 5.0—Inception, conception and perception. *Journal of Manufacturing Systems, 61*, 530–535.

Zeilina, L., & Toth, J. (2021). *Amplifying the "S" in ESG: Investor Myth Buster*. LA: International Sustainable Finance Centre.

Chapter 11

Foresight the future of Industry 5.0 in perspective of sustainable services in developing countries

Pankaj Singh
GLA University, Mathura, India

Ruchi Kushwaha and Jyoti Kushwaha
Jiwaji University, Gwalior, India

11.1 INTRODUCTION

Industry 4.0, a drive from Germany, has turned into a universally embraced term in the last 10 years. Numerous nations have given similar key drives, and a great deal of research and development work has gone into developing and implementing some of the Business 4.0 innovations (Sindhwani et al. 2022). At the 10-year mark since the inception of Industry 4.0, the European Commission declared Industry 5.0. Industry 4.0 is viewed as innovation-driven, while Industry 5.0 is esteem-driven (Venaik et al. 2023). The concurrence of two modern industrial revolutions welcomes questions and thus requests conversations and explanations. Industry 5.0, otherwise called the Fifth Industrial Revolution, is a new and arising period of industrialization that sees people working close by trend-setting innovation and artificial intelligence (AI)-fueled robots to upgrade working environment processes (Minculete et al., 2021). This is combined with a more human-driven center in addition to increased adaptability and an improved spotlight on manageability (Xu et al., 2021).

Enveloping something beyond assembling, this new stage expands upon the Fourth Modern Industrial Revolution (Industry 4.0) and is empowered by advancements in information communication technology (ICT) that incorporate the features, for example, man-made brainpower, computerization, Big Data examination, the Internet of Things (IoT), Artificial Intelligence, mechanical technology, brilliant frameworks, and virtualization (Ozdemir & Hekim, 2018; Al-Turzman et al., 2021). Expanding the ideas of Industry 4.0, this new industrial revolution is portrayed by the European Association as giving, "a dream of an industry that points past proficiency and efficiency as the sole objectives, and builds up the job and the commitment of industry to society" (Zong et al., 2021).

This is a significant differentiation from the methodology of Industry 4.0, as depicted by the EU, since "it puts the prosperity of the laborer at the focal

DOI: 10.1201/9781003489269-11

point of the creative interaction and uses new advances to give thriving past positions and development while regarding the creation furthest reaches of the planet." This is a shift away from an emphasis on monetary worth toward a more extensive idea of cultural worth and prosperity. While this idea has been addressed previously, through Corporate Social Obligation regarding the model, the thought of putting individuals and the planet before benefits makes another concentration for the industry (Saniuk et al., 2022).

Industry 5.0, also known as the Fifth Industrial Revolution, is the latest phase of industrial transformation that builds on Industry 4.0 by integrating human workers with advanced robotics and automation technologies that helps in better work–life balance (Kushwaha et al., 2022). While Industry 4.0 focuses on digitalization and automation, Industry 5.0 emphasizes the importance of human touch and creativity in the manufacturing process (Madsen & Berg, 2021).

Industry 5.0, simply known as the "human-centric" or "integrated" industry, emphasizes the integration of advanced technologies with human skills and values to achieve sustainable production and consumption (Ghobakhloo et al., 2022). In recent years, there has been growing interest in understanding how sustainable service production and consumption can be achieved in Industry 5.0 (Majerník et al., 2022). Notwithstanding, the possibility of Industry 5.0 goes past industry to envelop all associations and business systems to make a more extensive viewpoint than seen with Industry 4.0 (Table 11.1).

11.1.1 Industry 4.0 versus Industry 5.0

There are hybrids between these two transformations, yet while Industry 4.0 is innovation-driven, Industry 5.0 is esteem-driven (Kumar & Nayyar, 2020). Though 4.0 focuses on computerization and the expanded spotlight on innovation, 5.0 brings humankind back into the field, consolidating with innovation to give frameworks and cycles that serve individuals and our general surroundings first (Rajput et al., 2022). The distinction of the Fifth Industrial Revolution is the way current innovation is being utilized to drive helpful working among people and trend-setting innovations and simulated intelligence-empowered robots to upgrade working environment cycles to convey a more human-driven, strong, and earth-mindful, feasible future (Chander et al., 2022). Industry 5.0 has started to happen now, although it needs to build up progress to thrive (Leng et al., 2022).

Numerous organizations are centered on Industry 4.0, yet, as additional organizations move to consolidate the thoughts of Industry 5.0, we ought to see it grow further (Nayyar & Kumar, 2020). Industry 5.0 has started, with simulated intelligence-controlled frameworks taking up dreary errands and permitting individuals to zero in on more useful and esteem-adding assignments (Akundi et al., 2022). Industry 5.0 is still in its beginning phases, with numerous organizations actually zeroing in on Industry 4.0 but this is supposed to change as additional organizations adjust themselves to the

Table 11.1 Evolution and transition shift of Industrial Revolution from 1.0 to 5.0

S. No.	Stages of the Industrial Revolution	Major features in each phase
1	Industry 1.0	Starting in around 1780, this first upheaval zeroed in on modern creation in light of machines that were fueled by steam and water.
2	Industry 2.0	Nearly 100 years later, in 1870, this Second Industrial Revolution depended on electrification and occurred with large-scale manufacturing through sequential assembly line systems.
3	Industry 3.0	Venturing forward an additional 100 years, to 1970, Industry 3.0 saw robotization using PCs and hardware. This was upgraded by globalization (Industry 3.5), including offshoring of creation to minimal expense economies.
4	Industry 4.0	At present, we are living in the fourth modern revolution, which is based around the idea of digitalization and incorporates mechanization, man-made brainpower (man-made intelligence) advancements, associated gadgets, information examination, digital actual frameworks, computerized change, and that's just the beginning.
5	Industry 5.0	We are currently entering the fifth modern unrest with an emphasis on man and machines cooperating. In light of personalization and the utilization of cooperative robots, laborers are allowed to convey esteem-added undertakings for clients. This most recent emphasis goes past assembling cycles to incorporate expanded strength, a human-driven approach, and attention to supportability, which we investigate in more detail underneath.

objectives of Industry 5.0. Industry 5.0 calls out worldwide modern change. It expects to put people's prosperity at the focal point of assembling frameworks, accordingly accomplishing social objectives past businesses and development for the reasonable improvement of all mankind. Notwithstanding, the flow investigation of Industry 5.0 is still in its early stages where research discoveries are moderately scant and minimally efficient. The three driving attributes of Industry 5.0 are human-centricity, manageability, and versatility. The implication arrangement of Industry 5.0 is examined, and its broadened embodiment is broken down. Industry 4.0 has been accommodated for the last 10 years to help businesses and deficiencies; at last, the ideal opportunity for Industry 5.0 has shown up.

Brilliant manufacturing plants are improving business efficiency; accordingly, Industry 4.0 has impediments. Industry 5.0 is changing the worldview and brings the goal since it will diminish accentuation on the innovation and expect that the potential for progress depends on joint effort among the people and machines. The modern upheaval is further developing consumer loyalty by using customized items. In present-day business with the paid mechanical turns of events, Industry 5.0 is expected to acquire the upper hand as well as financial development for the plant.

11.1.2 Scope of Industry 5.0

Industry 5.0 is viewed as the following modern advancement, and its goal is to use the imagination of human specialists in a joint effort with proficient, smart, and precise machines, to get asset-effective and client-favored assembling arrangements contrasted with Industry 4.0. Various promising advancements and applications are supposed to help Industry 5.0 to increment creation and convey redid items in an unconstrained way. To give an absolute first conversation of Industry 5.0, in this chapter, we mean to give an overview put together instructional exercise with respect to possible applications and supporting advances of Industry 5.0. We initially present a few new ideas and meanings of Industry 5.0 according to the viewpoint of various industry professionals and scientists. We then, at that point, extravagantly examine the expected utilization of Industry 5.0, for example, keen medical services, cloud fabricating, inventory network the executives, and assembling creation. Consequently, we examine a few supporting innovations for Industry 5.0, for example, edge processing, computerized twins, cooperative robots, the Web of Each and Everything, blockchain, and 6G (Rupa et al., 2021). Finally, we feature a few examination difficulties and open issues that ought to be additionally evolved to acknowledge Industry 5.0.

11.1.2.1 Objectives of the chapter

To address the above-mentioned queries on the consequences of Industry 5.0, the following objectives are formulated:

- To foresee the future consequences of Industry 5.0 in the context of sustainable services;
- To analyze the role of Industry 5.0 in sustainable service production and consumption in developing countries;
- And, to detect the significance of Industry 5.0 in the attainment of Sustainable Development Goals (SDGs)

11.1.2.2 Organization of the chapter

The rest of the chapter is organized as follows. Section 11.2 provides a review of related studies in the area of Industry 5.0. Section 11.3 discusses technology enabled for Industry 5.0. In Section 11.4, the consequences of Industry 5.0 for major service sectors of developing economies are deliberated. Section 11.5 presents the significance of Industry 5.0 in attaining SDGs. And, finally, Section 11.6 concludes the chapter with future scope.

11.2 REVIEW OF LITERATURE

Industry 5.0 is a relatively new concept that has emerged as an evolution of Industry 4.0, with a focus on integrating human skills, creativity, and sustainability into the industrial production process. Here is a brief review of some of the literature on Industry 5.0.

11.2.1 Implications of Industry 5.0

Ivanov (2022) provided an overview of the Industry 5.0 concept, its main principles, and potential benefits. The author argues that Industry 5.0 could help address some of the challenges of Industry 4.0, such as the displacement of jobs and the neglect of human creativity and sustainability, by putting human needs and values at the center of the production process. Salgado and Rodríguez (2022) explored the role of human skills and creativity in Industry 5.0 and the potential for this new paradigm to enhance human well-being and creativity in the workplace. Authors argued that Industry 5.0 could lead to a more equitable and inclusive industrial production process that values the contributions of all workers. Ullah (2022) provided an overview of the main concepts and technologies associated with Industry 5.0, such as cyber-physical systems, artificial intelligence, and blockchain. Authors argued that Industry 5.0 has the potential to transform the industrial production process by enabling greater customization, flexibility, and sustainability.

Thakur and Sehgal (2021) discussed the potential impact of Industry 5.0 on the manufacturing sector, with a focus on the Indian context. Authors argued that Industry 5.0 could help India transform its manufacturing sector by improving efficiency, reducing waste, and promoting sustainability. Gorodetsky et al. (2020) explored the implications of Industry 5.0 for the future of work, with a focus on the role of human skills and creativity. Authors argued that Industry 5.0 could lead to a more diverse and inclusive workforce, with greater opportunities for creativity and innovation. Mourtzis et al. (2022) provided an overview of the current state of sustainability in Industry 5.0, including the challenges and opportunities for sustainable service production and consumption. Authors identified several key sustainability goals for Industry 5.0, such as reducing resource consumption, minimizing waste, and promoting social and environmental responsibility.

Nahavandi (2019) explored the challenges and opportunities of Industry 5.0, such as the need for new skills and competencies, the potential for job displacement, and the opportunities for innovation and entrepreneurship. Authors argued that Industry 5.0 requires a comprehensive approach that addresses the social, economic, and environmental dimensions of sustainable development. Zengin et al. (2021) explored the role of service-oriented computing in promoting sustainable service production and consumption in Industry 5.0. Authors highlighted that service-oriented computing can facilitate the integration of advanced technologies and human skills and values, enabling more sustainable and efficient production and consumption processes. Bryndin (2020) examined the current state of research on sustainable service systems in the industry, with a focus on Industry 4.0 and beyond, and identified several key themes in the literature, such as the importance of collaboration between stakeholders, the need for flexible and adaptable service systems, and the role of technology in promoting sustainability.

Adel (2022) examined the role of sustainable smart services in Industry 5.0, including the challenges and opportunities for promoting sustainability in service production and consumption. Authors elaborated that sustainable smart services can enable more efficient and sustainable production and consumption processes, while also promoting social and environmental responsibility. Aslam et al. (2020) provided an overview of the current state of research on sustainable service systems in Industry 5.0 and identified several key research directions for future studies. Authors deliberated that future research should focus on developing innovative service models and business models that promote sustainability, as well as exploring the role of advanced technologies in enabling sustainable service production and consumption.

11.2.2 Industry 5.0 consequences in developed and developing countries

Research on the possibilities of Industry 5.0 in developed and third-world countries is still in its early stages, but several studies have examined this topic. Here is a brief review of some of the literature on Industry 5.0 in both types of countries.

- Industry 5.0 in developed countries
  A study by the Organization for Economic Co-operation and Development (OECD) highlights the potential of Industry 5.0 in developed countries to create more flexible, human-centered production processes that can respond to changing customer demands and market conditions. Another study by the German Federal Ministry for Economic Affairs and Energy emphasizes the potential of Industry 5.0 to improve sustainability and reduce environmental impact through the use of advanced robotics and additive manufacturing. A study by McKinsey and Company highlights the potential of Industry 5.0 to create new business models and increase competitiveness in developed countries (Tiwari et al., 2022).
- Industry 5.0 in third-world countries
  A study by the United Nations Industrial Development Organization (UNIDO) explores the potential of Industry 5.0 in third-world countries to create more inclusive and sustainable industrialization,

improve productivity, and create employment opportunities. Another study by the World Economic Forum (WEF) emphasizes the potential of Industry 5.0 in third-world countries to promote economic development and reduce poverty through the adoption of advanced manufacturing technologies (Raja Santhi & Muthuswamy, 2023). Previous studies highlighted the potential of Industry 5.0 to create a more equitable and inclusive global industrial system by enabling third-world countries to participate in global value chains and access new markets.

Overall, the literature suggests that Industry 5.0 has the potential to transform manufacturing processes in both developed and third-world countries, creating new opportunities for economic growth, sustainability, and social inclusion by integrating human skills, creativity, and sustainability. However, realizing this potential will require significant investments in technology, infrastructure, and human capital development. Generally, these studies suggest that sustainable service production and consumption is a critical goal for Industry 5.0 and that achieving this goal will require collaboration between stakeholders, the development of innovative service models and business models, and the integration of advanced technologies with human skills and values. However, there is also a need to address potential challenges and risks associated with this new paradigm, such as the displacement of jobs and unequal distribution of benefits.

11.3 TECHNOLOGIES ENABLED FOR INDUSTRY 5.0

Industry 5.0 is viewed as the following modern advancement and its goal is to use the innovativeness of human specialists in a joint effort with productive, smart, and exact machines, and to obtain additional asset efficiency, that is, the client's preferred assembly arrangements, contrary to Industry 4.0. There are various promising developments in addition; applications are supposed to help Industry 5.0 to increment creation and convey modified items in an unconstrained way (Maddikunta et al., 2022).

11.3.1 Industry 5.0 benefits and burdens

The principal benefit of Industry 5.0 is the making of higher-worth positions that manage the cost of more prominent personalization for clients and further develop plan opportunities for laborers. By permitting fabricating cycles to be taken care of through robotization, human specialists can zero in a greater amount of their experience on conveying improved, custom-tailored administrations and items. At that time, this began with Industry 4.0, yet Industry 5.0 drives this further through better mechanization and input to make a help-based model where people can zero in on adding an incentive for end clients.

11.3.2 Futuristic applications of Industry 5.0

Basically, Industry 5.0 is supported by three systems, namely, human-centricity, versatility, and manageability.

(a) *Human-centricity*

Industry 5.0 incorporates a system that moves individuals from being viewed as assets to being real resources. As a result, this implies that instead of individuals serving associations, associations will serve individuals. Thus, rather than ability essentially being utilized to make an upper hand and an incentive for clients, Industry 5.0 pulls jointly to also create additional incentive for laborers to draw in and keep the best representatives.

(b) *Versatility*

As the world has become more signed up, over time, more people joined up; we have seen the far-reaching effect of worldwide issues like the Coronavirus pandemic and global inventory deficiencies. While numerous organizations hope to further develop efficiencies and advancing benefits, these variables don't further develop flexibility. As a matter of fact, there is a conviction that a focus on deftness and adaptability can make organizations less versatile, not more. As opposed to focusing on development, benefit, and effectiveness, stronger associations would hope to expect and respond to any emergency to guarantee soundness through testing times.

(c) *Manageability*

Industry 5.0 broadens maintainability from just lessening, limiting, or relieving environmental harm to effectively chasing after endeavors to make a positive change. At times alluded to as "Net Positive," this objective means to make the world a superior spot with organizations turning out to be important for the arrangement instead of being an issue or just offering empty talk to supportability objectives through "green washing."

11.3.3 Applications of Cobots in Industry 5.0

While robots have performed hazardous, dull, or truly debilitating work in assembling plants and different work environments, Industry 5.0 stretches out this to permit them to work cooperatively with human laborers (Batth et al., 2018). For instance, rather than being fenced off for security, another age of "Cobots" that can work securely close by individuals is setting out new open doors for organizations. Human and machine laborers working one next to the other permit individuals to zero in on esteem-adding cycles to take personalization of items to another level. Cobots, or collaborative robots, are robots designed to work alongside humans in a shared workspace.

They are equipped with sensors, cameras, and other safety features to detect human presence and ensure that they work safely and efficiently.

- *Manufacturing*: Cobots are commonly used in manufacturing environments to handle repetitive or dangerous tasks, such as assembling components, packaging, and machine tending. They can work alongside human workers to improve productivity, reduce errors, and increase safety.
- *Healthcare*: Cobots can assist healthcare workers by handling routine tasks such as delivering medications and supplies, taking vital signs, and cleaning hospital rooms. They can also be used in rehabilitation and therapy to assist patients with mobility and exercise.
- *Logistics and warehousing*: Cobots are increasingly used in warehouses and logistics facilities to handle tasks such as picking and packing, inventory management, and transportation of goods. They can work alongside human workers to improve efficiency and reduce errors.
- *Agriculture*: Cobots can be used in agriculture to automate tasks such as planting, weeding, and harvesting crops. They can work in tandem with human workers to increase productivity and efficiency in the field.
- *Education*: Cobots are being used in classrooms and educational settings to teach students about robotics and programming. They can help students learn about technology and automation in a hands-on way, preparing them for future careers in different fields.

For instance, the clinical calling could utilize this signed-up, helpful way to deal with making gadgets that are custom-fitted for an individual, for example, with a diabetes application that can follow your way of life and illuminate the production regarding a gadget to suit your singular requirements. Fitting items to suit individual necessities can be reached out to different businesses, including gadgets, and auto, and that's only the tip of the iceberg; adding an individual, human touch broadens the contributions made through Industry 5.0. The main applications of Cobots are shown in Figure 11.1.

11.4 CONSEQUENCES OF INDUSTRY 5.0 AS SMART SERVICES IN MAJOR SECTORS OF DEVELOPING ECONOMIES

Industry 5.0 is significant as it permits organizations and industries to effectively convey answers for society to protect assets, guarantee social dependability, and address environmental targets (Huang et al., 2022). With benefits zeroed in on the more extensive world, including workers, as opposed to just efficiency and benefit, Industry 5.0 transforms associated organizations

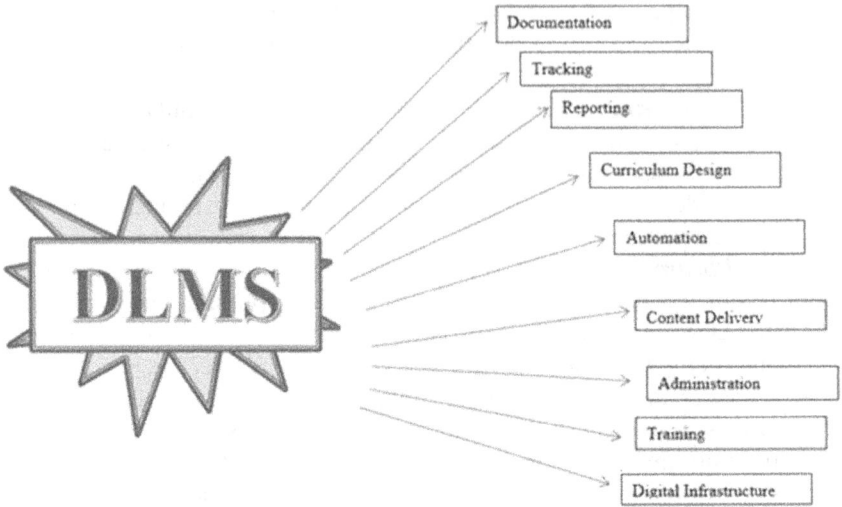

Figure 11.1 Major applications of Cobots in different sectors.

into part of the arrangement as opposed to presenting possible ecological and cultural issues. Major sectors affected by Industry 4.0 such as medical services, supply chains, and assembly are likely to profit from Industry 5.0 due to the use of massive data analysis, the IoT, cooperative robots, computerized twins, and, recently, an increased focus on personalization and maintainability (Mourtzis et al., 2022).

With the speedy improvement of development in the digitalization era, Industry 4.0 has transformed into an expression that transformed into a reference for imaginative work in the space of development in various regions. This continues to set off all people to encourage development to engage. Society 5.0 is an idea that gets a handle on the change in people's lives with the improvement of the fourth present-day upset. The possibility that should be looked into is how there is a disturbance in the public eye that both utilizes advancement and considers humanity's points of view. A few areas of work and needs are starting to enter digitalization that uses man-made brainpower, Big Data, mechanical technology, robotization, AI, and the Web of Things.

11.4.1 Smart healthcare practices

Industry 5.0 permits the assembling of a precisely tailored brilliant embed according to changing client necessities. In any case, this is reasonable. The field of clinical calling is moving toward automated reasoning that addresses a range of problems, for example, glucose levels in blood convey insulin into the blood and change the control framework. Industry 5.0 can change an existence with various innovations. It will, in general, robotize the assembling framework in a superior manner by making savvy plants, and fabricated item imparts carefully. The modern robot will computerize the entirety of the presentation cycle and proceed with the backward seating arrangement of human knowledge in this revolution (Sharma et al., 2020).

Industry 5.0 can assist with carrying out mass personalization by making inserts according to patient match which is an essential necessity for muscular health. It will change the conventional assembling techniques for patient inserts and is fit for updating various instruments and gadgets. Advances utilized in this unrest are likewise useful to carry out procedures in a more precise manner. It is likewise useful for a clinical understudy to give the best instruction, education, learning, treatment, and innovative work process (Carayannis & Morawska-Jancelewicz, 2022).

In muscular health, businesses need top-notch modified inserts with improved lifetimes that can be customized. Industry 5.0 is to tackle various difficulties such as overproduction, absence of straightforwardness, and wrong device choice. This unrest affects item dependability, item lifecycle, benefit, proficiency, administration, plan of action, IT security, climate, and machine and human safety. It will globally bring closer to worldwide businesses by creating great collaborations between humans and machines. It interfaces smart assembling frameworks through gadgets to make better mechanization by using human brainpower. Industry 4.0 creates digitization, where various advancements and gadgets such as brilliant machines, information examination, and man-made consciousness robotize the assembling system.

The primary utilization of Industry 5.0 is personalization. It centers on the co-activity between man and machine, prompting mass customization and personalization for the client. Humans smartly used mental processing and upskilled to provide some added value to the ongoing task. It adapts to the client's individual requirements. Industry 5.0 is useful for giving top-notch customized inserts, instruments, and gadgets with a necessary scope of detail according to the patient's prerequisites. In the future, this upheaval will be useful in muscular health by tackling issues over a drawn-out delicate life cycle. It can take information on illnesses from the different arrangements of patients, examine them, and effectively make changes in distinguishing the individual degree of sicknesses and aiding therapies without even understanding requesting something very similar.

In impending years, it will upgrade the job of machines that give imaginative and creative thoughts in clinical and muscular health fields. It gives excellent customized inserts and contraptions. The utilization of trend-setting innovations has expanded radically to keep up with any delicate records connected with schooling, well-being, or money. It assists with safeguarding the information from unapproved access by assailants. Notwithstanding, every one of the current cutting-edge innovations faces a few issues in light of their vulnerabilities. These innovations have a few omissions to give security, assault-free, straightforwardness, dependability, and adaptability. These attributes are fundamental while dealing with any touchy information such as instructive declarations or clinical testaments (Haleem & Javaid, 2019).

11.4.2 Smart online education practices

The possibility of Society 5.0 and Industry 5.0 is unquestionably not an essential successive continuation or choice as opposed to the Industry 4.0 perspective. Society 5.0 means to put individuals at the midpoint of advancement, exploiting the impact of development and Industry 4.0 results with the mechanical blend to chip away at individual fulfillment, social commitment, and viability. This notable perspective has typically centered on the objectives of the gathered sensible improvement targets for the nation. It moreover has huge repercussions on school modifications. Schools are called after transmitting information for fresh headways and social turn of events. Digitalization opens new perspectives for schools and can become one of the essential drivers of their change. Coordinating the speculations of Society 5.0 and Industry 5.0 into the school's practices and approaches will allow the universities and social orders to totally benefit from the cutting-edge change (Paschek et al., 2019).

Making the human-organized progression of the school's image name and developing new pleasing models will in like manner help with achieving prudent requirements. During the last 20 years, huge imaginative changes have happened around us, maintained by inconvenient advances, both on the item and hardware sides. A blend of data, correspondence, and computerized reasoning is occurring, as well as the cross-treatment of many ideas alluded to as advanced change. While the conversation on the most proficient method to operationalize the new keen frameworks of the fourth modern upheaval, Industry 4.0, is yet going on, the prevailing attributes of the fifth modern transformation, Industry 5.0 – going past delivering labor and products for benefit – expects all to think and act in an unexpected way. Because of the combination peculiarity, the limits between various disciplines are disintegrating, requiring an exhaustive conversation on what designing schooling ought to resemble from now on. Subsequent to thinking about the fifth modern transformation, significant patterns/impacting factors that will influence designing instruction essentially and dissecting the requirement for abilities from the future, we have recognized three procedures that might end up being useful to the advanced education foundations to overhaul their projects (Mitchell & Guile, 2022). These systems are as follows:

- deep-rooted learning and transdisciplinary training
- manageability, versatility, and human-driven plan modules
- involved information familiarity and the executives' courses

Contemporary advanced education is changing step by step. Gatherings of educators and understudies from auditoriums are progressively moving into the advanced space of the Web, embracing the recipe of distance learning. The coming of Society 5.0 and Economy 5.0 will infer further changes.

The need to coordinate the genuine and virtual worlds, expanded interest for the time being, assets, and the need to join proficient work with schooling will prompt advanced education to plan future residents to work in the space of sharing assets, to be compelled to additional versatile changes. Industry 5.0 turned into the reason for the creators to set up a proposition for a model of future instructive ways as per the Economy 5.0 pattern where the adaptability of spot and time, customization of the deal, collaboration, versatility of showing techniques and instruments, and the proactive job of the educator as a tutor and mentor comprise a bunch of set rules in the showing model representing things to come (Broo et al., 2022).

11.4.3 Smart logistic services

Given the significance of human centricity, versatility, and supportability, the arising idea of Industry 5.0 has pushed forward the examination boondocks of the innovation-centered Industry 4.0 to brilliant and amicable financial progress driven by innovations, where the job of the human in the mechanical change is transcendently centered around. A few examinations talk about the effects of problematic advances on smart strategic tasks in Industry 4.0. Nonetheless, since Industry 5.0 is another idea regardless of its outset, its suggestions for brilliant strategies have not been examined. The center components of Industry 5.0 show that following the innovation-driven change of Industry 4.0, the cultural, natural, and human viewpoints require more consideration, which will yield critical effects on coordinated factors tasks and the board. For example, the personalization of requests infers a customized conveyance framework. Integrating clients into the plan requires a profoundly insightful CPS and framework combination. Human–machine collaboration sets off the communication of different subjects such as security and the human way of behaving (Jafari et al., 2022).

Industry 5.0 gives potential chances to work on the astuteness and supportability of different strategies and activities through digitalization and human–machine coordinated effort. Notwithstanding, there is an absence of laid-out research and instructive modules in this field. The objective of the task is to empower industry 5.0 savvy operations from global perspective in order to lay out a stage for information building and move, foster new instructive modules, and advance understudy and staff trades (Qahtan et al., 2022).

11.4.4 Smart financial and insurance services

Innovation has created new possibilities for assembling enterprises with Industry 4.0 and has promoted the development of the assembling area. This chapter centers around the following stage, which is Industry 5.0, and the moves toward making mechanization powerful by expanding processes and functional effectiveness, as well as diminishing labor force size. In a modern economy, Industry 5.0 examines the mix of items, processes, machines, programming, and modern robots in acknowledging Industry 5.0. It covers the double combination of human insight with machine knowledge and audits the consequences of utilizing the Modern Web of Things (IoT) and computerized reasoning (simulated intelligence). The production of another classification of robots named Cooperative Robots (Cobots) explicitly intended to accelerate the assembling system and benefit is investigated.

Innovation, alongside political and financial variables, is one of the principal forces influencing future developments in banking (Nicoletti & Nicoletti, 2021). Banking chiefs direly need to know innovative patterns to pursue vital choices, know the future precisely, and take advantage of existing open doors. Industry 5.0 is the fantasy of present-day banking, in light of techniques for fruitful section into the field in something else entirely (Mehdiabadi et al., 2022).

The world has gone into the fifth modern upheaval which is portrayed as a human-driven upset as it depends on experts' imaginative and inventive capacities to direct quick, smart, and exact machines. The issues of breaking down and loss of efficiency during mechanization essentialized the utilization of people's ability to guide and control the assembling of merchandise and items. With regards to monetary establishments (banks), the use of I5.0 advances, for example, edge registering, computerized twins, Cobots, Web of Everything, Big Data examination, and 6G network, can work with banks to diminish the expense of their items and administrations, hold existing clients, keep up with classification and protection of clients, further develop their gamble the board rehearses, offer blunder-free monetary administrations, advance information-driven culture in the right direction, and enable effective joining of digitalization. This formative review is supposed to establish the groundwork for future investigations to investigate the progress of Industry 5.0 advancements inside banks experimentally.

Smart insurance services in Industry 5.0 play an important role in climate change resilience and weather risk mitigation, especially in rural areas (Al Faruqi, 2019; Jain and Singh, 2023). Weather index insurance is a type of insurance that provides protection against losses due to adverse weather events, such as droughts, floods, and hurricanes (Singh, 2022). Instead of indemnifying the actual losses, these policies pay out based on predetermined weather indexes, which can serve as proxies for the damage caused by weather-related risks (Singh & Agrawal, 2019). Over the past two decades, weather index insurance has gained popularity as a tool for managing weather-related risks among farmers, herders, and other vulnerable populations in developing countries (Singh & Agrawal, 2020). Weather index insurance can increase resilience and improve welfare among vulnerable populations (Singh & Agrawal, 2021). Weather index insurance can be an effective tool for managing weather-related risks (Singh & Agrawal, 2022).

Weather index insurance can provide significant benefits to farmers and other vulnerable populations (Singh et al., 2023). Weather index insurance was less likely to experience food insecurity during a drought and had higher levels of income and consumption (Singh et al., 2021a). Different studies prove that weather index insurance can be an effective tool for managing weather-related risks and improving welfare among vulnerable populations (Singh et al., 2024). However, the success of these insurance programs depends on careful design and implementation, especially for an unorganized sector (Singh, 2014).

11.4.5 Development practices in smart cities and smart villages

Savvy urban areas focus on an economical way of life with a greener and more secure metropolitan climate. The futurists have proactively begun examining Industry 5.0 as a subject of adding human contact or personalization in view of joint effort and collaboration of humans and robots. This chapter analyzes the idea of what Industry 5.0 can mean for brilliant urban communities and what sort of changes are expected – how it will make a different arrangement of chances (Kasinathan et al., 2022).

The quickly expanding lack of assets and energy, falling-apart foundation, ecological emergency, and interest in social advantages have prompted the improvement of the idea of a smart city or computerized city which bargains in applying brilliant advances to infrastructural administrations and parts. This incorporates the utilization of registering innovations to smooth-out city organization, training, medical services, public security, land, and transportation. In addition, various utilities in a smart city likewise make progress toward upgrading its assets by arranging its preventive support and security exercises. The principal point is to coordinate the physical, IT, social, and business foundations (Sharma et al., 2020).

The engineering of smart urban communities can be isolated into two sections: outside and interior. The outer design is utilized for observing roads, parks, recreation regions, and so on. The total city is observed through extended-run conventions for correspondence, for example, the utilization of sensors exposed to the daylight. Inward engineering is relatively less expensive to carry out and utilizes short-range conventions like Wi-Fi, Bluetooth, and RFID, for example, grasping the mathematical condition of the structures, accountability for, and so on. The smart urban communities are totally reliant upon IoT. The different sensors, gadgets, and applications are utilized for gathering the information which is subsequently examined to make innovation arrangements more viable (Visvizi & Lytras, 2019).

Smart urban communities are viewed as a venture that utilizes mechanical advances as an instrument to support and work on the nature of living. Information replaces the heart while planning and carrying out the idea of smart urban areas. The residents of a smart city are supposed to completely be mindful of its applications and follow protection, well-being, and security rehearses, which can be accomplished through suitable preparation and mindfulness crusades. The public authority is ought to be well-composed with the information and should have legitimate documentation and code-books as the onus of planning information approaches lies on the public authority (Ullah, 2022).

Savvy urban community's research has set up a good foundation for itself as one of today's most dynamic inter- and multidisciplinary centers. Research in this field is driven by the acknowledgment that advances in complex data and correspondence innovation (ICT), from one viewpoint, set out the freedom to ease a few difficulties that dynamic urbanization makes, and then again, may considerably add to the prosperity of urban areas' occupants. Unquestionably, basic voices exist and these ought to be noted. Smart towns remain a specialty idea, the connected discussion is yet to begin, and the exploration plan is actually open. Also, on account of the savvy urban communities' research, the brilliant town banter is driven by the subject of how and in which ways ICT can work on prosperity in pro-vincial regions. Regardless of whether wise significant contrasts between the two strands of exploration exist, the extensively imagined basis of maintainability is normal in the two discussions. By uniting these two dis-cussions, the target of this subject is twofold, i.e., to empower research on savvy urban areas and brilliant towns, individually, and, at the same time, to ponder the chance of building spans between the two discussions (Aslam et al., 2020).

Right away, the possibility of a Smart Metropolitan people group (metro-politan settlement) began from the IoT advancement, in any case, the usage of IoT development can contact the possibility of Splendid Towns (common settlement) as well, dealing with the presence of the occupants, and the orga-nizations overall (Fraga-Lamas et al., 2021).

Notwithstanding, the country settlements have fairly unforeseen essen-tials in contrast with the metropolitan settlements. If the utilization of IoT in the Splendid Metropolitan people group can be portrayed by the densifi-cation of IoT to regular daily existence, following metropolitan regions' essential characteristics of being thickly settled places, IoT-empowered Keen Towns are commonly a course of action of dispersing and need, as illus-trated in Figure 11.2.

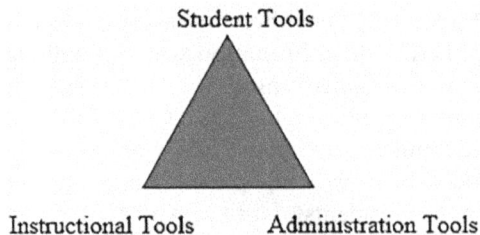

Student Tools

Instructional Tools   Administration Tools

Figure 11.2 Application of Internet of Things in smart city and village.

11.5 SIGNIFICANCE OF INDUSTRY 5.0 IN ATTAINING SUSTAINABLE DEVELOPMENT GOALS

Researchers accept that the recently presented Industry 5.0 can possibly move past the benefit-focused efficiency of Industry 4.0 and advance reasonable improvement objectives like human-centricity, socio-ecological maintainability, and versatility (Singh et al., 2021b). Nonetheless, little has been completed to understand how this badly characterized peculiarity might convey its indented maintainability values notwithstanding these speculative commitments. To address this information hole, the current review fostered a methodology guide that makes sense of the system by which Industry 5.0 conveys its planned manageable improvement capabilities (Chourasia et al., 2023). Authors previously created and presented the Business 5.0 reference model that portrays the specialized and utilitarian properties of this peculiarity. Huge changes in the public arena were underlined as being expected to accomplish Maintainable Improvement Objectives, a need which was additionally heightened with the development of the pandemic (Javaid et al., 2020). The imminent society ought to be coordinated toward a feasible turn of events, a cycle in which innovation assumes a critical part (Zengin et al., 2021).

11.5.1 SDG 1 (no poverty)

The major disruptive technological support for SDG 1 in Industry 5.0 is as follows:

- Developments upheld by troublesome innovations can make the items and administrations reasonable and open.
- Troublesome innovations can make a quick reskilling situation which, at the point when acknowledged by low-pay gatherings, would be profoundly advantageous to lift themselves from destitution through positions.
- Other digitalization approaches like portable banking and outstanding teleopenness can work on the availability and, thus, associate them with the possible openings on Earth.

11.5.2 SDG 2 (zero hunger)

The major disruptive technological support for SDG 2 in Industry 5.0 is as follows:

- Unstable advancements like artificial intelligence, machine learning, and the IoT may be able to reduce food waste and potentially connect the underprivileged and generous.
- The innovation can drastically improve and reinforce the food production network by giving flexibility, even in a debacle, for example, the Coronavirus pandemic.

- Consistent checking of natural variables through complex troublesome innovations can give continuous criticism to the food makers so that early activities can be carried out to forestall crop harm.

11.5.3 SDG 3 (good health and well-being)

The major disruptive technological support for SDG 3 in Industry 5.0 is as follows:

- Broadens medical care productivity and openness through digitalized processes. A few models incorporate telehealth and visit chat-bot facility.
- The fortified availability between organizations of specialists and patients would lean toward critical improvement in the prosperity of all age gatherings.

11.5.4 SDG 4 (quality education)

The major disruptive technological support for SDG 4 in Industry 5.0 is as follows:

- The digitalization-driven schooling components like smart classes, far-off schooling, and edu-tech organizations stretch the boundaries and approaches for instructing individuals.
- Then again, the development of troublesome innovations gives reasonable and quality training to all spaces of individuals. It makes the more youthful age totally ready for the quick changes in the mechanical area as well as to complete the changes in the forthcoming climate.

11.5.5 SDG 5 (gender equality)

The major disruptive technological support for SDG 5 in Industry 5.0 is as follows:

- Orientation uniformity is more articulated with regard to availability. In such a case, troublesome innovations assume an essential part in getting to the credits, food, and different assets. The gifted positions, on the other hand, ought to be guaranteed equivalent pay across sexes.

11.5.6 SDG 6 (clean water and sanitation)

The major disruptive technological support for SDG 6 in Industry 5.0 is as follows:

- With the guide of data and correspondence advances, more productive and powerful methods for gathering, screening, and improving water use information for public, business, cultural, and biological system needs are conceivable.

- Incorporated executives for water asset and brilliant home frameworks are a portion of the innovations that can profoundly lean toward ensuring clean water.

11.5.7 SDG 7 (affordable and clean energy)

The major disruptive technological support for SDG 7 in Industry 5.0 is as follows:

- With the development of ecologically friendly power, the significance of smart lattices and observation of ongoing energy requests are of extreme significance. These are the endless commitments from problematic advances.
- The consistent joining of different power age units and fulfillment of the unique burdens would open various potential doors. This makes society tough as well as makes more open positions.
- Digitalization can possibly grow admittance to clean energy access as well as upgrade its moderateness.

11.5.8 SDG 8 (decent work and economic growth)

The major disruptive technological support for SDG 8 in Industry 5.0 is as follows:

- Problematic advances support nature-comprehensive plans of action as well as circular economy approaches.
- The future worth or commitment of problematic advances in the economy.

11.5.9 SDG 9 (industry, innovation, and infrastructure)

The major disruptive technological support for SDG 9 in Industry 5.0 is as follows:

- Problematic innovations support the change to Industry 4.0 as well as Industry 5.0. This eventually improves efficiency, proficiency, network, versatility, and manageability.
- Problematic innovations open a wide new field and incorporate applications where the potential for advancement is high. It might likewise turn into a central substance in the public eye, like power in present society. Hence, progress in troublesome innovation can be valuable to the social order framework.

11.5.10 SDG 11 (sustainable cities and communities)

The major disruptive technological support for SDG 11 in Industry 5.0 is as follows:

- Savvy portability, rebuilt society, and Society 5.0 are a portion of the high-expected ideas for advancing maintainable urban communities that are brought into the world from problematic innovations.

11.5.11 SDG 12 (responsible consumption and production)

The major disruptive technological support for SDG 12 in Industry 5.0 is as follows:

- One of the remarkable advantages that disruptive advancements can give is enhancement. For diminished utilization of assets, streamlined creation and utilization are the best approaches.
- Diminish, reuse, and reuse can be supported by disruptive advances.

11.5.12 SDG 17 (partnerships for the goals)

The major disruptive technological support for SDG 17 in Industry 5.0 is as follows:

- Execution of troublesome advancements at a large scale requires systems administration of associations. This can support drawing in capitalization also, cultivating the advancement of different objectives.

As an example, for SDG 17, the case of Sweden is so great at reusing that, for quite a long time, it has imported junk from different nations to keep its reusing plants going. Under 1% of Swedish family squander was shipped off landfill last year or any year starting around 2011. Over the long haul, Sweden has executed a firm public reusing strategy so that, although privately owned businesses embrace the greater part of the matter of bringing in and consuming waste, the energy goes into a public warming organization to warm homes through the freezing Swedish winter. "That is a key explanation that how Sweden can utilize the warming from the waste plants" (Kushwaha et al., 2023a). In the southern part of Europe, they don't utilize the warming from the waste, it simply goes out the smokestack. Despite this fact, it will be utilized as a substitute for non-renewable energy sources.

11.6 CONCLUSION AND FUTURE SCOPE

Industry 5.0 is a concept that envisions a new paradigm of industrial production that combines the benefits of Industry 4.0 technologies, such as automation and artificial intelligence, with a renewed focus on human skills, creativity, and sustainability (Rozanec et al., 2022). While the full impact of Industry 5.0 is yet to be seen, it has the potential to transform the way we produce and consume goods and services, particularly in developing countries where sustainable development is a key priority. Some potential future consequences of Industry 5.0 in sustainable services production and consumption in developing countries are mentioned below.

- **Increased adoption of sustainable production practices:** Industry 5.0 technologies can enable more sustainable and resource-efficient production processes, such as 3D printing, robotics, and renewable energy. This could lead to increased adoption of sustainable production practices in developing countries, which could help reduce environmental impacts and promote more circular and low-carbon economies.
- **More inclusive and human-centered services:** Industry 5.0 emphasizes the importance of human skills and creativity in industrial production, which could lead to more inclusive and human-centered services in developing countries. For example, small-scale producers and artisans could use Industry 5.0 technologies to enhance their production processes and create unique and customized products that cater to local markets and cultures.
- **Increased access to digital services:** Industry 5.0 technologies can enable greater access to digital services, such as e-commerce, telemedicine, and education, particularly in rural and remote areas of developing countries. This could help bridge the digital divide and improve access to essential services, while also reducing the need for physical infrastructure and travel.
- **New business models and opportunities:** Industry 5.0 could enable new business models and opportunities in developing countries, particularly in the areas of sustainable production and consumption. For example, circular economy models, such as product-as-a-service or sharing economy platforms, could become more viable with the help of Industry 5.0 technologies (Martynov et al., 2019).

However, there are also potential challenges and risks associated with Industry 5.0 in developing countries, such as increased dependence on technology, displacement of jobs, and unequal distribution of benefits. It will be important to ensure that the benefits of Industry 5.0 are shared equitably and that the transition to this new paradigm is guided by principles of sustainability, inclusivity, and social responsibility.

Nonetheless, Industry 5.0 alludes to robots and savvy machines working close by individuals with added strength and maintainability objectives included. Where Industry 4.0 zeroed in on advancements, for example, the Web of Things and enormous information, Industry 5.0 looks to add human, ecological, and social viewpoints back into the situation. Because of the expanded information streams, sensors, and observation related to Industry

5.0, one of the biggest difficulties includes energy for the executives. This can be improved to permit huge brilliant gadgets to be associated with more astute energy utilization and energy gathering. Different difficulties for Industry 5.0 include insight and the eagerness, capacity, and monetary extent of organizations to embrace these new working techniques. Industry 5.0, while working at a full limit, will take into consideration the better mechanization of assembling processes, offering continuous information and yet permitting individuals to work close by hardware to further develop processes and give personalization to clients (Kushwaha et al., 2023b).

In such a manner, Industry 5.0 should be visible as supplementing the advances made in Industry 4.0 to help as opposed to overriding people. This permits people to intercede where expected and gets away from over-the-top computerization to consolidate decisive reasoning and flexibility, while as yet exploiting the accuracy and repeatability of machines. The future for Industry 5.0 incorporates the production of robots, including modern robots, with efforts toward advances in mental registration and computerized reasoning to further develop efficiencies and speed of conveyance, while simultaneously permitting individuals to zero in on different regions.

Innovation has made imaginative new possibilities for assembling enterprises and service providers with Industry 4.0 and has encouraged the development of the assembling area. This chapter centers around the following stage, which is Industry 5.0, and the moves toward making mechanization powerful by expanding processes and functional products, as well as lessening labor force size.

Further research on Industry 5.0 will examine the coordination of items, processes, machines, programming, and modern robots in acknowledging Industry 5.0 (Kolade & Owoseni, 2022). It will cover the double incorporation of human insight with machine knowledge and audit the aftereffects of utilizing the Modern Web of Things (IoT) and man-made consciousness (computer-based intelligence). The formation of another class of robots named Collaborative Robots (Cobots), explicitly intended to accelerate the assembling system and productivity, is investigated. This chapter likewise investigates the assembling system and offers more customized and tweaked items for clients.

REFERENCES

Adel, A. (2022). Future of industry 5.0 in society: human-centric solutions, challenges and prospective research areas. *Journal of Cloud Computing, 11*(1), 1–15.

Akundi, A., Euresti, D., Luna, S., Ankobiah, W., Lopes, A., & Edinbarough, I. (2022). State of Industry 5.0—Analysis and identification of current research trends. *Applied System Innovation, 5*(1), 27.

Al Faruqi, U. (2019). Future service in industry 5.0. *Jurnal Sistem Cerdas, 2*(1), 67–79.

Al-Turjman, F., Nayyar, A., Devi, A., & Shukla, P. K. (Eds.). (2021). *Intelligence of things: AI-IoT based critical-applications and innovations*. New York, US: Springer.

Aslam, F., Aimin, W., Li, M., & Ur Rehman, K. (2020). Innovation in the era of IoT and industry 5.0: Absolute innovation management (AIM) framework. *Information, 11*(2), 124.

Batth, R. S., Nayyar, A., & Nagpal, A. (2018, August). Internet of robotic things: driving intelligent robotics of future-concept, architecture, applications and technologies. In *2018 4th International Conference on Computing Sciences (ICCS)* (pp. 151–160). IEEE.

Broo, D.G., Kaynak, O. & Sait, S.M. (2022). Rethinking engineering education at the age of industry 5.0. *Journal of Industrial Information Integration, 25*, 100311.

Bryndin, E. (2020). Formation and management of Industry 5.0 by systems with artificial intelligence and technological singularity. *American Journal of Mechanical and Industrial Engineering, 5*(2), 24–30.

Carayannis, E. G., & Morawska-Jancelewicz, J. (2022). The futures of Europe: Society 5.0 and Industry 5.0 as driving forces of future universities. *Journal of the Knowledge Economy, 13*(4), 3445–3471.

Chander, B., Pal, S., De, D., & Buyya, R. (2022). Artificial intelligence-based internet of things for industry 5.0. In *Artificial intelligence-based internet of things systems* (pp. 3–45). Springer, Cham.

Chourasia, S., Pandey, S. M., Gupta, K., Murtaza, Q., & Walia, R. S. (2023). Industry 5.0 for sustainable manufacturing: New product, services, organizational and social information. In *Surface Engineering* (pp. 243–255). CRC Press.

Fraga-Lamas, P., Lopes, S. I. & Fernández-Caramés, T.M. (2021). Green IoT and edge AI as key technological enablers for a sustainable digital transition towards a smart circular economy: An industry 5.0 use case. *Sensors, 21*(17), 5745.

Ghobakhloo, M., Iranmanesh, M., Mubarak, M. F., Mubarik, M., Rejeb, A., & Nilashi, M. (2022). Identifying industry 5.0 contributions to sustainable development: A strategy roadmap for delivering sustainability values. *Sustainable Production and Consumption, 33*, 716–737.

Gorodetsky, V., Larukchin, V., & Skobelev, P. (2020). Conceptual model of digital platform for enterprises of Industry 5.0. In: Kotenko, I., Badica, C., Desnitsky, V., El Baz, D., Ivanovic, M. (eds) *Intelligent Distributed Computing XIII. IDC 2019. Studies in Computational Intelligence* (vol. 868). Springer, Cham. https://doi.org/10.1007/978-3-030-32258-8_4

Haleem, A., & Javaid, M. (2019). Industry 5.0 and its applications in orthopaedics. *Journal of Clinical Orthopaedics & Trauma, 10*(4), 807–808.

Huang, S., Wang, B., Li, X., Zheng, P., Mourtzis, D. & Wang, L., (2022). Industry 5.0 and Society 5.0—Comparison, complementation and co-evolution. *Journal of Manufacturing Systems, 64*, 424–428.

Ivanov, D. (2022). The Industry 5.0 framework: viability-based integration of the resilience, sustainability, and human-centricity perspectives. *International Journal of Production Research*, 1–13.

Jafari, N., Azarian, M. & Yu, H. (2022). Moving from Industry 4.0 to Industry 5.0: what are the implications for smart logistics? *Logistics, 6*(2), 26.

Jain, P., & Singh, P. (2023) 'Lethargic approach in expediting financial reporting system: a case of construction firm', *International Journal of Teaching and Case Studies, 14*(1), 1–10. https://doi.org/10.1504/IJTCS.2023.10055878

Javaid, M., Haleem, A., Singh, R.P., Haq, M.I.U., Raina, A. & Suman, R. (2020). Industry 5.0: Potential applications in COVID-19. *Journal of Industrial Integration and Management*, 5(4), 507–530.

Kasinathan, P., Pugazhendhi, R., Elavarasan, R.M., Ramachandaramurthy, V.K., Ramanathan, V., Subramanian, S., Kumar, S., Nandhagopal, K., Raghavan, R.R.V., Rangasamy, S. & Devendiran, R. (2022). Realization of sustainable development goals with disruptive technologies by integrating Industry 5.0, Society 5.0, smart cities and villages. *Sustainability*, 14(22), 15258.

Kolade, O., & Owoseni, A. (2022). Employment 5.0: The work of the future and the future of work. *Technology in Society*, 71, 102086.

Kumar, A., & Nayyar, A. (2020). si 3-Industry: A sustainable, intelligent, innovative, internet-of-things industry. A roadmap to Industry 4.0: Smart production, sharp business and sustainable development, 1–21.

Kushwaha, J., Sharma, A., & Singh, P. (2022). Exploration and prioritization of enablers to organization work–family balance planning for working sole Indian mothers integrating fuzzy Delphi and AHP. *International Journal of Social Economics*, 50(3), 398–418. https://doi.org/10.1108/IJSE-05-2022-0348

Kushwaha, J., Singh, P., & Kushwaha, R. (2023a), "Predicting working sole Indian mothers' satisfaction towards work-family balance integrating Kano model and weighted average method" Vol. ahead-of-print No. ahead-of print. https://doi.org/10.1108/IJSE-02-2023-0074

Kushwaha, J., Singh, P., & Sharma, A. (2023b) "Modelling the enablers of work-family balance among working single mothers integrating expert-mining and TISM-MICMAC", *Journal of Organizational Effectiveness: People and Performance*, Vol. ahead-of-print No. ahead-of print. https://doi.org/10.1108/JOEPP-05-2022-0106

Leng, J., Sha, W., Wang, B., Zheng, P., Zhuang, C., Liu, Q., Wuest, T., Mourtzis, D. & Wang, L. (2022). Industry 5.0: Prospect and retrospect. *Journal of Manufacturing Systems*, 65, 279–295.

Maddikunta, P. K. R., Pham, Q. V., Prabadevi, B., Deepa, N., Dev, K., Gadekallu, T. R., Ruby, R. & Liyanage, M. (2022). Industry 5.0: A survey on enabling technologies and potential applications. *Journal of Industrial Information Integration*, 26, 100257.

Madsen, D. Q., & Berg, T. (2021). An exploratory bibliometric analysis of the birth and emergence of industry 5.0. *Applied System Innovation*, 4(4), 87.

Majerník, M., Daneshjo, N., Malega, P., Drábik, P., & Barilová, B. (2022). Sustainable development of the intelligent industry from industry 4.0 to industry 5.0. *Advances in Science and Technology Research Journal*, 16(2), 12–18.

Martynov, V. V., Shavaleeva, D. N., & Zaytseva, A. A. (2019, September). Information technology as the basis for transformation into a digital society and industry 5.0. In *2019 International Conference "Quality Management, Transport and Information Security, Information Technologies" (IT&QM&IS)* (pp. 539–543). IEEE.

Mehdiabadi, A., Shahabi, V., Shamsinejad, S., Amiri, M., Spulbar, C., & Birau, R. (2022). Investigating Industry 5.0 and Its Impact on the Banking Industry: Requirements, Approaches and Communications. *Applied Sciences*, 12(10), 5126.

Minculete, G., Bârsan, G., & Olar, P. (2021). Conceptual Approaches of Industry 5.0. Correlative Elements with Supply Chain Management 5.0. *Revista de Management Comparat International*, 22(5), 622–635.

Mitchell, J., & Guile, D. (2022). Fusion skills and industry 5.0: conceptions and challenges. *Insights Into Global Engineering Education After the Birth of Industry 5(0)*, 53.

Mourtzis, D., Angelopoulos, J., & Panopoulos, N. (2022). A Literature Review of the Challenges and Opportunities of the Transition from Industry 4.0 to Society 5.0. *Energies*, 15(17), 6276.

Nahavandi, S. (2019). Industry 5.0—A human-centric solution. *Sustainability*, 11(16), 4371.

Nayyar, A., & Kumar, A. (Eds.). (2020). *A roadmap to industry 4.0: smart production, sharp business and sustainable development* (pp. 1–21). Berlin: Springer.

Nicoletti, B., & Nicoletti, B. (2021). Industry 5.0 and Banking 5.0. *Banking 5.0: How Fintech Will Change Traditional Banks in the'New Normal'Post Pandemic*, 13–53.

Ozdemir, V., & Hekim, N. (2018). Birth of industry 5.0: Making sense of big data with artificial intelligence, "the internet of things" and next-generation technology policy. *Omics: A Journal of Integrative Biology*, 22(1), 65–76.

Paschek, D., Mocan, A., & Draghici, A. (2019, May). Industry 5.0—The expected impact of next industrial revolution. In *Thriving on future education, industry, business, and Society, Proceedings of the MakeLearn and TIIM International Conference, Piran, Slovenia* (pp. 15–17).

Qahtan, S., Alsattar, H. A., Zaidan, A. A., Pamucar, D., & Deveci, M. (2022). Integrated sustainable transportation modelling approaches for electronic passenger vehicle in the context of industry 5.0. *Journal of Innovation & Knowledge*, 7(4), 100277.

Raja Santhi, A. & Muthuswamy, P. (2023). Industry 5.0 or industry 4.0 S? Introduction to industry 4.0 and a peek into the prospective industry 5.0 technologies. *International Journal on Interactive Design and Manufacturing (IJIDeM)*, 1–33.

Rozanec, J. M., Novalija, I., Zajec, P., Kenda, K., Tavakoli Ghinani, H., Suh, S., & Soldatos, J. (2022). Human-centric artificial intelligence architecture for industry 5.0 applications. *International Journal of Production Research*, 1–26.

Rupa, C., Midhunchakkaravarthy, D., Hasan, M.K., Alhumyani, H. & Saeed, R.A., (2021). Industry 5.0: Ethereum blockchain technology based DApp smart contract. *Mathematical Biosciences & Engineering*, 18(5), 7010–7027.

S. Rajput, E. Jain, A. Mehra, K. S. Thakur, O. Gupta and P. Singh, "Technology Driven Tourism: A PLS SEM Model of its Antecedents in Tourist Behavior Intention," *2022 IEEE Conference on Interdisciplinary Approaches in Technology and Management for Social Innovation (IATMSI)*, Gwalior, India, 2022, pp. 1–6. https://doi.org/10.1109/IATMSI56455.2022.10119265

Salgado-Reyes, A.P. & Rodríguez-Aguilar, R. (2022). Profile of the Business Science Professional for the Industry 4.0. In *Intelligent Computing & Optimization: Proceedings of the 4th International Conference on Intelligent Computing and Optimization 2021 (ICO2021)*, vol. 3 (pp. 820–831). Springer International Publishing.

Saniuk, S., Grabowska, S. & Straka, M. (2022). Identification of Social and Economic Expectations: Contextual Reasons for the Transformation Process of Industry 4.0 into the Industry 5.0 Concept. *Sustainability*, 14(3), 1391.

Sharma, I., Garg, I. & Kiran, D. (2020). Industry 5.0 and smart cities: A futuristic approach. *European Journal of Molecular & Clinical Medicine*, 7(8), 2515–8260.

Sindhwani, R., Afridi, S., Kumar, A., Banaitis, A., Luthra, S., & Singh, P. L. (2022). Can industry 5.0 revolutionize the wave of resilience and social value creation? A multi-criteria framework to analyze enablers. *Technology in Society, 68,* 101887.

Singh, P. (2014). Unorganized sector and India's informal economy: Challenges and prospects. *New Man International Journal of Multidisciplinary Studies,* 1(2), 209–218.

Singh, P. (2022). Weather index insurance viability in mitigation of climate change impact risk: A systematic review and future agenda. *Journal of Science and Technology Policy Management.* https://doi.org/10.1108/JSTPM-07-2021-0102

Singh, P., & Agrawal, G. (2019). Efficacy of weather index insurance for mitigation of weather risks in agriculture: An integrative review. *International Journal of Ethics and Systems,* 35(4), 584–616. https://doi.org/10.1108/IJOES-09-2018-0132

Singh, P., & Agrawal, G. (2020). Development, present status and performance analysis of agricultural insurance schemes in India: Review of evidence. *International Journal of Social Economics,* 47(4), 461–481. https://doi.org/10.1108/IJSE-02-2019-0119

Singh, P., & Agrawal, G. (2021). Modelling the barriers of weather index insurance service adoption integrating expert mining and ISM Fuzzy-MICMAC. *Benchmarking,* 29(8), 2527–2554. https://doi.org/10.1108/BIJ-04-2021-0183

Singh, P., & Agrawal, G. (2022). Mapping the customer centric weather index insurance service design using quality function deployment. *The TQM Journal,* 34(6), 1800–1822 https://doi.org/10.1108/TQM-08-2021-0236

Singh, P., Agrawal, G., & Kushwaha, R. (2021a, November). Evaluating customer satisfaction of weather index insurance service quality using Kano model. In *2021 IEEE Bombay Section Signature Conference (IBSSC)* (pp. 1–6). IEEE. https://doi.org/10.1109/IBSSC53889.2021.9673370

Singh, P., Kushwaha, R., & Kushwaha, J. (2021b, September). Assessing visitors satisfaction about museum service quality using Kano model. In *2021 IEEE 9th Region 10 Humanitarian Technology Conference (R10-HTC)* (pp. 1–6). IEEE.

Singh, P., Kushwaha, R., & Kushwaha, J. (2023). Evaluating farmer participation in weather index insurance in a socio-demographic perspective: Evidence from the crop insurance data of India. In *Handbook of Research on Managing the Urban-Rural Divide Through an Inclusive Framework* (pp. 69–88). IGI Global.

Singh, P., Kushwaha, R., & Kushwaha, J. (2024). Analyzing accountability of weather index insurance service in attainment of sustainable development goals: A sustainable accounting perspective. *In Springer Sustainable International Business: Smart strategies for business and society.* 978-3-031-43784-7, 531340_1_En, (Chapter 10), Contributions to Management Science, Pratik Arte et al. (Eds): Sustainable International Business. Springer Nature.

Thakur, P., & Sehgal, V. K. (2021). Emerging architecture for heterogeneous smart cyber-physical systems for industry 5.0. *Computers & Industrial Engineering,* 162, 107750.

Tiwari, S., Bahuguna, P. C., & Walker, J. (2022). Industry 5.0: A macroperspective approach. In *Handbook of Research on Innovative Management Using AI in Industry 5.0* (pp. 59–73). IGI Global.

Ullah, F. (2022). Smart Tech 4.0 in the built environment: Applications of disruptive digital technologies in smart cities, construction, and real estate. *Buildings,* 12(10), 1516.

Venaik, A., Jain, S., & Nayyar, A. (2023). Industry 4.0—Its Advancement and Effects on Security of Whistle-Blowers on Dark Web. In *Industry 4.0 and the Digital Transformation of International Business* (pp. 103–121). Singapore: Springer Nature Singapore.

Visvizi, A. & Lytras, M.D. (2019). Sustainable smart cities and smart villages research: Rethinking security, safety, well-being, and happiness. *Sustainability*, *12*(1), 215.

Xu, X., Lu, Y., Vogel-Heuser, B. & Wang, L. (2021). Industry 4.0 and Industry 5.0—Inception, conception and perception. *Journal of Manufacturing Systems*, *61*, 530–535.

Zengin, Y., Naktiyok, S., Kaygın, E., Kavak, O. & Topçuoğlu, E. (2021). An investigation upon industry 4.0 and society 5.0 within the context of sustainable development goals. *Sustainability*, *13*(5), 2682.

Zong, L., Memon, F. H., Li, X., Wang, H., & Dev, K. (2021). End-to-end transmission control for cross-regional industrial internet of things in industry 5.0. *IEEE Transactions on Industrial Informatics*, *18*(6), 4215–4223.

Chapter 12

Readiness of enterprises to implement Industry 5.0

Challenges and security concerns

Nguyen Huu Phuoc Dai

Can Tho Technical Economics College, Vietnam

FPT University, Vietnam

Pham Phi Giang

Can Tho Technical Economics College, Vietnam

Truong Hong Vo Tuan Kiet

FPT University, Vietnam

12.1 INTRODUCTION

Technologies have been playing an essential role in society and Industry for years recently. They influence changing processes or employees' behaviors and focus on holistic, sustainable, and human-centered value creation (Hein-Pensel et al., 2023). Since the beginning of the Industrial Revolution, many adjustments have been made to the production process. The term "innovation" may be used to refer to various things, from machines that are driven by steam or water (Industry 1.0) to ones that are powered by electricity or digital electronics (Industry 2.0), automation (Industry 3.0), and robotics (Industry 4.0), respectively. These improvements have extended the equipment's dependability, efficiency, and consistency while also simplifying and automating the production process. In addition, these improvements have shortened the manufacturing time. Furthermore, Industry 4.0 increases the complexity of digitalization when humans and machines collaborate during the implementation process. In this era, technologies support large companies and small- and medium-sized enterprises (SMEs) to work more effectively. In Industry 4.0, SMEs mainly focused on gaining economic objectives via digital transformation and automation of monotonous work procedures (Hein-Pensel et al., 2023). Regarding this industry, producers can reduce production, logistics, and quality management costs. Several approaches indicated that Industry 4.0 primarily emphasized replacing human workers' input with machines, artificial intelligence, and cyber–physical systems (TWI, 2023) to boost enterprises' efficiency, productivity, and revenue (AMFG, 2023). In another way, Industry 4.0 refers to intelligent systems,

DOI: 10.1201/9781003489269-12

innovative products, smart machines, and data storage as cyber–physical production systems (acatech – National Academy of Science & Engineering, 2013). Industry 4.0 enhanced the efficiency of enterprises by increasing mass production; however, it also skipped the human cost via process optimization (Maddikunta et al., 2022). As a result, it challenges employers to adopt this Industry. At this time, robots are essential to the day-to-day operations of Industry 4.0, while artificial intelligence (AI) will be used by Industry 5.0 in order to make mass customization possible (Prescient Technologies, 2022). The core of Industry 4.0 is centered on the concept of mass personalization. The industrial sector will shift significantly due to the enhanced mobility and decision-making capabilities of cooperative robots available in Industry 5.0. It can be seen that people may change from mass manufacturing to mass customization with the assistance of Industry 5.0. The incorporation of a personal touch is necessary for mass customization. Artificial intelligence, the digitization of industrial systems, and fast developments in production processes are the three pillars of Industry 5.0 that make mass customization possible. Besides, the Fifth Industrial Revolution (5.0) offers many benefits for our lives. Industry 5.0 paves the way for the production of personal devices that are less expensive, more covert, and more dependable.

Industry 5.0 is supposed to tackle this problem via human roles with machines. In another way, Industry 5.0 is an extension of Industry 4.0 with the primary purpose of putting the employee's well-being at the production process center. Likewise, the primary objective of Industry 5.0 is to liberate employees' time so that they may devote it to more fruitful pursuits, such as creative problem-solving and thinking outside the box. Taking advantage of all individuals' creative and analytical talents is of the utmost importance. For the manufacturing industry to transition from mass production to personalized manufacture, fast developments in manufacturing processes and the digital intelligence of its production systems are essential. The concept of "Industry 5.0" refers to the next generation of manufacturing, with the primary goal of which is to create an economy that places a higher priority on people and society than production (Sharp, 2021). It is essential to consider a company's resource footprint in today's climate when governments and investors are figuring out solutions to reduce their negative effects on the environment. For instance, various issues might be considered, including the procurement of raw materials, the management of garbage, the impact on the environment, and even the organization's capability to save energy and operate using renewable power sources. Renewable energy sources' usage is increasing among businesses, and many companies are also taking steps to reduce their overall carbon footprints. In order to achieve socio-ecological sustainability, it will be necessary to implement newly developed materials and composites in place of those based on petroleum, in addition to recycling and reusing the already available resources. The technological strategy of the firm was designed with Industry 4.0 serving as the cornerstone for its development. The market's overall state has improved in terms of productivity, competitiveness, and efficiency. In addition, with the development of a completely automated manufacturing facility supported by AI and other technologies, various factors need to be considered, including some further instances of how technology has advanced. Only through the collaboration of humans and machines can one hope to attain better results. It can be seen that Industry 5.0 will thrive now that the market has matured sufficiently to support its existence. A collaborative effort between humans and machines would minimize the amount of robotic manufacture and increase the amount of customized production. Consequently, we need not only technology and automated process advancements but also the originality and curiosity of human beings. Several companies were among the first to use the technologies that were developed for Industry 5.0. Apple, Tesla, and Boeing are three examples of organizations that have achieved extraordinary levels of success. In contrast, SMEs are now having difficulty with the adoption of Industry 4.0 while these global corporations can see beyond it. They are aware that, in addition to using cutting-edge technology, a successful firm is also required to keep its processes up to date and emphasize its people. The manufacturing sector needs the Industry 5.0 strategy to maintain its position as a sustainable and competitive industry. Therefore, organizations have the possibility of realizing their revolutionary potential. Models of technical governance that are innovative have the potential to increase the long-term profitability and stability of a corporation. Last but not least, a practical technique for Industry 5.0 would evaluate the strength and endurance of the existing industrial ecosystems. Providing workers with digital devices, endorsing a human-centric approach to technology, paving the way for environmentally sustainable technology usage, expanding the responsibility of businesses to include their entire value chains, and empowering workers are all components that will be necessary for a successful Industry 5.0 strategy. As a consequence, enterprises operating in the 21st century may be significantly affected by Industry 5.0. First, it fosters the expansion of value into new areas of the economy and across society as a whole. Second, it makes it easier to find realistic solutions to problems that arise in the political and economic spheres. Third, it contributes to consolidating the foundation upon which science and technology are constructed. A positive feedback loop is produced due to the resourceful use of the system's human, intellectual, and financial components. The present era of industrial growth is referred to as "Industry 5.0," and it is characterized by combining digital technology with human intelligence to produce methods of production that are individualized, favorable to the environment, and ethically acceptable. Two concepts that must be included in company plans are social responsibility and ecologically sound supplier chains. However, they face many challenges in achieving digitalization strategies in implementing restricted resources and security concerns related to new technology in the Industry 5.0 era.

Therefore, in this chapter, a meta-analysis of the literature review to describe the benefits of Industry 5.0 is conducted. In addition, several components and related technologies of Industry 5.0 are illustrated. Furthermore, the challenges of this technology to enterprises; for example, human–robot interaction, human factors, and security concerns are discussed to raise awareness and minimize the gaps between Industry 4.0 and 5.0. It is clear that Industry 5.0 is the future; however, its information is limited. The chapter aims to provide helpful information about Industry 5.0 via several aspects: main components of Industry 5.0, challenges of Industry 5.0 application for enterprise, outstanding opportunities of Industry 5.0, humans and Industry 5.0, and potential contribution of Industry 5.0 in future production and business. All of them lead to insight and knowledge about Industry 5.0, as well as readiness and confidence in applying it in the future.

12.1.1 Organization of the chapter

The rest of the chapter is organized as follows: Sections 12.2 and 12.3 present a general background of Industry 5.0, its components and related technologies; and the opportunities of Industry 5.0 for enterprises, respectively. Section 12.4 lists the challenges of Industry 5.0 for companies. Section 12.5 discusses the human factors in Industry 5.0. Section 12.6 expresses several representative examples. Section 12.7 explains future research directions. And, finally, Section 12.8 concludes the chapter with future scope.

12.2 BACKGROUND

12.2.1 A brief history of Industry 1.0 to 5.0

Industry 1.0, or the First Industrial Revolution, was first known as the extensive use of mechanized production and energy sources; for example, coal and steam & water power, during the late 18th century (1784) (Reddy et al., 2021). This revolution enhanced manufacturers' mass production; however, the drawbacks of this revolution were pollution, the time to operate, and a great demand for production machines (Vinitha et al., 2020). Industry 2.0 included electrical power and mass production in 1870 in many fields, such as iron, steel, rail, electrification, machine tools, and the like (Akundi et al., 2022), but it took a high cost to use electrical power, making people lose their jobs due to machine replacement (Vinitha et al., 2020). Industry 3.0 was related to electronics, automation, and information technologies in 1969 (Reddy et al., 2021). The notable achievements of this revolution were computers and programmable logic controllers in the factory processes, as well as the development of Internet and connectivity access. One of the disadvantages of this era was that automated systems wouldn't operate in special situations (Vinitha et al., 2020). Industry 4.0 combines the smart factory with the Internet of Things (IoT), artificial intelligence, machine learning, big data, and cognitive computing. Moreover, Industry 4.0 was established in 2011 using smart and connected technologies to decrease time in production processing (Gehlot et al., n.d.). Besides, the systems in this revolution were applied in different industrial processes in analyzing, guiding, and sharing smart actions or making devices smarter (Vinitha et al., 2020; Wang et al., 2017). This revolution enhanced higher productivity and quality, better efficiency and client experience, reduced costs (Azemi et al., n.d.), and increased profitability (Woboton, 2023). The main challenges of this revolution were data security, skilled human resources, team support, leveraging data (Mistry, 2023), reliability and stability for Machine to machine (M2M) communication, and so on (Happiestminds, 2023). Industry 5.0 enhances Industry 4.0 by applying innovation and research with human-centric, sustainable, flexible, advanced technologies (AI, robotics) (Nayyar & Kumar, 2020; Jha, 2023; George and George, 2020; Pramanik et al., 2020) (Figure 12.1). In other words, Industry 5.0 involves the combination of human workers and intelligent machines to develop the productivity of the factory industry and extend the efficiency of industrial production (Adel, 2022). In another way, several definitions related to Industry 5.0 were listed in Maddikunta et al. (2022). Adel (2022) offered five explanations of the meanings of the term Industry 5.0. This era opens new opportunities for companies to increase the overview of the maintenance plan (Stefanini, 2022), make factories sustainable, enhance human efficiency and productivity, manage environmental control inside the factory, and forecast line production efficiency (Cojocariu, 2023).

12.2.2 Industry 5.0 components

In this section, diverse components that make up Industry 5.0 are discussed as follows.

12.2.2.1 Robots

The term "Robot" was coined by a Czech novelist—Karel Capek, in 1920 in his play R.U.R (Rossum's Universal Robots) (Jordan, 2019; A brief history of robotics: The origin of the first humanoid robot, 2021; The History of Robotics, 2021; The WIRED Guide to Robots, 2020). In addition, a robot

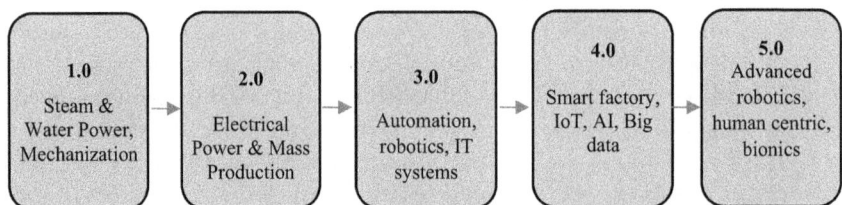

Figure 12.1 The revolution of industry.

can be defined as "a mechanical device which performs automated tasks, either according to direct human supervision, a pre-defined program or a set of general guidelines, using artificial intelligence techniques" (Goris, 2004). Likewise, robotics needs to follow three primary laws of robotics (Williams, 1991): (i) A robot may never hurt a human or not allow a human to come to harm via inaction, (ii) a robot has to follow human beings' orders, except when such orders conflict with the first law, and (iii) a robot needs to protect itself without breaking the previous rules. Robots bring many benefits to support and enhance our lives more conveniently and better. For instance, they can perform several dangerous and repetitive tasks in many areas which people can't do well, such as working 24/7, working in harsh environments (such as without the air, underwater, and in fire), working in industrial, military, and medical aspects, and the like. Industry 5.0 focuses on combining robots and human beings during the working process whenever and wherever (van den Bergh, 2017).

12.2.2.2 Artificial intelligent and big data

Artificial intelligence is related to using computers to simulate intelligent behavior with the least human involvement. It is frequently described as a scientific method for creating innovative devices that can make choices on their own (Khan et al., 2023). Big data is related to massive data collected from various sources that can't be analyzed and handled using conventional methods (Sydorenko, 2021). AI and big data have many advantages for enterprises, like lower costs, high productivity, improved operation and efficiency, competitive advantages, and better customer service (TP&P Technology, 2021) Indeed, the expected total market value is approximately USD 190.61 billion by 2025, and the global economy is expected to be USD 15.7 trillion in 2030 due to AI technology contribution (Woodward, 2023). AI positively impacts enterprises and businesses but has several negative consequences, including eliminating many jobs and requiring workers with high-tech skills. For instance, there were 1.8 million jobs removed globally because of AI. Likewise, big data also have critical effects on companies, industries, supply management (Bag et al., 2020), and management processes (Grover & Kar, 2017) by improving decision-making and effectiveness in many aspects, for instance, banks, information technology (IT), healthcare, financial services, and so on (Omoyiola, 2022). On the one hand, big data can increase revenue, profit, and value for companies; for example, its market is estimated at USD 274.3 billion (Zhaohao Sun, 2021), and the global big data analytics market's every-year revenue could be approximately USD 68.09 billion by 2025 (Palvia et al., 2021). Besides, according to Reinsel et al. (2018), the global data sphere is expected to grow from 33 Zettabytes (ZB) in 2018 to 175 ZB in 2025. As a result, big data strongly impacts developing companies and economic growth. Academics and entrepreneurs are discussing big data (Cheng et al., 2018). Industry 5.0 demands big data analytics. Big data analytics may assist "Industry 5.0" companies in understanding customers, enhancing pricing, production efficiency, overhead expenses, etc. Big data analytics may boost sales by tracking consumer happiness on Facebook, Twitter, and LinkedIn. Industry 5.0 relies on data, personalized production, and intelligent manufacturing—real-time big data analytics (Zhaohao Sun, 2021). According to Joglekar et al. (2023), big data analytics customizes Industry 5.0. To foresee and react adequately, AI, sensor-based networked systems, social media, and digital communication devices must evaluate massive amounts of data in real time. Big data began in the 1960s with the first database management system and data centers. IoT data is everywhere due to high-speed Internet. Since raw data is worthless, many firms do descriptive, exploratory, predictive, prescriptive, and preventive data studies (Bihani & Patil, 2014). This research helps major firms with client acquisition and retention, targeted advertising, new product creation, consumer behavior-based price optimization, supply chain management, risk reduction, and faster decision-making (Raja Santhi & Muthuswamy, 2022). Oracle showed that organizations underestimate data growth by 100% yearly. Zicari (2013) examined big data issues. Human-like computer "intelligence"—AI—humanizes robots-tech powers in Industry 4.0. Tankt and Hopfieldt (1987) pioneered AI, and Recurrent Graph Transformer, Deep Belief, and Convolutional Neural Networks enabled statistical learning (Vapnik, 1999) and the Greedy learning algorithm. Industry 4.0 employs AI in machine vision, industrial robots, supply chain management, inventory management, and predictive maintenance (Raja Santhi & Muthuswamy, 2022). AI-driven rotating equipment fault identification reduced machine downtime, maintenance costs, and safety risks. Since each solution has pros and cons in accuracy, speed, and robustness, they recommend a hybrid intelligent system for future problems. Wang et al. (2017) evaluated milling tool wear and gearbox/bearing failure using Recurrent Neural Network (RNN) models. Deutsch et al. (2017) investigated hybrid ceramic bearing lifespan. Raja Santhi and Muthuswamy (2022) studied AI-based industrial material flow solutions.

12.2.2.3 Smart machines

An intelligent machine refers to one that can have the capacity to think and make decisions on its own. It also functions as a cognitive computing device and solves problems without human help (Pereira, 2019). Besides, the smart machine is also a machine that can enhance connectivity, flexibility, efficiency, and safety (Beudert, 2018). In addition, it is integrated with several innovative technologies like AI and machine learning, or deep learning to figure out solutions, make decisions, and take action instead of human beings (Pratt, 2017). There are four categories of smart machines (Pereira, 2019). Smart devices play an essential role in boosting economic growth by adopting industrial robots, especially their impact on the Industry is about USD 153 billion by 2020, according to the Bank of America Merrill Lynch report (Matthews, 2015).

12.2.2.4 Cloud computing

Cloud computing is related to delivering services on the Internet, including data storage, servers, software, databases, and networking (Frankenfield, 2023). There are three main types of cloud computing: Software as a Service (SaaS), Platform as a Service (PaaS), Infrastructure as a Service (IaaS) (Frankenfield, 2023; Wesley Chai, 2023). In addition, cloud computing offers many benefits. For example, it serves companies, organizations, and industrial services. It allows users to create a virtual environment to back up their data (music, files, photos, data, etc.) and a virtual desktop to access data flexibly. Moreover, the impact of cloud computing is on many industrial aspects. It can enhance businesses by reducing costs, increasing productivity, improving customer service (Aparna, 2023), ensuring security, and improving performance, speed, and efficiency. According to predictions, the worldwide cloud computing market will develop at a 16.3% compound annual growth rate (CAGR) from USD 445.3 billion in 2021 to USD 947.3 billion in 2026 (Srushti Shah, 2022). The servers and information stored in the cloud are located in another location. Access to servers, data, and analytics/intelligence is provided through this system's "cloud" infrastructure. Data and software management can't compare to the reliability, scalability, and user-friendliness of cloud computing. On-demand ISPs facilitate communication between businesses and consumers. Unlike "cloud computing in manufacturing," which uses digital technology to simplify the process, "cloud manufacturing" refers to industrial sectors' indirect adoption of cloud computing. Production software, gear, and personnel are all watched by cloud computing. It was possible to "Design Anywhere and Manufacture Anywhere" (DAMA) using cloud manufacturing in the early 2000s. Productivity, effectiveness, and adaptability are all improved. Technologies like the IoT, radio frequency identification, sensors, global positioning systems, cyber–physical systems, and cloud computing make cloud manufacturing possible (Raja Santhi & Muthuswamy, 2023).

12.2.2.5 Cobotics and cyber–physical systems

12.2.2.5.1 Cobotics

"Cobot" refers to collaborative robots that can react to their surroundings parallel to humans to finish tasks or processes using sensor and machine learning technology (Essentracomponents, 2022). Cobots can help factories work more efficiently, cost-effectively, and accurately. They bring many benefits to enterprises, such as higher productivity, increased safety and flexibility, ease of use, fewer human resources, and better product quality (Essentracomponents, 2022). In the era of Industry 4.0, cobots seem to take on the human role in manufacturing, which could negatively impact sustainability development. The Fifth Industrial Revolution requires cobots to support human missions but work in sync so that the role of human workers becomes central, balancing with the part of cobots. In other words, in

Industry 5.0, cobots are designed to understand the needs of employees, determine if they need support, and predict risks and safety factors to protect the workers (Verma et al., 2022). In fact, according to Bi et al. (2021), researchers estimated the market size of cobots was USD 649 million in 2019 will increase to about 45% from 2019 to 2025. SMEs can use them to enhance working capabilities in autonomous production or assembly procedures. Furthermore, current cobots are sensor-driven, human-signal-forming, and no longer mechanical. Industry 5.0 cobots modify products. Tomography and 3D vision let surgical cobots perform complex surgeries in Healthcare 5.0. Another example is Telesurgery, where a surgeon instructs a cobot online. Cobots enhance output and automate tedious tasks, making them essential to the industrial workforce (Verma et al., 2022; Resende et al., 2021; Romero et al., 2016c). Automation needs humans. AI makes all computers cobots. Automation boosts teamwork. Cooperative robots fail. Cobots may aid humans and animals. Due to its sensors and fragility, cobots may shut down if employees find missing merchandise. Industrial robots are risky, and robots enjoy assembly lines. Product personalization may challenge supervised robots. Cooperative robots learn faster. Industry 5.0 may include healthcare cobots and smart applications that enable patients to combine medical and lifestyle objectives into a customized workout regimen—robotic surgery involves the use of mechanical tools to remove tissue. Da Vinci surgical robots improve precision. Gynecology and urology employ Da Vinci cobots. Human–machine cobots rule Industry 5.0. Cobots may improve competitive marketplaces (Maddikunta et al., 2022). German dryer-maker Stela Laxhuber says that adding a cobot to their welding cell has boosted production and component quality. Medicine, agriculture, food processing, electronics manufacturing, storage, transportation, vehicle production, metalworking, packaging, and logistics employ cobots more. Liability, public opinion, and job security must be addressed before cobots become mainstream (Raja Santhi & Muthuswamy, 2023).

12.2.2.5.2 Supply Chain Management

Supply chain (SC) is essential in industrial supply-chain ecosystems because of the various players and interactions. The supply chain is unreliable. The SC may automate cash transfers after supply chain peer agreements. SCs execute directives autonomously. SCs may be hosted on public, private, or permission BCs. User wallets hold contract native code. The parties' consent seals it. SCs can reliably preserve status or computation data in BCs with a few mouse clicks. Go, Solidity, or Serpent can program Ethereum. Ethereum runs contract bytecode. Ethereum contracts may express algorithms using branching and loops. Private hyperledger chain codes require instruction and cooperation. Verma et al. (2022) described a fifth-generation supply chain with human–robot cooperation utilizing cobots and AI. Hype personalization requires human–machine interaction.

Robots handle massive industrial supply networks. Supply chain Industry 5.0 may attract new careers and boost society's sustainability and intellect (Joglekar et al., 2023). Industry 5.0 widespread customization may optimize the supply chain. Industry 5.0 affects logistics. Industry 5.0 supply network models need conceptual breakthroughs. Human ingenuity with the Fourth Industrial Revolution (Digital Transformation (DT), cobots, 5G and beyond, machine learning, IoT, and the like) may help Industry 5.0 firms meet the need for personalized goods. Industry 5.0 requires SCM mass customization. DT digitizes logistics, stock, and assets. Design thinking may help SCM (Marmolejo-Saucedo et al., 2020). SCM-like DT collects IoT data. Logs, machine learning, big data, etc., may predict SCM issues. Simchenko et al. (2019) suggested simplifying the supply chain to speed up personalized product delivery. DT analyzes the production, service, inventory, and land prices. Moreover, DT also improves supply chain profitability. SCM proposes DT. Based on (Defraeye et al., 2019) advocate mango fruit thermal modeling. DT monitored plant climate. Logistics preserve fresh fruit. DT-optimized logistics improve refrigeration and waste. Greif et al. (2020) redesigned the construction DT. This research examined whether DT lowers SCM construction costs. This strategy was developed using several solvers, simulators, and analytical tools. Pharmaceutical distribution was predicted and studied. Drug analyses promote discussion. Cobots may simplify and help logistics. Simchenko et al. (2019) advocated SCM life cycle packaging, inspection, and lifting robots. Collaborative robots may help with material preparation, packing, quality assurance, shipping, customer service, and undesired item retrieval. Cobots simplify stocktaking, ordering, refunding, and quality monitoring (Maddikunta et al., 2022).

12.2.2.5.3 Augmented Reality and Virtual Reality (AR/VR) and Metaverse

- AR/VR
  The merging of digital data with the users' environment in real-time is called Augmented Reality (AR). On the other hand, using data such as text, graphics, audio, video, and others in real-time integrated with natural objects is recognized as AR. Unsimilar to AR, Virtual Reality (VR) creates an entirely virtual environment that helps users discover and interact with a 3D environment. This environment is created by using software and hardware to simulate virtual objects. AR/VR technologies can be used in various aspects, such as healthcare, education, entertainment, and visualization (Batth et al., 2019).
- Metaverse
  Metaverse refers to a combination of "meta" and "universe" in a sci-fi novel called Snow Crash in 1992, and it is linked to cyberspace, where people can experience life in different ways they couldn't do in the physical world (Tucci & Needle, 2023). There are two primary technologies to enhance the metaverse: VR and AR. The metaverse

will play an essential role in seven technologies: AI, IoT, extended reality, 3D modeling and reconstruction, spatial and edge computing, blockchain, and brain–computer interfaces. Moreover, it can be used in many industrial sectors like entertainment, real estate, and gaming.

12.2.2.6 6G networks and beyond

6G is the next generation and the successor to 5G cellular technology. Furthermore, 6G networks offer higher frequencies and capacity than 5G networks (Kranz & Christensen, 2022). Researchers believe that the 6G technology can improve imaging, presence technology, and location awareness. Besides, 6G can allow access points to handle multiple clients concurrently via Orthogonal Frequency-Division Multiple Access (OFDMA)—a feature of Wi-Fi 6 (802.11ax) (Kranz & Christensen, 2022). According to the EU Parliament report, 6G is one of the main trends for 2030 to enhance connectivity and autonomy in Industry and transportation (Angelica Mari, 2020). Several pioneer countries are working on 6G projects, like Finland, South Korea, China, the USA, Japan, Australia, and the EU. 6G promises a future way for workplaces by enhancing the way of online connectivity, using VR and AR in training workers, experiencing a new social media world (3D world), transforming healthcare, and managing the road (Marr, 2023).

In addition, the benefits of future 6G services to Industry 5.0 are promising. Tens of thousands of sensors, hardware nodes, and robots may complicate radio networking (Rajatheva et al., 2020). Smart infrastructure and apps will rule 4G and 5G. The 6G and beyond may aid Industry 5.0 because of its lower latency, enhanced service, calculable IoT infrastructure, and integrated AI (Chowdhury et al., 2020). Intelligent spectrum management, AI-powered mobile EC, and 6G networks help Industry 5.0 applications (Tariq et al., 2020). In a smart information society, Industry 5.0 applications will need 6G networks with rapid data transfer rates, low latency, high reliability, low energy consumption, huge traffic capacity, etc. Mobility and handover allow Industry 4.0 with 6G. 6G networks are large and complicated, requiring frequent handovers (Alwis et al., 2021): AI-predicted mobility and hand-off maintenance service. Industry 5.0's Internet speeds exceed existing software (Yang et al., 2020). Open-space 6G optical quantum communication may work. Industry 5.0's growing energy use and networked gadgets emphasize energy management. Novel approaches increase 6G energy harvesting and utilization. Digital worlds demand tools beyond digital and computational (Maddikunta et al., 2022). Telecoms want 6G by 2030. Smart gadgets may use 6G wireless connectivity, 6G capacity, and latency improvements. Nobody knows about 6G technology yet. Spectrum efficiency, coverage, bandwidths up to 100 GHz, power consumption, latency, stability, and data rates up to 1Tbps are some significant performance characteristics that may be improved. Strinati et al. (2019) briefly compare 5G and 6G key performance indicators (KPIs). The 6G and subsequent networks improve latency, quality, connected device infrastructure, and AI (Chowdhury et al., 2020) The 6G networks should use less energy due to linked devices and Industry 5.0 data volumes.

12.2.2.7 Blockchain

A distributed database or ledger shared by the nodes of a computer network is known as a blockchain (Hayes, 2023). However, they are not just used in cryptocurrency systems, where they play a vital function in keeping a secure and decentralized record of transactions. Any sector may utilize the blockchain to make data immutable, which is the phrase for the inability to be changed. This new technology can improve enterprises' performance and work more efficiently by tracking easily, reducing transactions, adding security, facilitating international trade, revolutionizing future business operations, and bringing future prospects in business operations (Lysak, 2023). Blockchain may help Industry 5.0. Industry 5.0 faces hardware diversity. Viriyasitavat and Hoonsopon (2019) said blockchain's safe peer-to-peer interactions enable permanent recording. Absolute ledgers enable Industry 5.0 transparency. Industry 5.0 smart contracts may analyze data and simplify procedures. Blockchain-based distributed and partitioned systems safeguard data and transact (Deepa et al., 2022). Blockchains may store digital identities in Industry 5.0. Businesses must manage network access. These changes may improve asset, commodity, and service management. Blockchain increases cloud manufacturing machine communication (Wang et al., 2022a). Industry 5.0 might benefit from blockchain technology. Industry 5.0 emphasizes device decentralization. Blockchain technology may boost trust in Distributed Ledger Technologies (DLTs) (Viriyasitavat & Hoonsopon, 2019). Blockchain-encrypted peer-to-peer interactions are here to stay (Viriyasitavat & Hoonsopon, 2019). Distributed ledgers make Industry 5.0 functions transparent and accountable. Industry 5.0 disputes require transparency (Hämäläinen & Inkinen, 2019). Smart contracts may help Industry 5.0 authentication and service-oriented behavior. Decentralizing and partitioning blockchains may increase data and transaction security (Xu et al., 2021). Blockchains may speed data collection and processing. Blockchain-stored IDs control Industry 5.0 subscribers. Check-in with staff and limit Internet access and digitally managed services (Mushtaq & Haq, 2019). Blockchain and smart contracts may simplify contracting. Blockchain-based cloud production may help machines share data (Tao et al., 2011). Blockchains record business network digital and physical asset transactions. Veracity, transparency, and traceability improve financial transactions. Block files archive recent transactions like ledger pages (Prabhadevi et al., 2021). Decentralized, unchangeable ledgers hold encrypted, time-stamped data. Blockchain is ideal for quickly, securely, and reliably exchanging data and information with stakeholders. Blockchain's decentralized, distributed, and irreversible characteristics allow organizations to transmit data securely (Raja Santhi & Muthuswamy, 2022).

12.3 INDUSTRY 5.0 OPPORTUNITIES

12.3.1 Smart manufacturing

The smart factory of the Fifth Industrial Revolution is based on IoT. Artificial intelligence is used to collect and evaluate cloud data. Edge computers, sensors, actuators, and ever-present robots on the factory floor hold the promise of fully automated factory processes. Edge computing allows for the integration of actuators and sensors. Smart manufacturing facilities, efficient production lines, and state-of-the-art consumer items all need high technical purity. Businesses today may be able to save expenses while minimizing their influence on the environment and the availability of natural resources, according to the opinions of (Sanchez et al., 2020). Environmentally friendly additive manufacturing produces thinner and more resilient layers. The realism of the 3D models improves. In smart additive manufacturing (SAM), AI and computer vision increase the quality and desirability of 3D printing and better supplies for 5D printing. Businesses and universities alike benefit from using intelligent manufacturing wares. The growth of smart manufacturing may be traced back to the incorporation of AI, IoT, the cloud, massive volumes of data, CPS, DT, and EC. Smart manufacturing improves both profitability and sustainability. In the recent decade, SAM has emerged as a subfield of smart manufacturing. Additive manufacturing, more commonly known as 3D printing, facilitates user input in Industry 4.0, and eco-friendly output in Industry 5.0 opens, automates, and provides access to valuable data. SAM is in charge of keeping an eye on the stuff in the layers. SAM reduces manufacturing's energy use, trash output, and material consumption. Supply chain management and product delivery in Industry 5.0 may benefit from SAM and integrated automation (Maddikunta et al., 2022).

12.3.2 Smart logistics

After Industry 4.0's technology-centric change, Industry 5.0 emphasizes social, environmental, and human views, which will impact logistics operations and management. Customizable distribution models are needed (Kumar et al., 2021a). Consumer inclusion requires intelligent CPS and system integration. Human–robot interactions raise security, psychological, and other difficulties (Gaiardelli et al., 2021). Thus, Industry 5.0 smart logistics issues may be handled from several aspects. "Smart logistics," which includes smart automation, devices, systems, and materials, focuses on human–technology interaction (Nahavandi, 2019).

12.3.3 Smart automation

Industry 5.0's system-centricity prioritizes humans. Industry 5.0 (Pathak, 2019) requires intelligent automation to balance human integration with automation (Nahavandi, 2019). Lean collaboration demands more planning and undermines the logistics system (Mekid et al., 2007). "Operator 4.0" was coined to characterize logistics system technologies in 2016. Assisted, collaborative, and enhanced work boost productivity in this notion (Romero et al., 2016a). The priority is human–technology interaction. In the second case, people and robots must collaborate. Third, employ technology to enhance hearing and sight. These applications improve production, storage, material handling, and information flow (Cimini et al., 2020a). Industry 5.0 emphasizes resilience and people. Romero and Stahre (2021) suggested an "Operator 5.0" who uses information and technology to unleash human imagination and resourcefulness in the face of adversity to develop creative, cost-effective solutions to keep production flowing and people safe. Industry 5.0 requires self- and system resilience to progress technologically. Self-resilience is centered on overcoming hardship (Romero & Stahre, 2021) and due to its resilience, humans and robots may share duties (Inagaki, 2003).

12.3.4 Smart devices

The purpose of Industry 5.0 is to leverage technology to improve logistics and supply chain management, with people as the primary emphasis. Wearable technologies are improving workers' cognitive and physical performance and are becoming more popular in the workplace (Longo et al., 2020). Unmanned aerial vehicles (UAVs) have profoundly affected intralogistics, material handling, and personalized shipment. A substantial investigation has been devoted to auto-ID and RFID in smart logistics and supply chains (Fraga-Lamas et al., 2021) because they improve traceability, warehouse operations, and inventory management.

12.3.5 Smart systems

Industry 5.0 requires improved supply chain decision-making and information sharing for customized and case-based production. Stakeholder data and information transmission improve smart logistics system responsiveness and savviness. It needs to transmit and analyze massive sensor data (Kumar, 2019). Smart Cyber–Physical Systems (SCPS) in Industry 5.0 might improve data transmission and logistics networks. Green processes like green manufacturing, recycling/disposal, and G-IoT are vital; however, the digital revolution may only enable a small circular economy (CE). Information technology revolutionized blockchain technology (Carayannis et al., 2022). Recommender systems assist the supply chain in targeting specialized markets by analyzing social media, language, and analytics. Online tracking and verification may enhance output (Kumar et al., 2021b). Real-time decision-making and high-quality visualization enable smart digital twins for logistics systems in Industry 5.0's virtual intelligent logistics system (Hakanen & Rajala, 2018).

12.3.6 Smart materials

Industry 5.0 developed smart materials. The malleability and versatility of these cutting-edge materials may affect logistics and respond differently to different temperatures, lighting conditions, stress levels, and the like (Li et al., 2017). Most obviously, it can be used in 4D printing (Javaid & Haleem, 2020). 4D printing, like 3D printing, relies on the adherence of successive layers of material (Pei et al., 2017). The material used is the most crucial factor (Li et al., 2017). Smart materials allow things to modify their behavior in response to their surroundings and increase their robustness, flexibility, and dependability, such as medicine, aircraft, and electronics.

12.4 CHALLENGES OF INDUSTRY 5.0 FOR ENTERPRISES

The core of Industry 5.0 is about sustainability, human-centricity, and resilience, and human-centric manufacturing has been and is one of the much-discussed topics. For fulfilling this kind of manufacturing, the role of human workers must be pivotal and as equitable as possible to the machine's position in manufacturing production. A crucial question is how humans and machines can work together effectively. Genuinely, human–machine collaboration is currently one of the most challenging aspects of Industry 5.0. Within the scope of this chapter, this challenge will be discussed in the following three approaches: human workers, machine teammates, and intelligent manufacturing.

12.4.1 Human workers

The cognitive capacity of humans, in general, and factory workers, in specific, is vital and constantly evolves in the historical timeline of the industrial revolution. In the Fifth Industrial Revolution, the understanding and skills of human workers are the elementary requirements for discovering new knowledge. Due to the development of advanced technology in Industry, human workers face limitations on job positions and roles in the manufacturing environment, such as the competency to interact and communicate with modern machines in high-tech factories. Accordingly, to adapt to technological changes, human workers must raise their sense of initiative in learning and training, which seems to be a challenge worth considering.

In fact, associated with focus conversion from machines to humans, the active learning approach is an essential strategy enabling humans to actively improve and enhance craftsmanship, which could increase and preserve the central human-worker position (Güğerçin, 2021). A valuable forecast is that the labor force will be widely reallocated so that job guarantees will be highly competitive (Yücebalkan, 2020). To take advantage of the future, either active learning or self-learning of human workers is feasible not only for the age of discovery but also for lifelong benefits (Komiyama & Yamada, 2018). Once human workers take ownership of knowledge, human well-being can be achieved through care, satisfaction, motivation, and self-expression. Learning would bring necessary knowledge to the workforce, whereas training could help enhance their competency skills. Human strength is the ability to reason, creativity, adaptability, etc., which could improve employees' ability to handle new and diverse situations, especially technical circumstances. Therefore, the problem-solving ability of workers is precious in Industry 5.0, while machines do not automatically respond to unexpected technical situations. For the cyber–physical-system vision, the problem-solving competency of human workers is expected to contribute to the development of smart manufacturing. In intelligent factories equipped with complex problem-solving processes based on digital systems, out-standing human worker skills are still essential such as creativity, analysis, and design (Bellet et al., 2013). The core of Industry 5.0 is about advanced technology rapidly evolving, which has made the global industrial environment full of technical engineering systems. Working in such an environment requires the competency to use and develop technology. Typically, interact-ing and communicating with machines and technical systems would enable human workers to enhance controlling, monitoring, recovering, operating, etc., and raise the level of work safety proactively. Advanced technology skills have been in great demand for prospective job positions relating to data analytics, big data analytics, AI, and machine learning (Güğerçin, 2021). In addition, the digital transformation era always needs the digi-tization capacity of human resources, which organizations and businesses are looking for and expecting in that workforce (Knudsen et al., 2021). In addition, researching and inventing new techniques or improving them is also essential to serving technology development. With the experience and creativity of humans, the improvement and upgrading of production pro-cesses and production methods are expected to both improve productivity and ensure flexibility. On the other hand, technology development capacity can positively impact the workforce transition between manual and mental labor-related job positions (Morgan-Thomas et al., 2020). The development of technology is also associated with the social revolution, where human life is always appreciated. Thus, human workers, in addition to work and technical skills, need to adapt to changes and impacts outside the factory. Normal problems of modern life, such as work pressure and family respon-sibilities, require human workers to have the ability to deal with them. Thus, stress management skills are necessary and valuable for humans to maintain life balance (Wu et al., 2013). In addition, to adapt to various work and life conditions, time and functional flexibility are considered relevant skills for human workers, which could make an advantage in choosing the type of work, such as full-time, part-time, remote, and job positions, such as IT, Marketing, Sales (Güğerçin, 2021). Overall, the knowledge and competency skills mentioned above toward human challenges in Industry 5.0 would promise to contribute to enhancing and ensuring good collaboration with machine teammates to increase productivity and efficiency for manufacturing production.

12.4.2 Machine teammates

In the context of teamwork with human workers, cobots are called machine teammates.

There are three categories of human–robot interaction in industrial robotic systems (García Olaizola et al., 2022). In reality, ensuring labor safety, providing jobs for humans, and enhancing production efficiency are considered urgent challenges to maintain and improve the centric role of humans (Cimini et al., 2020b). To understand human behavior, machine intelligence applied in cyber–physical systems requires deep learning and advanced training to acquire human understanding and skills directly or indirectly (Nahavandi, 2019). Implementing automated learning systems is a complex requirement due to human biological characteristics such as cognitive, physical, and social factors (Munir et al., 2013). The ability of high-tech machines would help replace the limitations of human abilities and calculation, and it also supports human strengths such as intelligence, creativity, and problem-solving. With this approach, the role of human workers is increased in manufacturing and has become a central goal for technology applications (Pathak, 2019).

12.4.3 Intelligent manufacturing

While Industry 4.0 focuses on intelligent automation, Industry 5.0 aims at the intelligent collaboration of humans and machines. This collaboration is known as new-generation intelligent manufacturing or human–robot collaboration (Zhou et al., 2018; Rachel, 2018). The main challenge of smart manufacturing is to create a working environment that fosters a good relationship between humans, machines, and organizations in terms of both technology and culture. As for the technology aspect, notable achievements such as AI, big data, cloud computing, IoT, and digital twins are being widely applied. Of those, the big expectation is reducing the gap between AI training results and the real world of human–robot collaboration systems (Wang et al., 2021). Quality of AI training would support automated learning systems that are well-adapted to real and human tasks. In addition, the effectiveness of human–machine cooperation could be preserved and enhanced by mutual self-learning with digital twin-based training (Liu et al., 2021). Behaviors, skills, operational tasks of human workers and technical space, and physical elements of smart factories could be digitalized by virtual modeling and visualization for digital twin applications. Human workers could train and enhance their expertise with digital twin space, while machine intelligence could use mine scenario data from the digital twin operation (Pan & Zhang, 2021). The development of mutual learning could contribute to the collaborative intelligence of humans and machines, promising flexible application of AI and human intelligence to resilient production (Daugherty, 2018). Through this harmonious intellectual integration, handling new, complex, unexpected problems benefits from leadership between humans and machine teammates, whose strengths are used harmoniously and effectively (Leng et al., 2022). While the technical aspect is seen as the backbone of this new generation of smart manufacturing, the organizational culture would play the role of the heart of the administration. The well-being of human workers, central to smart factories, could not be preserved without the impact of organizational culture. Creativity, innovation, investment in technology, and resource management are always strongly influenced by administration policies, which could bring satisfaction, such as income, promotion, and passion, to human workers, employees, and other stakeholders. This vision is unlikely to be a simple business strategy when critical analysis of return on investment, time, revenue, and benefits management is always a glaring challenge for any organization (Cimini et al., 2017). Besides management culture, education plays a crucial role in educating and training a generation of humans with modern skills and knowledge to meet the requirements of the new industrial revolution. Along with the rapid change of technology, education engineering must constantly develop, as well as from educational methods, goals, facilities, and teaching staff. For instance, knowledge-based and outcome-based methods could be flexibly combined to enhance professional and soft skills rather than focusing on academic understanding (Diaz Lantada, 2020). Ethical issues have always been of particular concern in a hi-tech society, where the educational environment must address critical issues such as gender and racial bias. In addition, transdisciplinary education is considered a potential challenge in this generation of Industry because the requirements for job positions and the workforce's competitiveness interfere with many fields and areas. Education engineering outcomes and lifelong learning would contribute to the well-being of humans and the binding relationship between humans, technology, and society (Gürdür Broo et al., 2022).

12.4.4 Wage disparity

Technological advancements have reduced the salary gap between various workforce groups over the last several decades. The median salary of middle-educated and experienced workers has fallen over the 1990s, while the pay gap has grown, according to 2016 World Bank research. The information economy and cognitive jobs have raised the demand for knowledgeable workers. Acemoglu and Restrepo (2018) discovered that low-educated workers lost the most employment to automation. Pervasive automation and AI have revived the idea of a universal basic income (Raja Santhi & Muthuswamy, 2023) for everyone. Industry 5.0 builds on the human-centered concepts of "4.0." Human–robot collaboration would boost the economy. Tech businesses targeting Gen Z and Millennial workers must accommodate their diverse interests. Community activities, scheduling flexibility, and promotion of historically underrepresented groups retain younger employees. Human-centered design requires rethinking workplace robot–human collaboration. Frontline workers may benefit from digital twins and collaborative robots in industrial operations. Valued personnel work harder for the company (Raja Santhi & Muthuswamy, 2023).

12.4.5 Loss of craftsmanship

Technocrats have witnessed computer programs beat grandmasters since the 1980s. AI excels in visual recognition, verbal comprehension, strategic games, and autonomous cars. AI outperforms attorneys in contract flaw detection and songwriting (Miley, 2018). In 2016, "The Next Rembrandt" used AI to generate Rembrandt-like paintings. Researchers used AI to reproduce Rembrandt's style pixel-by-pixel (NPR Staff, 2016). AI can compete with humans in many categories requiring creativity, subtlety, and expertise (Raja Santhi & Muthuswamy, 2023).

12.4.6 New role for the Industry Worker

Employee roles and histories will evolve as a consequence of Industry 5.0. Workers contribute to the success of a company and their own personal development; they are not an expense. The organization cares deeply about its employees and is committed to their professional and personal development. This technique prioritizes human capital over monetary resources in favor of a more conventional cost–benefit analysis. In Industry 5.0, humans should take precedence over robots. The needs of those who work in the manufacturing sector should drive technological development, not the other way around. The employee-employer relationship becomes closer. To achieve this goal, workers might take part in developing industrial technologies that rely on artificial intelligence and robots.

12.4.7 Unemployment

The Industrial Revolution, the Great Depression, and the World Wars increased unemployment. Due to industrialization and population concentration, rural and urban employment markets differed. Humans developed and created during the First Industrial Revolution, while machines and robots handled the dirty jobs. AI in Industry 4.0 can tackle complex problems and replace human intelligence. AI will help experienced and educated people, but not others. Wage gaps may grow. AI-enabled robots may replace software engineers, doctors, and architects (Gay et al., n.d.). The United Nations anticipates 6 billion working-age people by 2050, whereas AI and robotics are progressively infiltrating human-only fields (Raja Santhi & Muthuswamy, 2023).

12.4.8 Security concerns

- Phishing
  Phishing is an attack in which scammers send messages to the victims, showing them as a prestige site but leading to a trap. In this way, attackers can steal all credentials of victims. This attack attempts to gain sensitive data from victims, which can cause financial losses (Gupta et al., 2017). Five stages of a phishing cycle and various types of this attack were described in Gupta et al. (2017). They indicated that these attacks have significant impacts on enterprises' financial losses. Indeed, a phishing attack is one of the most popular cyber crimes nowadays (approximately more than USD 4 million of the average cost for an organization's data breach, based on AAG (2023a)).
- Intellectual Property (IP) theft
  Intellectual Property (IP) is related to creative expressions/works, designs, inventions, trade secrets, or valuable information known as company property (De Groot, 2021; Proofpoint, 2023; Simpson, 2022; Thales, 2023). In another way, it is also the intangible property of a company. IP theft is a threat when someone unauthorized exploits or uses company property, and it can lead to serious financial issues for individuals, businesses, and governmental offices. There are several types of intellectual theft, including copyright, trademark, design, patent infringement, and trade secret theft. These types can influence financial loss, remediation cost, and reputational damage for enterprises.
- Ransomware
  Ransomware is a malicious code that prevents users from accessing their data until it is paid (AAG, 2023b). There are four types of the latest ransomware: crypto, locker, scareware, and double extortion ransomware (Blackfog, 2023). Ransomware is becoming one of the most dangerous attacks on enterprises, even big companies, and SMEs; in fact, a big meat processing company (JBS) in Brazil was under attack by ransomware and forced to shut down temporarily several branches of it in Australia, Canada, and the USA from May 30, 2021, to June 2, 2021, with USD 11 million payment in bitcoin to resume regular activity (Jones, 2022). This attack can impact enterprises in many ways; for example, operation interruption, company reputation damage, sensitive data exposure, and financial problems. According to global statistics (Figure 12.2), the

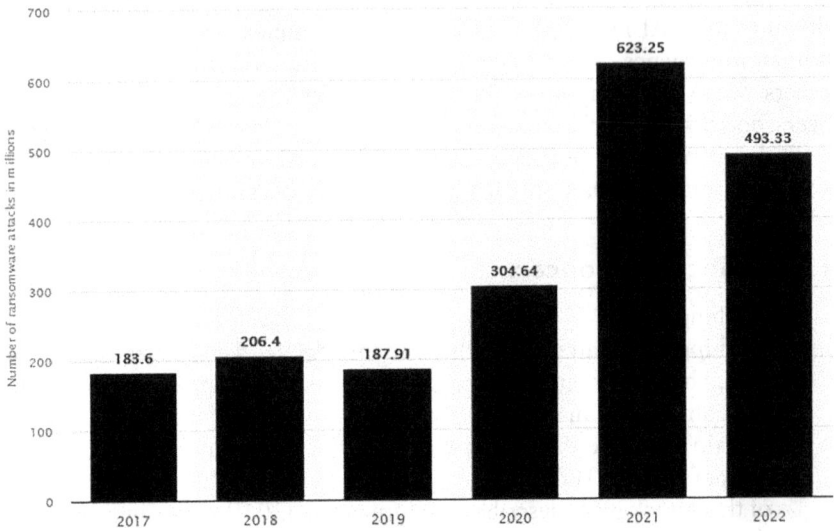

Figure 12.2 Total number of global ransomware attacks from 2017 to 2022 (in millions) (Statista, 2023a).

number of ransomware attacks slightly increased in 2017 and 2018 and marginally decreased in 2019. However, it was nearly 1.5 times rising in 2020 and approximately double that in 2021 compared to 2020, but a slight decrease in 2022. Besides, Figure 12.3 shows that the percentage of companies under attack by ransomware increased continuously from 2018 to 2022; particularly, over 71 percent of companies were victimized by ransomware. This leads to significant challenges for enterprises and SMEs in adopting new technologies in Industry 5.0.

- Supply chain attack
  This attack targets the weak links in an enterprise's supply chain involving producing and selling product processes (Gillis, 2023). Moreover, cyber hackers often exploit the companies' trust in third-party vendors by using integrated malware in hardware and software while delivering materials from the supplier to the producers' procedure. Supply chain attacks are similar to Advanced Persistence Threats (APT) by using complex methods and taking a long time; even organizations' systems are quite good (Helpnet Security, 2021). This attack can cause massive damage to enterprises. Recently, there have been several critical supply chain attacks, such as Solarwinds, Kaseya, Codecov, NotPetya, Atlassian, and British Airways (Checkpoint, 2023). There are many categories of supply chain attacks with the primary purpose of creating or taking benefits of security vulnerabilities by malicious code in software (Fortinet, 2023).

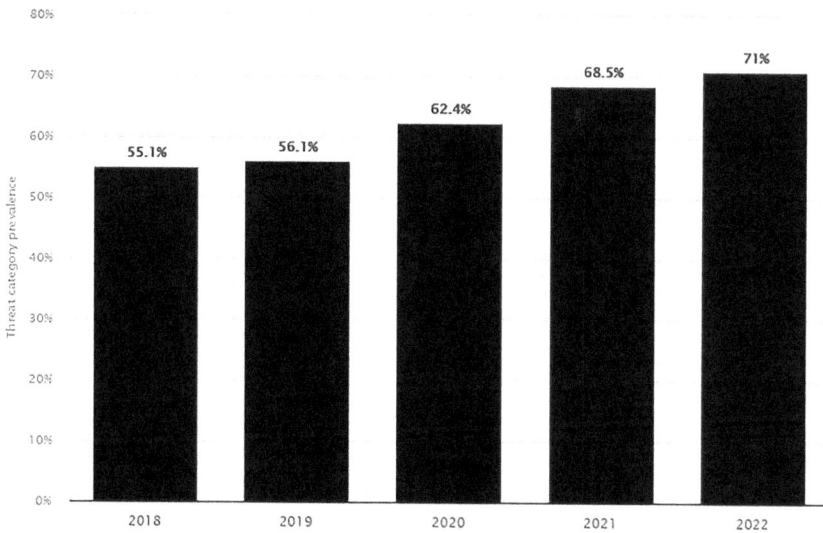

Figure 12.3 Total percentage of global organizations under attack by ransom-ware from 2018 to 2022 (Statista, 2023b).

12.5 HUMAN FACTORS IN INDUSTRY 5.0

The Fifth Industrial Revolution places the human factor at the heart of the evolutionary strategy of both technology and society, where the human factor is not only a clear premise but also a key driving force for changes and development toward total success. This vision is also differentiating, supplementing, and perfecting the achievements of the Fourth Industrial Revolution because the opposite impact of high-tech machines and technology-driven one-way management overshadows the role of human workers. In fact, it seems that in any stage of science–technology evolution, human factors engineering is always a top concern. This is obvious because its goal is to help humans overcome their own capacity limits and promote their specific strengths to achieve comfort, safety, and efficiency in work (Kadir et al., 2019). As human worker satisfaction increases, so does their creativity and dedication to production goals. Therefore, in intelligent manufacturing, replacing human workers completely with machines is not recommended or expected. Furthermore, other focus areas of Industry 5.0, such as healthcare, infrastructure, mobility, fintech, and more, human factors engineering must be considered. For example, in the healthcare sector, human factors engineering and human–computer interaction application continues to be of practical importance in the future through human-driven design and evaluation for clinical informatics systems (Holden et al., 2022). During industrialization, the machine components and hardware elements of engineering are often called physical systems. These have also been integrated with human factors toward sustainability and resilience for production.

Human–physical systems are the popular model in traditional factories, where workers work with the support of machines in certain stages of production (Wang et al., 2022b). As technology advanced, machines with digital and computerized integration, known as cyber–physical systems, gradually replaced workers in many specific jobs in smart factories (Wang et al., 2022a). The priority of modern machinery in the factory has affected human job security, adversely affecting society's sustainable development. Considered an inevitable achievement, the human factor is once again placed at the heart of the production process, known as the human–cyber–physical system (HCPS), which is the harmonious integration of the role of workers in the cyber–physical system (Fantini et al., 2020). The central role of workers is actively demonstrated in operation, control, and decision-making in HCPS, in which human inputs are always ensured and maintained (Tehrani et al., 2019). In special situations, the decision and action to solve the problem are due to the reasonable flexibility between workers and machines. For example, in human–machine collaborative systems in the healthcare sector, safety is enhanced through both operational efficiency and reduced errors related to human decisions (Fu et al., 2017). In addition, human input also plays a key role in personalization, one of the essential goals of the human-oriented Industry 5.0. Unlike in Industry 4.0, in the new generation of Industry, personalization enables humans to be directly or indirectly involved in production through the design, trial, and maintenance phases (Zhang et al., 2017; Zheng et al., 2017). Typically, in HCPS, the application of metaverse and digital twin using human input for product personalization both contribute to improved human satisfaction and is an enjoyable experience of the human-centric role (Wang et al., 2022a). In order to meet the resiliency of the Industry, the human factor is indispensable in leveraging human intelligence and perception in handling failures and unexpected situations. Thanks to this strong human support of the cyber–physical system, HCPS can ensure stability and adaptability in production, limiting or avoiding unnecessary interruptions affecting work efficiency (Chen et al., 2019). Moreover, the knowledge and understanding of humans at every position in organizations is a precious asset, having high value when contributing to collective wisdom (Tuptuk & Hailes, 2018). Making full use of collective intelligence is an indispensable requirement in human resource policy to ensure sustainable development, not only in the immediate but also in the long term.

12.6 REPRESENTATIVE EXAMPLES

12.6.1 Industrial Internet of Things (IIoT)-based solution for human-centric Digital Twins (DTs)

Most DTs are currently made with models that don't support reusability and scalability well. In addition, modeling humans as data-driven agents and visualizing their interactions in factories remains limited. An Industrial IoT-based platform, named Clawdite, can help customize data visualization of human factors and manufacturing systems (Montini et al., 2022). Besides, it can also provide a flexible mechanism for easy DT initialization and acceleration. Clawdite's design is suitable for specific purpose applications, supporting data flows of varying volumes, velocities, and variations. The flexibility of the reference model allows the representation of diverse data structures (time series, JSON, XML). Different types of input and output data streams in special situations are also well supported by this solution. Typical scenarios that have been successfully evaluated are estimating worker location, optimizing autonomous Mobile Robot fleet, estimating worker fatigue, detecting worker activities, and profiling human workers, as well as skills assessment.

12.6.2 Augmented Reality (AR) enhancing human–machine collaboration

AR is a prominent technology that enables real-time visual interaction between the real world and virtual space on computers. AR applications can help improve awareness, train workers, train robots, and enhance the ability to evaluate blueprints related to products and manufacturing systems. However, widespread use of this tool is hampered by the complexity of creating and deploying AR content. The ARContent tool (Moya et al., 2023) is a potential solution to the challenge. The ARContent tool offers extended usability for non-experts as no programming skills are required and faster access using the Internet interface, which can enhance the active participation of employees in creating context-aware AR content. As a result, the benefit is less cognitive and mental effort in engineering operations, pilot-testing design, human–robot training, and maintaining production quality.

12.7 FUTURE RESEARCH DIRECTIONS

The ideal of Industry 5.0 is human well-being at the heart of the manufacturing world, sustainable development, and guaranteed resilience. Contributing to the realization of that vision, from a technical perspective, to ensure the central role of humans in production, solutions to improve the efficiency of cooperation between humans and machines must be studied in-depth. Potential strategies and methods of enhancing mutual understanding, empathy, and learning between humans and machines will be essential in the future. AI, IoT, big data, human-centric digital twins, and Augmented Reality are expected to play a key role in serving the specification of a human-driven industry. These technologies need to be researched for ease of use, deployment, and accessibility to a wide range of people, helping to improve human satisfaction both inside and outside the factory. In addition, multidisciplinary researchers should pay attention to the harmony of ecological, economic, and social characteristics toward the overall goal of not only Industry 5.0 but also of the coming era.

People's trust in modern technology is essential to be part of the path to Industry 5.0. When machines act as colleagues, they must create empathy, understanding, and interest in the factory space in order to facilitate and enhance human trust. Therefore, solutions for improving and enhancing the characteristics and requirements in human cognitive science for smart technology, smart manufacturing, and smart society will be one of the most studied topics. This, very clearly, could push AI into a new stage of development where deep learning and cognitive computing can rely on cognitive or unsupervised models instead of structured or predefined ones. Besides, the transparency and explainability of AI-based decision-making need further investigation. The achievements of this field will enhance the quality of mutual learning between humans and machines, making it possible to customize and personalize the communication between human life and the engineering space.

On the other hand, Industry 5.0 will open a new door to a smart society where people can enjoy life to the best of their ability. To achieve this vision, lifelong learning is an effective tool for maintaining and enhancing social and technical human capacity. Therefore, researchers should seek answers on how to make lifelong learning a habit and a common culture and custom as possible in a smart life.

The Fifth Industrial Revolution warrants skepticism. Many worries about the economy, job losses, and the value of "humans" eroding. Industry 5.0 brings humans and robots/AI together to innovate. Not replacing humans, it inspires creativity. Industry 5.0 must balance production and efficiency. Industry 5.0 seeks perfection. Future manufacturing complexity and personalization will necessitate a more robotized and collaborative human–machine approach.

Industry 5.0 aims to enhance people's lives worldwide by changing people's views on AI and supporting people-centered digital technology development. The Skills Agenda and Digital Education Action Plan should prepare workers for AI and robotics. If we extend our knowledge base and provide upskilling opportunities, people can take advantage of Industry 5.0's imaginative advantages and employ their creative skills in high-value industrial jobs. Human-centered solutions based on the IoT and artificial intelligence will improve healthcare by identifying, treating, and assessing contemporary society's many diseases. To deliver mass customization with a human touch, the healthcare sector needs business 5.0's networked machines' productivity, efficiency, safety, and waste reduction. Industry 5.0 is a once-in-a-lifetime opportunity to improve the world. Sustainability underpins Industry 5.0. UN-supported companies get financing. Industry 5.0 may inspire sustainable government, business, and climate initiatives. A new idea will succeed if it improves the economy, culture, and ecology. Industry 5.0 will change everything. We must include it in our technological sphere as it expands worldwide. We're only starting a significant transformation that might lead to a more optimistic, creative, and prosperous future for everyone.

12.8 CONCLUSION AND FUTURE SCOPE

Industry 5.0 offers many advantages for all criteria. It is changing the way workers behave in their jobs by applying innovative technologies inherited from Industry 4.0; for example, AI, machine learning, big data, blockchain, and cobots. Industry 5.0 is an option for businesses. Meaning "predictive maintenance" is the natural progression from "preventive maintenance." Smart sensors, IoT gadgets, and computer programs lessen inaccuracies. They are repairing the most broken devices. In Industry 5.0, manufacturing is centered on meeting the client's demands. Business models may shift to accommodate human–machine interaction. Local economies benefit from manufacturing and employment. The revival of the manufacturing sector is facilitated by technological re-employment. With the help of collaborative robots, people may focus on more complex mental activities without risk or boredom. Skilled workers who are invested in their jobs accomplish more. The human–robot collaboration will accelerate the next industrial revolution. Industry 5.0 excels at customization. Industry 5.0 rapidly develops virtual worlds, supercomputers, and other information technologies. "Industry 5.0" reduces complexity. Industry 5.0 includes big data, AI, IoT, cloud, collaborative robots, and inventive problem-solving. Industry 5.0 improves money and freedom. Individualization and productivity improve. Preparing employees for increasingly automated situations is difficult. Industry 5.0's broad usage of standard communication protocols makes critical industrial systems and production lines more vulnerable. Industry 5.0 will give robots more independence, but humans will face moral dilemmas. Industry 5.0 favors human–robot manufacturing.

This chapter comprehensively overviews Industry 5.0 technology by critically reviewing scientific papers and publications. Furthermore, several features and opportunities of Industry 5.0 for enterprises are discussed in this chapter. Although large enterprises and SMEs can benefit from technologies in Industry 5.0, they face many challenges, such as human–machine collaboration, intelligent manufacturing, human factors, and security concerns. Therefore, this chapter plays an essential role in raising awareness for companies and motivates them to figure out solutions in order to mitigate these issues.

BIBLIOGRAPHY

A brief history of robotics: The origin of the first humanoid robot. (2021). https://www.lucarobotics.com/blog/first-humanoid-robot

AAG. (2023a). *The Latest 2023 Phishing Statistics (updated May 2023)*. https://aag-it.com/the-latest-phishing-statistics/

AAG. (2023b). *The Latest 2023 Ransomware Statistics (updated May 2023)*. https://aag-it.com/the-latest-ransomware-statistics/

acatech – National Academy of Science, & Engineering. (2013). *Recommendations for implementing the strategic initiative INDUSTRIE 4.0. Final report of the Industrie 4.0 Working Group. April.*

Acemoglu, D., & Restrepo, P. (2018). The race between man and machine: Implications of technology for growth, factor shares, and employment. In *American Economic Review* (Vol. 108, Issue 6, pp. 1488–1542). American Economic Association. whttps://doi.org/10.1257/aer.20160696

Adel, A. (2022). Future of industry 5.0 in society: human-centric solutions, challenges and prospective research areas. *Journal of Cloud Computing, 11*(1). https://doi.org/10.1186/s13677-022-00314-5

Akundi, A., Euresti, D., Luna, S., Ankobiah, W., Lopes, A., & Edinbarough, I. (2022). State of Industry 5.0—Analysis and Identification of Current Research Trends. *Applied System Innovation, 5*(1), 1–14. https://doi.org/10.3390/asi5010027

Alwis, C. De, Kalla, A., Pham, Q. V., Kumar, P., Dev, K., Hwang, W. J., & Liyanage, M. (2021). Survey on 6G Frontiers: Trends, Applications, Requirements, Technologies and Future Research. *IEEE Open Journal of the Communications Society, 2*, 836–886. https://doi.org/10.1109/OJCOMS.2021.3071496

AMFG. (2023). *Industry 4.0 vs. Industry 5.0: Understanding the Real Difference.* https://amfg.ai/2023/03/17/industry-4-0-vs-industry-5-0-understanding-the-real-difference/

Angelica Mari. (2020). *6G, European internet, censorship: EU Parliament sets out vision for digital services.* https://www.computerweekly.com/news/252484423/6G-European-Internet-Censorship-EU-Parliament-sets-out-vision-for-digital-services

Aparna. (2023). *The Impact Of Cloud Computing On Business Efficiency.* https://stefanini.com/en/insights/articles/the-impact-of-cloud-computing-on-business-efficiency

Azemi, F., Šimunović, G., Lujić, R., Tokody, D., & Rajnai, Z. (n.d.). The Use of Advanced Manufacturing Technology to Reduce Product Cost. In *Acta Polytechnica Hungarica* (Vol. 16, Issue 7).

Bag, S., Wood, L. C., Xu, L., Dhamija, P., & Kayikci, Y. (2020). Big data analytics as an operational excellence approach to enhance sustainable supply chain performance. *Resources, Conservation and Recycling, 153.* https://doi.org/10.1016/j.resconrec.2019.104559

Batth, R. S., Nayyar, A., & Nagpal, A. (2019). Internet of Robotic Things: Driving Intelligent Robotics of Future - Concept, Architecture, Applications and Technologies. *Proceedings - 4th International Conference on Computing Sciences, ICCS 2018*, 151–160. https://doi.org/10.1109/ICCS.2018.00033

Bellet, T., Hoc, J. M., Boverie, S., & Boy, G. (2013). From Human-Machine Interaction to Cooperation: Towards the Integrated Copilot. In *Human-Computer Interactions in Transport* (pp. 129–155). John Wiley and Sons. https://doi.org/10.1002/9781118601907.ch5

van den Bergh, A. (2017). The pros and cons of robotics and its implementation. *Quint Wellington Redwood*, 1–9. https://doi.org/10.1016/S1473-3099(13)70172-5

Beudert, R. (2018). *Understanding Smart Machines: How They Will Shape the Future.* https://www.techbriefs.com/component/content/article/tb/pub/features/technology-leaders/28863

Bi, Z. M., Luo, M., Miao, Z., Zhang, B., Zhang, W. J., & Wang, L. (2021). Safety assurance mechanisms of collaborative robotic systems in manufacturing. *Robotics and Computer-Integrated Manufacturing*, 67(June 2020). https://doi.org/10.1016/j.rcim.2020.102022

Bihani, P., & Patil, S. (2014). *A Comparative Study of Data Analysis Techniques.*

Blackfog. (2023). *4 Key Types of Ransomware – And How to Guard Against Them.* https://www.blackfog.com/4-types-of-ransomware-how-to-recognize-them/

Carayannis, E. G., Christodoulou, K., Christodoulou, P., Chatzichristofis, S. A., & Zinonos, Z. (2022). Known Unknowns in an Era of Technological and Viral Disruptions—Implications for Theory, Policy, and Practice. *Journal of the Knowledge Economy*, 13(1), 587–610. https://doi.org/10.1007/s13132-020-00719-0

Checkpoint. (2023). *Supply Chain Attacks Are Surging.* https://www.checkpoint.com/cyber-hub/threat-prevention/what-is-a-supply-chain-attack/

Chen, B., Wan, J., Lan, Y., Imran, M., Li, D., & Guizani, N. (2019). Improving Cognitive Ability of Edge Intelligent IIoT through Machine Learning. *IEEE Network*, 33, 61–67. https://doi.org/10.1109/MNET.001.1800505

Cheng, Y., Chen, K., Sun, H., Zhang, Y., & Tao, F. (2018). Data and knowledge mining with big data towards smart production. *Journal of Industrial Information Integration*, 9, 1–13. https://doi.org/10.1016/j.jii.2017.08.001

Chowdhury, M. Z., Shahjalal, M. D., Ahmed, S., & Jang, Y. M. (2020). 6G Wireless Communication Systems: Applications, Requirements, Technologies, Challenges, and Research Directions. *IEEE Open Journal of the Communications Society*, 1, 957–975. https://doi.org/10.1109/ojcoms.2020.3010270

Cimini, C., Lagorio, A., Romero, D., Cavalieri, S., & Stahre, J. (2020a). *Product Avatar View project SO SMART-Socially Sustainable Manufacturing for the Factories of the Future View project Smart Logistics and The Logistics Operator 4.0.* https://www.researchgate.net/publication/340952295

Cimini, C., Pinto, R., Pezzotta, G., & Gaiardelli, P. (2017). The transition towards industry 4.0: business opportunities and expected impacts for suppliers and manufacturers. *IFIP Advances in Information and Communication Technology*, 513, 119–126. https://doi.org/10.1007/978-3-319-66923-6_14

Cimini, C., Pirola, F., Pinto, R., & Cavalieri, S. (2020b). A human-in-the-loop manufacturing control architecture for the next generation of production systems. *Journal of Manufacturing Systems*, 54, 258–271. https://doi.org/10.1016/j.jmsy.2020.01.002

Cojocariu, O. (2023). *Industry 5.0 opportunities and challenges: bring your factory into the future.* https://digitalya.co/blog/industry-5-opportunities-and-challenges/

Daugherty, P. R. (2018). *Collaborative Intelligence: Humans and AI Are Joining Forces.*

De Groot, J. (2021). *IP Theft: Definition and Examples.* https://www.digitalguardian.com/blog/ip-theft-definition-and-examples

Deepa, N., Pham, Q. V., Nguyen, D. C., Bhattacharya, S., Prabadevi, B., Gadekallu, T. R., Maddikunta, P. K. R., Fang, F., & Pathirana, P. N. (2022). A survey on blockchain for big data: Approaches, opportunities, and future directions. In *Future Generation Computer Systems* (Vol. 131, pp. 209–226). Elsevier B.V. https://doi.org/10.1016/j.future.2022.01.017

Defraeye, T., Tagliavini, G., Wu, W., Prawiranto, K., Schudel, S., Assefa Kerisima, M., Verboven, P., & Bühlmann, A. (2019). Digital twins probe into food cooling and biochemical quality changes for reducing losses in refrigerated supply chains.

Resources, Conservation and Recycling, 149, 778–794. https://doi.org/10.1016/j.resconrec.2019.06.002

Deutsch, J., He, M., & He, D. (2017). Remaining useful life prediction of hybrid ceramic bearings using an integrated deep learning and particle filter approach. *Applied Sciences (Switzerland), 7*(7). https://doi.org/10.3390/app7070649

Diaz Lantada, A. (2020). Engineering Education 5.0: Continuously Evolving Engineering Education. *International Journal of Engineering Education, 36,* 1814–1832.

Essentracomponents. (2022). *How are cobots benefiting manufacturers?* https://www.essentracomponents.com/en-gb/news/trends/industry-40/how-are-cobots-benefiting-manufacturers

Fantini, P., Pinzone, M., & Taisch, M. (2020). Placing the operator at the centre of Industry 4.0 design: Modelling and assessing human activities within cyber-physical systems. *Computers and Industrial Engineering, 139.* https://doi.org/10.1016/j.cie.2018.01.025

Fortinet. (2023). *Supply Chain Attacks: Examples and Countermeasures.* https://www.fortinet.com/resources/cyberglossary/supply-chain-attacks

Fraga-Lamas, P., Varela-Barbeito, J., & Fernandez-Carames, T. M. (2021). Next Generation Auto-Identification and Traceability Technologies for Industry 5.0: A Methodology and Practical Use Case for the Shipbuilding Industry. *IEEE Access, 9,* 140700–140730. https://doi.org/10.1109/ACCESS.2021.3119775

Frankenfield, J. (2023). *What is Cloud Computing? Pros and Cons of Different Types of Services.* https://www.investopedia.com/terms/c/cloud-computing.asp

Fu, Z., Guo, C., Ren, S., Ou, Y., & Sha, L. (2017). *Modeling and Integrating Human Interaction Assumptions in Medical Cyber-Physical System Design.* 373–378. https://doi.org/10.1109/CBMS.2017.50

Gaiardelli, S., Spellini, S., Lora, M., & Fummi, F. (2021). Modeling in industry 5.0: What is there and what is missing: Special session 1: Languages for industry 5.0. *Forum on Specification and Design Languages,* 2021-September. https://doi.org/10.1109/FDL53530.2021.9568371

García Olaizola, I., Quartulli, M., Garcia, A., & Barandiaran, I. (2022). *Artificial Intelligence from Industry 5.0 perspective: Is the Technology Ready to Meet the Challenge?*

Gay, D., Ng, L., Cheng, H. W., Vergara, S., Ernst, E., Tobin, S., Milasi, S., Pin-Heiro, V., Kraft, C., Women, U. N., Fernandez, A., Renner, S., Roig, M., Jimenez, M., Perry, J., Mukherjee, S., Roehrl, R. A., & Freire, C. (n.d.). *Substantive contributions in the form of draft text for sections and boxes and research assistance were made by of the Division of Social Policy and Develop-ment (DESA/DSPD); and.*

Gehlot, A., Duggal, A. S., Malik, P. K., Gaba, G. S., Masud, M., & Al-Amri, J. F. (n.d.). *Original Research Paper A sequential roadmap to Industry 6.0: Exploring future manufacturing trends.* https://doi.org/10.1049/cmu2.12284

George, A. S., & George, A. S. (2020). Industrial Revolution 5.0: The transformation of the modern manufacturing process to enable man and machine to work hand in hand. *Seybold Report, 15,* 214–234. https://doi.org/10.5281/zenodo.6548092

Gillis, A. S. (2023). *Supply chain attack.* https://www.techtarget.com/searchsecurity/definition/supply-chain-attack

Goris, K. (2004). *Eindwerk ingediend tot het behalen van de academische graad burgerlijk werktuigkundig-elektrotechnisch ingenieur.*

Greif, T., Stein, N., & Flath, C. M. (2020). Peeking into the void: Digital twins for construction site logistics. *Computers in Industry, 121.* https://doi.org/10.1016/j.compind.2020.103264

Grover, P., & Kar, A. K. (2017). Big Data Analytics: A Review on Theoretical Contributions and Tools Used in Literature. *Global Journal of Flexible Systems Management, 18*(3), 203–229. https://doi.org/10.1007/s40171-017-0159-3

Güğerçin, S. (2021). HOW EMPLOYEES SURVIVE IN THE INDUSTRY 5.0 ERA: IN-DEMAND SKILLS OF THE NEAR FUTURE. *International Journal of Disciplines In Economics and Administrative Sciences Studies (IDEAstudies),* 7(31), 524–533. https://doi.org/10.26728/ideas.452

Gupta, B. B., Tewari, A., Jain, A. K., & Agrawal, D. P. (2017). Fighting against phishing attacks: state of the art and future challenges. In *Neural Computing and Applications* (Vol. 28, Issue 12, pp. 3629–3654). Springer London. https://doi.org/10.1007/s00521-016-2275-y

Gürdür Broo, D., Kaynak, O., & Sait, S. M. (2022). Rethinking engineering education at the age of industry 5.0. *Journal of Industrial Information Integration, 25,* 100311. https://doi.org/10.1016/j.jii.2021.100311

Hakanen, E., & Rajala, R. (2018). Material intelligence as a driver for value creation in IoT-enabled business ecosystems. *Journal of Business and Industrial Marketing, 33*(6), 857–867. https://doi.org/10.1108/JBIM-11-2015-0217

Hämäläinen, E., & Inkinen, T. (2019). Industrial applications of big data in disruptive innovations supporting environmental reporting. *Journal of Industrial Information Integration, 16.* https://doi.org/10.1016/j.jii.2019.100105

Happiestminds. (2023). *Industry 4.0.* https://www.happiestminds.com/insights/industry-4-0/

Hayes, A. (2023). *Blockchain Facts: what is it, How it works, and How it can be used.* https://www.investopedia.com/terms/b/blockchain.asp

Hein-Pensel, F., Winkler, H., Brückner, A., Wölke, M., Jabs, I., Mayan, I. J., Kirschenbaum, A., Friedrich, J., & Zinke-Wehlmann, C. (2023). Maturity assessment for Industry 5.0: A review of existing maturity models. *Journal of Manufacturing Systems, 66*(December 2022), 200–210. https://doi.org/10.1016/j.jmsy.2022.12.009

Helpnet Security. (2021). *Supply chain attacks expected to multiply by 4 in 2021.*

Holden, R., Abebe, E., Hill, J., Brown, J., Savoy, A., Voida, S., Jones, J., & Kulanthaivel, A. (2022). *Human Factors Engineering and Human-Computer Interaction: Supporting User Performance and Experience* (pp. 119–132). https://doi.org/10.1007/978-3-030-93765-2_9

Inagaki, T. (2003). *Chapter 8 of the Handbook of Cognitive Task Design pp Adaptive Automation: Sharing and Trading of Control.*

Javaid, M., & Haleem, A. (2020). Critical components of industry 5.0 towards a successful adoption in the field of manufacturing. *Journal of Industrial Integration and Management, 5*(3), 327–348. https://doi.org/10.1142/S2424862220500141

Jha, A. (2023). *The Evolution of Industry: From 1.0 to 6.0.* https://www.linkedin.com/pulse/evolution-industry-from-10-60-arnav-jha

Jones, N. (2022). *5 Ways Ransomware Can Negatively Impact Your Business.* https://www.egnyte.com/blog/post/5-ways-ransomware-can-negatively-impact-your-business

Jordan, J. M. (2019). *The Czech Play That Gave Us the Word 'Robot'.* https://thereader.mitpress.mit.edu/origin-word-robot-rur/

Kadir, B., Broberg, O., & Conceição, C. (2019). Current Research and Future Perspectives on Human Factors and Ergonomics in Industry 4.0. *Computers & Industrial Engineering, 137,* 106004. https://doi.org/10.1016/j.cie.2019.106004

Khan, M., Haleem, A., & Javaid, M. (2023). Changes and improvements in Industry 5.0: A strategic approach to overcome the challenges of Industry 4.0. *Green Technologies and Sustainability*, 1(2), 100020. https://doi.org/10.1016/j.grets.2023.100020

Knudsen, E. S., Lien, L. B., Timmermans, B., Belik, I., & Pandey, S. (2021). Stability in turbulent times? The effect of digitalization on the sustainability of competitive advantage. *Journal of Business Research*, 128, 360–369. https://doi.org/10.1016/j.jbusres.2021.02.008

Komiyama, H., & Yamada, K. (2018). *New Vision 2050: A Platinum Society*. https://doi.org/10.1007/978-4-431-56623-6

Kranz, G., & Christensen, G. (2022). *What is 6G? Overview of 6G networks & technology*. https://www.techtarget.com/searchnetworking/definition/6G

Kumar, P., Mangal, P., Ahmed, S., Al-Absi, A., & Kumar, P. (2021a). *Lecture Notes in Networks and Systems 149 Proceedings of International Conference on Smart Computing and Cyber Security Strategic Foresight, Security Challenges and Innovation (SMARTCYBER 2020)*. http://www.springer.com/series/15179

Kumar, R. (2019). *Sustainable Supply Chain Management in the Era of Digitialization* (pp. 446–460). https://doi.org/10.4018/978-1-5225-8933-4.ch021

Kumar, R., Gupta, P., Singh, S., & Jain, D. (2021b). Human Empowerment by Industry 5.0 in Digital Era: Analysis of Enablers. *Lecture Notes in Mechanical Engineering*, 401–410. https://doi.org/10.1007/978-981-33-4320-7_36

Leng, J., Sha, W., Wang, B., Zheng, P., Zhuang, C., Liu, Q., Wuest, T., Mourtzis, D., & Wang, L. (2022). Industry 5.0: Prospect and retrospect. *Journal of Manufacturing Systems*, 65, 279–295. https://doi.org/10.1016/j.jmsy.2022.09.017

Li, X., Shang, J., & Wang, Z. (2017). Intelligent materials: A review of applications in 4D printing. *Assembly Automation*, 37(2), 170–185. https://doi.org/10.1108/AA-11-2015-093

Liu, Q., Leng, J., Yan, D., Zhang, D., Wei, L., Yu, A., Zhao, R., Zhang, H., & Chen, X. (2021). Digital twin-based designing of the configuration, motion, control, and optimization model of a flow-type smart manufacturing system. *Journal of Manufacturing Systems*, 58, 52–64. https://doi.org/10.1016/j.jmsy.2020.04.012

Longo, F., Padovano, A., & Umbrello, S. (2020). Value-oriented and ethical technology engineering in industry 5.0: A human-centric perspective for the design of the factory of the future. *Applied Sciences (Switzerland)*, 10(12), 1–25. https://doi.org/10.3390/APP10124182

Lysak, A. (2023). *Blockchain and its Impact on Business Operations*. https://planetcompliance.com/blockchain-and-its-impact-on-business-operations/

Maddikunta, P. K. R., Pham, Q. V., B, P., Deepa, N., Dev, K., Gadekallu, T. R., Ruby, R., & Liyanage, M. (2022). Industry 5.0: A survey on enabling technologies and potential applications. *Journal of Industrial Information Integration*, 26(August 2021). https://doi.org/10.1016/j.jii.2021.100257

Marmolejo-Saucedo, J. A., Hurtado-Hernandez, M., & Suarez-Valdes, R. (2020). Digital Twins in Supply Chain Management: A Brief Literature Review. *Advances in Intelligent Systems and Computing*, 1072, 653–661. https://doi.org/10.1007/978-3-030-33585-4_63

Marr, B. (2023). *6G Is Coming: What Will Be The Business Impact?* https://www.forbes.com/sites/bernardmarr/2023/03/17/6g-is-coming-what-will-be-the-business-impact/?sh=7efa0c242f10

Matthews, D. (2015). *Industry 4.0: smart machines are new industrial revolution*. https://www.raconteur.net/industry-4-0-smart-machines-are-new-industrial-revolution/

Mekid, S., Schlegel, T., Aspragathos, N., & Teti, R. (2007). Foresight formulation in innovative production, automation and control systems. *Foresight*, *9*(5), 35–47. https://doi.org/10.1108/14636680710821089

Miley, J. (2018). *11 Times AI Beat Humans at Games, Art, Law and Everything in Between.* https://interestingengineering.com/innovation/11-times-ai-beat-humans-at-games-art-law-and-everything-in-between

Mistry, P. (2023). *Industrial Internet of things (IIoT) and Industry 4.0: Are you ready for it?* https://radixweb.com/blog/what-is-industry-4-0

Montini, E., Cutrona, V., Bonomi, N., Landolfi, G., Bettoni, A., Rocco, P., & Carpanzano, E. (2022). An IIoT Platform For Human-Aware Factory Digital Twins. *Procedia CIRP*, *107*, 661–667. https://doi.org/10.1016/j.procir.2022.05.042

Morgan-Thomas, A., Dessart, L., & Veloutsou, C. (2020). *Digital ecosystem and consumer engagement: a socio-technical perspective author details digital ecosystem and consumer engagement: a socio-technical perspective digital ecosystem and consumer engagement: a socio-technical perspective.*

Moya, A., Bastida, L., Aguirrezabal, P., Pantano, M., & Abril-Jiménez, P. (2023). Augmented Reality for Supporting Workers in Human–Robot Collaboration. *Multimodal Technologies and Interaction*, *7*(4). https://doi.org/10.3390/mti7040040

Munir, S., Stankovic, J., Lin, S., Stankovic, J. A., & Liang, C.-J. M. (2013). *Cyber Physical System Challenges for Human-in-the-Loop Control Cognitive Assistant Systems for Emergency Response View project breadcrumb system View project Cyber Physical System Challenges for Human-in-the-Loop Control.* https://www.researchgate.net/publication/259933586

Mushtaq, A., & Haq, I. U. (2019, May 9). Implications of blockchain in industry 4.O. *2019 International Conference on Engineering and Emerging Technologies, ICEET 2019.* https://doi.org/10.1109/CEET1.2019.8711819

Nahavandi, S. (2019). Industry 5.0-a human-centric solution. *Sustainability (Switzerland)*, *11*(16). https://doi.org/10.3390/su11164371

Nayyar, A., & Kumar, A. (2020). *A Roadmap to Industry 4.0: Smart Production, Sharp Business and Sustainable Development* (A. Nayyar & A. Kumar, Eds.). Springer International Publishing. https://doi.org/10.1007/978-3-030-14544-6

NPR Staff. (2016). *A "New" Rembrandt: From The Frontiers Of AI And Not The Artist's Atelier.* https://www.npr.org/sections/alltechconsidered/2016/04/06/473265273/a-new-rembrandt-from-the-frontiers-of-ai-and-not-the-artists-atelier

Omoyiola, B. O. (2022). The social implications, risks, challenges and opportunities of big data. *Emerald Open Research*, *4*, 23. https://doi.org/10.35241/emeraldopenres.14646.1

Palvia, P., Ghosh, J., Jacks, T., & Serenko, A. (2021). Information technology issues and challenges of the globe: the world IT project. *Information and Management*, *58*(8). https://doi.org/10.1016/j.im.2021.103545

Pan, Y., & Zhang, L. (2021). A BIM-data mining integrated digital twin framework for advanced project management. *Automation in Construction*, *124*, 103564. https://doi.org/10.1016/j.autcon.2021.103564

Pathak, P. (2019). Fifth revolution: applied AI and human intelligence with cyber physical systems. In *International Journal of Engineering and Advanced Technology (IJEAT)* (Issue 3). https://www.researchgate.net/publication/331966435

Pei, E., Loh, G. H., Harrison, D., De, H., Almeida, A., Domingo, M., Verona, M., & Paz, R. (2017). *A Study of 4D Printing and Functionally Graded Additive Manufacturing.*

Pereira, A. (2019). *What are smart machines?* https://www.careerinstem.com/what-are-smart-machines/

Prabadevi, B., Deepa, N., Pham, Q. V., Nguyen, D. C., Maddikunta, P. K. R., Reddy, T., Pathirana, P. N., & Dobre, O. (2021). *Toward Blockchain for Edge-of-Things: A New Paradigm, Opportunities, and Future Directions.* https://doi.org/10.1109/IOTM.0001.2000191

Pramanik, P. K. D., Mukherjee, B., Pal, S., Upadhyaya, B. K., & Dutta, S. (2020). Ubiquitous Manufacturing in the Age of Industry 4.0: A State-of-the-Art Primer. In *Advances in Science, Technology and Innovation* (pp. 73–112). Springer Nature. https://doi.org/10.1007/978-3-030-14544-6_5

Pratt, M. K. (2017). *Smart machines.* https://www.techtarget.com/searchcio/definition/smart-machines

Prescient Technologies. (2022). *Benefits of Industry 4.0.* https://www.pre-scient.com/knowledge-center/industry-4-0/benefits-of-industry-4-0.html

Proofpoint. (2023). *What Is Intellectual Property Theft?* https://www.proofpoint.com/us/threat-reference/intellectual-property-theft

Rachel. (2018). *7 Advantages of Robots in the Workplace.* https://www.roboticstomorrow.com/story/2018/08/7-advantages-of-robots-in-the-workplace/12342/

Raja Santhi, A., & Muthuswamy, P. (2022). Influence of Blockchain Technology in Manufacturing Supply Chain and Logistics. In *Logistics* (Vol. 6, Issue 1). MDPI. https://doi.org/10.3390/logistics6010015

Raja Santhi, A., & Muthuswamy, P. (2023). Industry 5.0 or industry 4.0S? Introduction to industry 4.0 and a peek into the prospective industry 5.0 technologies. *International Journal on Interactive Design and Manufacturing.* https://doi.org/10.1007/s12008-023-01217-8

Rajatheva, N., Atzeni, I., Bjornson, E., Bourdoux, A., Buzzi, S., Dore, J.-B., Erkucuk, S., Fuentes, M., Guan, K., Hu, Y., Huang, X., Hulkkonen, J., Jornet, J. M., Katz, M., Nilsson, R., Panayirci, E., Rabie, K., Rajapaksha, N., Salehi, M., ... Xu, W. (2020). *White Paper on Broadband Connectivity in 6G.* http://arxiv.org/abs/2004.14247

Reddy, P. K., Pham, V. Q., & Deepa, N. (2021). *Industry 5.0: A Survey on Enabling Technologies and Potential Applications Industry 5.0: A Survey on Enabling Technologies and Potential Applications.* July. https://doi.org/10.1016/j.jiii.2021.100257

Reinsel, D., Gantz, J., & Rydning, J. (2018). *The Digitization of the World From Edge to Core.*

Resende, A., Cerqueira, S., Barbosa, J., Damasio, E., Pombeiro, A., Silva, A., & Santos, C. (2021). Ergowear: An ambulatory, non-intrusive, and interoperable system towards a Human-Aware Human-robot Collaborative framework. *2021 IEEE International Conference on Autonomous Robot Systems and Competitions, ICARSC 2021,* 56–61. https://doi.org/10.1109/ICARSC52212.2021.9429796

Romero, D., & Stahre, J. (2021). Towards the Resilient Operator 5.0: The Future of Work in Smart Resilient Manufacturing Systems. *Procedia CIRP, 104,* 1089–1094. https://doi.org/10.1016/j.procir.2021.11.183

Romero, D., Stahre, J., Wuest, T., Noran, O., Bernus, P., Fast-Berglund, Å., & Gorecky, D. (2016a). *Towards An Operator 4.0 Typology: A Human-Centric Perspective On The Fourth Industrial Revolution Technologies.* 29–31.

Romero, D., Stahre, J., Wuest, T., Noran, O., Bernus, P., Fast-Berglund, Å., & Gorecky, D. (2016c). *Towards an Operator 4.0 Typology: A Human-Centric Perspective on the Fourth Industrial Revolution Technologies Demonstrating and testing smart digitalisation for sustainable human-centred automation in production View project Additive Manufacturing: Sensors and Electronics Integration for 3D printed Structural Electronics View project TOWARDS AN OPERATOR 4.0 TYPOLOGY: A HUMAN-CENTRIC PERSPECTIVE ON THE FOURTH INDUSTRIAL REVOLUTION TECHNOLOGIES.* 29–31. https://www.researchgate.net/publication/309609488

Sanchez, M., Exposito, E., & Aguilar, J. (2020). *Autonomic Computing in Manufacturing Process Coordination in Industry 4.0 Context.* https://www.elsevier.com/open-access/userlicense/1.0/

Sharp, N. (2021). *Industry 5.0 and the future of sustainable manufacturing.* https://www.escatec.com/blog/industry-5.0-and-the-future-of-sustainable-manufacturing

Shweta Joglekar, Sachin Kadam, Sonali Dharmadhikari, & Bharati Vidyapeeth. (2023). *Industry 5.0: Analysis, Applications And Prognosis.*

Simchenko, N. A., Tsohla, S. Y., & Chyvatkin, P. P. (2019). IoT & Digital Twins Concept Integration Effects on Supply Chain Strategy: Challenges and Effects. In *Int. J Sup. Chain. Mgt* (Vol. 8, Issue 6). http://excelingtech.co.uk/

Simpson, A. (2022). *What is Intellectual Property (IP) Theft?* https://www.code42.com/blog/what-is-intellectual-property-theft/

Srushti Shah. (2022). *Impact of Cloud computing in Different Industries.* https://www.business2community.com/cloud-computing/impact-of-cloud-computing-in-different-industries-02451160

Statista. (2023a). *Annual number of ransomware attacks worldwide from 2017 to 2022.* https://www.statista.com/statistics/494947/ransomware-attacks-per-year-worldwide/

Statista. (2023b). *Percentage of organizations victimized by ransomware attacks worldwide from 2018 to 2022.* https://www.statista.com/statistics/204457/businesses-ransomware-attack-rate/

Stefanini. (2022). *What Is Industry 5.0? How Shifting Objectives Enables Transformation.* https://stefanini.com/en/insights/news/what-is-industry-50-how-shifting-objectives-enables-transformation

Strinati, E. C., Barbarossa, S., Gonzalez-Jimenez, J. L., Kténas, D., Cassiau, N., & Dehos, C. (2019). *6G: The Next Frontier.* http://arxiv.org/abs/1901.03239

Sydorenko, I. (2021). *Big Data and Its Business Impacts: Benefits, Challenges, and Use Cases.* https://labelyourdata.com/articles/big-data-business-impact

Tankt, D. W., & Hopfieldt, J. J. (1987). *Neural computation by concentrating information in time (neural network/connectionist/speech recognition/parallel processing)* (Vol. 84). https://www.pnas.org

Tao, F., Zhang, L., Venkatesh, V. C., Luo, Y., & Cheng, Y. (2011). Cloud manufacturing: A computing and service-oriented manufacturing model. *Proceedings of the Institution of Mechanical Engineers, Part B: Journal of Engineering Manufacture,* 225(10), 1969–1976. https://doi.org/10.1177/0954405411405575

Tariq, F., Member, S., Khandaker, M. R., Wong, K.-K., Imran, M. A., Bennis, M., & Debbah, M. (2020). *A Speculative Study on 6G* (Vol. 27, Issue 4).

Tehrani, B. M., Wang, J., & Wang, C. (2019, May). *Review of Human-in-the-Loop Cyber-Physical Systems (HiLCPS): The Current Status from Human Perspective.* https://doi.org/10.1061/9780784482438.060

Thales. (2023). *What is Intellectual Property Theft & Why it Matters.* https://cpl.thalesgroup.com/software-monetization/what-is-intellectual-property-theft

The History of Robotics. (2021). https://www.thomasnet.com/articles/automation-electronics/history-of-robotics/

The WIRED Guide to Robots. (2020). https://www.wired.com/story/wired-guide-to-robots/

TP&P Technology. (2021). *The Impact Of Big Data On Businesses, Workforce, and Society*. https://www.tpptechnology.com/blog/the-impact-of-big-data-on-businesses-workforce-and-society/

Tucci, L., & Needle, D. (2023). *What is the metaverse? An explanation and in-depth guide*. https://www.techtarget.com/whatis/feature/The-metaverse-explained-Everything-you-need-to-know

Tuptuk, N., & Hailes, S. (2018). Security of smart manufacturing systems. *Journal of Manufacturing Systems, 47*, 93–106. https://doi.org/10.1016/j.jmsy.2018.04.007

TWI. (2023). *What is Industry 5.0? (Top 5 Things You Need To Know)*. https://www.twi-global.com/technical-knowledge/faqs/industry-5-0#EvolutionoftheIndustrialRevolutionIndustry10to50

Vapnik, V. N. (1999). An Overview of Statistical Learning Theory. In *IEEE transactions on neural networks* (Vol. 10, Issue 5).

Verma, A., Bhattacharya, P., Madhani, N., Trivedi, C., Bhushan, B., Tanwar, S., Sharma, G., Bokoro, P. N., & Sharma, R. (2022). Blockchain for Industry 5.0: Vision, Opportunities, Key Enablers, and Future Directions. *IEEE Access, 10*, 69160–69199. https://doi.org/10.1109/ACCESS.2022.3186892

Vinitha, K., Ambrose Prabhu, R., Bhaskar, R., & Hariharan, R. (2020). Review on industrial mathematics and materials at Industry 1.0 to Industry 4.0. *Materials Today: Proceedings, 33*(xxxx), 3956–3960. https://doi.org/10.1016/j.matpr.2020.06.331

Viriyasitavat, W., & Hoonsopon, D. (2019). Blockchain characteristics and consensus in modern business processes. *Journal of Industrial Information Integration, 13*, 32–39. https://doi.org/10.1016/j.jii.2018.07.004

Wang, B., Tao, F., Fang, X., Liu, C., Liu, Y., & Freiheit, T. (2021). Smart manufacturing and intelligent manufacturing: A comparative review. In *Engineering* (Vol. 7, Issue 6, pp. 738–757). Elsevier Ltd. https://doi.org/10.1016/j.eng.2020.07.017

Wang, B., Zheng, P., Yin, Y., Shih, A., & Wang, L. (2022a). Toward human-centric smart manufacturing: A human-cyber-physical systems (HCPS) perspective. *Journal of Manufacturing Systems, 63*, 471–490. https://doi.org/10.1016/j.jmsy.2022.05.005

Wang, B., Zhou, H., Yang, G., Li, X., & Yang, H. (2022b). Human Digital Twin (HDT) Driven Human-Cyber-Physical Systems: Key Technologies and Applications. *Chinese Journal of Mechanical Engineering (English Edition), 35*(1). https://doi.org/10.1186/s10033-022-00680-w

Wang, J., Zhao, R., Wang, D., Yan, R., Mao, K., & Shen, F. (2017). Machine health monitoring using local feature-based gated recurrent unit networks. *IEEE Transactions on Industrial Electronics, 65*(2), 1539–1548. https://doi.org/10.1109/TIE.2017.2733438

Wesley Chai, S. J. B. (2023). *Cloud computing*. https://www.techtarget.com/searchcloudcomputing/definition/cloud-computing

Williams, B. (1991). An Introduction to Robotics in Catalonia. *Catalònia, 23*, 28–29.

Woboton. (2023). *5 key benefits of Industry 4.0 for factories*. https://woboton.com/5-key-benefits-of-industry-40-for-factories/

Woodward, M. (2023). *Artificial Intelligence Statistics For 2023*. https://www.searchlogistics.com/learn/statistics/artificial-intelligence-statistics/

Wu, G., Feder, A., Cohen, H., Kim, J. J., Calderon, S., Charney, D. S., & Mathé, A. A. (2013). Understanding resilience. In *Frontiers in Behavioral Neuroscience* (Issue January 2013). https://doi.org/10.3389/fnbeh.2013.00010

Xu, L. Da, Lu, Y., & Li, L. (2021). Embedding Blockchain Technology into IoT for Security: A Survey. *IEEE Internet of Things Journal, 8*(13), 10452–10473. https://doi.org/10.1109/JIOT.2021.3060508

Yang, H., Alphones, A., Xiong, Z., Niyato, D., Zhao, J., & Wu, K. (2020). Artificial-Intelligence-Enabled Intelligent 6G Networks. *IEEE Network, 34*(6), 272–280. https://doi.org/10.1109/MNET.011.2000195

Yücebalkan, B. (2020). Endüstri 4.0'dan endüstri 5.0'a geçiş sürecine genel bakiş. *IEDSR Association, 5*(9), 241–250. https://doi.org/10.46872/pj.181

Zhang, Y., Ren, S., Liu, Y., Sakao, T., & Huisingh, D. (2017). A framework for Big Data driven product lifecycle management. *Journal of Cleaner Production, 159*. https://doi.org/10.1016/j.jclepro.2017.04.172

Zhaohao Sun. (2021). *Intelligent Analytics with Advanced Multi-industry Applications*. IGI Global.

Zheng, P., Xu, X., Yu, S., & Liu, C. (2017). Personalized product configuration framework in an adaptable open architecture product platform. *Journal of Manufacturing Systems, 43*, 422–435. https://doi.org/10.1016/j.jmsy.2017.03.010

Zhou, J., Li, P., Zhou, Y., Wang, B., Zang, J., & Meng, L. (2018). Toward New-Generation Intelligent Manufacturing. In *Engineering* (Vol. 4, Issue 1, pp. 11–20). Elsevier Ltd. https://doi.org/10.1016/j.eng.2018.01.002

Zicari, R. V. (2013). *3 Big Data: Challenges and Opportunities*.

Index

Pages in *italics* refer to figures and pages in **bold** refer to tables.

For Product Safety Concerns and Information please contact our EU
representative GPSR@taylorandfrancis.com
Taylor & Francis Verlag GmbH, Kaufingerstraße 24, 80331 München, Germany

www.ingramcontent.com/pod-product-compliance
Lightning Source LLC
Chambersburg PA
CBHW060425220326
41598CB00021BA/2292

9 781032 787572